The Photosynthetic Bacterial Reaction Center II

Structure, Spectroscopy, and Dynamics

NATO ASI Series

Advanced Science Institutes Series

A series presenting the results of activities sponsored by the NATO Science Committee, which aims at the dissemination of advanced scientific and technological knowledge, with a view to strengthening links between scientific communities.

The series is published by an international board of publishers in conjunction with the NATO Scientific Affairs Division

A	**Life Sciences**	Plenum Publishing Corporation
B	**Physics**	New York and London
C	**Mathematical and Physical Sciences**	Kluwer Academic Publishers
D	**Behavioral and Social Sciences**	Dordrecht, Boston, and London
E	**Applied Sciences**	
F	**Computer and Systems Sciences**	Springer-Verlag
G	**Ecological Sciences**	Berlin, Heidelberg, New York, London,
H	**Cell Biology**	Paris, Tokyo, Hong Kong, and Barcelona
I	**Global Environmental Change**	

Recent Volumes in this Series

Volume 231—Formation and Differentiation of Early Embryonic Mesoderm
edited by Ruth Bellairs, Esmond J. Sanders, and James W. Lash

Volume 232—Oncogene and Transgenics Correlates of Cancer Risk Assessments
edited by Constantine Zervos

Volume 233—T Lymphocytes: Structure, Functions, Choices
edited by Franco Celada and Benvenuto Pernis

Volume 234—Development of the Central Nervous System in Vertebrates
edited by S. C. Sharma and A. M. Goffinet

Volume 235—Advances in Cardiovascular Engineering
edited by Ned H. C. Hwang, Vincent T. Turitto, and Michael R. T. Yen

Volume 236—Rhythms in Fishes
edited by M. A. Ali

Volume 237—The Photosynthetic Bacterial Reaction Center II: Structure, Spectroscopy, and Dynamics
edited by Jacques Breton and André Verméglio

Series A: Life Sciences

The Photosynthetic Bacterial Reaction Center II

Structure, Spectroscopy, and Dynamics

Edited by

Jacques Breton

CEN Saclay
Gif-sur-Yvette, France

and

André Verméglio

CEN Cadarache
Saint Paul lez Durance, France

Springer Science+Business Media, LLC

Proceedings of a NATO Advanced Research Workshop on
The Photosynthetic Bacterial Reaction Center:
Structure, Spectroscopy, and Dynamics,
held May 10–15, 1992,
at the Centre d'Etudes Nucléaires de Cadarache, France

NATO-PCO-DATA BASE

The electronic index to the NATO ASI Series provides full bibliographical references (with keywords and/or abstracts) to more than 30,000 contributions from international scientists published in all sections of the NATO ASI Series. Access to the NATO-PCO-DATA BASE is possible in two ways:

—via online FILE 128 (NATO-PCO-DATA BASE) hosted by ESRIN, Via Galileo Galilei, I-00044 Frascati, Italy

—via CD-ROM "NATO-PCO-DATA BASE" with user-friendly retrieval software in English, French, and German (©WTV GmbH and DATAWARE Technologies, Inc. 1989)

The CD-ROM can be ordered through any member of the Board of Publishers or through NATO-PCO, Overijse, Belgium.

Library of Congress Cataloging-in-Publication Data

The Photosynthetic bacterial reaction center II : structure, spectroscopy, and dynamics / edited by Jacques Breton and André Verméglio.
p. cm. -- (NATO ASI series. Series A, Life sciences ; v. 237)
"Published in cooperation with NATO Scientific Affairs Division."
"Proceedings of a NATO advanced research workshop on The Photosynthetic Bacterial Reaction Center: Structure, Spectroscopy, and Dynamics, held May 10-15, 1992, at the Centre d'etudes nucléaires de Cadarache, France"--T.p. verso.
Includes bibliographical references and index.
ISBN 978-1-4613-6330-9 ISBN 978-1-4615-3050-3 (eBook)
DOI 10.1007/978-1-4615-3050-3
1. Photosynthetic reaction centers--Congresses. 2. Photosynthetic bacteria--Congresses. I. Breton, Jacques, 1942- II. Verméglio, André. III. Series.
QR88.5.P482 1992
589.9'013342--dc20 92-36001
CIP

ISBN 978-1-4613-6330-9

Originally published by Plenum Press in 1992
Softcover reprint of the hardcover 1st edition 1992

PREFACE

The NATO Advanced Research Workshop entitled "The Photosynthetic Bacterial Reaction Center: Structure, Spectroscopy, and Dynamics" was held May 10–15, 1992, in the Maison d'Hôtes of the Centre d'Etudes Nucléaires de Cadarache near Aix–en–Provence in the south of France. This workshop is the most recent of a string of meetings which started in Feldafing (Germany) in March 1985, soon after the three–dimensional structure of the bacterial reaction center had been elucidated by X–ray crystallography. This was followed, in September 1987, by a workshop in Cadarache and, in March 1990, by a second meeting in Feldafing.

Although one of the most important processes on Earth, photosynthesis is still poorly understood. Stimulated by the breakthrough of solving the bacterial reaction center structure at atomic resolution, the field of relating this structure to the function of the reaction center, i.e. the remarkably efficient conversion and storage of solar energy, has been developing vigorously. Once the general organization of the cofactors and some details of the protein–cofactor interactions were known, it became possible to combine a variety of spectroscopic techniques with the powerful tool of site–directed mutagenesis in order to address increasingly incisive questions about the specific role of some amino acid residues in the electron transfer process. Still another promising tool is being developed, namely the exchange of a number of the native bacteriochlorophyll and bacteriopheophytin cofactors by chemically modified pigments. In order to understand the electron transfer mechanism in the reaction center, one needs to know not only the atomic structure provided by the X–ray studies but also the electronic structure of the cofactors within their binding site as well as the electronic coupling among the pigments and the relative energetics. Several groups are currently attempting such calculations and comparing their results to spectroscopic observations. Furthermore, the X–ray picture is static and reflects neither the fluctuation of structure, which might play an essential role in fine–tuning the energy levels of the pigments and in assisting the electron transfer, nor the changes of structure that stabilize the separated charges. Thus, the combination of X–ray crystallography, theoretical calculations, and spectroscopy is needed to achieve a complete understanding of the dynamic processes of photosynthesis. In addition, due to its detailed structural characterization, the photosynthetic bacterial reaction center offers unprecedented opportunities for the study of basic questions which go much beyond photosynthesis, such as electron transfer theories, protein dynamics, proton transfer mechanisms, membrane protein assembly and folding, development of artificial donor–acceptor systems...

All of these issues have been addressed during the workshop and are discussed in these proceedings which contain the invited papers. We take this opportunity to thank all the participants for their contributions and the authors for the timely preparation and the high scientific quality of the manuscripts.

We are greatly indebted to Bernadette Fournal for her dedicated assistance with the secretarial correspondence and the preparation of this volume and to Frédéric Sarrey for his efficient contribution to the smooth running of the meeting. We would also like to express our sincere thanks to all our coworkers for their most valuable assistance and to F. Guichot for promoting the pleasant and stimulating atmosphere of the medieval castle of Cadarache.

In line with the "Cadarache I meeting", the organization of this workshop was made possible by a grant from the NATO Scientific Affairs Division which is gratefully acknowledged. Additional financial support was also provided by the Direction des Sciences du Vivant du Commissariat à l'Energie Atomique, by the Conseil Régional Provence–Alpes–Côte d'azur and by the Conseil Général des Alpes de Haute Provence et des Bouches du Rhone.

Jacques BRETON and André VERMEGLIO

September, 1992

CONTENTS

THE 3-D STRUCTURE OF THE REACTION CENTER FROM *RHODOPSEUDOMONAS VIRIDIS*

Johann Deisenhofer[1] and Hartmut Michel[2]

[1]Howard Hughes Medical Institute and Department of Biochemistry University of Texas Southwestern Medical Center, Dallas, Texas 75235, U.S.A.

[2]Max-Planck-Institut für Biophysik, Abteilung Molekulare Membranbiochemie, Frankfurt/Main, F.R.G.

INTRODUCTION

The photosynthetic reaction center (RC) from the purple bacterium *Rps. viridis* is a complex of 4 protein subunits and 14 cofactors. The protein subunits are, in order of decreasing size, cytochrome (336 amino acids), subunit M (323 a.a.), subunit L (273 a.a.), and subunit H (258 a.a.); the complete amino acid sequences of these subunits were derived from the sequences of the genes coding for them[1-3]. The cofactors are four heme groups, covalently linked to the cytochrome subunit, four bacteriochlorophyll-b (BChl-b), two bacteriopheophytin-b (BPh-b), one menaquinone-9, one ubiquinone–9, one ferrous iron ion, and the carotenoid 1,2–dihydroneurosporene[4,5].

In this contribution we briefly describe structure analysis, and atomic model of the RC, based on X-ray diffraction data at 2.3 Å resolution. We emphasize structural details important for the function of the RC; structural properties that characterize the RC as an integral membrane protein have been described elsewhere[6,7].

STRUCTURE DETERMINATION

Crystallization

Well ordered three-dimensional crystals of the RC from *Rps. viridis* were obtained from RCs solubilized with the detergent N,N-dimethyl-dodecylamine-N-oxide (LDAO); the small amphiphile heptane-1,2,3-triol was an essential additive, and ammonium sulphate was the precipitant in a vapor diffusion experiment[8,9]. The crystals have the symmetry of space group $P4_32_12$. The unit cell dimensions are a = b = 223.5 Å, c = 113.6 Å, with 1 RC molecule per asymmetric unit[8,10]; about 70% of the unit cell's volume is occupied by solvent. The crystals diffract X-rays to at least 2.3 Å resolution.

The Photosynthetic Bacterial Reaction Center II
Edited by J. Breton and A. Verméglio, Plenum Press, New York, 1992

X-ray analysis

The structure of the RC crystals was initially determined at 3.0 Å resolution using the method of multiple isomorphous replacement with heavy atom compounds[10]. Phase information from five heavy atom derivatives was used to calculate an electron density map, which the method of solvent flattening[11] further improved. On the basis of this map, an atomic model of the RC was built for the cofactors[10] and subsequently for the protein subunits[12]. The combination of sequence information and crystallographic data allowed the description of protein cofactor interactions[13]. Crystallographic refinement at 2.3 Å resolution (J. Deisenhofer, O. Epp, I. Sinning, H. Michel in preparation) led to a an atomic model consisting of 10288 atoms, including e.g. 201 bound water molecules, 7 bound anions, and 1 firmly bound LDAO molecule[6,14,15]. The model has a crystallographic R-value of 0.19 for 95762 unique reflections to 2.3 Å resolution ($R = \sum|F_{obs}-F_{calc}|/\sum F_{obs}$; F_{obs}, F_{calc} are observed and calculated structure factor amplitudes, respectively; the summation includes all unique reflections used in refinement); the upper limit for the average error in atomic coordinates[16] was estimated to 0.266 Å[6,15]. In a low-resolution neutron diffraction study, the distribution of the detergent in the crystal could be determined[17].

THE ATOMIC MODEL

Overview

Figure 1 shows an overall view of the model of the RC from *Rps. viridis*. In its longest dimension, the complex measures ~130 Å; the maximum width perpendicular to that direction is ~70 Å. The closely associated subunits L and M, together with the bound BChl, BPh, quinones, nonheme iron, and carotenoid, form the central part of the RC. The most prominent structural features of each of the central subunits are five long hydrophobic helices that are assumed to span the bacterial membrane. This assumption is supported by the distribution of detergent in the crystal[17], by properties of the model[6], and by functional considerations. The orientation of RCs with respect to the cytoplasmic and periplasmic sides of the membrane was determined e.g. by proteolysis experiments[18-20] which showed that the NH_2 termini of subunits L and M, and subunit H are accessible at the cytoplasmic membrane surface. The polypeptide backbones of subunits L and M and the attached cofactors display a high degree of approximate local two-fold symmetry; the local two-fold symmetry axis is oriented perpendicular to the membrane plane[21]. On either side of the membrane-spanning region of the L-M complex, a peripheral subunit is attached: the cytochrome with its four bound heme groups at the periplasmic side, and the globular domain of the H-subunit at the cytoplasmic side of the membrane. The H-subunit contributes a single membrane-spanning helix. Neither the cytochrome nor the H-subunit obey the local symmetry found in the central part of the RC; the cytochrome has internal local symmetry of its own.

Subunit structure

Figure 2 shows a schematic drawing of the polypeptide chain folding of subunit M; subunit L is folded very similarly. Structurally similar segments in both subunits include the transmembrane helices (labeled A, B, C, D, E) and a large fraction of the polypeptide chains connecting them. In total, 216 α–carbons

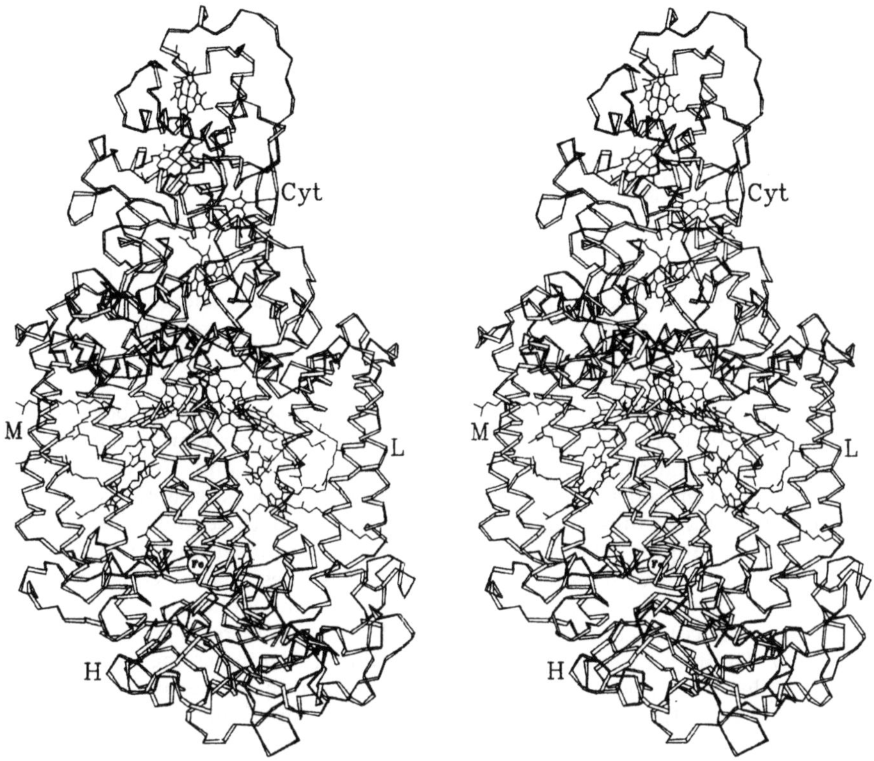

Figure 1 (stereo pair): Overall view of the RC from *Rps. viridis*. The polypeptide chains of the protein subunits are represented by ribbons; co-factors are drawn as wire models. The membrane plane is approximately perpendicular to the plane of projection; the approximate positions of the periplasmic (upper line) and the cytoplasmic membrane surfaces are indicated. Figures 1 to 7 were prepared with a computer program by A.M. Lesk and K.D. Hardman[22].

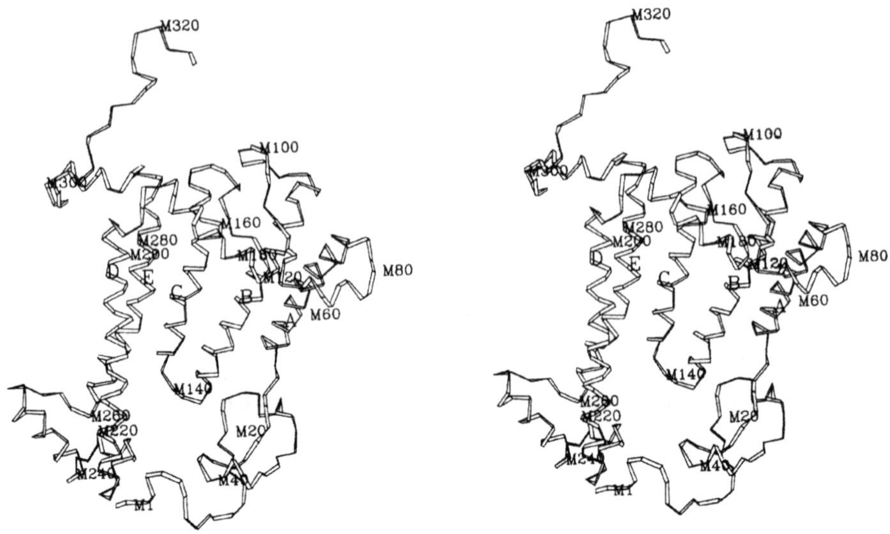

Figure 2 (stereo pair). M-subunit of the RC from *Rps. viridis* in ribbon representation. The transmembrane helices are labeled A, B, C, D, and E; selected residue numbers are indicated.

from the M subunit can be superimposed onto corresponding α–carbons of the L subunit by a rotation of almost exactly 180° to a root mean square (rms) deviation of 1.22 Å. The transmembrane helices are between 21 and 28 residues long; additional, shorter helices can be found in the connecting polypeptide chains. The M subunit is 50 residues longer than the L subunit. Both sequence alignments[2], and structural alignments[6,12] show that the sequence insertions in M reside on either side of the transmembrane helices (20 residues at the NH_2 terminus, 7 residues in the connection between transmembrane helices MA and MB, 7 residues in the connection between MD and ME, and 16 residues at the COOH terminus). Through these insertions, the M subunit dominates the contacts of the L-M complex to the peripheral subunits. The insertion between transmembrane helices MD and ME, containing another small helix, is of importance for binding of the nonheme iron and for the different conformations of the quinone binding sites in L and M.

The H-subunit can be divided into three structural regions with different characteristics. The amino terminal segment, beginning with formylmethionine[1], contains the only transmembrane helix of subunit H (residues H12 to H35). Near the end of the transmembrane helix, the polypeptide chain consists of seven consecutive charged residues (H33 to H39). Residues H47 to H53 are disordered in the crystal. Following the disordered region, the H chain forms an extended structure along the surface of the L-M complex, apparently deriving structural stability from that contact. The surface region contains a short helix and two two-stranded antiparallel β-sheets. The third structural segment of the H-subunit, starting at about residue H105, forms a globular domain. This domain contains an extended system of parallel and antiparallel β–sheets (between residues H134 and H203), and an α–helix (residues H232 to H248).

The cytochrome is the largest subunit in the reaction center complex[3]. Its structure consists of an NH_2-terminal segment, two pairs of heme binding segments, and a segment connecting the two pairs. Each heme binding segment

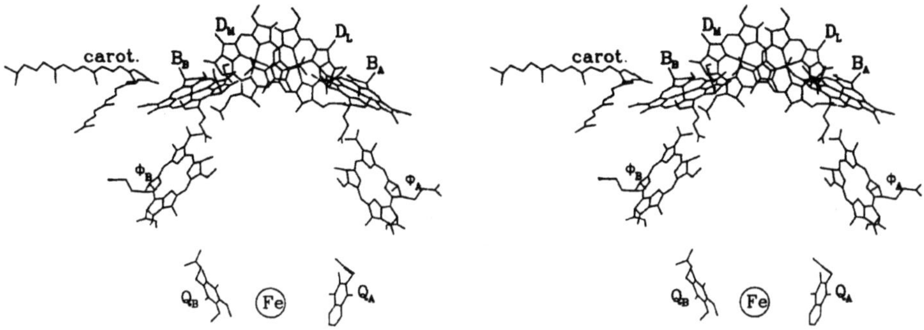

Figure 3: Model of the co-factors associated with subunits L and M in the RC from *Rps. viridis*. The molecules are labeled according to the nomenclature used in the text; "carot." is the carotenoid 1,2 dihydroneurosporene. The phytyl chains of the BChl-b and BPh-b, and the isoprenoid chains of the quinones are not shown.

consists of a helix with an average length of 17 residues followed by a turn and the Cys-X-X-Cys-His sequence typical for c-type cytochromes. The hemes are connected to the cysteine residues via thioether linkages in such a way that the heme planes are parallel to the helix axes. The Nε2 atoms of the histidine residues are the fifth ligands to the heme irons (the pyrrole nitrogens are ligands 1 to 4). The sixth ligands to three of the four heme irons are the sulphur atoms of methionine residue within the helices. The exception in this scheme is heme 4 where histidine C124, located within heme binding segment 2, serves with its Nε2 atom as a sixth ligand for the iron. The two pairs of heme binding segments, containing hemes 1, 2 and 3, 4, respectively, are related by local twofold symmetry; 65 residues within each of these regions obey this local symmetry with an rms deviation between corresponding α–carbon atoms of 0.93 Å. The local symmetry of the cytochrome is not related to the local symmetry in the center of the RC (see above). Lack of electron density for the last four residues, C333 to C336, indicates disorder. Also disordered is the lipid moiety bound to the NH_2 terminal cysteine residue[23].

The cofactors

Figure 3 shows arrangement and nomenclature of the cofactors associated with the protein subunits L and M, excluding phytyl side chains or isoprenoid side chains for clarity. A closely associated pair of BChl-bs, the special pair, resides at the origin of two branches of cofactors, each of which consists of another BChl-b (the accessory BChl-b), a BPh-b, and a quinone. The nonheme iron sits between the quinones and is bound to five amino acid residues, four histidines (L190, L230, M217, M264), and one glutamic acid (M232)[12,13]. The Mg^{2+} ions of the BChl-b are five-coordinated with two histidine residues each from the subunits L and M forming metal-nitrogen bonds (L153–>B_A, L173–>D_L, M180–>B_B, M200–>D_M). D_L, D_M, Φ_A, Φ_B, Q_A, and Q_B form hydrogen bonds to the protein[13]. We use D for special pair, B for the accessory BChl, Φ for the BPh, and Q for the quinones. The branches are denoted by subscripts A and B. Because D belongs to both branches, its two BChl, whenever they are mentioned individually, are distinguished by subscripts L and M according to the protein subunit to which their magnesium atoms are linked.

The tetrapyrrole rings of BChl-b, BPh-b, and the quinone head groups follow the same approximate local symmetry displayed by the L and M chains. This local two-fold symmetry is most perfect in the arrangement of the special pair BChl-bs. The two molecules overlap with their pyrrole rings I. The orientation of the rings leads to a close proximity between the ring I acetyl groups and the Mg^{2+} ions, but the acetyl groups do not act as ligands to the Mg^{2+}. The pyrrole rings I from D_L and D_M are nearly parallel to each other, and to the symmetry axis; they are ~3.2 Å apart. However, the remainders of the tetrapyrrole ring systems of D_L and D_M are not parallel to each other: Planes through the pyrrole nitrogens of D_L and D_M form an angle of 11.3°. Individual atoms of the rings can deviate from these planes by more than 0.5 Å. The D_M ring is considerably more deformed than the D_L ring.

The tetrapyrrole rings of the BChl-b and BPh-b of D and of the A and B branches can be superimposed using a single coordinate transformation with the reasonably low rms deviation of 0.38 Å between the positions of equivalent atoms. Close inspection, however, shows significant differences between the local symmetry operations, individually superimposing D_M to D_L (180° rotation), B_B to B_A (176° rotation), and Φ_B to Φ_A (173° rotation)[6]. Imperfect symmetry causes the interatomic distances and inter-planar angles to differ in the two branches. For example, the closest distance of atoms involved in double bonds in D and in Φ_A is shorter by 0.7 Å than the corresponding distance between D and Φ_B. These structural differences, together with different environments of the cofactors[13] may account in part for different electron transfer properties in both branches (see below).

Large deviations from the local symmetry are manifest in the structures of the phytyl chains of the B and Φ cofactors, in the different chemical nature and different occupancy of the quinones (Q_B is bound to only ~ 30% of the RCs in the crystal), and in the presence of a carotenoid molecule near B_B only. Differences in the crystallographically refined B-values of Φ and of the phytyl chains of B indicate a higher mobility of the cofactors of the B branch.

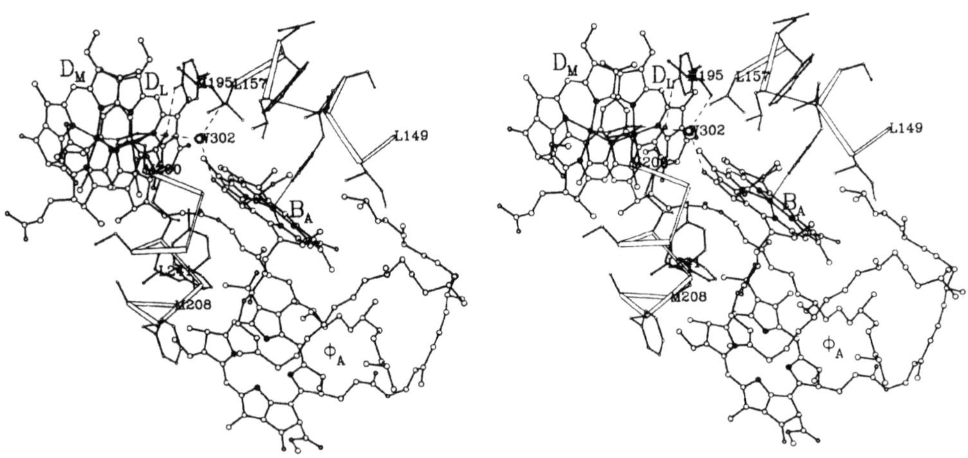

Figure 4. The D-B_A-Φ_A region. The phytyl chain of D_M is not shown. The two contiguous segments of polypeptide chain (residues M200 to M208, and L149 to L157) are drawn as ribbons, together with selected amino acid side chains. H-bonds are indicated by dashed lines. W302 is a localized water molecule.

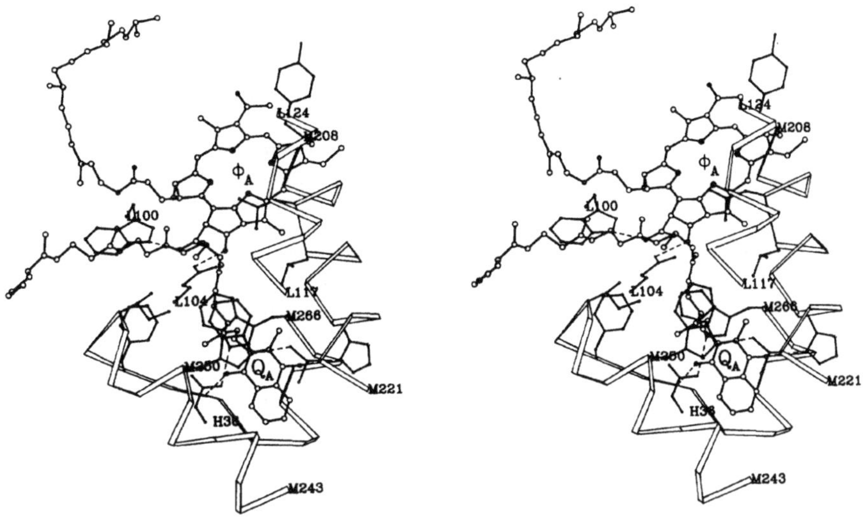

Figure 5. The Φ_A-Q_A region. Only part of the isoprenoid chain of Q_A is shown. Three contiguous segments of polypeptide chains (residues L117 to L124, M208 to M221, and M243 to M266) are drawn as ribbons. Selected amino acid side chains are also shown. H-bonds are indicated by dashed lines.

Electron transfer pathways

Functional aspects of RCs are the topic of most of the articles in this volume. Here we describe the main structural features along the electron tranfer pathways through the RC.

In spite of the structural symmetry, the RC is functionally highly asymmetric: Absorption of a photon or energy transfer from light-harvesting complexes in the membrane raises the special pair D to its first excited singlet state, D*, which is the starting point for a series of electron transfer reactions across the membrane, leading to Φ_A[24-26], Q_A, and Q_B. In the first step $D^+\Phi_A^-$ is formed from $D^*\Phi_A$ with a time constant of ~3 ps at room temperature, and ~1.1 ps at 10 K[27-29]. Figure 4 shows the RC-region along this pathway with D, B_A, Φ_A, selected protein residues, and a bound water molecule. The BChl-bs of D, in particular D_M, are in van der Waals contact to B_A; B_A is in turn in contact with Φ_A. The phytyl chains of D_L, B_A, and Φ_A participate significantly in these contacts, with the phytyl chain of D_L filling a cleft between B_A and Φ_A. The protein structure shown in the figure includes residues M200 to M208 from the D-helix of subunit M, residues L149 to L157 from the peripheral CD helix of subunit L, and phenylalanine L241. The histidines M200 and L173 are ligands to the Mg^{2+} ions of D_M, and B_A, respectively. The hydroxyl group of tyrosine M195 forms an H-bond to the ring I acetyl group of D_M. A localized water molecule forms H-bonds to B_A, the side chain of histidine M200, and to the backbone carbonyl oxygen of tyrosine M195. Without being involved in H-bonds, the side chain of tyrosine M208 occupies a central position in the arrangement of D, B_A, and Φ_A. Recent experimental results suggest electron transfer in two-steps, first from D* to B_A, followed by electron transfer from B_A^- to Φ_A[30].

From Φ_A^-, the electron is transferred to Q_A with a time constant of ~200 ps at room temperature[30-32]; this time constant shortens to ~100 ps at temperatures below 100 K[31]. Figure 5 shows the RC structure in the Φ_A - Q_A region. Φ_A is bound in a pocket formed by residues from the helices B, C and E of subunit L, and from helix D of subunit M.

The side chains of residues tryptophane L100, and glutamic acid L104 interact by H-bonds with carbonyl oxygens of Φ_A. The Q_A binding pocket includes residues from helices D and E, and from their connection. The head group and the first isoprenoid units of Q_A are flanked by the side chains of tryptophanes M250 and M266; tryptophane M250 points towards Φ_A. The two carbonyl oxygens of Q_A are H-bonded to the side chain of histidine M217 (one of the ligands of the non-heme iron), and to the peptide nitrogen of alanine M258. The "bottom" of the Q_A binding pocket is mostly hydrophobic and shielded from the cytoplasm by the subunits M and H.

Electron transfer from Q_A^- to Q_B happens with a time constant of ~25 μs[33]. Figure 6 shows a simplified view of the binding regions of Q_A, the non-heme iron, and Q_B. The non-heme iron is bound by four histidines (L190, L230, M217, and M264), and by the two side chain oxygens of glutamic acid M232. The Q_B binding site is formed by residues from the L-subunit which, in analogy to the Q_A site, are from helices D and E, and from their connection. The Q_B binding pocket differs from the Q_A pocket by a high content in polar or charged residues, such as glutamic acid L212, asparagine L213, arginine L217, serine L223, aspartic acid M43, and glutamic acid H177. Bound water molecules indicate possible pathways for protons from the cytoplasm to Q_B.

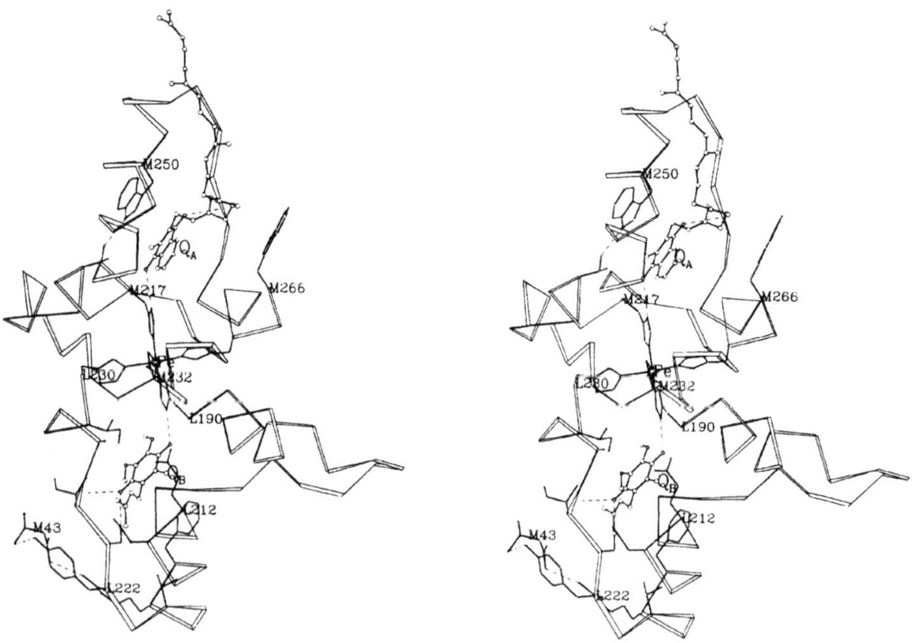

Figure 6. The Q_A-Fe-Q_B region. Only part of the isoprenoid chain of Q_A is shown. Two contiguous segments of polypeptide chain (residues L190 to L230, and M217 to M266) are drawn as ribbons. Selected amino acid side chains are also shown. H-bonds are indicated by dashed lines.

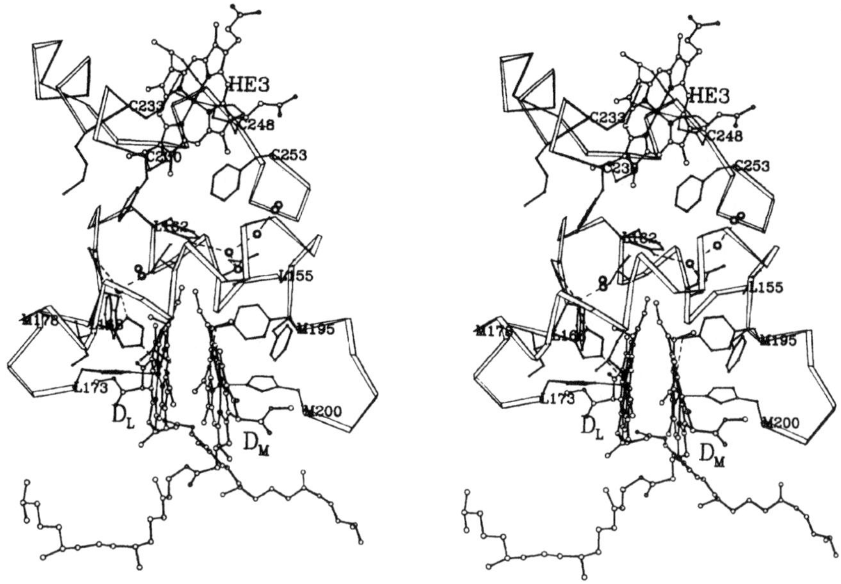

Figure 7. The region between heme 3 of the cytochrome and D. Three contiguous segments of polypeptide chain (residues C230 to C253, L155 to L173, and M178 to M200) are drawn as ribbons. Selected amino acid side chains and localized water molecules are also shown. H-bonds are indicated by dashed lines.

Electrons and protons leave the RC when the quinol dissociates; the electrons return to the RC via the cytochrome b/c1 complex, and the small soluble cytochrome c2, which transfers them to the RC's own cytochrome subunit. Electron transfer to D^+ from the heme groups in the cytochrome follows complicated, temperature dependent kinetics with an overall $t_{1/2}$ of 120 ns at room temperature, and 1.2 ms at 10 K[34]. Figure 7 shows the region between D and the closest heme group (HE3) in the cytochrome subunit. Residue tyrosine L162 is located directly between these two groups. The region is also rich in bound water molecules, some of which forms chains H-bonded to the protein, and to each other.

Acknowledgements

Our colleagues Otto Epp, Kunio Miki, and Irmi Sinning made significant contributions to the X-ray crystallographic studies of the RC from *Rps. viridis.* We thank Diana Diggs for technical assistance with the figures.

REFERENCES

1. H. Michel, K.A. Weyer, H. Gruenberg, and F. Lottspeich, The 'heavy' subunit of the photosynthetic reaction centre from Rhodopseudomonas viridis: isolation of the gene, nucleotide and amino acid sequence, *EMBO J.* 4:1667 (1985).
2. H. Michel, K.A. Weyer, H. Gruenberg, I. Dunger, D. Oesterhelt, and F. Lottspeich, The 'light' and 'medium' subunits ofthephotosynthetic reaction centre from Rhodopseudomonas viridis: isolation of the genes, nucleotide and amino acid sequence, *EMBO J.* 5:1149 (1986).

3. K.A. Weyer, F. Lottspeich, H. Gruenberg, F.S. Lang, D. Oesterhelt, and H. Michel, Amino acid sequence of the cytochrome subunit of the photosynthetic reaction centre from the purple bacterium Rhodopseudomonas viridis, *EMBO J.* 6:2197 (1987).
4. J.P. Thornber, R.J. Cogdell, R.E.B. Seftor, and G.D. Webster, Further Studies on the Composition and Spectral Properties of the Photochemical Reaction Centers of Bacteriochlorophyll-b Containing Bacteria, *Biochim. Biophys. Acta* 593:60 (1980).
5. P. Gast, T.J. Michalski, J.E. Hunt, and J.R. Norris, Determination of the amount and the type of quinones present in single crystals from reaction center protein from the photosynthetic bacterium Rhodopseudomonas viridis, *FEBS Lett.* 179:325 (1985).
6. J. Deisenhofer and H. Michel, The photosynthetic reaction centre from the purple bacterium *Rhodopseudomonas viridis,EMBO J.* 8:2149 (1989).
7. Michel, H. and J. Deisenhofer. 1990. The Photosynthetic Reaction Center from the Purple Bacterium *Rhodopseudomonas viridis*: Aspects of Membrane Protein Structure. In Current Topics in Membranes and Transport Volume 36: Protein-Membrane Interactions. T. Claudio, editor. Academic Press, San Diego. 53-69.
8. H.Michel, Three-dimensional Crystals of a Membrane Protein Complex The Photosynthetic Reaction Centre from Rhodopseudomonas viridis, *J. Mol. Biol.* 158:567 (1982).
9. H. Michel, Crystallization of membrane proteins, *Trends Bioch. Sci.* 8:56 (1983).
10. J. Deisenhofer, O. Epp, K. Miki, R. Huber, and H. Michel, X-Ray Structure Analysis of a Membrane Protein Complex: Electron Density Map at 3 A Resolution and a Model of the Chromophores of the Photosynthetic Reaction Center from Rhodopseudomonas viridis, *J. Mol. Biol.* 180:385 (1984).
11. B.-C. Wang, Resolution of Phase Ambiguity in Macromolecular Crystallography, *Meth. Enzymol.* 115:90 (1985).
12. J. Deisenhofer, O. Epp, K. Miki, R. Huber, and H. Michel, Structure of the protein subunits in the photosynthetic reaction centre of Rhodopseudomonas viridis at 3 A resolution, *Nature* 318:618 (1985).
13. H. Michel, O. Epp, and J. Deisenhofer, Pigment-protein interactions in the photosynthetic reaction centre from Rhodopseudomonas viridis, *EMBO J.* 5:2445 (1986).
14. Deisenhofer, J. and H. Michel. 1988. The Crystal Structure of the Photosynthetic Reaction Center from Rhodopseudomonas viridis. In The Photosynthetic Bacterial Reaction Center Structure and Dynamics. J. Breton and A. Vermeglio, editors. Plenum Press, New York. 1-3.
15. Deisenhofer, J. and H. Michel. 1989. The Photosynthetic Reaction Centre from the Purple Bacterium Rhodopseudomonas viridis. In Les Prix Nobel 1988. T. Frangsmyr, editor. Nobel Foundation, Stockholm. 134-188.
16. P.V. Luzzati, Traitement statistique des erreurs dans la determination des structures cristallines,Acta Crystallogr. 5:802 (1952).
17. M. Roth, A. Lewit-Bentley, H. Michel, J. Deisenhofer, R. Huber, and D. Oesterhelt, Detergent structure in crystals of a bacterial photosynthetic reaction centre, *Nature* 340:659 (1989).
18. M.H. Tadros, R. Frank, B. Doerge, N. Gad'on, J.Y. Takemoto, and G. Drews, Orientation of the B800-850, B870, and Reaction CenterPolypeptides on the Cytoplasmic and Periplasmic Surfaces of Rhodobacter capsulatus Membranes, *Biochemistry* 26:7680 (1987).
19. M.H. Tadros, R. Frank, J.Y. Takemoto, and G. Drews, Localization of reaction center and B800-850 antenna pigment proteins in membranes of *Rhodobacter sphaeroides, J. Bacteriol.* 170:2758 (1988).
20. M.H. Tadros, D. Spormann, and G. Drews, The localization of pigment-binding polypeptides in membranes of *Rhodopseudomonas viridis, FEMS Microbiology Letters* 55:243 (1988).
21. J. Breton, Orientation of the chromophores in the reaction center of *Rhodopseudomonas viridis*. Comparison of low-temperature linear dichroism spectra with a model derived from X-ray crystallography, *Biochim. Biophys. Acta* 810:235 (1985).
22. A.M. Lesk and K.D. Hardman, Computer-Generated Pictures of Proteins, *Meth. Enzymol.* 115:381 (1985).
23. K.A. Weyer, W. Schafer, F. Lottspeich, and H. Michel, The Cytochrome Subunit of the Photosynthetic Reaction Center from Rhodopseudomonas viridis Is a Lipoprotein, *Biochemistry* 26:2909 (1987).
24. W. Zinth, W. Kaiser, and H. Michel, Efficient Photochemical Activity and Strong Dichroism of Single Crystals of Reaction Centers from Rhodopseudomonas viridis, *Biochim. Biophys. Acta* 723:128 (1983).
25. W. Zinth, E.W. Knapp, S.F. Fischer, W. Kaiser, J. Deisenhofer, and H. Michel, Correlation of Structural and Spectroscopic Properties of a Photosynthetic Reaction Center, *Chem. Phys. Lett.* 119:1 (1985).

26. E.W. Knapp, S.F. Fischer, W. Zinth, M. Sander, W. Kaiser, J. Deisenhofer, and H. Michel, Analysis of optical spectra from single crystals of Rhodopseudomonas viridis reaction centers, *Proc. Natl. Acad. Sci. USA* 82:8463 (1985).
27. J.-L. Martin, J. Breton, A.J. Hoff, A. Migus, and A. Antonetti, Femtosecond spectroscopy of electron transfer in the reaction center of the photosynthetic bacterium Rhodopseudomonas sphaeroides R-26: Direct electron transfer from the dimeric bacteriochlorophyll primary donor to the bacteriopheophytin acceptor with a time constant of 2.8+-0.2 psec, *Proc. Natl. Acad. Sci. USA* 83:957 (1986).
28. J. Breton, J.-L. Martin, A. Migus, A. Antonetti, and A. Orszag, Femtosecond spectroscopy of excitation energy transfer and initial charge separation in the reaction center of the photosynthetic bacterium Rhodopseudomonas viridis, *Proc. Natl. Acad. Sci. USA* 83:5121 (1986).
29. G.R. Fleming, J.-L. Martin, and J. Breton, Rates of primary electron transfer in photosynthetic reaction centres and their mechanistic implications, *Nature* 333:190 (1988).
30. W. Holzapfel, U. Finkele, W. Kaiser, D. Oesterhelt, H. Scheer, H.U. Stilz, and W. Zinth, Initial electron-transfer in the reaction center from *Rhodobacter sphaeroides*, *Proc. Natl. Acad. Sci. USA* 87:5168 (1990).
31. C. Kirmaier, D. Holten, and W.W. Parson, Temperature and detection-wavelength dependence of the picosecond eletron-transfer kinetics measured in Rhodopseudomonas sphaeroides reaction centers. Resolution of new spectral and kinetic components in the primary charge-separation process, *Biochim. Biophys. Acta* 810:33 (1985).
32. C. Kirmaier and D. Holten, Primary Photochemistry of Reaction Centers from the Photosynthetic Purple Bacteria, *Photosynth. Res.* 13:225 (1987).
33. P. Mathis, I. Sinning, and H. Michel, Kinetics of electron transfer from the primary to the secondary quinone in *Rhodopseudomonas viridis*, *Biochim. Biophys. Acta Bio-Energetics* 1098:151 (1992).
34. J.M. Ortega and P. Mathis, Effect of temperature on the kinetics of electron transfer from the tetraheme cytochrome to the primary donor in *Rhodopseudomonas viridis*, *FEBS Lett.* 301:45 (1992).

CORRELATION BETWEEN THE POLARIZED LIGHT ABSORPTION AND THE X-RAY STRUCTURE OF SINGLE CRYSTALS OF THE REACTION CENTER FROM *RHODOBACTER SPHAEROIDES* R-26

Harry A. Frank and Mila L. Aldema

Department of Chemistry
University of Connecticut
215 Glenbrook Road
Storrs, CT 06269-3060

INTRODUCTION

The crystallization and X-ray diffraction analysis of the reaction center from the bacterium *Rb. sphaeroides* R-26 has revealed the three-dimensional structure of the protein and bound cofactors to atomic resolution (Allen et al., 1986; Allen et al., 1987; Yeates et al., 1987; Yeates et al., 1988; Allen et al., 1988; Komiya et al., 1988; Chang et al., 1991; El-Kabbani et al., 1991). With this information available, it is of interest to ask how the spectroscopic properties of the reaction center correlate with the structural features. Ultimately this correlation will allow an elucidation of the molecular details that control the spectral features and relate to the primary photochemical events carried out by the reaction center. Polarized light absorption is one technique for correlating the spectroscopic features with its molecular structure (Breton, 1985). In order to make the correlation more precise, it is distinctly advantageous to carry out the spectroscopic experiments directly on the crystalline samples used in the X-ray diffraction analyses. In this way the clearest link between the structure of the complex and its photochemical properties will emerge.

A specific question that can be asked concerning polarized light measurements on single crystals of the reaction center is: Does the existing X-ray structure predict a linear dichroism (defined as a polarization ratio) consistent with that measured on the single crystals? The present work seeks to answer this question by carrying out polarized light absorption measurements directly on the single crystals of *Rb. sphaeroides* R-26 reaction centers and comparing the results to the polarization ratios calculated from the X-ray structure coordinates.

MATERIALS AND METHODS

The reaction centers were isolated from *Rhodobacter sphaeroides* strain R-26 whole cells according to a procedure similar to those of Feher (1971) and Clayton and Wang (1971). This was as follows: The *Rb. sphaeroides* whole cells were diluted to an

The Photosynthetic Bacterial Reaction Center II
Edited by J. Breton and A. Verméglio, Plenum Press, New York, 1992

OD of ~ 50 cm^{-1} at 860 nm with 15 mM Tris buffer and 1.0 mM EDTA at pH 8.0. Small amounts of DNAase (~ 20 mg) and $MgCl_2$ (~ 0.1 g) were mixed into the solution. The homogenized cells were broken by high pressure (~ 1250 psi) at a dropwise flow rate through a French Pressure cell. The effluent was centrifuged at 8000 rpm (SS-34 rotor) for 10 minutes. The supernatant, containing chromatophores, from the low-speed spin was centrifuged for 90 minutes at 50,000 rpm (55.2 Ti rotor) at 4°C. This pelleted the chromatophores.

The chromatophores were then solubilized with 150 mM NaCl and 0.6% LDAO in 15 mM Tris buffer at pH 8 and ultracentrifuged for 90 minutes at 50,000 rpm. This separated the reaction center-rich supernatant from the light-harvesting complex rich pellet. The reaction center-rich supernatant was collected and subjected to further purification using successive high and low cut extractions with ammonium sulfate. Final purification was achieved by DEAE sephacel column chromatography. The purified protein complex was eluted from the column using 15 mM Tris buffer, pH 8.0 containing 300 mM NaCl, 1 mM EDTA and 0.06% LDAO. The NaCl was removed by dialysis of the complex in Spectra/Por membrane tubing (12,000 - 14,000 molecular weight cut-off) against 15 mM Tris buffer containing 0.06% LDAO, pH 8.0 at 4°C in the dark. The dialyzed sample was concentrated to a final absorbance of up to ~50 at 802 nm using an Amicon microconcentrator.

Crystallization of the complex was accomplished using the vapor diffusion technique at 20°C in the dark according to the procedure of Allen et al., (1986). 30 ml of a solution containing 3.8 % 1,2,3-heptanetriol (Oxyl, 8903, Bobingen, high m. p. isomer) as the amphiphile, 12 % of polyethylene glycol (MW 4000) and 0.32 M NaCl as the precipitating agents, 0.06 % LDAO as the detergent, and the complex at an optical density of ~10 at 802 nm, in 15 mM Tris-HCl buffer with 0.1% NaN_3 and 1 mM EDTA at pH 8.0 $\pm$ 0.02 were placed in a crystal growing well and allowed to equilibrate with a solution containing 22% PEG, 0.6M NaCl, 1 mM EDTA, and 0.1% NaN_3 at pH 8. After one

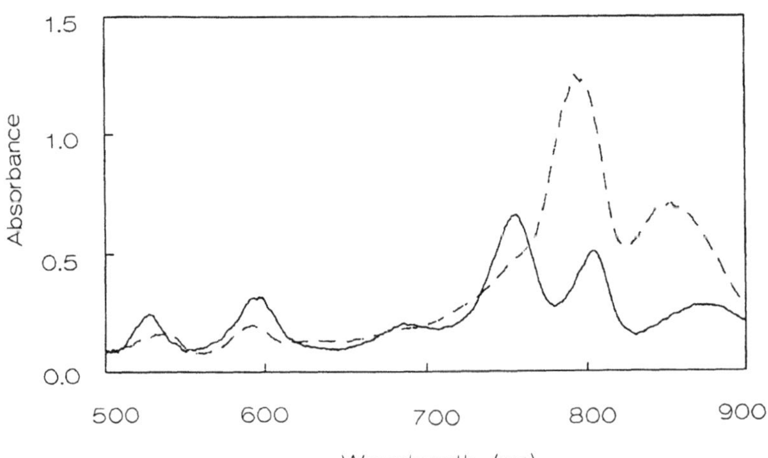

Figure 1. *Polarized absorption spectra of a single crystal of Rb. sphaeroides R-26. The solid line corresponds to the light polarization parallel to the long axis of the crystal. The dashed line corresponds to the light polarization perpendicular to the long axis of the crystal.*

week, several crystals were observed. Crystal growth was complete after 3-4 weeks. The dimensions of typical crystals used in the present experiments were 2-4 mm x 0.5-1 mm x 0.3 mm.

For the spectroscopic experiments, crystals from the well, with the equilibrated solution, were carefully deposited on a cover slip, around which was a thin ring of silicon vacuum grease. A second cover slip was placed on top to make a good seal. A mylar spacer was used between the cover slips to prevent the crystal from being pressed. The sample was then mounted in the microspectrometer previously described (Frank et al., 1991). For the polarized light experiments, an HN-7 polarizer sheet from Polaroid was placed in a rotating calibrated circular stage in the measuring beam and aligned relative to the morphological axes of the crystals. A video camera attached to the microspectrometer facilitated crystal selection and alignment.

The crystallographic notation used in the present work follows that of Allen et al., (1986). These authors found that for $P2_12_12_1$ crystals grown using LDAO as the detergent, the unit cell dimensions were a = 142.4 Å, b = 75.5 Å and c = 141.8 Å; *i. e.* $a > c > b$. Subsequent studies by this group (Allen et al., 1987) showed that upon exchanging the detergent in the crystals for β-octyl-glucoside the unit cell dimensions changed to a = 138.0 Å, b = 77.5 Å and c = 141.8 Å; *i. e.* $c > a > b$. The crystals used in this study are those of the former variety.

RESULTS AND DISCUSSION

The polarized light absorption spectra from a single crystal of *Rb. sphaeroides* R-26 are given in Fig. 1. The major features of the spectra are as follows:

Primary donor region

As the polarization of the measuring beam is rotated from parallel to perpendicular with respect to the long morphological axis of the crystal there is an increase in the polarized absorption intensity from 0.24 ± 0.03 to 0.70 ± 0.03 for the primary donor transition at ~860 nm. This results in a negative polarization ratio (defined below) of -0.48 ± 0.03 for this transition. A negative polarization ratio was also observed for the primary donor of *Rb. sphaeroides* wild type strain 2.4.1 previously reported at room and low temperatures (Frank et al., 1989; Frank et al., 1991). The negative polarization ratio indicates that the primary donor transition moments are preferentially oriented perpendicular to the long morphological axes of the crystal.

Bacteriochlorophyll monomer region

There is a clear resolution of at least two bacteriochlorophyll absorptions near 800 nm (Q_Y transitions). When the direction of polarization of the measuring beam is parallel to the long axis of the crystal, the absorption maximum is located at 804 nm. Rotation of the direction of polarization to perpendicular results in a build-up of absorption intensity from 0.50 ± 0.03 to 1.2 ± 0.03 and a broadening of the spectral feature. Also, there is an ~ 13 nm shift to shorter wavelength suggesting the presence of more than one transition moment. This pronounced enhancement of absorption gives rise to a large negative polarization ratios of -0.55 ± 0.02 and -0.36 ± 0.02 for the two peaks.

The bacteriochlorophyll absorption peaks near 600 nm (the Q_X transitions) have also been resolved in this work. Rotation of the polarizer from parallel to perpendicular produces an ~ 4 nm shift of the band to shorter wavelength. The long wavelength absorbing bacteriochlorophyll has been assigned to the bacteriochlorophyll (B_B) bound to the subunit that is inactive in electron transport (Maroti et al., 1985). The positive polarization ratio of 0.25 ± 0.04 implies that the Q_X transition moment for this molecule is

Table 1. *Summary of the experimentally observed polarization ratios for the various chromophores in the reaction center of Rb. sphaeroides R-26. BChl A and B are the monomeric bacteriochlorophylls and ϕ_A and ϕ_B are the bacteriopheophytins.*

	Transition	Wavelength (nm)	Polarization ratio
Primary donor	Q_Y	860	-0.48 ± 0.03
BChl A	Q_Y	791	-0.55 ± 0.03
BChl B	Q_Y	804	-0.36 ± 0.02
ϕ_A	Q_X	543	-0.15 ± 0.03
	Q_Y	not resolved	n. d.
ϕ_B	Q_X	528	0.30 ± 0.03
	Q_Y	755	0.16 ± 0.03

preferentially directed along the long axis of the crystal which also compliments the large negative polarization ratio of its orthogonal Q_Y transition described above.

Bacteriopheophytin region

Rotation of the direction of the polarization of the measuring beam from parallel to perpendicular results in almost complete disappearance of the ~760 nm peak (Q_Y transitions). The upper limit of the polarization ratio is 0.16 ± 0.03 and implies that the preferential orientation for this (presumably ϕ_B) bacteriopheophytin Q_Y transition moment is along the long morphological axis of the crystal.

The bacteriopheophytin Q_X region is very clearly resolved in these experiments. When the polarized light is rotated from parallel to perpendicular to the long axis of the crystal, the spectral maximum shifts from ~530 nm to ~545 nm. It has been reported that the bacteriopheophytin A (ϕ_A) absorbs at longer wavelengths than the bacteriopheophytin B (ϕ_B) (Kirmaier et al., 1985). The polarization ratio measured at 528 nm and 543 nm are $+0.30 \pm 0.03$ and -0.15 ± 0.03 respectively. This implies that ϕ_B (which absorbs at ~ 530 nm) with positive polarization ratio is preferentially oriented parallel whereas the ϕ_A (which absorbs at ~ 545 nm) with negative polarization ratio is oriented preferentially perpendicular to the long morphological axes of the crystal. Table 1 summarizes the polarization ratios for all the transitions.

The question posed above was whether the X-ray structure of the reaction center from *Rb. sphaeroides* R-26 predicted a linear dichroism consistent with that measured on the single crystals. In order to see if the experimentally observed linear dichroism (defined as a polarization ratio) is consistent with that predicted from the X-ray diffraction data, we shall focus on the spectral features of the bacteriopheophytin Q_X transitions near 540 nm.

Mathematical Model

The observed linear dichroism will be an average of the contributions from all four of the reaction center sites. The individual contributions are determined using the crystal site permutation matrices that depend on the space group symmetry of the unit cell. For the $P2_12_12_1$ space group these are as follows (Frank et al., 1989):

$$\hat{S}_1 = \begin{vmatrix} 1 & 0 & 0 \\ 0 & 1 & 0 \\ 0 & 0 & 1 \end{vmatrix} \qquad \hat{S}_2 = \begin{vmatrix} 1 & 0 & 0 \\ 0 & -1 & 0 \\ 0 & 0 & -1 \end{vmatrix}$$

$$\hat{S}_3 = \begin{vmatrix} -1 & 0 & 0 \\ 0 & 1 & 0 \\ 0 & 0 & -1 \end{vmatrix} \qquad \hat{S}_4 = \begin{vmatrix} -1 & 0 & 0 \\ 0 & -1 & 0 \\ 0 & 0 & 1 \end{vmatrix}$$

These matrices, denoted $\hat{S}_n$, are used to transform a vector from one site in the unit cell into another. In this case the vectors to be transformed are molecular structure segments that define Q_X transitions of the bacteriopheophytins. These structure segments are obtained from normalized vector differences between the X-ray diffraction atomic coordinates of individual nitrogens within the bacteriopheophytin ring systems. See Table 2.

This model assumes that the transition moments lie along these molecular structure segments. Within the limits of precision of this analysis, this is most likely a good approximation. The individual site, bacteriopheophytin, Q_X transition moment directions are denoted $\vec{\mu}(\phi_A)_i$ and $\vec{\mu}(\phi_B)_i$ where i represents the site index and ϕ_A and ϕ_B denote the two different bacteriopheophytins in the reaction center.

Once $\vec{\mu}(\phi_A)_i$ and $\vec{\mu}(\phi_B)_i$ are obtained for each site, their projections in the unit cell (a, b, c) axes must be transformed into the macroscopic crystalline morphological axis system (a, b, c). There are six different possible orientations for this unit cell to arrange itself in the crystal. See Fig. 2. We shall denote these "unit cell configuration" matrices by $\hat{P}_m$ where m indexes one of the six possible orientations of the unit cell with respect to the crystalline morphological axes. The $\hat{P}_m$ matrices are given as follows:

$$\hat{P}_1 = \begin{vmatrix} 1 & 0 & 0 \\ 0 & 1 & 0 \\ 0 & 0 & 1 \end{vmatrix} \qquad \hat{P}_2 = \begin{vmatrix} 1 & 0 & 0 \\ 0 & 0 & 1 \\ 0 & 1 & 0 \end{vmatrix}$$

$$\hat{P}_3 = \begin{vmatrix} 0 & 0 & 1 \\ 1 & 0 & 0 \\ 0 & 1 & 0 \end{vmatrix} \qquad \hat{P}_4 = \begin{vmatrix} 0 & 0 & 1 \\ 0 & 1 & 0 \\ 1 & 0 & 0 \end{vmatrix}$$

$$\hat{P}_5 = \begin{vmatrix} 0 & 1 & 0 \\ 1 & 0 & 0 \\ 0 & 0 & 1 \end{vmatrix} \qquad \hat{P}_6 = \begin{vmatrix} 0 & 1 & 0 \\ 0 & 0 & 1 \\ 1 & 0 & 0 \end{vmatrix}$$

Finally, the crystal must be oriented in the laboratory axis system (X, Y, Z). The experiment is carried out with light polarized parallel to the long axis of the crystal (denoted X) in the lab frame $\vec{E}_{\parallel} = (1,0,0)$ and (perpendicular to the long axis of the crystal (denoted either the Y or Z) in the lab frame. $\vec{E}_{\perp} = (0,1,0)$ or $\vec{E}_{\perp} = (0,0,1)$. The intensity of polarized absorption for a given transition is proportional to the sum (over the four unit cell sites) of the squares of the site-projected and unit cell-oriented transition moments in the laboratory axis system (X, Y, Z).

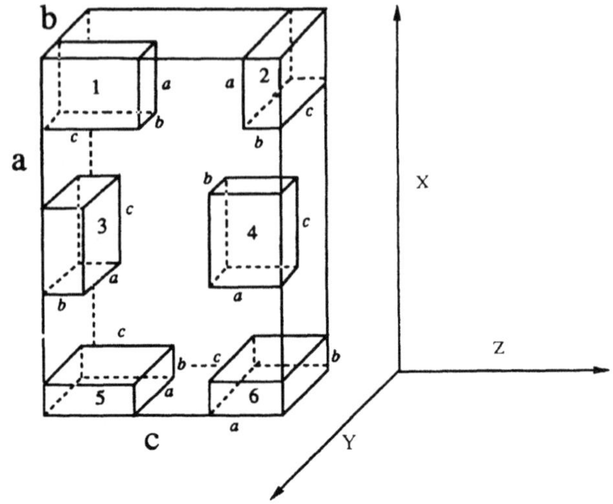

Figure 2. *The six different configurations of the unit cell axes (a, b, c) with respect to the crystalline macroscopic morphological axes* (a, b, c) *in the laboratory frame* (X, Y, Z).

Table 2. *Atomic coordinates (in Å units) of the central nitrogens of the bacteriopheophytin molecules in the reaction center of Rb. sphaeroides R-26. a, b and c refer to the unit cell axis system. N1, N2, N3, N4 refer to the nitrogens in the bacteriopheophytin rings 1-4, respectively. The vector differences representing the Q_X and Q_Y transition moments have been normalized.*

	a	*b*	*c*
ϕ_A			
N1	28.975	35.011	50.825
N2	29.035	35.937	53.698
N3	27.662	33.504	54.457
N4	27.641	32.513	51.653
Q_X (N2-N4)	0.33	0.81	0.48
Q_Y (N1-N3)	0.32	0.36	-0.88
ϕ_B			
N1	48.423	44.305	61.250
N2	48.066	41.494	60.436
N3	49.070	40.616	63.044
N4	49.484	43.369	63.832
Q_X (N2-N4)	-0.34	-0.45	-0.82
Q_Y (N1-N3)	-0.16	0.89	-0.43

$$A_{\parallel} \propto \sum_{n=1}^{4}\left[\hat{S}_n \cdot \hat{P}_m \cdot \vec{\mu}(\phi_{AB})_1 \cdot \vec{E}_{\parallel}\right]^2$$

$$A_{\perp} \propto \sum_{n=1}^{4}\left[\hat{S}_n \cdot \hat{P}_m \cdot \vec{\mu}(\phi_{AB})_1 \cdot \vec{E}_{\perp}\right]^2$$

Here $\vec{\mu}(\phi_{AB})_i$ represents the site 1 transition moment of either ϕ_A or ϕ_B. The linear dichroism is expressed as an anisotropy (or polarization) ratio as

$$p = \frac{A_{\parallel} - A_{\perp}}{A_{\parallel} + A_{\perp}}$$

The calculated ratios are then compared with the experimental results. Table 3 shows the values and signs of the bacteriopheophytin Q_X polarization ratios calculated from the X-ray diffraction coordinates based on each of the six different configurations of the unit cell in the crystal.

Among all the calculated polarization ratios, only the cases corresponding to $\hat{P}_3$ ($\vec{E}_{\perp}$ = (0,0,1)) and $\hat{P}_4$ ($\vec{E}_{\perp}$ = (0,1,0)), give the same sign combination (negative for the Q_X transition of ϕ_A and positive for the Q_X transition of ϕ_B) as the experimental values. These are the only orientations of the unit cell that are consistent with the observed polarization ratios. Both of these correspond to the case where the unit cell *c* axis is parallel to the crystal morphological a axis. The previous X-ray diffraction data indicate precisely the same alignment; *i. e.* that the *c* unit cell axis is parallel to the long morphological a axis of the crystal. Hence, the linear dichroism predicted from the X-ray structure is consistent with that experimentally observed.

Another interesting question that can be asked is whether the data from polarized light measurements on single crystals can be used to predict the orientations of the transition moments with respect to the molecular axis systems of the cofactors. For this one needs the individual-site projections of the transition moments in the unit cell axis system. Once these are obtained, the transition moment vectors can be overlaid with the atomic coordinates specified in the same axis system. Deconvolving the experimental linear dichroism into its individual-site transition moment projections is possible only with

Table 3. *Values of the ϕ_A and ϕ_B polarization ratios determined from the X-ray coordinates. These were calculated for the two cases where $\vec{E}_{\perp}$ = (0,1,0) and $\vec{E}_{\perp}$ = (0,0,1).*

	ϕ_A (Q_X) 543 nm		ϕ_B (Q_X) 528 nm	
$\hat{P}_m$	$\vec{E}_{\perp}$ = (0,1,0)	$\vec{E}_{\perp}$ = (0,0,1)	$\vec{E}_{\perp}$ = (0,1,0)	$\vec{E}_{\perp}$ = (0,0,1)
1	-0.715	-0.365	-0.272	-0.703
2	-0.365	-0.715	-0.703	-0.272
3	0.365	-0.474	0.703	0.533
4	-0.474	0.365	0.533	0.703
5	0.715	0.474	0.272	-0.533
6	0.474	0.715	-0.533	0.272

the aid of additional experimental data; *e. g.* from magnetophotoselection and single crystal EPR measurements. Such an analysis has been carried out for reaction centers from *Rb. sphaeroides* wild type strain 2.4.1 (Frank et al., 1989), and is in progress for the reaction centers from *Rb. sphaeroides* R-26.

ACKNOWLEDGMENTS

The authors wish to thank Professor D. C. Rees for supplying the atomic coordinates of the *Rb. sphaeroides* R-26 reaction center for this analysis. This work was supported by grants from the National Institutes of Health (GM-30353) and the University of Connecticut Research Foundation.

REFERENCES

Allen, J. P., Feher, G., Yeates, T. O., Komiya, H., and Rees, D. C., 1987, Structure of the reaction center from *Rhodobacter sphaeroides* R-26: The protein subunits. *Proc. Natl. Acad. Sci. U. S. A.* 84:6162-6166.

Allen, J. P., Feher, G., Yeates, T. O., Komiya, H., and Rees, D. C., 1988, Structure of the reaction center from *Rhodopseudomonas sphaeroides* R-26: Protein-cofactor (quinones and Fe^{2+}) interactions. *Proc. Natl. Acad. Sci. U. S. A.*, 85:8487-8491.

Allen, J. P., Feher, G., Yeates, T. O., Rees, D. C., Deisenhofer, J., Michel, H., and Huber, R., 1986, Structural homology of reaction centers from *Rhodopseudomonas sphaeroides* and *Rhodopseudomonas viridis* as determined by X-ray diffraction. *Proc. Natl. Acad. Sci. U. S. A.*, 83:8589-8593.

Breton, J., 1985, Orientation of the chromophores in the reaction center of Rhodopseudomonas viridis. Comparison of low temperature linear dichroism spectra with a model derived from X-ray crystallography. *Biochim. Biophys. Acta*, 810:235-245.

Chang, C. H., El-Kabbani, O., Tiede, D., Norris, J., and Schiffer, M., 1991, Structure of the membrane-bound protein photosynthetic reaction center from *Rhodobacter sphaeroides. Biochemistry*, 30:5352-5360.

Clayton, R. K. and Wang, R. T., 1971, Photochemical reaction centers from *Rhodopseudomonas spheroides. Methods Enzymol.* 23:696-704.

El-Kabbani, O., Chang, C. H., Tiede, D., Norris, J., and Schiffer, M., 1991, Comparison of reaction centers from *Rhodobacter sphaeroides* and *Rhodopseudomonas viridis*: Overall architecture and protein-pigment interactions. *Biochemistry*, 30:5361-5369.

Feher, G., 1971, Some Chemical and physical properties of a bacterial reaction center particle and its primary photochemical reactants. *Photochem. Photobiol.* 14:373-387.

Frank, H. A., Aldema, M. A., Violette, C. A. and Parot, P. H., 1991, Low temperature polarized absorption microspectroscopy of single crystals of the reaction center from *Rhodobacter sphaeroides* wild type strain 2.4.1. *Photochem. Photobiol.* 54:151-155.

Frank, H. A., Violette, C. A., Taremi, S. S., and Budil, D., 1989, Linear dichroism of single crystals of the reaction center from *Rhodobacter sphaeroides* wild type strain 2.4.1. *Photosyn. Res.* 21:107-116.

Kirmaier, C., Holten, D., and Parson, W. W., 1985, Picosecond-photodichroism studies of the transient states in *Rhodopseudomonas sphaeroides* reaction centers at 5 K. Effects of electron transfer on the six bacteriochlorin pigments. *Biochim. Biophys. Acta* 810:49-61.

Komiya, H., Yeates, T. O., Rees, D. C., Allen, J. P., and Feher, G., 1988, Structure of the reaction center from *Rhodopseudomonas sphaeroides* R-26 and 2.4.1: Symmetry relations and sequence comparisons between different species. *Proc. Natl. Acad. Sci. U. S. A.*, 85:9012-9016.

Maroti, P., Kirmaier, C., Wraight, C., Holten, D., and Pearlstein, R. M., 1985, Photochemistry and electron transfer in borohydride-treated photosynthetic reaction centers. *Biochim. Biophys. Acta*, 810:132-139.

Yeates, T. O., Komiya, H., Chirino, A., Rees, D. C., Allen, J. P., and Feher, G., 1988, Structure of the reaction center from *Rhodopseudomonas sphaeroides* R-26 and 2.4.1: Protein-cofactor (bacteriochlorophyll, bacteriopheophytin, and carotenoid) interactions. *Proc. Natl. Acad. Sci. U. S. A.*, 85:7993-7997.

Yeates, T. O., Komiya, H., Rees, D. C., Allen, J. P., and Feher, G., 1987, Structure of the reaction center from *Rhodobacter sphaeroides* R-26: Membrane-protein interactions. *Proc. Natl. Acad. Sci. U. S. A.*, 84:6438-6442.

SYMMETRICAL INTER-SUBUNIT SUPPRESSORS OF THE BACTERIAL REACTION CENTER cd-HELIX EXCHANGE MUTANTS

Steven J. Robles, Teresa Ranck, and Douglas C. Youvan

Department of Chemistry
Massachusetts Institute of Technology
Cambridge, MA 02139

INTRODUCTION

The L and M subunits of the reaction center (RC) show sequence homology and structural symmetry, yet paradoxically electron transfer is unidirectional. We have attempted to determine the structural unit of the RC responsible for the unidirectionality of electron transfer by switching homologous regions of the L and M subunits. In the past we reported the mutagenic exchange and duplication of the D helices in the RC of *Rhodobacter capsulatus*.[1,2] The exchange of the D helices did not produce any obvious wrong-way electron transfer mutants and, therefore, differences between the two subunits in this region do not appear to be the only ones responsible for undirectionality. We have now switched the regions which include the amphipathic cd-helices, each forming a major portion of a monomeric bacteriochlorophyll binding pocket and having several residues in the vicinity of the special pair of bacteriochlorophylls (P).[3,4]

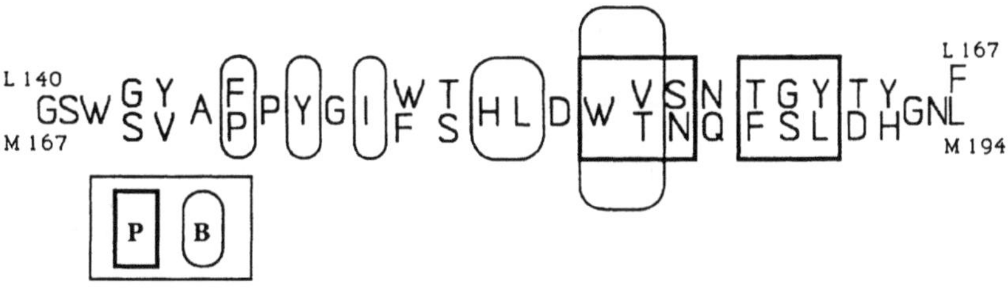

Figure 1. Amino acid sequence of the exchanged region of the *Rb. capsulatus* RC. Thirteen differences are present between the L and M subunit sequences (top and bottom respectively). The key on the bottom left of the figure indicates the interaction of these residues with the special pair (P), and the monomeric bacteriochlorophylls (B). The active or A-branch prosthetic groups interact with the L subunit residues in this region and those on the B-branch with the M subunit. Note that His residues L153 and M180 are the axial ligands for B_A and B_B, respectively.

The Photosynthetic Bacterial Reaction Center II
Edited by J. Breton and A. Verméglio, Plenum Press, New York, 1992

RESULTS

The amino acid sequences of the regions of the L and M subunits that were exchanged are shown in Figure 1. These regions encode the amphipathic cd helices which include the axial histidine ligands to the monomeric bacteriochlorophylls.[3] The methods for mutagenesis in the *Rb. capsulatus* RC have previously been described.[2]

Three mutants were constructed, cd_{LL}, cd_{MM} and cd_{LM}. The cd_{LL} mutant has an L-subunit cd-helix in the M-subunit, and the cd_{MM} mutant has an M-subunit cd-helix in the L-subunit. Both of these mutants express reduced levels of RCs, and neither appear to be significantly photochemically active. The mutant cd_{LM} which has the two cd-helices switched does not appear to produce RCs. None of these three mutants are capable of photosynthetic growth. Therefore a photosynthetic selection was used to obtain revertants for both cd_{LL} and cd_{MM}. Various attempts to isolate photosynthetically competent revertants for cd_{LM} were unsuccessful. Sixteen and eight individually isolated revertants for cd_{LL} and cd_{MM}, respectively, were sequenced (Table 1). Colonies of several revertants for the cd_{LL} mutant are blue or gray in color, and therefore probably have altered carotenoid expression. When the plasmids of these revertants were isolated and returned to the *Rb. capsulatus* strain U43, the colony pigmentation appeared normal. These reconstructed revertants are still capable of impaired photosynthetic growth. A reversion was found in both of two independent isolates of this type of revertant that were sequenced, at residue M188 which substituted Gly with Ser.

Table 1. Revertant Genotypes

Parent Strain	Reversion	Compensatory mutations	Occurrence[1]	PS assay[2]
cd_{LL}[3]	M188 G→S		8	+
cd_{LL}	M188 G→S M173(P) F→V		2	++
cd_{LL}	M188 G→S	L177 I→V	1	++
cd_{LL}	M188 G→S	L245 A→S	1	++
cd_{LL}	M188 G→S	L161 G→S	2	++
cd_{MM}		M271 A→P	5	++
cd_{MM}		L93 V→A	1	+++
cd_{MM}		L237 A→P	2	+

[1]The "Occurrence" column lists the number of times the sequence was independently isolated.
[2]The growth rates of the revertants are expressed relative to each other under this column, where cd_{LL} or cd_{MM} < + < ++ < +++ < wild type.
[3]All revertants for cd_{LL} have this reversion, including two not listed that have altered carotenoid expression.

DISCUSSION

As shown in Table 1, all photosynthetic revertants of cd_{LL} have a reversion at residue M188 which substituted the mutagenic Gly with Ser. This residue is in the vicinity of the B-branch bacteriochlorophyll of the special pair, P_B. In addition to M188 Gly to Ser, two cd_{LL} revertants had a second substitution in the mutagenized region of the M subunit which changed the Phe at M173 to Val. Pro occurs at this residue in WT. The remaining cd_{LL} revertants have suppressors or compensatory mutations at non-mutagenized residues in the L subunit. Two of these are L177 Ile to Val, and L245 Ala to Ser. These residues are in the D and E helices respectively, and both are in the vicinity of the A-branch bacteriochlorophyll of the dimer, P_A. A third suppressor for the cd_{LL} mutant occurs in the L-subunit cd-helix at L161 which is also near P_A. Interestingly, L161 is homologous to M188 at which the Gly to Ser reversion is simultaneously present. Like the M188 reversion, the compensatory mutation at L161 substitutes Ser for Gly.

Analogous to the cd_{LL} mutant, symmetrical suppressors were also obtained for the cd_{MM} mutant. One suppressor for the cd_{MM} mutant resulted in the substitution L237 Ala to Pro. This residue is in the E helix near H_A. L237 is outside of the mutagenized region, but occurs in the mutagenized subunit. Another compensatory mutation that occurred independently five times is M271 Ala to Pro. M271 is homologous to L237, and in the vicinity of H_B. In the cd_{MM} mutant the symmetrical suppressors occur independently in separate revertants, where as in the cd_{LL} mutant they occur simultaneously.

Previously we found that several compensatory mutations could suppress the photosynthetically impaired phenotypes of the D helix exchange mutants.[1,2] These are all intragenic or intra-subunit suppressors, that is, they occur in the same subunit or gene as the original mutation. Although some are in other helices, these D helix suppressors are structurally in the vicinity of the mutagenized region. Such intragenic suppressors are not unexpected, and have been isolated for mutations in other proteins, such as staphylococcal nuclease.[5] For the cd helix mutants, not only have we isolated an intra-subunit suppressor (i.e. L237 Ala to Pro), but also inter-subunit suppressors (i.e. M271 Ala to Pro, L245 Ala to Ser, L177 Ile to Val and L161 Gly to Ser). In addition, two of these inter-subunit suppressors occur at residues which are homologous to an intra-subunit suppressor or a reversion in the other subunit. Neither the inter- nor the intra-subunit suppressors are in the immediate vicinity of the original mutations.

These results are difficult to interpret but they do lead to interesting hypotheses. Since the RC has been symmetrized, one possibility is second site mutations in either subunit could produce similar effects that stabilize the RC structure. A speculative but more interesting possibility is that the symmetrization of the RC causes the two chromophore branches to have the same potential for electron transfer. If the two chromophore branches in the RC are structurally and functionally similar in these helix exchange mutants, either branch could be activated by a secondary mutation. Because of the symmetry, a compensatory mutation activating the A-branch would be homologous to a mutation that activates the B-branch. Future experiments will be aimed at detecting such aberrant electron transfer in these mutants.

REFERENCES

1. S. J. Robles, J. Breton, and D. C. Youvan (1990) Science 248, 1402-1405.
2. S. J. Robles, J. Breton, and D. C. Youvan (1990) in "Reaction Centers of Photosynthetic Bacteria", Vol. 6 (M.-E. Michel-Beyerle, ed.) Springer Series in Biophysics, pp. 283-289.
3. J. Deisenhofer, O. Epp, K. Miki, R. Huber, and H. Michel (1985) Nature 618-624.
4. D. M. Tiede, D. E. Budil, J. Tang, O. El-Kabbani, J. Norris, C.-H. Chang, and M. Schiffer (1988) in "The Photosynthetic Reaction Center: Structure and Dynamics", Vol. 149 (J. Breton, and A. Vermeglio eds.) Plenum Press, pp. 13-20.
5. D. Shortle, and B. Lin (1985) Genetics 110: 539-555.

MUTATIONS THAT AFFECT THE DONOR MIDPOINT POTENTIAL IN REACTION CENTERS FROM *RHODOBACTER SPHAEROIDES*

J. C. Williams, N. W. Woodbury, A. K. W. Taguchi, J. M. Peloquin, H. A. Murchison, R. G. Alden, and J. P. Allen

Department of Chemistry and Biochemistry
and Center for the Study of Early Events in Photosynthesis
Arizona State University, Tempe, AZ 85287-1604, USA

INTRODUCTION

Modulation of the redox midpoint potential of the initial electron donor is key for achieving electron transfer with high yields in photosynthetic systems. The initial electron donor in reaction centers of purple nonsulfur bacteria, P, has a much lower potential than the donor in photosystem II reaction centers, which are capable of oxidizing water (for general reviews of chlorophylls and their oxidation potentials, see ref. 1). Isolated bacteriochlorophyll and chlorophyll have similar oxidation potentials of approximately 700 mV and 800 mV, respectively. However, *in vivo* the midpoint oxidation potential of the donor in bacterial systems, approximately 490 mV, is much lower than that of monomer bacteriochlorophyll, while that of the donor in photosystem II is higher than 1.0 V. This dramatic alteration of the oxidation potentials in the reaction center must be due to interactions between chlorophylls or between the donor and the surrounding protein. This is supported by the observation of changes of approximately 100 mV in the oxidation potentials for bacteriochlorophylls and chlorophylls due to changes in solvent[1]. In this paper we address the role of specific protein-donor interactions in modulating the potential.

The Photosynthetic Bacterial Reaction Center II
Edited by J. Breton and A. Verméglio, Plenum Press, New York, 1992

RELATIONSHIP BETWEEN HYDROGEN BONDS AND P/P+ MIDPOINT POTENTIAL

One of the most direct ways that a protein can interact with a bacteriochlorophyll is through hydrogen bonds to the carbonyl groups that are part of the conjugated system. The influence of hydrogen bonds on the dimer midpoint potential has been tested by the construction of three single site mutations in *Rb. sphaeroides* (Table I). Two mutations, Leu to His at M160 and Leu to His L131, were each designed to introduce a hydrogen bond to the ring V keto group of one of the bacteriochlorophylls of the electron donor[2]. Both of these mutations resulted in a substantial increase in the midpoint potential of the electron donor as measured by chemical titration experiments (Table I). Another mutation, Phe to His at L168, results in the loss of the hydrogen bond between His L168 and one of the ring I acetyl groups of the donor[3]. This mutation resulted in a decrease in the midpoint potential of P by about 80 mV (Table I).

Table I. Alteration of hydrogen bonds to P by mutagenesis

Strain name	Mutation	Midpoint potential[2,3] compared to wild type	Putative change in hydrogen bonding
LH(L131)	Leu to His at L131	+ 80 mV	Addition to 9-keto of P_A
LH(M160)	Leu to His at M160	+ 55 mV	Addition to 9-keto of P_B
HF(L168)	His to Phe at L168	− 80 mV	Loss from 2a-acetyl of P_A

The introduction or removal of hydrogen bonds could change the oxidation potential of the P/P+ couple in the reaction center in at least two general ways. First, hydrogen bonding could directly affect the electronic structure of P through the bonding interaction, altering the conjugated π system. Second, hydrogen bonding could cause distortions of the bacteriochlorophyll ring system or change the relative orientation of the two bacteriochlorophylls. Significant structural changes due to hydrogen bonds to the 9-keto groups are unlikely. However, rotation of the acetyl group of ring I due to the formation of a hydrogen bond could significantly change the electronic structure[4]. Detailed studies are in progress of the state P^+ in these mutants in terms of spin delocalization *via* ENDOR measurements (see Rautter, *et al.* in these proceedings) and C=O vibrational frequencies using difference FTIR measurements (see Nabedryk, *et al.* these proceedings).

RELATIONSHIP BETWEEN P/P+ MIDPOINT POTENTIAL AND ELECTRON TRANSFER RATES

Alteration of the energy level of the oxidized electron donor, P^+, results in changes in the energies of the charge separated states $P^+H_A^-$, $P^+Q_A^-$, and $P^+B_A^-$. Due to the changes in energy levels, the mutants have altered driving forces, *i.e.* altered free energy differences between the initial and final states, for both the forward charge separation $P^* \rightarrow P^+H_A^-$ and the charge recombination $P^+Q_A^- \rightarrow PQ_A$. These altered driving forces should lead to changes in the associated rates of electron transfer. The results of experimental measurements of this relationship for both reactions are described below.

Relative changes of the $P^*/P^+H_A^-$ standard energy difference can be estimated from P/P^+ oxidation/reduction measurements. Due to the proximity of the mutations to P, the change in the P/P^+ redox potential is likely to be the dominant perturbation to the driving force for initial electron transfer in these mutants[2,3]. The P* decay times, as measured using a femtosecond transient laser system, were determined to be 4.5, 5.7, and 12.2 ps for the HF(L168), LH(M160), and LH(L131) mutants, respectively, compared to 3.4 ps for the wild type[2,3]. Slower rates are observed for mutants that have higher midpoint potentials and thus lower driving forces than wild type, while a rate comparable to wild type was observed for the HF(L168) mutant that has a lower potential and hence a larger driving force than wild type (Figure 1).

The charge recombination from $P^+Q_A^-$ is driven by the $P^+Q_A^-/PQ_A$ energy difference. The mutations are approximately 20 Å from Q_A and no significant change for the Q_A^-/Q_A energy difference is expected. This is consistent with the observation that the rate of electron transfer from H_A to Q_A does not change substantially in the mutants[2,3]. The charge recombination rate was measured by monitoring the change in absorption at 865 nm after excitation by a flash lamp. The charge recombination characteristic times determined for the HF(L168), LH(M160), and LH(L131) mutants were 220, 75, and 70 ms, respectively, compared to 100 ms for the wild type[2,3]. A correlation is found between the rates and the driving force in these mutants (Figure 2). In contrast with the data of Figure 1, driving forces that are higher than that of wild type result in increased rates.

COMPARISON OF DATA WITH MODELS AND OTHER MEASUREMENTS

Modeling of the relationship between the initial rate of electron transfer and the $P^*/P^+H_A^-$ energy difference is difficult due to uncertainty concerning the role of B_A and the energy of the state $P^+B_A^-$ [5,6]. The three mutations are near P and experiments indicate that the principal effect of the changes is to alter the P/P^+ potential and not the P/P^* energy difference[2,3]. If one assumes that the P/P^+ redox potential is the only thermodynamic parameter affected by these mutations, then the $P^*/P^+B_A^-$ and $P^*/P^+H_A^-$ energy differences should be altered from the wild type to the same extent in the mutants and the $P^+B_A^-/P^+H_A^-$ energy difference should be the same as wild type. Under this assumption, the $P^*/P^+B_A^-$ energy difference should change by the same amount as the P/P^+ midpoint potential, *i.e.* 80 meV for the LH(L131) mutant. Preliminary experiments show that the P* decay rate in the LH(L131) mutant has a weak inverse temperature dependence. If B_A serves as a real

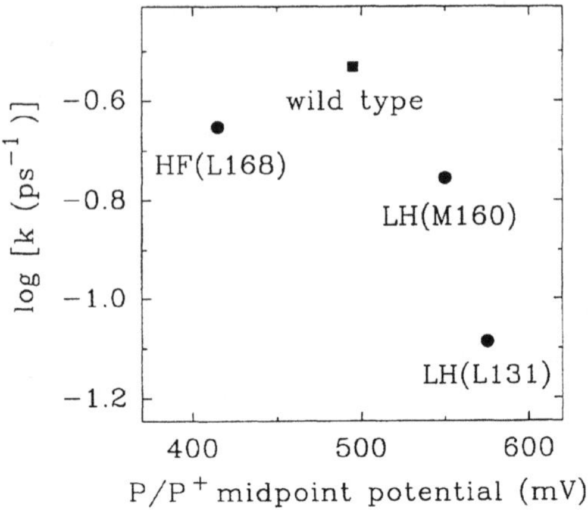

Figure 1. Relationship between the rate of electron transfer from the excited electron donor to the intermediate electron acceptor and the P/P$^+$ midpoint potential at 295 K for mutants and wild type in reaction centers from *Rb. sphaeroides*. Details concerning the measurements have been presented[2,3].

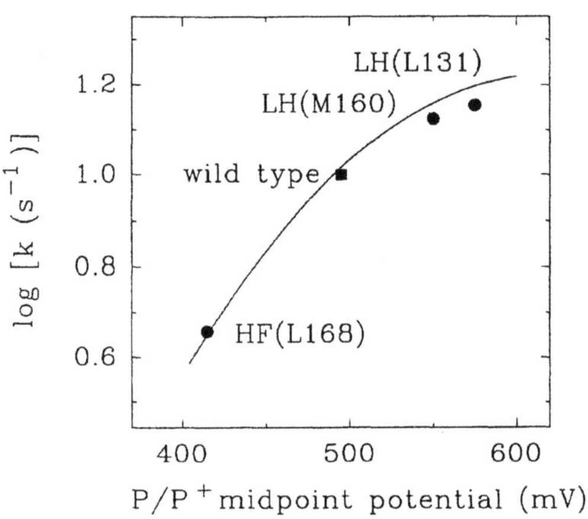

Figure 2. Correlation between the rate of charge recombination from Q_A^- to the electron donor P$^+$ and the P/P$^+$ midpoint potential at 295 K for mutants and wild type in reaction centers from *Rb. sphaeroides*. Details concerning the measurements have been presented[2,3]. The curve is a fit to the data using Eq. 1 with the parameters $\lambda = 635$ meV and $\nu = 360$ cm^{-1}.

populated acceptor, then for the $P^+B_A^-$ state to remain below the P* state in the mutants it should lie at least 80 meV below P* in the wild type. If B_A serves as a virtual acceptor, then the combined effect of decreasing the $P^*/P^+H_A^-$ energy difference while increasing the $P^*/P^+B_A^-$ activation energy by 80 meV results in only a 4 fold change in electron transfer rate. This is a surprisingly small effect since these changes are comparable to the $P^*/P^+H_A^-$ energy difference[7].

Charge recombination from the primary quinone Q_A to the electron donor P occurs in 100 ms by a direct non-activated process in wild type reaction centers[8-10]. The $P^+Q_A^-/PQ_A$ energy difference which drives this process has been measured to be essentially equivalent to the P/P^+ midpoint potential in wild type reaction centers[8]. This is expected to be true also for the mutants. The relationship between the rate and midpoint potential can be modeled assuming that the electron coupling constant does not change due to the mutations. In particular, we consider a widely-used general model that assumes the donor and the acceptor are coupled to a single vibrational mode in a harmonic approximation[11]. For a system with a temperature T and a driving force ΔG, the dependence of the rate on driving force is:

$$k \propto \left(\frac{v+1}{v}\right)^{P/2} \exp[-S(2v+1)]\, I_P\!\left(2S\sqrt{v(v+1)}\right) \qquad (1)$$

where $v = 1/[\exp(h\nu/k_BT) - 1]$ which is the thermal population of a mode with vibrational frequency ν, k_B is the Boltzmann constant, $P = -\Delta G/h\nu$, $I_P(\ )$ is a modified Bessel function of order P, and $S = \lambda/h\nu$ where λ is the reorganizational energy. The data shown in Figure 2 was modeled using the parameters $\lambda = 635$ meV and $\nu = 360\ \mathrm{cm}^{-1}$. The dependence of the recombination rate upon the $P^+Q_A^-/PQ_A$ energy difference has been investigated by several groups. Our data agrees well with the results from experiments that altered the energy difference by replacing the native ubiquinone with quinones of other redox potentials[10]. An alternate approach has been to alter the energy difference by the application of an electric field across the reaction centers. Our results are in qualitative agreement with the studies of Feher, *et al.*[9] and Franzen, *et al.*[12] but significant differences are found between our data and studies using reaction centers embedded in Langmuir-Blodgett films[13]. Possible reasons for these differences, including protein dehydration in the films and local field corrections, have been discussed previously[9,12,13].

CONCLUSIONS

It thus appears that the gain or loss of a hydrogen bond between the protein matrix and P can dramatically alter the midpoint potential of the electron donor. It is clear from the data that the redox potential has been set by interactions with the protein to both maximize the forward electron transfer from P^* to $P^+H_A^-$ and minimize the backward transfer from $P^+Q_A^-$ to PQ_A. This has been accomplished by approximately matching the driving force with the reorganizational energy for the forward reaction but having a large difference between the reorganizational energy and free energy for the back reaction. The question of whether even higher potentials can be achieved by multiple hydrogen bonds is being addressed by double mutants. As structural studies of photosystem II progress, it will be interesting to see whether hydrogen bonding plays a major role in achieving the strongly elevated redox potential of the donor.

ACKNOWLEDGMENTS

We wish to thank V. H. Coryell for assistance with the preparation of the reaction centers. This work was supported by grants GM41300 and GM45902 from the N. I. H., grants DMB89-177729 and DMB91-58251 from the N. S. F., a Postdoctoral Fellowship in Plant Biology DIR-9104322 from the N. S. F. and a Shell Oil graduate fellowship. Instrumentation was purchased with funds from N. S. F. grant DIR-8804992 and D. O. E. grants DE-FG-05-88-ER75443 and DE-FG-05-87-ER75361. This is publication #110 from the Arizona State University Center for the Study of Early Events in Photosynthesis. The Center is funded by D. O. E. grant DE-FG-88-ER13969 as a part of the USDA/DOE/NSF Plant Science Centers Program.

REFERENCES

1. H. Scheer, ed. "Chlorophylls", CRC Press, Boca Raton (1991).
2. J. C. Williams, R. G. Alden, H. A. Murchison, J. M. Peloquin, N. W. Woodbury, and J. P. Allen, Effects of mutations near the bacteriochlorophylls in reaction centers from *Rhodobacter sphaeroides*, *Biochemistry*, submitted.
3. H. A. Murchison, R. G. Alden, J. P. Allen, J. M. Peloquin, A. K. Taguchi, N. W. Woodbury, and J. C. Williams, manuscript in preparation.
4. M. Plato, E. Trankle, W. Lubitz, F. Lendzian, and K. Mobius, Molecular orbital investigation of dimer formations of bacteriochlorophyll a. Model configurations for the primary donor of photosynthesis, *Chem. Phys.* 107: 185 (1986).
5. W. Holzapfel, U. Finkele, W. Kaiser, D. Oesterhelt, H. Scheer, H. U. Stilz, and W. Zinth, Initial electron-transfer in the reaction center from *Rhodobacter sphaeroides*, *Proc. Natl. Acad. Sci. USA* 87, 5168 (1990)
6. C. Kirmaier and D. Holten, An assessment of the mechanism of initial electron transfer in bacterial reaction centers, *Biochemistry* 30, 609 (1991).
7. N. W. T. Woodbury and W. W. Parson, Nanosecond fluorescence from isolated photosynthetic reaction centers of *Rhodopseudomonas sphaeroides*, *Biochim. Biophys. Acta* 767: 345 (1984).

8. H. Arata and W. W. Parson, Delayed fluorescence from *Rhodopseudomonas sphaeroides* reaction centers enthalpy and free energy changes accompanying electron transfer from P-870 to quinones, Biochim. Biophys. Acta 638: 201 (1981).
9. G. Feher, T. R. Arno, and M. Y. Okamura The effect of an electric field on the charge recombination rate of $D^{+}Q_{A}^{-} \rightarrow DQ_{A}$ in reaction centers from *Rhodobacter sphaeroides* R-26, in: "The Photosynthetic Bacterial Reaction Center" J. Breton and A. Vermeglio, eds., Plenum, New York, 271 (1988).
10. M. R. Gunner and P. L. Dutton, Temperature and $-\Delta G^{o}$ dependence of the electron transfer from Bph^{-} to Q_A in reaction center protein from *Rhodobacter sphaeroides* with different quinones as Q_A, *J. Am. Chem. Soc.* 111: 3400 (1989).
11. J. Jortner, Temperature dependent activation energy for electron transfer between biological molecules, *J. Chem. Phys.* 64: 4860 (1976).
12. S. Franzen, R. F. Goldstein, and S. G. Boxer, Electric field modulation of electron transfer reaction rates in isotropic systems: long-distance charge recombination in photosynthetic reaction centers, *J. Phys. Chem.* 94: 5135 (1990).
13. Z. D. Popovic, G. J. Kovacs, P. S. Vincett, G. Alegria, and P. L. Dutton, Electric field dependence of recombination kinetics in reaction centers of photosynthetic bacteria, *Chem. Phys.* 110:227 (1986).

SUGGESTIONS FOR DIRECTED ENGINEERING OF REACTION CENTERS: METAL, SUBSTITUENT AND CHARGE MODIFICATIONS

Jack Fajer,[1] Louise Karle Hanson,[1] Michael C. Zerner,[2] and Mark A. Thompson[1,2,3]

[1]Department of Applied Science, Brookhaven National Laboratory Upton, NY 11973

[2]Quantum Theory Project, University of Florida, Gainesville, FL 32611

[3]Molecular Science Research Center, Pacific Northwest Laboratory, Richland WA 99352

INTRODUCTION

The multifaceted biological roles fulfilled by porphyrin derivatives repeatedly illustrate the control of electronic and chemical properties exerted by variations in the porphyrin macrocycles as well as their metals, peripheral substituents, and protein microenvironments.[1,2] The rapid advances in mutation technology[3,4,5] and the recent developments in chromophore exchanges[6] offer unprecedented opportunities to probe the structure-function interactions that control the photophysics and photochemistry of photosynthetic reaction centers (and antennas). We discuss here experimental and theoretical models that predict or rationalize the consequences of altering the metals and substituents of the chromophores as well as their protein environment. Specifically, we consider the following: 1) The substitution of zinc or nickel for the magnesium of the bacteriochlorophylls (BChl) in the reaction center (RC), or the insertion of these metals into the bacteriopheophytins (BPheo), both as redox modifiers and as EXAFS probes of their immediate coordination sphere. 2) Modifications of the peripheral substituents of the chromophores and their observed and predicted effects on the optical properties of RCs with particular attention to the large differences in phototrap energies in organisms comprised of BChls b and g. 3) The theoretical basis for electrochromic effects induced by the generation of cations and anions in RCs, or by the directed introduction of "charged" or highly polar residues.

METAL SUBSTITUTIONS

Mutations of axial histidines that induce the loss of the magnesium from BChl or its introduction into BPheo have already yielded RCs with significantly altered properties. Examples are heterodimers in which one magnesium is lost in the special pair of BChls,[7,8] or the insertion of a magnesium converting the native BPh electron acceptor into a BChl.[9]

The Photosynthetic Bacterial Reaction Center II
Edited by J. Breton and A. Verméglio, Plenum Press, New York, 1992

Recent reconstitution experiments[6] open the possibility of selectively introducing other metallo bacteriochlorins (or chlorins) at the acceptor BChl or BPheo sites. Zinc substitutions are obvious choices since the optical and coordination properties of Zn complexes resemble those of Mg derivatives but the former are easier to reduce (and harder to oxidize) than the latter.[10,11] Perhaps a less obvious candidate for metal replacement is nickel which offers several advantages: Ni complexes are also easier to reduce and harder to oxidize than Mg compounds but, in addition, redox potentials of Ni porphyrins and chlorins are close to those of the metal-free derivatives; the optical properties of Ni chlorins are similar to those of Mg ones; they possess short-lived excited states as a result of which they have proved to be valuable for resonance Raman studies.[11,12] A more practical advantage of Ni complexes is their enhanced stability towards demetallation, compared to Mg or Zn complexes, a property of merit if the chromophores are to be used in reconstitution reactions. (Ni protoporphyrins IX have been successfully exchanged for the histidine-bound hemes in hemoglobin.)[13]

Selective reconstitution of native RCs with Zn or Ni affords the possibility of probing the immediate structural environment of the modified chromophores by synchrotron radiation techniques such as XANES and EXAFS, and, eventually, time-resolved EXAFS, without the need for crystals. Ni-N distances have been shown to be sensitive diagnostic probes of macrocycle conformations in Ni porphyrins, chlorins and tetrapyrroles. Short Ni-N distances are typically found in nonplanar conformations whereas long distances signal planar macrocycles.[14-16] When such data are available, comparison of EXAFS and crystallographic data for several Ni tetrapyrroles agree within 0.02 Å.[15,16] A further advantage of using Ni substitutions is that the coordination number of the Ni is revealed almost on inspection of the x-ray absorption pre-edge features. In square planar complexes, a distinct peak, attributed to a 1s → $4p_z$ electronic transition is present at approximately 6 eV below the main absorption edge. Introduction of one axial ligand results in attenuation of this peak together with the enhancement of a weakly allowed 1s → 3d peak, arising from the reduction in symmetry. Addition of a second ligand to form hexaccordinated Ni results in the disappearance of the preedge peaks.[16] Several examples of these changes in Ni chlorins and other hydroporphyrins have recently been reported.[16] The structural consequences of redox changes of Ni tetrapyrroles have also been investigated.[15-17] Formation of the π anion radicals of a chlorin and a porphycene (a porphyrin isomer) leaves the metal environment undisturbed. As well, the close similarity between EXAFS spectra of the parents and the radicals indicates that no significant changes occur in the overall conformations of the compounds in the two states. Extrapolation of these results to BPheo or BChl in vivo suggests that formation of their π anions during charge separation does not induce large structural changes. Obviously, the in vitro results do not address possible changes due to chromophore-protein interactions.

PERIPHERAL MODIFICATIONS

Modifications of peripheral substituents in natural photosynthetic chromophores are common. Examples are Chls a, b, d and BChls c, d, e among chlorins, and BChls a, b and g among bacteriochlorins.[18] At the chlorin saturation level, the effects of variations at the common C3 site of ring I are readily demonstrated by a series of Ni(II) pyropheophorbide a derivatives synthesized by Smith and Shiau in which the usual 3-vinyl group found in Chl a is replaced.[19] The Qy transitions for the chromophores red-shift in the order: ethyl (640 nm), hydroxymethyl (644 nm), hydroxyethyl (646 nm), vinyl (654 nm), acetyl (666 nm), cyano (668 nm) and formyl (680 nm). Clearly, the more polar groups that can also conjugate with the chlorin π system cause the larger red shifts. Theoretical calculations suggest a similar trend for bacteriochlorins.[20,21] These simple models thus readily predict the optical shifts that may be expected by synthetic (or biosynthetic) manipulation of the molecules.

Particularly striking examples of the influence of peripheral substituents are the recently discovered BChls g found in the Heliobacteria Heliobacterium chlorum, H. gestii, H. fasciculum, and Heliobacillus mobilis.[22] The molecular structure of BChl g is similar to that of the more common BChl b, except for the substitution of a vinyl group for the 3-acetyl function found on ring I of BChl b. Optical spectra of BChls b and g do not differ significantly in vitro. The low energy Qy absorption maxima are found at 791 and 762 nm in acetone, for b and g, respectively.[23] In vivo, however, RCs of Rhodopseudomonas viridis, that contain BChls b, exhibit low energy maxima at 960 nm[6] whereas the first RC red bands of H. chlorum and Hb. mobilis,[22] comprised of BChls g, are found at 798-800 nm, a difference of 2000 cm^{-1} in the energies of the phototraps or primary electron donors.

We have recently presented INDO/s calculations that considered whether the large optical difference between RCs comprised of BChls b and g reflected a different architecture for the primary donor in Heliobacteria or whether the observed differences could be still explained in terms of the special pairs found in RCs containing BChls b or a.[21] The results are summarized in Table 1.

Calculations of BChls b, based on the coordinates[24] of the two monomer subunits of the special pair in Rps. viridis, yield Qy values of 12484 cm^{-1} (801 nm) and 12789 cm^{-1} (782 nm).[25] The differences are due to variations in the conformations of the porphyrin cores and in the orientations of the acetyl groups. The average of the calculated Qy transitions of the two BChls b conformers equals the value observed in acetone solution,[23] 12642 cm^{-1} or 791 nm. For BChl g calculations, the BChl b coordinates were modified by simply replacing the ring I acetyl groups with vinyl groups and maintaining the same dihedral angles relative to the porphyrin plane. The Qy values obtained for the BChls g are 12922 cm^{-1} (774 nm) and 13471 cm^{-1} (742 nm) for an average value of 13197 cm^{-1} or 758 nm, in good agreement with the experimental value in acetone of 13123 cm^{-1} (762 nm).[23] As noted before, the calculated Qy transitions are sensitive to the ruffling of the macrocycles and to the orientation of the acetyl (or vinyl) groups.[20] Rolling the latter into the plane of the porphyrin ($\theta = 180°$ or $0°$) will maximize conjugation and result in a red shift whereas rolling the groups out of plane ($\theta = 90°$) will minimize their effect and cause blue shifts. Nonetheless, the Rps. viridis coordinates yield reasonable values for the BChl b and g monomers. Since INDO/s calculations of the BChl b special pair of Rps. viridis, based solely on the crystallographic coordinates, afford good agreement with the experimental Qy transitions, linear dichroism, and positions of the exciton bands,[25] initial calculations of BChl g dimers started with the architecture of the BChl b special pair in Rps. viridis with the acetyl groups replaced with vinyl groups. The latter were oriented as in Rps. viridis, i.e. the dihedral angles are -174° and -156° for the L and M monomers, respectively, and they point into the space between the two monomer subunits. In addition, other orientations of the vinyl groups into and out of the porphyrin planes were considered. These calculations were prompted by crystallographic, NMR and EPR data for chlorophylls, pheophytins and protoporphyrins which clearly show that the vinyl groups of the chlorins and porphyrins can adopt a wide range of orientations relative to the macrocycle planes.[21] Steric constraints limit rotation of the vinyl groups into the BChl g dimer. However, the groups will swivel out of the dimer to dihedral angles of 120° without crowding.

Table 1 presents the results of the BChl g dimer calculations for various configurations. Note that simply introducing vinyl groups and changing their orientations yields BChl g dimers with Qy bands that are blue shifted 700-1300 cm^{-1} relative to the BChl b dimer which is calculated at 10786 cm^{-1}, and observed at 10417 cm^{-1}. Previous calculations of special pairs predicted[25] that increasing the interplanar separation in BChl dimers would cause blue shifts, and the different absorption maxima of Rb. sphaeroides and Rps. viridis have been attributed to the crystallographic observed interplanar differences of ~0.2 Å in the two organisms.[25] The calculations for the BChl g dimer as a function of interplanar separation illustrate the blue shifts that can be induced by widening the spacing within the dimers. Starting with the Rps. viridis configuration at ~3.3 Å, increasing the separations by 0.4 Å to 3.7 Å yields a blue shift of ~1000 cm^{-1} within the BChl g series to 12497 cm^{-1} or 800 nm. Clearly, reasonable combinations of interplanar separations and vinyl rotations can yield BChl g dimers with the desired spectral properties. A separation of ~3.5 Å and vinyl orientations of 150° also result in a Qy transition at 800 nm. Furthermore, inclusion of the histidines found as Mg axial ligands in Rps. viridis and Rb. sphaeroides causes a further blue shift to 789 nm.

We emphasize that we are not attempting to establish a unique structural configuration for the BChl g dimer. Clearly, less distorted conformations than are found for the BChls b in the Rps. viridis structure would also cause blue-shifts in the dimers and therefore require smaller interplanar separations and/or different vinyl group orientations. Rather, the major point here is that quite "standard" dimers of BChl with only small variations on the Rps. viridis configuration and its ligands, are entirely compatible with the observed optical properties of Heliobacteria.

The concept of special pairs was originally invoked to explain the properties of P700, the primary electron donor of photosystem I in green plants and algae comprised of Chl a.[27] In vitro, isomerization of the ethylidene group of BChl g leads directly to Chl a formation, and extensive analogies between Heliobacteria Rcs and photosystem I components are indeed observed.[22] Calculations for Chl a dimers with configurations similar to those considered above for the Bchls g yield Qy transitions at 680-690 nm in reasonable agreement with the experimental value of 700 nm.

Given the many peripheral variations already uncovered in photosynthetic chromophores, we suggest that there may exist an as-yet undiscovered bacterium with a hypothetical BChl "h" in which the 3-acetyl group of BChl a has been changed to a vinyl group, as in the BChl b to g transformation. Calculations for such a BChl "h" dimer with the same configuration as described above for BChl g predict a Qy transition at 12515 cm^{-1}, a value comparable to that of the BChls g. Clearly, the introduction or removal of acetyl function and other polar substituents that can conjugate with the chlorin π system provide simple synthetic avenues to modulate the optical properties of porphinoid monomers; effects that are further enhanced by dimerization, as evidenced by the large differences in the BChls g and b in vivo.

Table 1. Calculated Qy transitions[a]

Species	Dihedral Angle, Degrees	Interplanar Separation, Å	Qy, cm^{-1}	Qy 2,[b] cm^{-1}
BChl b (L)	-174[24]	-	12484[25]	-
BChl b (M)	-156[24]	-	12789[25]	-
BChl (aver.)	-	-	12637	-
BChl g (L)	-174	-	12922	-
BChl g (M)	-156	-	13471	-
BChl g (aver.)	-	-	13197	-
BChl b dimer	-174,-156	~3.3[24]	10786[25]	12345
BChl g dimer[c]	-174,-156	3.3	11505	12726
BChl g dimer	-174,-156	3.5	12150	13090
BChl g dimer	-174,156	3.6	12351	13197
BChl g dimer	-174,156	3.7	12497	13273
BChl g dimer	120,120	3.3	12072	13376
BChl g dimer	150,150	3.3	11931	13176
BChl g dimer	180,180	3.3	11702	12899
BChl g dimer	150,150	3.5	12495	13435
BChl g dimer + histidines	150,150	3.5	12676	13706

[a]The INDO/s calculations were performed with the program ARGUS using the coordinates of Rps. viridis obtained from the Brookhaven Protein Data Bank. Configuration interactions included all single-excited configurations from the 14 (28) HOMOS into the 14 (28) LUMOS for the monomers and (dimers).
[b]Upper Qy exciton component in the dimers.
[c]BChl g′, the 13^2 epimer of BChl g, has been suggested as a possible component of P798 in H. chlorum and Hb. mobilis.[26] The isolated epimers display identical Qy transitions in vitro[27] and the epimerization is therefore unlikely to significantly affect the Qy transitions of the BChl g dimers.

ELECTROCHROMIC EFFECTS REVISITED

The primary electron transfer events of bacterial photosynthesis create cations and anions that lie in close proximity to each other. Our previous INDO/s calculations[28] of the electrochromic effects of the charged species on the neighboring BChls and BPhs b yielded trends consistent with the spectral changes observed as primary charge separation evolves in Rps. viridis. The calculations are extended here to consider electrostatic effects on the properties of BChls a, b and g as well as Chl a. The results should also be relevant to the electrostatic effects of incorporating site specific or regiospecific "charged" or polar residues near the chromophores, modifications now possible because of the recent advances in mutagenesis and reconstitution of RCs and antennas. The calculated effects on optical and redox properties are explained by a qualitative model that considers the distribution of electron density in the ground and excited states and the dipole moment difference between these states.

Table 2 presents the electrochromic shifts of the Qy transitions induced by point charges placed 3.5 Å above various positions of BChls a, b and g and pyroChl a. (The latter was used instead of Chl a to calibrate the calculations against experimental results, vide infra. Very similar effects are induced in Chl a itself.)

As seen from Table 2, a distinctive pattern of Qy shifts emerges which is common to all the chromophores. The electrochromic responses generally follow the order BChl g> Bchl b> BChl a> Chl a.[29] The magnitude and sign of the shifts depend on the placement and sign of the point charges: positive charges situated near rings III or V cause large blue shifts whereas reversing the charges results in large red shifts. The effect of a charge reverses along a line passing through the molecular x-axis, which passes through rings II and IV, and the magnitude of the shifts is related to the y-axis along rings I and III. A charge near the x-axis causes a much smaller shift than one situated on or near the y-axis. In accord with this pattern, experimental results by Davis et al.[30] show a small blue shift of 90 cm^{-1} for a derivatized pyroChl protonated at C7 to be compared with a calculated shift of 135 cm^{-1} for a positive charge placed near C7. Similar calculations for synthetic protonated porphyrin and (bacterio)chlorin Schiff bases[31] also yield good agreement with experiment.

The electrochromic behavior of the (B)Chl Qy bands can be understood in terms of the distribution of electron density within the ground and first excited states and their respective dipole moments. Table 3 lists the calculated dipole moments for the ground (μ_o) and Qy states (μ_{Qy}) for the (B)Chls in the absence of point charges along with the difference in dipole moment between the two states ($\Delta\mu$).

The electrochromic effects result from the strong y-polarization of the Qy transition and the large difference between the calculated ground and Qy dipole moments. In the ground state the electron density is higher in the ring III region than the ring I region. The Qy transition results in a net shift in electron density along the y-axis from ring III to ring I which is reflected in the large positive μ_y value for the ground state and the small μ_y value for Qy. (Although the dipole moment differences calculated by INDO/s tend to be larger than estimated from Stark effects,[32] the predicted trends between the ground and excited states appear valid.[20,31])

Figure 1 schematically shows the changes in the calculated energies of the ground and Qy states when a point charge is placed 3.5 Å above rings I and III. A positive

Table 2. Effect of Point Charges on the Qy Transitions of BChls a, b and g, and PyroChl a[a]

Charge (a.u.)	Position[b]	Shift (cm^{-1})[c]			
		BChl g	BChl b	BChl a	PyroChl a
+1	NI	-1510	-1440	-1053	-716
-1		2291	2297	2014	591
+1	NII	-205	-239	-51	34
-1		21	109	-5	-84
+1	NIII	1502	1329	1579	623
-1		-1109	-895	-962	-635
+1	NIV	-551	-564	-224	41
-1		439	440	182	-159
+1	C20	-1341	-1272	-866	-634
-1		2066	1993	1669	303
+1	C2	-1754	-1603	-1229	-1187
-1		2919	2872	2617	782
+1	C8	796	712	519	-26
-1		-1204	-1059	-645	-294
+1	C12	2772	2579	2709	995
-1		-1472	-1160	-1079	-728
+1	O1	1937	1796	1958	1002
-1		-1110	-1000	-895	-583

[a]To simplify comparisons, the geometries of all the chromophores are based on the crystal structure of ethyl chlorophyllide a[33] with appropriate adjustments for the different ring substituents.

[b]See the generic (bacterio)chlorin skeleton for numbering system. Charges are placed 3.5 Å above the designated atom.

[c]Shift in frequency relative to no charge. Positive values denote blue shifts; negative values, red shifts. The values are calculated for a dielectric constant of 1. Increasing the dielectric constant to 2 to simulate a hydrophobic environment in the RCs halves the values shown.

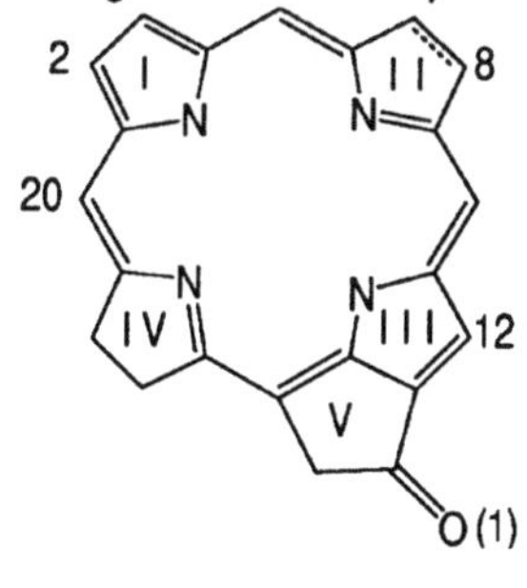

Table 3. Calculated State Dipoles[a]

Molecule	State	μ_x [b]	μ_y	μ_z
BChl g	ground state	-0.1256	13.6710	0.3203
	Qy	1.7287	1.2652	0.3391
	$\Delta\mu$ [c]	1.8543	-12.4058	0.0188
BChl b	ground state	-1.3464	10.8295	4.0070
	Qy	0.5019	- 0.8447	3.9985
	$\Delta\mu$	1.8483	-11.6742	-0.0085
BChl a	ground state	-0.0860	10.3297	3.9951
	Qy	0.2572	- 0.7976	3.9351
	$\Delta\mu$	0.3432	-11.1273	-0.0600
pChl a	ground state	0.6811	6.4990	-0.4502
	Qy	-1.5782	1.5312	-0.3418
	$\Delta\mu$	-2.2593	-4.9678	0.1084

[a] Values listed are obtained from INDO/s calculations of the unperturbed molecules.
[b] Debyes
[c] Excited stated (Qy) minus the ground state dipole moments.

charge lowers the energy of both states. When placed near ring I, the "head" of the dipole moment vectors, the charge depresses the energy of the Qy state more than the ground state. The Qy state gains electron density near ring I relative to the ground state and therefore its stabilizing interaction with a positive charge is larger, and $E^+_I < E_o$ results in a red shift (Figure 1). Conversely, a positive charge near ring III, the "tail" of the dipole moment vectors depresses the energy of the ground state more than the Qy state because the ground state has relatively greater electron density near the ring III region, and thus $E^-_{III} > E_o$ results in a blue shift. Negative charges induce shifts opposite to those of positive charges because of their energetically destabilizing interactions with regions of high electron density. A negative charge raises the energies of both states, but the state with relatively more electron density in the vicinity of the negative charge is raised the most. Thus, a negative charge near the ring I region raises the energy of the Qy state more than the ground state ($E^-_I > E_o$, Figure 1) resulting in a blue shift. A negative charge near ring III, however, raises the energy of the ground state more than the Qy state ($E^-_{III} < E_o$) and the calculated transition is red shifted. Thus, it is the placement of the point charge relative to μ_o and μ_{Qy} that determines whether the electrochromic shift is blue or red, and therefore explains why a given charge causes electrochromic shifts of opposite sign on different sides of the molecular x-axis.

Note that in Figure 1, a positive charge always lowers the energies of both states while a negative charge raises these energies, albeit differentially. This trend is particularly relevant to our discussion of modifying the immediate environment of the chromophores in RCs.[34] Because the Qy transition in BChls is predominantly a

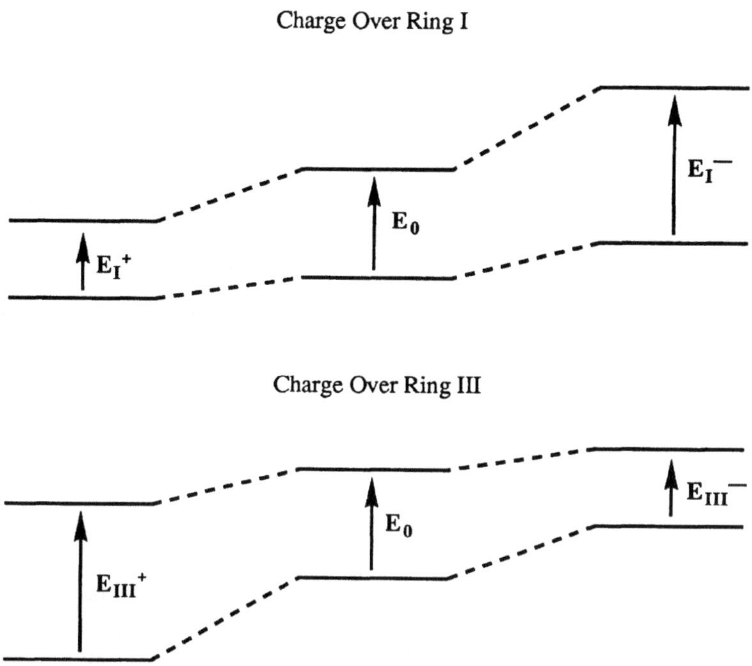

Figure 1. Schematic representation of the effect of point charges placed over rings I and III on the calculated energy of the Qy transition of (B)Chls. Abbreviations are: E_o, the Qy transition energy in absence of point charges,$E^{+(-)}_{I}$, the Qy transition energy in the presence of a positive (negative) point charge over ring I, and $E^{+(-)}_{III}$, the Qy transition energy in the presence of a positive (negative) point charge over ring III.

HOMO to LUMO excitation, Figure 1 also represents the effects of charges on the HOMOs and LUMOs of the chromophores. Thus the introduction of neighboring residues with increased positive character will render the molecules harder to oxidize and easier to reduce, while groups with more negative character will cause the BChls to be easier to oxidize and harder to reduce. Whether such "charged" groups induce blue or red shifts will depend on their placement, as discussed above.

ACKNOWLEDGMENTS

This work was supported by the Division of Chemical Sciences, U.S. Department of Energy, under Contracts No. DE-AC02-76CH00016 at BNL and DE-AC06-76RLD1830 at PNL, and a grant from Eastman Kodak Company at U.F.

REFERENCES

1. "The Porphyrins," D. Dolphin, ed., Academic Press, New York (1978).
2. J. Fajer, Chemistry & Industry, 869 (1991).

3. W.J. Coleman and D.C. Youvan, Annu. Rev. Biophys. Biophys. Chem. 19:333 (1990).
4. "Reaction Centers of Photosynthetic Bacteria," M. E. Michel-Beyerle, ed., Part III, Springer-Verlag, Berlin (1990).
5. See also the several articles in these Proceedings by J. Norris, D. Oesterhelt, C. Schenck, M. Schiffer, N. Woodbury, and D. Youvan and coworkers.
6. A. Struck, E. Cmiel, I. Katheder, and H. Scheer, FEBS Lett. 268:180 (1990).
7. E.J. Bylina and D.C. Youvan, Proc. Nat'l. Acad. Sci. U.S.A. 85:7226 (1988).
8. C.C. Schenck, D. Gaul, M. Steffen, S.G. Boxer, L. McDowell, C. Kirmaier, and D. Holten, In ref. 4, p. 229.
9. C. Kirmaier, D. Gaul, R. Debey, D. Holten, and C.C. Schenck, Science 251:922. (1991).
10. R.H. Felton in ref. 1, Vol. 5, p. 53
11. R.L. Heald and T.M. Cotton, J. Phys. Chem. 94:3968 (1990).
12. N.J. Boldt, R.J. Donohue, R.R. Birge, D.F. Bocian, J. Am. Chem. Soc. 109:2284 (1987).
13. R.G. Alden, M.R. Ondrias, and J.A. Shelnutt, J. Am. Chem. Soc. 112:691 (1990).
14. C. Kratky, R. Waditschatka, C. Angst, J.E. Johansen, J.C. Plaquevent, J. Schreiberg, and A. Eschenmoser, Helv. Chim. Acta. 68:1312 (1985).
15. M.W. Renner, L.R. Furenlid, K.M. Barkigia, A. Forman, H.K. Shim, D.J. Simpson, K.M. Smith, and J. Fajer, J. Am. Chem. Soc. 113:6891 (1991).
16. L.R. Furenlid, M.W. Renner, and J. Fajer, J. Am. Chem. Soc. 112:8987 (1990).
17. L.R. Furenlid, M.W. Renner, K.M. Smith, and J. Fajer, J. Am. Chem. Soc. 112:1634 (1990).
18. "Chlorophylls," H. Scheer, ed., CRC Press, Boca Raton (1991).
19. K.M. Smith and F.Y. Shiau, private communication.
20. E. Gudowska-Nowak, M.D. Newton, and J. Fajer, J. Phys. Chem. 94:5795 (1990).
21. M.A. Thompson and J. Fajer, J. Phys. Chem. 96:2933 (1992).
22. For recent reports see: J.T. Trost and R.E. Blankenship, Biochemistry 28:9898 (1989) and W. Nitschke, P. Setif, U. Liebl, U. Feiler, and A.W. Rutherford, Biochemistry 29:11079 (1990).
23. A. Hoff and J. Amesz in ref. 18, p. 723.
24. J. Deisenhofer and H. Michel, Science 245:1463 (1989).
25. M.A. Thompson, M.C. Zerner, and J. Fajer, J. Phys. Chem. 95:5693 (1991).
26. M. Kobayashi, E.J. Van de Meet, C. Erkelens, J. Amesz, I. Ikegami, and T. Watanabe, Biochim. Biophys. Acta 1057:89 (1991).
27. J.R. Norris, R.A. Uphaus, H.L. Crespi, and J.J. Katz, Proc. Nat'l. Acad. Sci. U.S.A. 68:625 (1971).
28. L.K. Hanson, J. Fajer, M.A. Thompson, and M.C. Zerner, J. Am. Chem. Soc. 109:4728 (1987).
29. J. Eccles and B. Honig, Proc. Nat'l. Acad. Sci. U.S.A. 80:4959 (1983), calculated similar trends using a CNDO method.
30. R.C. Davis, S.L. Ditson, A.F. Fentiman, and R.M. Pearlstein, J. Am. Chem. Soc. 103:6823 (1981).
31. L.K. Hanson, C.K. Chang, B. Ward, P.M. Callahan, G.T. Babcock, and J.D. Head, J. Am. Chem. Soc. 106: 3950 (1984).
32. D.J. Lockhart and S.G. Boxer, Proc. Nat'l. Acad. Sci. U.S.A. 85:107 (1988).
33. H.C. Chow, R. Serlin, and C.E. Strouse, J. Am. Chem. Soc. 97:7230 (1975).
34. See also W.W. Parson, Z.T. Chu, and A. Warshel, Biochim. Biophys. Acta 1017:251 (1990).

POTENTIAL ENERGY FUNCTION FOR PHOTOSYNTHETIC REACTION CENTRE CHROMOPHORES : ENERGY MINIMISATIONS OF A CRYSTALLINE BACTERIOPHEOPHYTIN A ANALOG

N. Foloppe[1], J. Breton[2], and J.C. Smith[1],

[1]Section de Biophysique des protéines et des membranes

[2]Section de Bioénergétique
Département de Biologie Cellulaire et Moléculaire
CE-Saclay, 91191 Gif-sur-Yvette, France

INTRODUCTION

The elucidation of the three-dimensional structure of two bacterial photosynthetic reaction centres (RC) [1-3] has revealed the geometrical arrangement of the pigments relative to each other and to the protein. An improved understanding of the factors determining the pigment orientations should result from molecular mechanics calculations. However, the usefulness of the calculations depends on the accuracy of the potential function, particularly concerning the nonbonded interactions. Whereas several force fields devoted to protein investigations exist, there is a need to develop a reliable energy function for the bacterial photosynthetic pigments bacteriochlorophyll (BChl) and bacteriopheophytin (BPhe). In what follows we present a preliminary parameterisation and testing of a molecular mechanics force field for a BPhe A analog, methyl-bacteriopheophorbide A (Me-BPh). This molecule differs from BPhe only in that it lacks the peripheral phytyl chain. The advantage of using Me-Bph for parameterisation purposes is that, unlike for BPhe, an accurate crystal structure is known for Me-BPh [4]. This allows a test of the potential function to be made by performing energy minimisations of the molecule in the presence of its crystal environment.

The Photosynthetic Bacterial Reaction Center II
Edited by J. Breton and A. Verméglio, Plenum Press, New York, 1992

METHODS

Form of the Energy Function

All our calculations were performed with the CHARMM program [5] with the potential energy function **V** :

$$\mathbf{V} = \sum_{bonds} k_r(r - r_0)^2 + \sum_{angles} k_\theta(\theta - \theta_0)^2 + \sum_{dihedrals} k_\phi \, [1 + cos(n\phi - \delta)$$

$$+ \sum_{impropers} k_\omega(\omega - \omega_0)^2 + \sum_{ij} 4\epsilon_{ij}[(\sigma_{ij}/r_{ij})^{12} - (\sigma_{ij}/r_{ij})^6] \; + \sum_{ij} \frac{q_i q_j}{r_{ij}}$$

In the above equation, r, θ, and ω are the bond lengths, valence angles and improper torsion angles with force constants k_r, k_θ, k_ω and reference values r_0, θ_0, ω_0, respectively. ϕ are dihedral angles of periodicity n with force constants k_ϕ and phase angles δ. The nonbonded terms consist of a 12-6 Lennard-Jones potential for van der Waals interactions and a Coulombic term representing electrostatic interactions between atomic partial point charges q_i and q_j. r_{ij} and ϵ_{ij} are the nonbonded distance and the Lennard-Jones well depth. $\sigma_{ij} = 2^{1/6} r_{min}$, where r_{min} is the distance between atoms i and j at their van der Waals energy minimum. The Coulombic interactions are brought smoothly to zero at a cut-off distance of 13 Å using the 'shift' truncation method [5]. The dielectric constant was taken as 1.0. There is no explicit hydrogen-bonding term in the potential function; hydrogen bonds are described by a combination of van der Waals and electrostatic interactions [6]. All atoms were explicitly included.

Parameter Derivation

Initial parameters for the macrocyclic core were taken from a CHARMM 22 parameter set for the haem group [7]. Some minor adjustments of the internal geometric terms were made and the resulting haem group potential was energy minimised. The minimised geometry has an RMS coordinate deviation of 0.109 Å from the high resolution crystal geometry of (2,3,7,8,12,13,17,18 - octaethylporphyrinato) iron (II) [8]. The resulting haem group potential function was transferred as far as possible to Me-BPh; some further internal geometry adjustments were needed to take into account the added fifth ring and the reduction of rings 2 and 4 (Fig. 1). The additional geometric parameters were taken from the high resolution Me-BPh crystal geometry [4]. Geometric parameters and torsional barriers for the ketone and ester groups were transferred from previous work [9-11].

Peripheral aliphatic groups were assigned standard parameters available for alkanes in CHARMM 22 [10], including partial charges.

The partial charges q_i were derived from semi-empirical quantum - mechanical calculations using MOPAC 6.0 with the AM1 Hamiltonian [12]. Due to computational limitations, the partial charges of the macrocyclic core and of the ester substituents were calculated separately. The final charge set is shown in fig. 1. The benzene of crystallisation was included using CHARMM 22 parameters [13]. A full list of the parameters will be published elsewhere.

Energy Minimisation

Three different types of energy minimisation were performed. The first involves minimisation of the isolated Me-BPh molecule starting from the crystallographic coordinates but without including the crystal environment. The second method involves minimising the molecule including the crystal environment and keeping the lattice constants at their experimental (200 K) values. The final calculations take into account that a lower potential energy can be found by letting the lattice constants vary; they are therefore optimised along with the molecular geometry.

Crystallographic coordinates of Me-BPh were obtained from the Cambridge structural data base [14]. The Me-BPh crystal is triclinic, space group P1, a=7.184 Å, b=8.073 Å, c=17.071 Å, α = 91.04°, β = 93.50°, γ = 110.06° [4]. The crystal was simulated by replicating the unit cell using periodic boundary conditions. In the crystal the Me-BPh associates in one dimensional chains by stacking of rings 1 and 3.

The energy function was employed to calculate minimum energy geometries using steepest descent (SD), adopted basis Newton-Raphson (ABNR) and Newton-Raphson (NRAP) minimisation algorithms. The energy optimisation of isolated Me-BPh was pursued to an RMS energy gradient $< 10^{-6}$ kcals/mol Å. Me-BPh in the crystal was minimised to an RMS energy gradient $< 5\times10^{-4}$ kcals/mol Å.

RESULTS AND DISCUSSION

The RMS coordinate deviation between Me-BPh energy minimised in vacuum and the crystalline reference is 0.491 Å, with 0.135 Å for the macrocyclic core (cyclic moiety and ketone groups) and 0.647 Å for the flexible peripheral substituents.

The model structure is in better agreement with the experimental structure when minimised in its crystalline environment. When the lattice parameters were not optimised, the overall RMS deviation is 0.187 Å (0.101 Å for the macrocyclic core and 0.211 Å for the peripheral substituents).

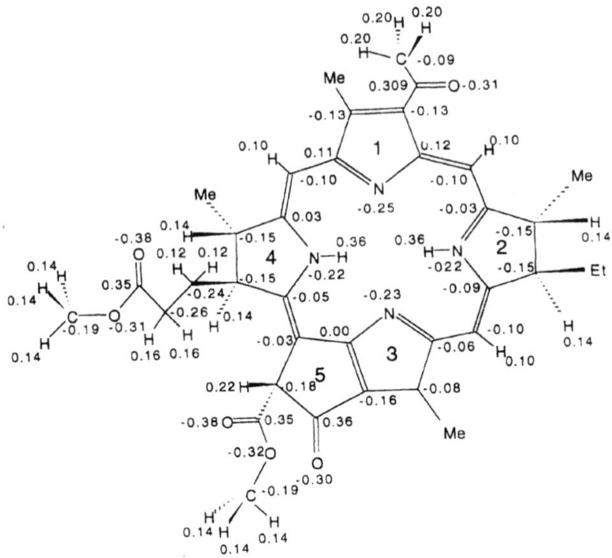

Figure 1. Chemical structure, partial charges and ring numbers for Me-BPh.

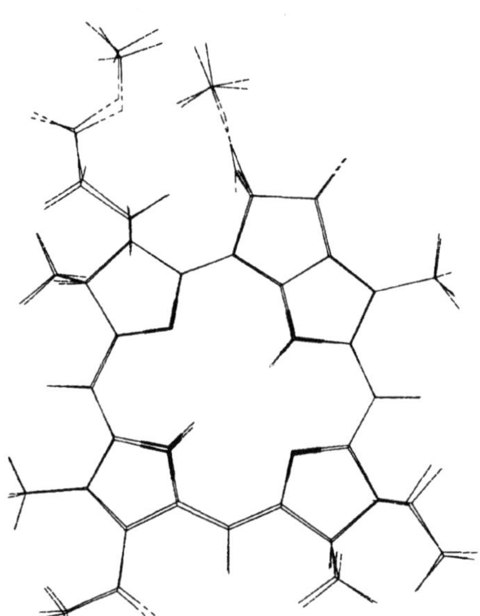

Figure 2. Experimental and energy minimised crystalline Me-Bph.

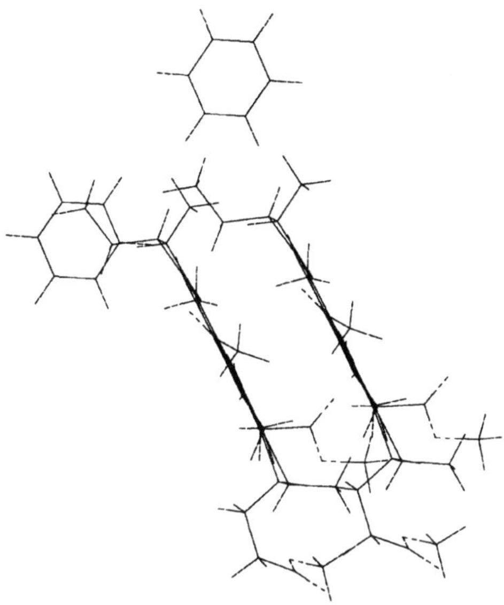

Figure 3. Two adjacent Me-Bph molecules stacked in the crystal, after energy minimisation. The two co-crystallised benzene molecules are also shown.

Fig. 2 shows the experimental geometry of Me-BPh superimposed on its modeled geometry obtained in these conditions. Figs. 3 shows the stacking of two successive crystalline Me-BPh after this energy minimisation.

When the lattice is optimised the RMS deviation is 0.192 Å for the overall structure (0.101 Å for the macrocyclic core and 0.228 Å for peripheral substituents). The unit cell volume reduces by 4.1 per cent from 927.43 $Å^3$ to 890.89 $Å^3$ (a=7.240 Å, b=7.865 Å, c=16.580 Å, $\alpha = 92.520^o$, $\beta = 90.315^o$, $\gamma = 109.168^o$); this small reduction may be due to thermal contraction from 200 K to 0 K (corresponding to the energy-minimum geometry) resulting from anharmonic effects.

CONCLUDING DISCUSSION

The work described here represents the determination of a ground-state potential energy function for Me-BPh. Internal geometrical parameters were derived from high-resolution small-molecule crystallography and the partial charges from quantum-mechanical calculations. Energy minimisations of Me-BPh with and without the crystalline environment indicate a better agreement with experiment when the environment is included; crystal minimisations provide a more realistic test of the force field. The reasonable agreement between the crystal-minimised and experimental geometries gives confidence in the parameterisation. Future work will extend the parameterisation to Mg^{2+}-containing chromophores and to protein calculations.

REFERENCES

[1] J. Deisenhofer, O. Epp, K. Miki, R. Huber, and H. Michel, *Nature*, **318**, 618, (1985).

[2] J.P. Allen, G. Feher, T.O. Yeates, H. Komiya, and D.C. Rees, *Proc. Natl. Acad. Sci. USA*, **84**, 6162, (1987).

[3] C.H. Chang, O. El-Kabbani, D. Tiede, J. Norris, and M. Schiffer, *Biochemistry*, **30**, 5352, (1991).

[4] K.M. Barkigia, D.S. Gottfried, S.G. Boxer, and J. Fajer, *J. Am. Chem. Soc.*, **111**, 6444, (1989).

[5] B.R. Brooks, R.E. Bruccoleri, B.D. Olafson, D.J. States, S. Swaminathan , and M. Karplus, *J. Comp. Chem.*, **4**, 187, (1983).

[6] W. Reiher, Ph. D. Thesis, Harvard University, (1985).

[7] K. Kuczera, J. Kuriyan, and M. Karplus, *J. Mol. Biol.*, **213**, 351, (1990).

[8] S.H. Strauss, M.E. Silver, KM. Long, R.G. Thompson, R.A. Hudgens, K Spartalian, and J.A. Ibers, *J. Am. Chem. Soc.*, **107**,4207, (1985).

[9] N.L. Allinger, K. Chen, M. Rahman, and A. Pathiaseril, *J. Am. Chem. Soc.*, **113**, 4505, (1991).

[10] J.C. Smith, and M. Karplus, *J. Am. Chem. Soc.*, **114**, 801, (1992).

[11] G.D. Smith, and R.H. Boyd, *Macromolecules*, **23**, 1527, (1990).

[12] M.J.S. Dewar, E.G. Zoebisch, E.F. Healy, and J.J.P. Stewart, *J. Am. Chem. Soc.*, **107**, 3902, (1985).

[13] A.D. Mac Kerrell, J. Straub, and M. Karplus, in preparation.

[14] F.H. Allen, S.A. Bellard, M.D. Brice, B.A. Cartwright, A. Doubleday, H. Higgs, T. Hummelink, B.G. Hummelink-Peters, O. Kennard, W.D.S. Motherwell, J.R. Rodgers, and D.G. Watson, *Acta Crystallogr.*, **B35**, 2331, (1979).

BACTERIAL REACTION CENTERS WITH PLANT-TYPE PHEOPHYTINS

Hugo Scheer, Michaela Meyer, and Ingrid Katheder

Botanisches Institut der Universität München,
Menzinger Str. 67, D-8000 München 19

SUMMARY

The exchangeability of the bacteriopheophytins at sites H_A and H_B with modified (bacterio)pheophytins (=**(B)Phe**)*) was tested in reaction centers (RC) of *Rhodobacter spheroides* R26. An exchange at both sites is possible with Pyro-**BPhe a** lacking the 13^2-$COOCH_3$ group, and with three plant-type pheophytins: **Phe a** (which contains a 3-vinyl-group), 13^2-hydroxy-**Phe a** (which contains in addition a hydroxy group), and [3-acetyl]-**Phe a** (which differs from **BPhe a** only by the unsaturated ring II). In all cases, the exchange appears to be easier at the H_B-site. An exchange only at this site, was obtained with **BPhe** $\mathbf{a}_{gg}$ in which the esterifying phytol is replaced by geranyl-geraniol, and with 13^2-hydroxy-**Phe a**. Environment-induced red-shifts (EIRS) are observed with all pigments, and they are in the range of the ones known for the native **BPhe a**. Strong optical activity is induced in most pigments. Shifts in the absorptions of the monomeric **BChl**s at sites $B_{A,B}$ indicate an interaction with the **BPhe**s sites $H_{A,B}$, or an indirect structural effect.

INTRODUCTION

In reaction centers (RC) of *Rhodobacter spheroides* (Rb.), the tetrapyrrole pigments at the sites $B_{A,B}$ ("monomeric" bacteriochlorophylls,

---

*) **Abbreviations: Chl** = chlorophyll, **Phe** = pheophytin, **BChl** = bacteriochlorophyll, **BPhe** = bacteriopheophytin; the subscripts refer to the esterifying alcohols ("p" or none for phytol, "gg" for geranylgeraniol), RC = reaction centers, *Rb.* = *Rhodobacter*, cd = circular dichroism, P = primary donor site, B = site of monomeric BChl, H = site of BPhe in RC. The subscripts "A" and "B" refer to the active ("L") and inactive branch ("M"), respectively, of the electron transport chain.

The Photosynthetic Bacterial Reaction Center II
Edited by J. Breton and A. Verméglio, Plenum Press, New York, 1992

BChl-$B_{A,B}$) and $H_{A,B}$ (bacteriopheophytins, **BPhe**-$H_{A,B}$) are exchangeable against chemically modified pigments. **BChl**-$B_{A,B}$ could be exchanged with a variety of modified **BChls**, but neither with plant-type chlorophylls (**Chl**), nor with bacteriopheophytins (**BPhe**) or plant-type pheophytins (**Phe**) (Struck *et al.*, 1990). The exchangeability of the bacteriopheophytins at sites $H_{A,B}$ has been studied less. Struck (1990) has shown, that they are exchangeable against some modified **BPhes**, but not against any Mg-containing **(B)Chls**. The selectivity of the sites according to the presence or absence of the central Mg-atom, rsp., corroborates results from site-directed mutagenesis (Coleman and Youvan, 1990; Woodbury *et al.*, 1990; Schenck *et al.*, 1990). However, little is presently known on the influence of the reduction level (chlorin *vs.* bacteriochlorin) or the peripheral substituents of **(B)Phes** on the exchangeability. We wish to report exchange experiments with **BPhes** modified at C-13^2 and C-17^4, and in particular with plant-type **Phes** containing an unsaturated ring II.

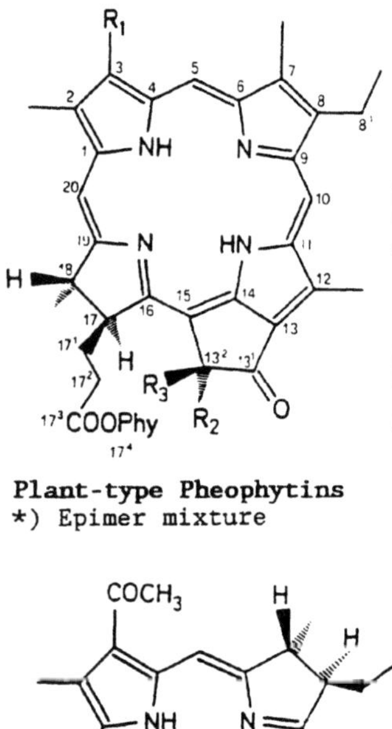

Pigment	R_1	R_2	R_3
Phe a	$CHCH_2$	$COOCH_3$	H
Phe a'	$CHCH_2$	H	$COOCH_3$
[3-Acetyl]-Phe a	$COCH_3$	$COOCH_3$	H
13^2-OH-Phe a*	$CHCH_2$	$COOCH_3$	OH

Plant-type Pheophytins
*) Epimer mixture

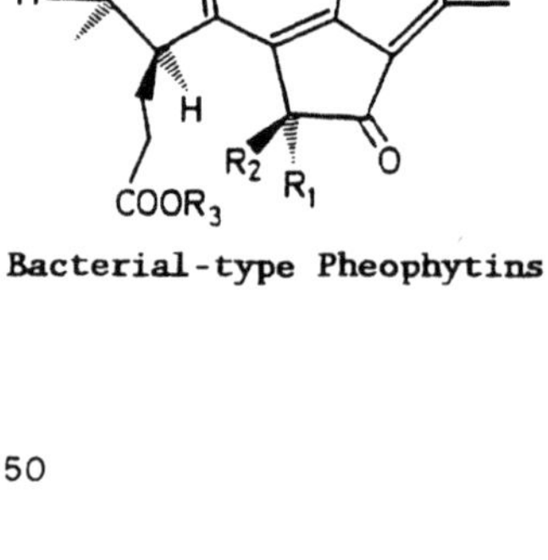

Pigment	R_1	R_2	R_3
BPhe a_p	$COOCH_3$	H	$C_{20}H_{39}$
BPhe a_p'	H	$COOCH_3$	$C_{20}H_{39}$
BPhe a_{gg}	$COOCH_3$	H	$C_{20}H_{33}$
Pyro-BPhe a	H	H	$C_{20}H_{39}$

Bacterial-type Pheophytins

MATERIALS AND METHODS

RC of *Rb. spheroides* R26 were prepared from chromatophores by repeated solubilization with increasing concentrations of LDAO and NaCl (modified after Feher and Okamura, 1978) and purified on DEAE-cellulose (Struck, 1990).

BPhe $\mathbf{a}_p$, **BPhe** $\mathbf{a}_{gg}$ and **Phe a** were extracted from *Rb. spheroides* 2.4.1, *Rhodospirillum rubrum* G9 and *Spirulina gleitleri* (SOSA Texcoco), rsp., by standard procedures and purified on DEAE-cellulose (Satoh and Murata, 1978). Demethoxycarbonylation of **BPhe** $\mathbf{a}_p$ to Pyro-**BPhe** $\mathbf{a}_p$ was done according to Pennington *et al.* (1963). [3-acetyl]-**Phe a** was made from **Phe a** (Smith and Calvin, 1966). 13^2-hydroxy-**Chl a** was obtained as a by-product during the isolation of **Chl a** and pheophytinized according to Rosenbach-Belkin (1988). Structures of the pigments were verified by VIS-NIR absorption, ^{1}H-NMR and mass spectroscopy.

The conditions for the exchange experiments of **BPhe a** against the modified pigments as described by Struck (1990), were optimized. The incubation temperature of the RC was increased to 43.5°C. The modified pigments were added in a 10-fold excess, the solvent for the pigments was 100% acetone, its final concentration 10%. After incubation, the excess of free pigments was removed by repeated chromatography on DEAE-cellulose.
Extraction of the pigments from RC was done with $CHCl_3/CH_3OH$ = 5:1 (v/v). The extract was dried under a stream of argon, dissolved in toluene, and then subjected without delay to HPLC-analysis according to Watanabe *et al.* (1984).

RESULTS

Three plant-type pheophytins were tested: i) [3-acetyl]-**Phe a**, which differs from **BPhe a** only by the unsaturation of ring II. ii) **Phe a**, which has in addition the 3-acetyl- replaced by a vinyl-group. iii) 13^2-hydroxy-**Phe a**, which contains furthermore an OH-group instead of the enolizable 13^2-proton.

The absorption spectra of **Phe a** and 13^2-hydroxy-**Phe a** are essentially identical (Fig. 1). In comparison to **BPhe a**, there are two characteristic blue-shifted Q_X-bands for the two plant-type pheophytins, and the Q_Y-bands are blue-shifted by 70-80 nm, too. Unlike **Phe a**, [3-acetyl]-**Phe a** shows a split Soret-band as do the bacteriopheophytins. Providing the same assignments, they are red-shifted by 24 nm (B_X)and 28 nm (B_Y). The main peak of the Soret-band of **Phe a** is red-shifted by ≈52 nm compared to B_X of **BPhe a.**

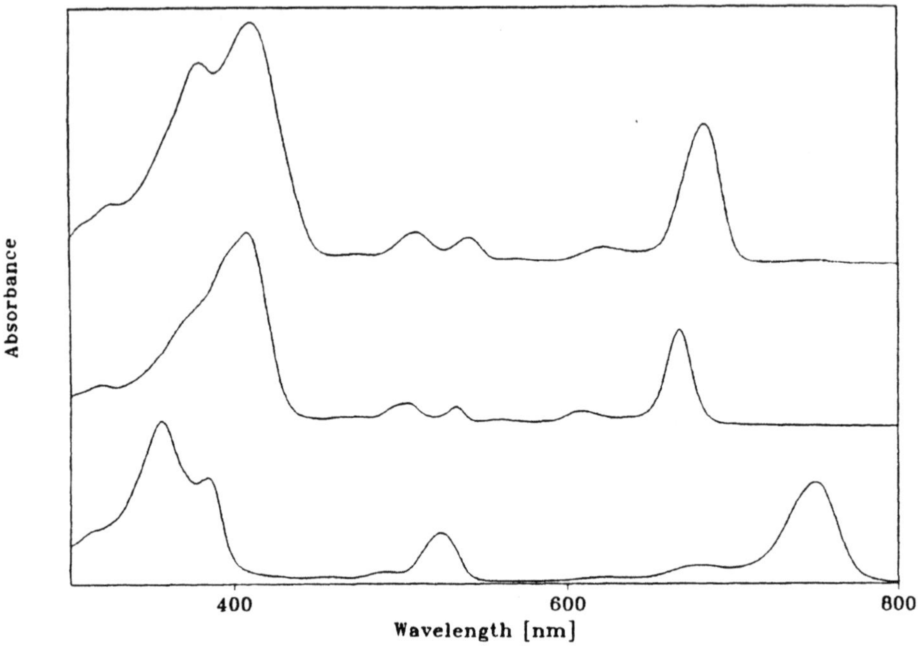

Fig. 1. Absorption spectra of pigments in ether solution: Bacteriopheophytin a (**BPhe a**, bottom), pheophytin a (**Phe a**, center) and [3-acetyl]-pheophytin a ([3-acetyl]-**Phe a**, top).

Table 1. Exchange rates of some plant-type pheophytins and bacterial-type pheophytins

Pigment	exchange-rates [%]	
BPhe a_{gg}	38	
Pyro-BPhe a	35 (1)	68 (2)
Phe a	80 (1)	95 (2)
[3-Acetyl]-Phe a	68	
13^2-OH-Phe a	35	

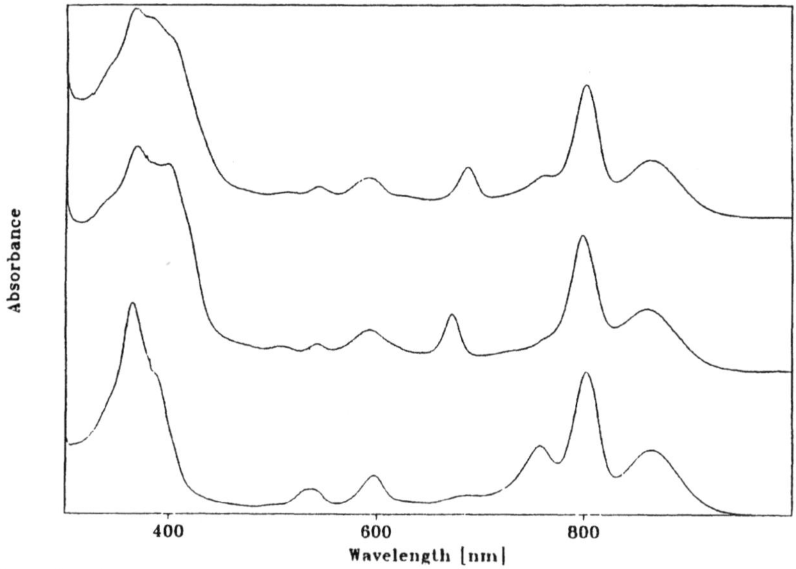

Fig. 2. Absorption spectra of reaction centers with modified pheophytins: in tris-HCl buffer (20 mM, pH 8) containing LDAO (0.1%). Native RC from *Rhodobacter spheroides* R26 (bottom), RC after double exchange of **BPhe a** against **Phe a** (center), and RC after single exchange of **BPhe a** against [3-acetyl]-**Phe a** (top). Spectra were normalized to the same absorption at the dimer band (≈865 nm).

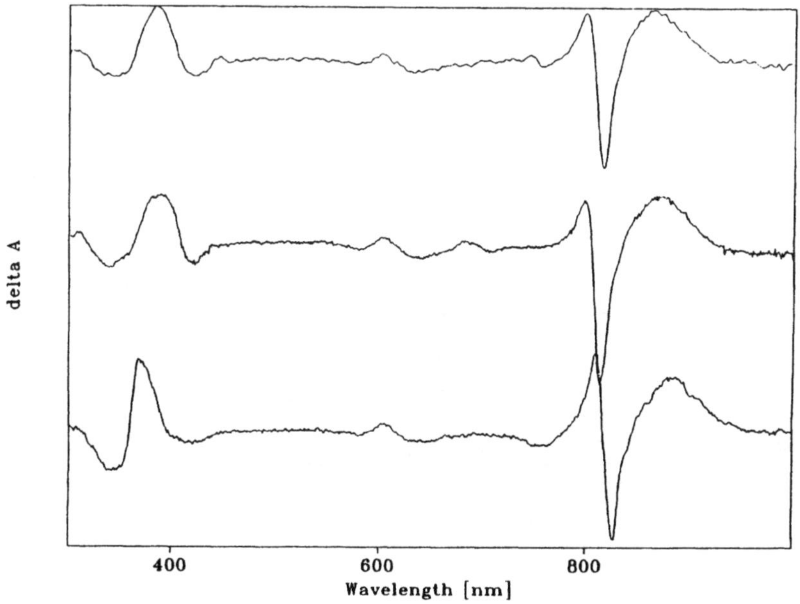

Fig. 3. Circular dichroism spectra of reaction centers with modified pheophytins, in tris-HCl buffer (20 mM, pH 8) containing LDAO (0.1%). Native RC from *Rhodobacter spheroides* R26 (bottom), RC after exchange of **BPhe a** against **Phe a** (center), and RC after single exchange of **BPhe a** against [3-acetyl]-**Phe a** (top). Spectra were normalized to the same **absorption** at the dimer band (≈865 nm).

The absorption spectra of the three bacteriopheophytins tested (**BPhe** $\mathbf{a}_p$, **BPhe** $\mathbf{a}_{gg}$ and Pyro-**BPhe** $\mathbf{a}_p$) are nearly identical and not shown.

The exchange-rates of these **(B)Phes** are summarized in Table 1. **Phe a** and [3-acetyl]-**Phe a** exchange readily to >50%, e.g. they exchange both in H_A- and H_B-sites of the RC. The HPLC-chromatogram of RC after repeated (double) exchange of **BPhe a** against **Phe a** (Fig. 4), shows only traces of the former (> 90% exchange). Because of the dehydrogenation of ring II, **Phe a** has a shorter retention time than **BPhe a**.

The absorption spectra of the RC modified with plant-type pheophytins (exchange rates >90%) show distinct changes as compared to the native ones. These changes follow the differences in the solution spectra of the respective pigments (Fig. 2). The Q_X(0-1)-band shows a blue-shift of about 22 nm (**Phe a**) or 14 nm ([3-acetyl]-**Phe a**) and the Q_X(0-0)-band a red-shift of about 7 nm (**Phe a**) or 17 nm ([3-acetyl]-**Phe a**), as compared to the center of the **BPhe a**-band Q_X(0-0) in native RC. The Q_Y-band of **BPhe a** ($\lambda_{max} \approx$ 758 nm in native RC) is replaced by a strongly blue-shifted one. Comparing RC containing **Phe a** and [3-acetyl]-**Phe a**, the relative band positions of the solution spectra are preserved. The Q_Y-band of the latter is in particular red-shifted compared to the former.

In the cd-spectra, the bands assigned to **BPhe a** are diminished. RC containing **Phe a** show instead a distinct, s-shaped feature at the position of

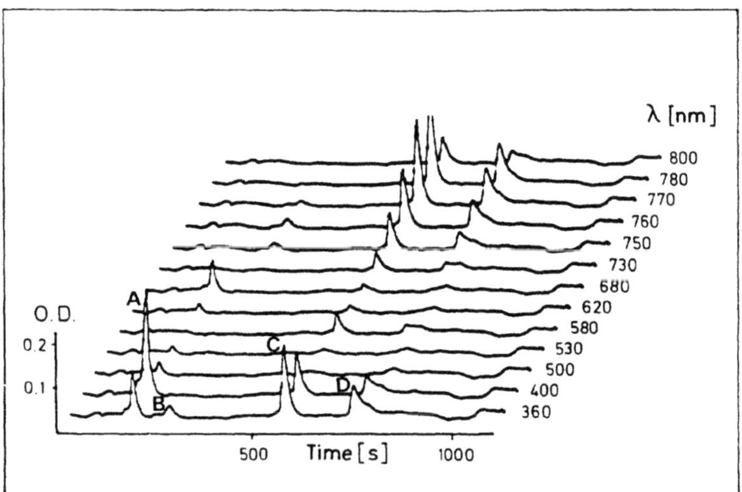

Fig. 4. HPLC chromatogram of RC after repeated exchange of BPhe a against Phe a: (chromatography system of Watanabe *et al.* (1984), detection with HP diode array). The detection wavelengths are given on the right-hand side of the individual traces. Peak assignments: A: **Phe a**, B: **BPhe a**, C: **BChl a**, D: 13^2-hydroxy-**BChl a**

Table 2. Absorption maxima [nm] of (bacterio)pheophytins in ether solution, of the same pigments in the H-sites of RC, and the resulting environment-induced red-shifts (EIRS) (nm and cm^{-1}).

Pigment	Ether [nm]		Protein [nm]		red-shift [nm] (cm^{-1})	
	Q_X	Q_Y	Q_X	Q_Y	Q_X	Q_Y
BPhe a_P	524	750	537	758	13 (462)	8 (141)
BPhe a_{gg}	524	749	539	758	15 (531)	9 (159)
Pyro-BPhe a	527	749	538	757	11 (388)	8 (141)
Phe a	504/533	667	509/542	674	5/9 (195/312)	7 (156)
[3-Acetyl]-Phe a	510/541	680	516/544	689	6/3 (228/102)	9 (192)
13^2-OH-Phe a	502/531	667	506/539	673	4/8 (157/280)	6 (134)

the Q_Y-band of the newly introduced pigment, but the latter is not obvious in the spectrum of RC containing [3-acetyl]-**Phe a**. It is noteworthy, that there is also an effect of the cd assigned to the monomeric **BChl**, e.g. a decrease of the exciton band of the monomeric **BChls** in the Q_X-region and at about 380 nm (Fig. 3).

DISCUSSION

All **(B)Phes** investigated, exchanged selectively into the H-binding site(s) of **BPhe a**. The presence or absence of the central Mg-atom again (Struck *et al.*, 1990) then seems to determine whether the pigment is accepted in $H_{A,B}$ or $B_{A,B}$, rsp. This complements site directed mutagenesis of amino acids: **BPhe** replaces **BChl** if a suitable ligand (his, glu, ser, thr) is introduced, and *vice versa* (Schenck *et al.*, 1990; Woodbury *et al.*, 1990; Coleman and Youvan, 1990).

Compared to the structural variations allowed for exchange of **BChls** into the $B_{A,B}$ binding sites, the results indicate that the $H_{A,B}$ sites allow for considerably more extensive structural changes. It is particularly noteworthy, that there is a ready exchange possible with the plant-type **Phes** in the $H_{A,B}$ binding sites, because the plant-type **Chls** (= Mg-complexes) were not accepted in previous experiments in the **BChl**-binding sites (Struck, 1990). The efficiency of the exchange with **Phe a** is greater than with 13^2-hydroxy-**Phe a** and [3-acetyl]-**Phe a** (Table 1). Remarkable is also the preference of **Phe a** (= 3-vinyl) over all other pigments containing the 3-acetyl-group, which is characteristic for the native **BPhe a**. Although exchangeability is strictly an operational criterion, these results indicate a greater structural plasticity at the H- than at the B-sites.

Most of the pigments investigated, showed exchanges amounting to >50%, which by repeated incubation led to replacements ≤95%. This clearly shows that both binding sites are accessible to these pigments. The asymmetry introduced by glu-100, has then generally not a strongly selecting influence. **BPhe** a_{gg} is an exception. This pigment has a more unsaturated esterifying alcohol (four double bonds instead of one), which changes the flexibility, polarity and spatial structure of this part of the molecule. It should be noted, that the different arrangement of the esterifying alcohols of **BPhe**-H_A and **BPhe**-H_B is one of the distinctive symmetry braking elements in RC, which indicates a specific function of the alcohol in binding. No such specificity was observed in **BChl** exchanges at the $B_{A,B}$ sites (Struck, 1990), and neither is there a comparable asymmetry.

At the $B_{A,B}$-sites, differential exchange kinetics were observed, with B_B exchanging more rapidly that B_A. A similar difference was seen in earlier **BPhe** exchange experiments. The present results with plant-type pheophytins indicate no obvious difference with these pigments, but due to band-overlap in the Q_X-region this result needs further studies, e.g. at low temperature.

Comparing the pigments in ether solution and in the RC environment, an environment-induced red-shift (EIRS) of the Q_X and Q_Y-bands in the protein is found for all pigments. This shift shows only relatively small variations (Table 2). It is not clear if this is a result of protein-chromophore or protein-protein-interactions (Scherz *et al.*, 1990), or both. At least in **Phe a**, there is also a concomitant increase in optical activity. Both effects are compatible with a non-planar distortion of the macrocyclic system.

ACKNOWLEDGEMENTS

Work was supported by the Deutsche Forschungsgemeinschaft (SFB 143, "Elementarprozesse der Photosynthese"). ^{1}H-NMR spectra were kindly recorded for us by E. Cmiel (Technische Universität München, Garching), mass spectra by W. Schäfer (Max-Planck Institut für Biochemie, Martinsried).

REFERENCES

Coleman W.J. and Youvan D.C., 1990, *Ann. Rev. Biophys.* 895:63

Feher G. and Okamura M.T., 1978, *Nature* 339:111

Pennington F.C., Strain H.H., Svec W.A., and Katz J.J., 1963, *J. Am. Chem. Soc.* 86:1418

Rosenbach-Belkin V., 1988, Ph. D. thesis; The Weizmann Institute of Science, Rehovot, Israel

Schenck C.C., Gaul D., Steffen M., Boxer S.G., McDowell L., Kirmaier C., and Holten D., 1990, in: "Reaction Centers of Photosynthetic Bacteria", ed. Michel-Beyerle M.-E., Springer, Berlin, pp. 229

Scherz A., Fischer J.R.E., and Braun P., 1990, in: "Reaction Centers of Photosynthetic Bacteria", ed. Michel-Beyerle M.-E., Springer, Berlin, pp. 377

Smith J.R.L. and Calvin M., 1966, *J. Am. Chem. Soc.* 88:4500

Struck A., 1990, Ph. D. thesis, Ludwig-Maximilians-Universität, München, Germany

Struck A., Beese D., Cmiel E., Fischer M., Müller A., Schäfer W., and Scheer H., 1990, in: "Reaction Centers of Photosynthetic Bacteria", ed. Michel-Beyerle M.-E., Springer, Berlin, pp. 313

Struck A., Cmiel E., and Katheder I., 1990, *FEBS Lett.* 268:180

Watanabe T., Hongu A., Konda K., Nakazato M., Konno M., and Saitoh S., 1984, *Anal. Chem.* 56:251

Woodbury N.W., Taguchi A.K., Stocker J.W., and Boxer S.G., 1990, in: "Reaction Centers of Photosynthetic Bacteria, ed. Michel-Beyerle M.-E., Springer, Berlin, pp. 303

TRAPPING OF A STABLE FORM OF REDUCED BACTERIOPHEOPHYTIN AND BACTERIOCHLOROPHYLL IN *ECTOTHIORHODOSPIRA SP.* PHOTOREACTION CENTER

Ted Mar and Gabriel Gingras

Département de Biochimie
Université de Montréal
Montréal, Québec
CANADA H3C 3J7

INTRODUCTION

The pathway of electron transfer from the primary electron donor, P, to the primary electron acceptor, Q_A, in the photoreaction centers of various organisms has been unraveled by two main approaches : very fast kinetic measurements and phototrapping. While the first technique has been the main source of information about the identity of the intermediates and about the sequence of events they undergo, the latter has advantages that have made it an indispensable complement. Its main value resides in its greater sensitivity that permits the identification of products that are formed with quantum yields too low to be detectable by flash photolysis. Perhaps the principal asset of phototrapping is that it renders amenable the intermediary states it captures to detailed spectroscopic scrutiny. The identification of one of the pheophytin molecules (Φ_A) as an intermediary of the primary reaction owes a great deal to phototrapping [1-3]. The following is an illustration of the power of phototrapping as applied to the photoreaction center of *Ectothiorhodospira sp.*

This photoreaction center which contains the three usual H, L and M polypeptides, bacteriochlorophyll a and bacteriopheophytin a, menaquinone and ubiquinone as prosthetic groups and one atom of iron, also contains a four-heme c cytochrome [4]. This cytochrome plays an essential role in the phototrapping experiments that are reported below: when it and the primary quinone acceptor are kept in the reduced state, it furnishes electrons to the primary donor rapidly enough to allow the forward reactions to overcome the back reactions.

Reduction of Φ_B could be observed only indirectly by flash photolysis because of the low quantum yield of this reaction [5,6]. In the photoreaction center from *Ectothiorhodospira sp.*, we were able [7] to trap not only Φ_A^- and Q_A^{2-} but also Φ_B^- although at a rate 275 times slower than Φ_A^-. We have also shown the complete reduction of both pheophytin molecules as a result of two successive electron donations. Although trapping B_A^- or B_B^- could not so far be achieved, we show a donation of two electrons to each of the the four monomeric bacteriochlorin molecules of the *Ectothiorhodospira* photoreaction center. This implies singly reduced monomeric bacteriochloropyll as intermediary in the single electron transfer from P to B_A. This has been difficult to show directly by femtosecond flash photolysis [8,9] due to the complications introduced by vibrational relaxation processes [10,11].

The Photosynthetic Bacterial Reaction Center II
Edited by J. Breton and A. Verméglio, Plenum Press, New York, 1992

EXPERIMENTAL CONDITIONS

All the experiments were carried out anaerobically in the presence of sodium dithionite as a reductant and under selective illumination of only the 870 nm band to prevent possible side reactions. More details are found in the figure captions.

THE DIFFERENT CHARGE ACCUMULATED STATES

Trapping of Q_AH_2 and Φ_A^-

Before illumination, the *Ectothiorhodospira sp.* photoreaction center displays an EPR signal that is typical of Q_A^-. As illumination goes on, this signal disappears due to the double reduction of Q_A [12]. This double reduction is followed by the appearance of a new EPR signal and of a broad absorption band at 645 nm [12], both typical of anionic bacteriopheophytin [13]. The trapping of Φ_A^- is preceded by a time lag that corresponds to the reduction of Q_A^- to doubly reduced Q_A (presumably Q_AH_2). The appearance of these signals is concomitant with bleaching of the 540 nm (Q_x) and 755 nm (Q_y) bands of Φ_A. When the illumination is turned off, the Φ_A^- species decays with a lifetime of 7 - 8 s at room temperature; this is very short compared to the 15 - 20 min observed with the *Rb. sphaeroides* photoreaction center [3].

Trapping of doubly reduced Φ_A

Continued illumination under reducing (and anaerobic) conditions of the photoreaction center in the Φ_A^- - trapped state results in the disappearance of the Φ_A^- EPR signal and of the 645 nm band. Another result of continued illumination is the decline of the response of the Φ_A absorption bands to light and dark. As shown in Fig 1, the amount of this decline equals the amount of unrecoverable (under reducing conditions) bleaching of the Φ_A 755 nm absorption band. The loss of the light-dark recoverable bleaching under these conditions means that Φ_A^- is further reduced into a stable form. We infer from these experiments that Φ_A has been captured in a doubly reduced state.

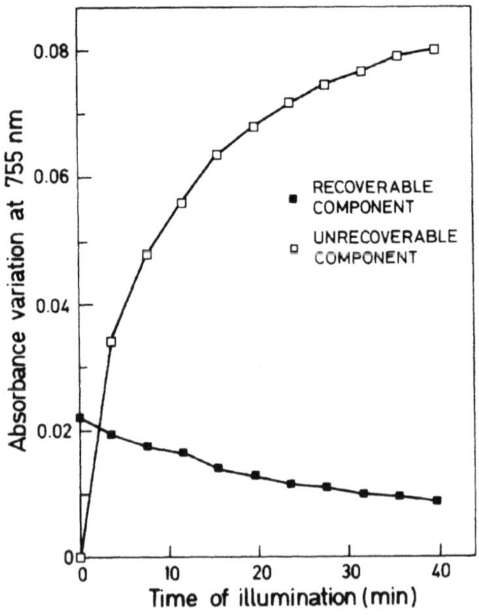

Figure 1. Kinetics of the light-dark recoverable (open circles) and unrecoverable (closed circles) components of the light-induced absorbance changes at 755 nm in *Ectothiorhodospira sp.* photoreaction center at 29 °C. The photoreaction center (1.7 μm) was suspended in 50 mM Tris-HCl (pH 8.0) containing 0.1% W/V Triton X-100 and 0.5 mM Na dithionite. Illumination was provided by a tungsten lamp filtered with a Baird Atomic wide-band interference filter centered at 930 nm. Intensity of illumination was 1.48 ergs cm^{-2} s^{-1}. The kinetics of the unrecoverable component was resolved into two components (dotted line).

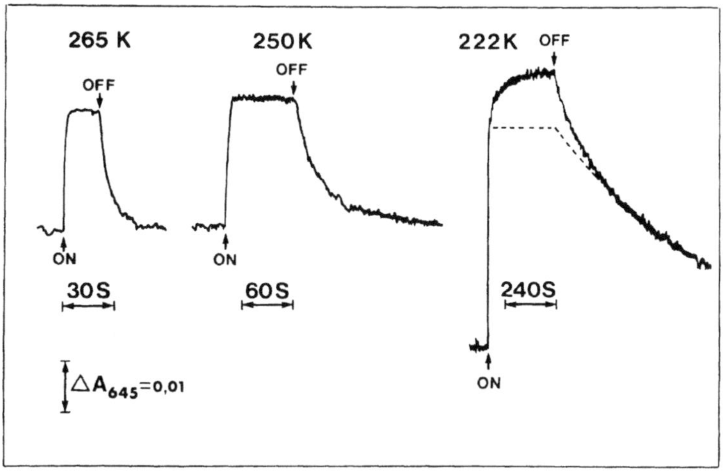

Figure 2. Kinetics of light-induced absorbance change at 645 nm in *Ectothiorhodospira sp.* photoreaction center measured at three different temperatures. The photoreaction center was suspended in 50 mM Tris-HCl (pH 8.0) containing glycerol 50% W/V, 0.1% W/V Triton X-100 and 0.5 mM Na dithionite. Illumination was provided by a tungsten lamp filtered with a Baird Atomic wide-band interference filter centered at 930 nm. Intensity of illumination was 1.48 ergs cm^{-2} s^{-1}.

Trapping of doubly reduced Φ_B

The rate of accumulation of the unrecoverable (under reducing conditions) bleaching measured at 755 nm has two components (Fig 1). The slower component, which was discussed in the previous paragraph, corresponds to the trapping of doubly reduced Φ_A. The faster initial component is due to the trapping of a reduced state of Φ_B [12]. The absence of a corresponding fast component for the EPR signal and for the 645 nm band indicates that this reduced form is not an anion but most probably doubly reduced Φ_B.

Although trapped Φ_B^- was not observed at room temperature, clear evidence for its existence was brought forth by EPR and optical spectroscopy [7]. Fig 2 shows that at temperatures above 222 K, the absorbance change at 645 nm has a single component corresponding to the trapping of Φ_A^-. At 222 K, however, another component appears that corresponds to the trapping of Φ_B^-. As seen in Fig 2, the lifetime of trapped Φ_B^- is much shorter and its rate of trapping much slower than that of Φ_A^-. In fact, at 217 K the rate of trapping of Φ_B^- is 275 times slower than that of Φ_A^- [7]. Even at 217 K, only small amounts of Φ_B- were trapped under steady state conditions because its rate of decay is faster than that of trapping. For opposite reasons, large amounts of Φ_A^- can be trapped above 222 K. The trapped doubly reduced Φ_B was observed because this state is very stable under reducing and anaerobic conditions and can be accumulated slowly over a long period of time.

Trapping of doubly reduced B_B and B_A

Figure 3 shows the spectra of a photoreaction center preparation measured after different times of illumination that lead to unrecoverable (under reducing conditions) bleaching of the Q_x and Q_y bands of both molecules of bacteriopheophytin. These spectra show, in fact, the time course of the double reduction of Φ_A and Φ_B as monitored at 755 nm. The absence of bleaching at 600 nm indicates that no reduced bacteriochlorophyll was trapped. The partial bleaching and band shift of the 802 nm band under those conditions is interpreted as being due to a strong interaction between the B and Φ molecules. After 7.5 minutes, the trapping of doubly reduced Φ_A and Φ_B was

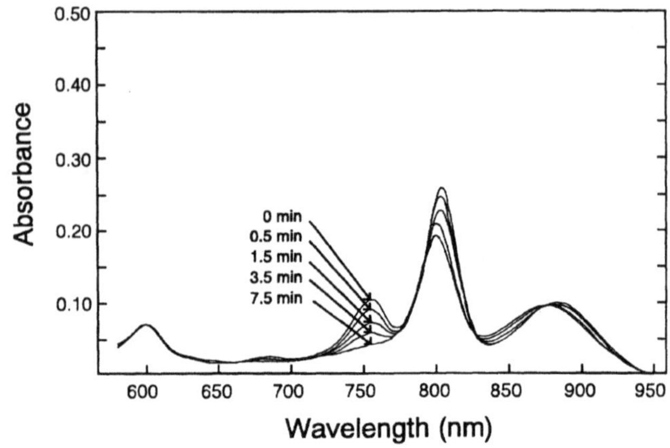

Figure 3. Bleaching as a function of the time of illumination in photoreaction center from *Ectothiorhodospira sp.* The photoreaction center (0.82 μm) was suspended in 50 mM Tris-HCl (pH 8.0) containing 0.1% W/V Triton X-100 and 0.5 mM Na dithionite. Illumination was provided by a tungsten lamp filtered with a Baird Atomic wide-band interference filter centered at 930 nm. Intensity of illumination was 1.5 x 10^4 ergs cm^{-2} s^{-1}. The experiment was carried out at 20 ^oC. Illumination periods are indicated.

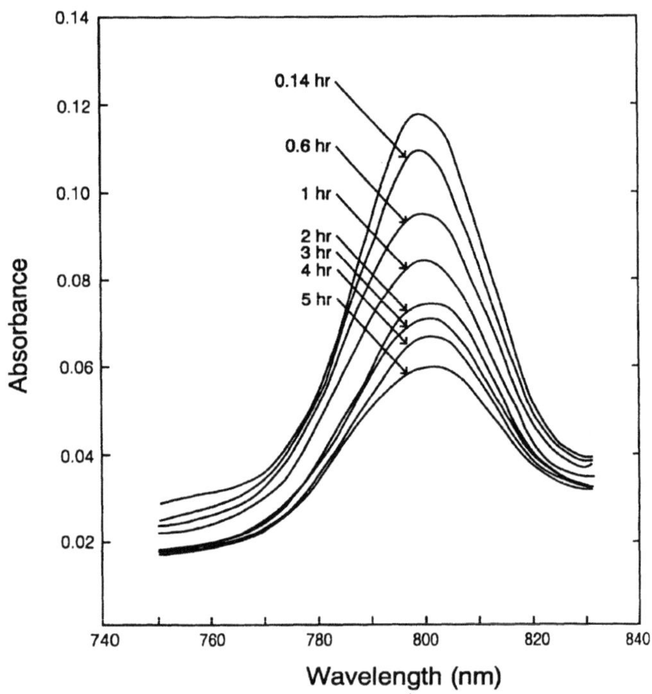

Figure 4. Bleaching of the 800 nm band as a function of the time of illumination in photoreaction center from *Ectothiorhodospira sp.* The photoreaction center (0.51 μm) was suspended in 50 mM Tris-HCl (pH 8.0) containing 0.1% W/V Triton X-100 and 0.5 mM Na dithionite. Illumination was provided by a tungsten lamp filtered with a Baird Atomic wide-band interference filter centered at 930 nm. Intensity of illumination was 1.5 x 10^4 ergs cm^{-2} s^{-1}. Illumination periods are indicated.

completed. If the illumination was prolonged up to 5 hours, as shown in Fig 4, the 802 nm band of bacteriochlorophyll was almost completely bleached and the 600 nm band approximately half bleached (not shown). The bleaching of these two bands was not accompanied by the appearance of a band at 645 nm nor by a new EPR signal that would testify to anionic bacteriochlorophyll. The bleaching of the 600 and 802 nm bands under the same reducing conditions that lead to the trapping of doubly reduced Φ_A and Φ_B indicate, therefore, that doubly reduced B_A and B_B were trapped. We assume that, under our experimental conditions, the lifetime of B_A^- and B_B^- is too short for these species to be accumulated in any detectable quantity.

It is noteworthy that the rate of bleaching of the 800 nm band is not uniform as a function of wavelength : the faster bleached component has a peak that is blue-shifted to 798 nm whereas the slower component's peak is red-shifted to 802 nm. The attribution of these components to B_A or to B_B is impossible at this time.

Absorption spectrum of the primary donor

The trapping of all four bacteriochlorin monomers in their doubly reduced state results in the stable bleaching of their Q_x and Q_y absorption bands. This leaves intact the spectrum of the primary donor which thus can be measured experimentally. Fig 5 clearly shows a large splitting of the Q_y transition between a low and a high energy band. This is consistent with a large excitonic interaction between the two bacteriochlorophylls of the special pair. The low oscillator strength of the high energy band compared with the stronger oscillator strength of the low energy band is consistent with excitonic interaction between two not quite parallel bacteriochlorophyll molecules, as shown by X-ray diffraction [14-16]. The bandwidths of the high and of the low energy bands are not equal but are respectively narrower and broader than that of a monomer, in line with the theory attributing this band broadening to the coupling of the excitonic states with charge transfer states [17]. Detailed calculations of such coupling within the special pair predict that the coupling is weak for the high energy transition and strong for the low energy transition [18]. The experimental spectrum of Fig 5 gives credence to those calculations.

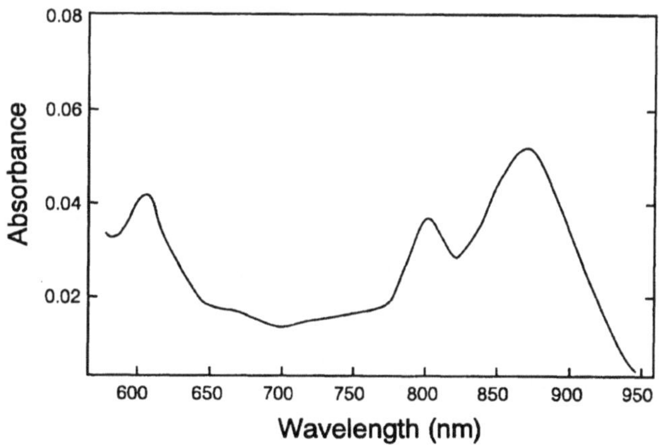

Figure 5. Absorption spectrum of the special pair in *Ectothiorhodospira sp.* photoreaction center. The spectrum was taken after bleaching of the absorption bands was maximal. The photoreaction center (0.54 μm) was suspended in 50 mM Tris-HCl (pH 8.0) containing 0.1% W/V Triton X-100 and 0.5 mM Na dithionite. Illumination was provided by a tungsten lamp filtered with a Baird Atomic wide-band interference filter centered at 930 nm. Maximum bleaching was attained after 70.5 minutes with an intensity of illumination of 1.2×10^5 ergs cm^{-2} s^{-1}.

CONCLUSION

This work provides strong evidence for the trapping of all four monomeric bacteriochlorins in their doubly reduced state. In the case of the bacteriopheophytins, the trapping technique showed the accumulation of Φ_A^- and Φ_B^- upon illumination. Continued illumination under reducing and anaerobic conditions led to the disappearance of these species and to a stable bleaching of the Q_x and Q_y bands of Φ_A and Φ_B. Prolonged illumination finally led to the bleaching of the Q_x and Q_y bands of B_A and B_B leaving intact the absorption spectrum of the primary donor. Addition of an ocidant such as oxygen or ferricyanide restored the original spectrum of the untreated photoreaction center. Allowing the double reduction of B_A and B_B, and taking into account that the primary donor is a single electron donor leads to the inescapable conclusion that singly reduced B_A and B_B must be in the path of this reaction. The implication of B_A^- as an obligatory intermediate is in accord with the femtosecond kinetic experiments which also indicate that it is a conventional intermediate in the primary act of photosynthesis.

ACKNOWLEDGEMENTS

This work was supported financially by a grant from the Natural Science and Engineering Council of Canada to G.G.

REFERENCES

1. Shuvalov, V.A. and Klimov, V.V. The primary photoreactions in the complex P-890 - P-760 (bacteriopheophytin P_{760}) of *Chromatium minutissimum* at low redox potentials. Biochim. Biophys. Acta 440, 587-599 (1976)
2. Tiede D.M., Prince, R.C., Dutton, P.L. EPR and optical spectroscopic properties of the electron carrier intermediate between the reaction center bacteriochlorophylls and the primary acceptor in *Chromatium vinosum*. Biochim Biophys Acta 449, 447-467 (1976)
3. Okamura, M.Y., Issacson, R.A. and Feher, G. Spectroscopic and kinetic properties of the transient intermediate acceptor in reaction centers of Rhodopseudomonas sphaeroides. Biochim. Biophys. Acta 546, 394-417 (1979)
4. Lefebvre, S., Picorel, R., Cloutier, Y. and Gingras, G. Photoreaction center of *Ectothiorhodospira sp*.: pigment, heme, quinone, and polypeptide composition. Biochemistry 23, 5279-5288 (1984)
5. Schenck, C.C., Parson, W.W., Holten, D. and Windsor, N.W. Transient states in reaction centers containing reduced bacteriopheophytin. Biochim. Biophys. Acta 663, 383-392 (1981)
6. Kellogg, E.C., Kolaczkowski, S., Wasielewski, M.R., Tiede, D.M. Measurement of the extent of electron transfer to the bacteriopheophytin in the M-subunit in reaction centers of Rhodopseudomonas viridis. Photosynthesis Research 22, 47-59 (1989)
7. Mar, T. and Gingras, G. Relative phototrapping rates of the two bacteriopheophytins in the photoreaction center of *Ectothiorhodospira sp.* Biochim. Biophys. Acta 1017, 112-117 (1990)
8. Holzapfel, W., Finkele, U., Kaiser, W., Oesterhelt, D., Scheer, H., Stilz, H.U. and Zinth, W. Initial electron-transfer in the reaction center from *Rhodobacter-sphaeroides*. Proc. Natl. Acad. Sci. U.S.A. 87, 5168-5172 (1990)
9. Chan, C.K., Dimagno, T.J., Chen, L.X.Q., Norris, J.R., and Fleming, G.R. Mechanism of the initial charge separation in bacterial photosynthetic reaction centers. Proc. Natl. Acad. Sci. U.S.A. 88, 11202-11206 (1991)
10. Vos, M.H., Lambry, J.C., Robles, S.J., Youvan, D.C., Beton, J. and Martin, J.L. Direct observation of vibrational coherence in bacterial reaction centers using femtosecond absorption spectroscopy. Proc. Natl. Acad. Sci. U.S.A. 88, 8885-8889 (1991)

11. Vos, M.H., Lambry, J.C., Robles, S.J., Youvan, D.C., Breton, J. and Martin, J.L. Femtosecond spectral evolution of the excited state of bacterial reaction centers at 10-K. Proc. Natl. Acad. Sci. U.S.A. 89, 613-617 (1992) .
12. Mar, T.and Gingras, G. Evidence for the photoreductive trapping of doubly reduced bacteriopheophytin in the photoreaction center of *Ectothiorhodospira sp.* Biochim. Biophys. Acta 1056, 190-194 (1991)
13. Fajer, J., Brune, D.C., Davis, M.S., Forman, A, and Spaulding, L.D. Primary charge separation in bacterial photosynthesis : oxidized chlorophylls and reduced pheophytin. Proc. Natl. Acad. Sci. U.S.A. 72, 4956-4960 (1975)
14. Deisenhofer, J., Epp, O., Miki, K., Huber, R. and Michel, H. Structure of the protein subunits in the photosynthetic reaction centre of *Rhodopseudomonas viridis* at 3A resolution. Nature (London) 318, 618-624 (1985)
15. Allen, J.P., Feher, G, Yeates T.O., Komiya, H. and Rees, D.C.Structure of the reaction center from *Rhodobacter sphaeroides* R 26: the cofactors. Proc, Natl. Acad. Sci. U.S.A. 84, 5730-5734 (1987)
16. Chang, C.H., Tiede, D., Tang, J., Smith, U., Norris, J.R. and Schiffer, M. Structure of Rhodopseudomonas sphaeroides R26 reaction center. Febs Lett 205, 82-86 (1986)
17. Won, Y. and Friesner, R.A. Simulation of photochemical hole-burning experiments on photosynthetic reaction centers. Proc. Natl. Acad. Sci. U.S.A. 84, 551-5515 (1987)
18. Parson, W.W. and Warshel, A. Spectral properties of photosynthetic reaction centers. 2. Application of the theory to *Rhodopseudomonas viridis*. J. Am.Chem. Soc. 109, 6152-6163 (1987)

TRIPLET-MINUS-SINGLET ABSORBANCE DIFFERENCE SPECTROSCOPY OF *HELIOBACTERIUM CHLORUM* MONITORED WITH ABSORBANCE-DETECTED MAGNETIC RESONANCE

J. Vrieze, E.J. van de Meent and A.J. Hoff

Department of Biophysics, Huygens Laboratory, Leiden University
Leiden, The Netherlands

INTRODUCTION

Since the discovery of the photosynthetic bacterium *Heliobacterium (H.) chlorum* [1], the first identified species of the family of the Heliobacteriaceae, this species has been investigated intensively. Membranes of *H. chlorum* contain bacteriochlorophyll (BChl) *g* as major photosynthetic pigment [2]. They have about 35 antenna BChl *g* molecules per reaction center [3], which absorb close to 790 nm (Q_Y transition) and 575 nm (Q_X transition). At temperatures below 77 K the Q_Y band can be roughly resolved into three spectral components, centered at 778, 793 and 808 nm [4].

Excitation energy is believed to be transferred to antenna pigments absorbing at the longest wavelength, BChl *g*-808, and subsequently to the reaction center [4-6]. The primary donor, P-798, in the reaction center is presumably a dimer of BChl *g'*, the 13^2 epimer of BChl *g* [7]. In the reaction center charge separation is induced between the primary donor and the primary acceptor A_o, identified as 8^1-hydroxy Chl *a* [8]. When electron transport to the secondary acceptor(s) is blocked, $P798^+A_o^-$ recombines to the triplet state of P-798 with a yield of 30% at low temperatures [9]. The acceptor side of the heliobacterial reaction center seems to be similar to that of Photosystem I (PSI) of plants. It contains a Chl *a* derivative as primary acceptor and secondary electron transport involves at least one iron-sulfur cluster and possibly a quinone molecule [10-14]. However, there still remains substantial uncertainty about the order of electron transfer from the primary acceptor to the secondary acceptors.

The Photosynthetic Bacterial Reaction Center II
Edited by J. Breton and A. Verméglio, Plenum Press, New York, 1992

At low temperatures two triplets have been observed, using EPR [12,13] and optical spectroscopy [5,9,12]: one located on the primary donor in the reaction center, formed by radical pair recombination, and another on antenna pigments absorbing at 814 nm, formed by intersystem crossing. The zero-field splitting (ZFS) parameters obtained by EPR are subject to discussion [12,13].

In this paper we will report results of Absorbance-Detected Magnetic Resonance (ADMR) measurements of membranes of *H. chlorum* at 1.2 K. With this method the ZFS parameters are more accurately determined than with high-field EPR because of the optical selectivity of the ADMR method and the much higher resolution of this zero-field technique. Using the microwave selectivity of the ADMR method, Triplet-minus-Singlet (T-S) spectra have been obtained, under different reducing conditions, from which we have been able to ascertain the origin of the triplets of BChl *g* in *H. chlorum*. The ZFS parameters and the decay rates of the triplet sublevels are compared with those obtained for BChl *g* in ethanol.

MATERIALS AND METHODS

The ADMR set-up was described in [15,16]. The T-S spectra were recorded at 1.2 K using amplitude modulation of the microwaves (312 Hz) and lock-in detection. Microwaves were generated by a HP8350B sweep oscillator with HP83525A plug-in, using its maximal output without further amplification.

Decay rates were measured by the pulse method [17], using microwave pulses with a width of 20 μsec, and a repetition rate of 33 Hz (EG&G 9650 pulse generator and HP33190B microwave switch), and were amplified by a IFI wideband amplifier (M5580). The microwaves were frequency modulated with 100 kHz through the resonance band. The change in transmittance induced by the microwave pulses was measured by a peltier-cooled photodiode (RCA 30842), whose output was fed into a differential amplifier (Tektronix AM 502). The signals were averaged by a Nicolet 527.

BChl *g* and BChl *g'* were obtained as described before [7]. The solvent was repeatedly degassed by freezing and pumping vacuum to remove all traces of oxygen. Solutions of BChl *g* and BChl *g'* were handled under nitrogen gas in the dark.

Heliobacterial membranes were prepared as described in Ref. 3. The optical absorption spectrum had an A_{788}/A_{670} ratio of 6.7 or better at room temperature, checked before and after each measurement.

Strongly reducing conditions were obtained by diluting the sample in 30 mM CAPS buffer (pH=10.5), containing 10 mM sodium ascorbate and 10 mM sodium dithionite. The sample was frozen under illumination from 240 K down to 77 K. Weakly reducing conditions were obtained by diluting the membranes in 10 mM TRIS buffer (pH=8), containing 10 mM sodium ascorbate and freezing in the dark. The absorbance at 788 nm was about 0.5 in a 1 mm perspex cuvet. All samples contained 65% (v/v) glycerol to obtain a clear glass upon freezing.

RESULTS AND INTERPRETATION

The triplet state of BChl *g* in ethanol

Zero-field splitting parameters. For BChl *g* in ethanol two zero-field (zf-) transitions were found at a detection wavelength of 760 nm: one centered at 480 MHz, and another, with lower intensity, centered at 940 MHz (not shown). The center frequency depended on the detection wavelength in the Q_Y transition, varying 12 MHz for the low-frequency transition and about 20 MHz for the high-frequency transition. For BChl *g'*, identical frequencies and relative intensities were observed. As for BChl *a* and BChl *b* we take D>0 [18]. The sign of E is not known and is also taken positive. Thus the 480 and 940 MHz transitions are ascribed to the |D|-|E| and |D|+|E| transitions, respectively. The 2|E| transition was not observed. It follows that $|D| = 237 \pm 4\ 10^{-4}\ cm^{-1}$ and $|E| = 77 \pm 2\ 10^{-4}\ cm^{-1}$. The relative intensities of the two zf-transitions were comparable with those found for other BChls *in vitro* [19].

Decay rates measured at detection wavelengths within the Q_Y transition were 3800 (± 400), 2700 (± 500) and 800 (± 100) s^{-1} for k_x ,k_y and k_z, respectively, the slowest decay being assigned to the lowest (*z*-) sublevel. The same decay rates were measured within the error limits, varying the detection wavelength over the Q_Y band.

The two microwave transitions correspond to an increase in transmittance at the maximum of the Q_Y band, i.e. the concentration of the triplet state increases when resonant microwaves are applied. Thus the slow-decaying *z*-sublevel is less populated than the faster decaying sublevels, similar to what has been observed for BChl *a* in methyltetrahydrofuran [19].

T-S spectrum. The ADMR detected T-S spectrum shows bands at 760, 697 and 595 nm (Fig. 1). They can be ascribed to the Q_Y, a vibronic and the Q_X transition, respectively, similar to the bands observed in the absorption spectrum of BChl *g* at room temperature [7]. The broad positive contributions to the T-S spectrum, between 600 and 670 nm and above 800 nm, may be assigned to triplet-triplet absorptions.

For comparing the *in vitro* T-S spectrum with the *in vivo* T-S spectrum (see further) we have inverted the sign in Fig. 1. Thus, for the *in vitro* spectrum a bleaching in the T-S spectrum corresponds to a decrease in absorbance, whereas for the *in vivo* T-S spectrum it corresponds to an increase in absorbance upon application of resonant microwaves.

In the absorption spectrum taken at room temperature the Q_X region shows two bands centered at 595 nm and 575 nm (not shown) with equal amplitude. At 1.2 K the 595 nm band dominates the Q_X region, both in the absorption spectrum (not shown) and in the T-S spectrum. This might be explained by a different coordination of the solvent to the central magnesium atom at room temperature and 1.2 K, as ethanol tends to coordinate stronger at low temperatures [20]. This is supported by the absorbance of BChl *g* in diethylether [7], which has a Q_X band at 565 nm in accordance with 5-coordination in diethylether. The Q_Y transition is less sensitive to this effect, absorbing at 756 nm and 760 nm at room temperature and 1.2 K, respectively, in agreement with results obtained for BChl *a* [20]. Therefore we ascribe the measured ZFS parameters to the BChl *g* species whose central magnesium atom is 6-coordinated.

Triplet states in membranes of *H. chlorum*

In membranes of *H. chlorum* the relative intensities of the ADMR transitions (not shown) are dependent on the chosen wavelength of detection in the Q_Y band. This indicates the existence of several triplets in the photosynthetic apparatus of *H. chlorum*. Different reducing conditions also result in different relative intensities of the zf-transitions, again indicating the presence of different triplets, formed via different pathways.

In order to assign the zf-transitions to specific pigments we recorded the ADMR-detected T-S spectra at the resonant frequencies under different reducing conditions. By comparing the wavelengths of maximum bleachings for different reducing conditions we can assign the ZFS parameters to triplets residing in the reaction center or in antenna pigments. Furthermore the decay rates for each zf-transition were measured.

The results are summarized in Table 1, together with the results obtained for BChl *g* in ethanol.

The reaction center triplet. Fig. 2 shows the T-S spectra recorded at 473, 509 and 982 MHz for samples under strongly reducing conditions (see Materials and Methods). Samples frozen in the dark in the absence of dithionite (pH=8) resulted in a 5-6 times lower intensity at the wavelength of maximum bleaching. For samples frozen in the dark in the presence of dithionite (pH=10.5) this factor was about 5 (not shown).

The T-S spectra have a maximum at 794 nm and a shoulder around 785 nm (Fig. 2 inset). Additionally, in the T-S spectra recorded at 509 and 982 MHz a positive signal contributes between 805 and 815 nm. In the T-S spectrum recorded at 473 MHz an extra bleaching at 814 nm is present due to a triplet on pigments absorbing at 814 nm (see further). Apart from the 814 nm component, the T-S spectra are equal in shape, from which we conclude that the zf-transitions at 473, 509 and 982 MHz belong to the same triplet state.

The intensity of the shoulder at 785 nm depends on the sample and on the amount of reduction; under weakly reducing conditions the shoulder at 785 nm is higher as compared to the maximum at 794 nm than under strongly reducing conditions (not shown). Therefore, it can not be ascribed to the same pigment as the 794 nm absorbing pigment, on which the triplet is located.

The 473, 509 and 982 MHz spectra show two positive peaks around 670 nm, superimposed on a broad positive signal between 600-700 nm. This can be ascribed to a shift caused by an interaction between the primary acceptor, a Chl *a* species [8], and the triplet-carrying molecule absorbing at 794 nm.

The bleaching at 573 nm is attributed to a bleaching of the Q_X band of BChl-*g* or

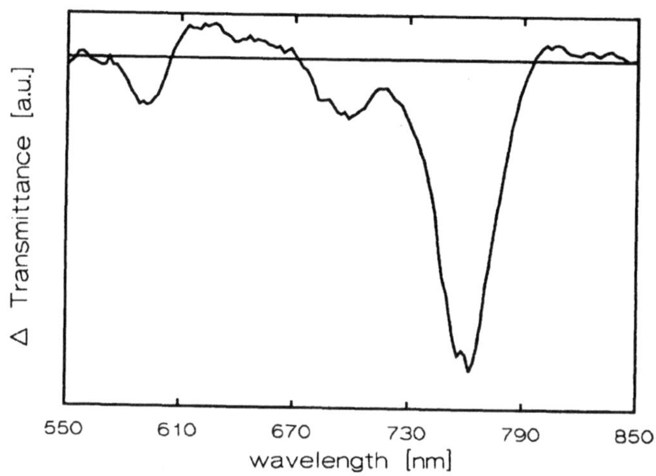

Fig.1. T-S spectrum of BChl *g* in ethanol for excitation at the |D|-|E| transition at 480 MHz.

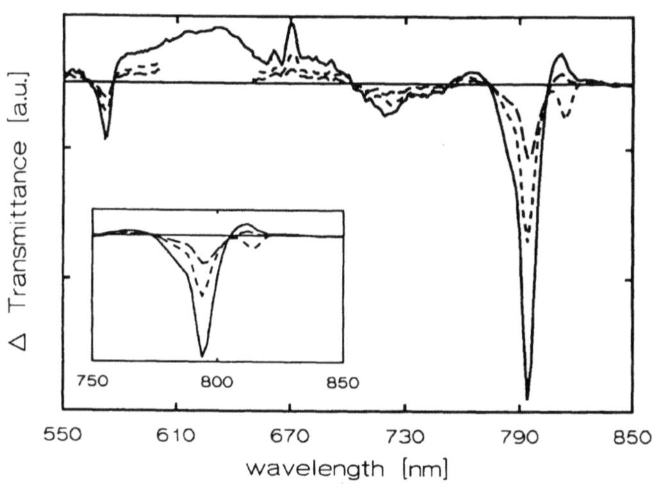

Fig.2. T-S spectra of membranes of *H. chlorum*, under strongly reducing conditions, recorded at the |D|+|E|, |D|-|E| and 2|E| transition of the reaction center triplet, 982 MHz (—), 473 MHz (- - -) and 509 MHz (– – –), respectively. Inset: Q_Y region (750-850 nm).

BChl g'. At 725-730 nm a bleaching of a vibronic transition is observed, similar to the T-S spectrum of BChl g in ethanol (see Fig. 1). The positive band at 765 nm can be ascribed to monomeric BChl g absorption.

Comparison of the flash-induced T-S spectrum [9] with the ADMR-detected T-S spectrum at 509 or 982 MHz, shows that both spectra have an identical shape around the Q_Y maximum, including the shoulder at 785 nm. The positive contribution between 805 and 815 nm in the ADMR-detected T-S spectrum is not seen in the flash-induced T-S spectrum. The ADMR-detected T-S spectrum shows more structure than the flash-induced T-S spectrum, which can easily be explained by the fact that the selectivity with resonant microwaves used by ADMR allows better discrimination than flash spectroscopy when several triplets are present with triplet lifetimes in the same order of magnitude. Smit *et al.* [5] and Kleinherenbrink *et al.* [9] attributed the flash-induced T-S spectrum to the triplet state of the primary donor, formed by radical recombination. As the ADMR signal increases when the sample is frozen under illumination our measurements confirm this conclusion. The appearance of the acceptor signal at 670 nm, which is only seen in the spectra recorded at 473, 509 and 982 MHz, is further evidence that these transitions belong to the reaction center triplet.

The decay rates for the individual triplet sublevels are similar to those found for BChl g in ethanol, being 3700 (± 300), 2800 (± 400) and 870 (± 50) s^{-1} for the k_x, k_y and k_z, respectively. Therefore, we assign the transition at 473 MHz to the $|D|-|E|$ transition, and that at 509 MHz to the $2|E|$ transition. It is not possible to assign the 473 MHz resonance to the $2|E|$ transition; in that case the resulting value of $|D|$ is much higher than the value found with high-field EPR [12,13].

Antenna triplets formed by intersystem crossing. The ADMR transitions that do not depend on the redox state of the reaction center will be described here.

Zf-transitions at 935 and 925 MHz are observed (not shown), of which the intensities do not vary with different reducing conditions. At 935 MHz two well-resolved bleachings of comparable amplitude are detected at 798 and 814 nm (Table I). The decay rates measured at these frequencies are 650 s^{-1} (k_z) and about 4500 s^{-1} (k_x).

For the 798 nm component mentioned here, the corresponding $|D|-|E|$ transition was at 450 MHz. This is in agreement with the observation that the k_z values are equal for 450 and 935 MHz.

For the 814 nm component zf-transitions are observed, apart from 925 and 935 MHz, at 450, 466 and 476 MHz. We did not observe a bleaching at 798 nm for 466 and 476 MHz. The decay rates measured at the 450, 466 and 476 MHz all contain a slow component (Table I), ascribed to the decay rate of a z-sublevel. Apparently, these frequencies constitute $|D|-|E|$ transitions. The corresponding $|D|+|E|$ transitions are at 925 MHz (with $|D|-|E|$= 466 MHz) and at 935 MHz (with $|D|-|E|$= 450 MHz and 476 MHz). The k_z value obtained at 935 MHz only matches the k_z value measured at 450 MHz; probably two $|D|+|E|$ transitions, one of which is dominant, contribute to 935 MHz. Thus, the 814 nm component originates from at least three triplets, formed by ISC. The difference in ZFS parameters can be ascribed to different sites.

Table 1. Zero field-splitting parameters and sublevel decay rates of triplet states of BChl *g in vitro* and *in vivo*

	λ (nm)	|D|-|E| (MHz)	|D|+|E|	|D| ($\times 10^{-4}$ cm^{-1})	|E|	k_x	k_y (s^{-1})	k_z
BChl *g* in ethanol	760	480	940	237 (4)	77 (2)	3800 (400)	2700 (500)	800 (100)
Triplet I	794	473	982	242.5 (0.5)	84.8 (0.2)	3700 (300)	2800 (400)	870 [1] (50)
Triplet II	798	450	935	231	81	4500 (500)	4500 (800)	650 (100)
	814	466	925	232	76.5			850 [2] (100)
		476	935	235	76.5			1200 [2] (100)
		450	935	231 (1)	81 (0.5)			690 [2] (100)

The estimated error is given in parentheses.

[1] The decay rates are measured at three frequencies: 509, 473, 982 MHz.

[2] Decay rates measured at the indicated |D|-|E| transitions and 814 nm; at the |D|+|E| transition k_z= 650 s^{-1} (see text). The k_x and k_y values are both about 4500 s^{-1}.

DISCUSSION AND CONCLUSIONS

ZFS parameters and decay rates of triplets of BChl g in ethanol and in *H. chlorum*

The value of the ZFS parameters reflect the structure and the environment of the molecule or molecular system on which the triplet state is located. The value and sign of D especially reflect the extense of the triplet wavefunction and its shape (oblete or prolate). For purple bacteria D is positive [18] and 18% (*Rb. sphaeroides*) to 25% (*Rps. viridis*) lower than the D-value of BChl *a* and BChl *b*, respectively [21]. EPR and ADMR experiments have shown that the triplet state of the primary donor of *Rps. viridis* is located on the L monomer of the BChl *b* dimer [22,23]. The lower value of D compared to that of BChl *b in vitro* was explained by the admixture of 23% charge transfer character in the triplet wavefunction [22]. The primary donor of *Rb. sphaeroides* appears to be significantly delocalized [22]. Taking into account the geometry of the dimer, the lower value of D was explained as resulting from exciton interaction between the two monomers of the dimer, in combination with 13% charge transfer admixture [22].

Considering the values of |D| and |E| for the triplet state of BChl *g* in ethanol and *H. chlorum* we note that they are very close (Table I). The higher |E| value of the reaction center triplet as compared to those of BChl *g* in ethanol can easily be explained by different coordination of the central magnesium atom. Accepting the observation of Angerhofer [20] that |D| and |E| decrease about 3% and 15%, respectively, going from 5- to 6-coordinated species, the |E| value is in reasonable agreement with 5-coordinated BChl *g* whereas the |D| value is 11% reduced. This reduction can be explained by environmental effects or structural pecularities as ring puckering, (partial) delocalization and/or admixture of charge transfer character. Further data are needed to chose between these possibilities. In principle, the decay rates can afford a discrimination [24]. In practice, however, both for *Rps. viridis* and *Rb. sphaeroides* the decay rates are practically indistinguishable from those of BChl *a* and BChl *b in vitro* [25], even when the (de)localization of the triplet state is quite different in the two species. The same situation prevails for BChl *g* in ethanol and in *H. chlorum*. Thus, the ZFS parameters and the decay rates alone provide insufficient data to decide on the (de)localization of the triplet state and multiplicity (dimer or monomer) of the primary donor of *H. chlorum*. A similar situation is present in the two plant photosystems [26].

A notable feature of both BChl *g* in ethanol and in *H. chlorum* is the high value of |E|. It is in fact close to 1/3 |D|. This is quite unlike the values of |E| of other (B)Chls; it must be explained by the presence of a vinyl group at position 3 on ring I, which is not present in BChl *a* and BChl *b* [7]. Another notable feature is the high intensity of the 2|E| transition for the reaction centert triplet compared to that of BChl *g* and other BChls. Compared to BChl *a* and BChl *b* this difference is due to the larger difference between k_x and k_y. Compared to BChl *g* in ethanol, whose 2|E| transition is not detected, the relative high intensity of the 2|E| transition of the reaction center triplet must be due to a difference in populating probabilities. This is likely, as in the reaction center triplets are formed via radical recombination, whereas *in vitro* triplets are formed via ISC. Taking the populating probabilities determined for the reaction center triplet in *Rb. sphaeroides* [27], the calculated relative intensities of the zf-transitions are in reasonable agreement with our measurements.

Antenna and reaction center triplets in *H. chlorum*

For the reaction center triplet we found |D|= 242.5 10^{-4} cm^{-1} and |E|= 84.8 10^{-4} cm^{-1}. Using high-field EPR, Nitschke *et al.* [12] ascribed the triplet with this |D| value to a triplet of antenna pigments absorbing at 814 nm. The interpretation of the EPR data, however, is complicated by the fact that |D| equals about 3|E| for all triplets in *H. chlorum*. Clearly, the advantage of ADMR of providing optical selectivity is crucial in the assignment of the two triplet states.

Three triplets, formed by ISC, can be assigned to antenna pigments absorbing at 814 nm with ZFS parameters |D|= 235 10^{-4} cm^{-1} and |E|= 76.5 10^{-4} cm^{-1}, |D|= 232 10^{-4} cm^{-1} and |E|= 76.5 10^{-4} cm^{-1} and |D|= 231 10^{-4} cm^{-1} and |E|= 81 10^{-4} cm^{-1}. These slight differences can be explained by structurally or environmentally different antenna pigments (different sites). As the triplet yields are not influenced by the state of the reaction center, we may conclude that the antenna absorbing at 814 nm do not transfer energy to the reaction center.

Because the triplet with |D|+|E|= 935 MHz, located on the pigments absorbing at 798 nm, is not influenced by the redox state of the reaction center, we also attribute this triplet to an antenna triplet formed by ISC. This is in agreement with ps measurements, which revealed a fast-decaying component at 793 nm, that was attributed to the decay of the excited state of antenna pigments absorbing around 793 nm [28].

T-S spectrum of the reaction center triplet

Analogously to the 685 nm component in the T-S spectrum of P-700 in PSI [29] we may ascribe the 785 nm shoulder to accessory BChl *g*. A positive band is observed around 765 nm, which is due to monomeric BChl *g* absorption. Taking the exciton interaction comparable to that proposed for P-700 (about 500 cm^{-1} [29]), the high-energy exciton band is expected around 735 nm, which is hidden under the vibronic band at 725-735 MHz, and which shifts together with the accessory BChl *g* absorbing at 785 nm to 765 nm, the same value obtained for BChl *g* in ethanol. It can not be excluded however that the shoulder at 785 nm has a contribution from the high-energy exciton band, although the analogue in PSI at 685 nm has not been explained in that way. In the latter case the exciton interaction between the two monomers forming the dimer would be much weaker than obtained for reaction centers of purple bacteria (about 70 cm^{-1} vs. approximately 1000 cm^{-1} for *Rps. viridis* [23,30] and 600 cm^{-1} for *Rb. sphaeroides* [31]). In this context we like to point to the P^+X^- spectrum [5,9,12], which has a positive signal at 780 nm ascribed to an electrochromic shift of an adjacent BChl *g* [5]. As the EPR linewidth of P^+ has been explained by a positive charge delocalized on two BChl *g* molecules [32], the component at 785 nm can not be due to one of the monomers of the dimer.

The positive band at the red side is also seen in the P^+X^- spectrum [9]. An analogous band is observed for the triplet state of PSI and has been ascribed to triplet-triplet absorption with admixture of charge transfer states. The broad positive signal between 600-650 nm is similar to that found for BChl *g* in ethanol, which we attribute to triplet-triplet absorptions in the T-S spectrum of BChl *g*.

Around 670 nm two bands are observed, which can be interpreted as a red shift to 670 nm and a blue shift to 662 nm due to the presence of two 8^1-hydroxy Chl *a* molecules, one of which serves as the primary acceptor that shows a maximum at 666 nm in the $P^+A_0^-$ spectrum [9,28]. This is analogous to the T-S spectrum of *Rps. viridis*, which also shows a blue and a red shift for the two bacteriopheophytins in the reaction center [23].

Thus, from comparison with the T-S spectrum of BChl *g* in ethanol and with T-S spectra reported earlier, the T-S spectrum of the primary donor indicates a dimeric structure of the triplet carrying molecule.

In conclusion, we have definitively assigned the reaction center triplet state, with |D|= 242.5 10^{-4} cm^{-1} and |E|= 84.8 10^{-4} cm^{-1}, and several antenna triplets, with closely spaced |D| and |E| values. The ZFS parameters of all triplets are comparable in magnitude to the ZFS parameters of BChl *g* ethanol. The T-S spectrum of the reaction center triplet, compared to that of the triplet of BChl *g* in ethanol, supports the notion that the primary donor is a dimer.

REFERENCES

[1] H. Gest and J.L. Favinger, *Arch. Microbiol.* 136, 11 (1983).

[2] H. Brockmann and A. Lipinski, *Arch. Microbiol.* 136, 17 (1983).

[3] E.J. Van de Meent, F.A.M. Kleinherenbrink and J. Amesz, *Biochim. Biophys. Acta* 1015, 223 (1990).

[4] R.J. Van Dorssen, H. Vasmel and J. Amesz, *Biochim. Biophys. Acta* 809, 199 (1985).

[5] H.W.J. Smit, R.J. Van Dorssen and J. Amesz, *Biochim. Biophys. Acta* 973, 212 (1989).

[6] F.A.M. Kleinherenbrink, G. Deinum, S.C.M. Otte, A.J. Hoff and J. Amesz, *Biochim. Biophys. Acta* 1099, 175 (1992).

[7] M. Kobayashi, E.J. Van de Meent, C. Erkelens, J. Amesz, I. Ikegami and T. Watanabe, *Biochim. Biophys. Acta* 1057, 89 (1991).

[8] E.J. Van de Meent, M. Kobayashi, C. Erkelens, P.A. Van Veelen, J. Amesz and T. Watanabe, *Biochim. Biophys. Acta* 1058, 356 (1991).

[9] F.A.M. Kleinherenbrink, T.J. Aartsma and J. Amesz, *Biochim. Biophys. Acta* 1057, 346 (1991).

[10] A. Hiraishi, *Arch. Microbiol.* 151, 378 (1989).

[11] H.W.J. Smit, J. Amesz and M.F.R. Van der Hoeven, *Biochim. Biophys. Acta* 893, 232 (1987).

[12] W. Nitschke, P. Setif, U. Liebl, U. Feiler and A.W. Rutherford, *Biochemistry* 29, 11079 (1990).

[13] M.R. Fisher, *Biochim. Biophys. Acta* 1015, 471 (1990).

[14] M. Brok, H. Vasmel, J.T.G. Horikx and A.J. Hoff, *FEBS Lett.* 194, 322 (1986).

[15] H.J. Den Blanken and A.J. Hoff, *Biochim. Biophys. Acta* 681, 365 (1982).

[16] R. Van der Vos, D. Carbonera and A.J. Hoff, *Appl. Magn. Res.* 2, 179 (1991).

[17] W.G. Van Dorp, W.H. Schoemaker, M. Soma and J.H. Van der Waals, *Mol. Phys.* 30, 1701 (1975).

[18] M.C. Thurnauer and J.R. Norris, *Chem. Phys. Lett.* 47, 100 (1977).

[19] H.J. Den Blanken and A.J. Hoff, *Chem. Phys. Lett.* 96, 343 (1983).

[20] A. Angerhofer, PhD Thesis, Universität Stuttgart (1987).

[21] A.J. Hoff in: Triplet State ODMR Spectroscopy (R.H. Clarke ed.), 367 (1982).

[22] J.R. Norris, D.E. Budil, P. Gast, C.-H. Chang, O. El-Kabbani and M. Schiffer, *Proc. Natl. Acad. Sci. USA* 86, 4335 (1989).

[23] E.J. Lous and A.J. Hoff, *Proc. Natl. Acad. Sci. USA* 84, 6147 (1987).

[24] W.U. Hägele, PhD Thesis, Universität Stuttgart (1977).

[25] H.J. Den Blanken, A.P.J.M. Jongenelis and A.J. Hoff, *Biochim. Biophys. Acta* 725, 472 (1983).

[26] A.J. Hoff in: Light Emission by Plants and Bacteria (Govindjee, J. Amesz, D.C. Fork, eds.) Acad. Press, Orlando, 225 (1986).

[27] A. Angerhofer, R. Speer, J. Ullrich, J.U. von Schütz and H.C. Wolf, *Appl. Magn. Res.* 2, 203 (1991).

[28] P.J.M. Van Kan, T.J. Aartsma and J. Amesz, *Photosynth. Res.* 21, 61 (1989).

[29] H.J. Den Blanken and A.J. Hoff, *Biochim. Biophys. Acta* 724, 52 (1983).

[30] E.W. Knapp, P.O.J. Scherer and S.F. Fischer, *Biochim. Biophys. Acta* 852, 295 (1986).

[31] P.O.J. Scherer and S.F. Fischer, *Biochim. Biophys. Acta* 891, 157 (1987).

[32] R.C. Prince, H. Gest and R.E. Blankenship, *Biochim. Biophys. Acta* 810, 377 (1985).

MID- AND NEAR-IR ELECTRONIC TRANSITIONS OF P+: NEW PROBES OF RESONANCE INTERACTIONS AND STRUCTURAL ASYMMETRY IN REACTION CENTERS

William W. Parson#, Eliane Nabedryk% and Jacques Breton%

#Dept. of Biochemistry, Univ. of Washington, Seattle, WA 98195 USA

%SBE/DBCM, CEN-Saclay, 91191 Gif-sur-Yvette Cedex France

Semiempirical molecular orbital treatments provide a useful framework for relating the reaction center's structure to its unusual spectroscopic properties and to the interaction matrix elements that govern the rates of electron transfer. Such treatments encounter a number of obstacles, but perhaps the most difficult of these is to evaluate the interactions of atoms that are relatively far apart, where interatomic resonance integrals are not well calibrated. Alternative approaches to this problem have led to diverging opinions concerning the resonance interactions that mix intramolecular transitions with charge-transfer transitions of the special pair of BChls.* Many of the RC's spectroscopic properties can be rationalized either by strong interactions with a charge-transfer state that lies considerably higher in energy than the Q_y transitions of the BChls, or by weaker interactions with a state that lies closer in energy.[1-3]

The resonance energies and charge-transfer transition energies of P are not directly measurable, and attempts to calculate them on the basis of the crystal structure are subject to major uncertainties. However, the oxidized special pair (P^+) should have a series of eigenstates whose separations depend largely on resonance interactions between the two BChls.[4,5] We

*Abbreviations: BChl, bacteriochlorophyll; c.i., configuration-interaction; HOMO, highest occupied molecular orbital; LUMO, lowest normally unoccupied molecular orbital; P, special pair of BChls (P_L and P_M).

The Photosynthetic Bacterial Reaction Center II
Edited by J. Breton and A. Verméglio, Plenum Press, New York, 1992

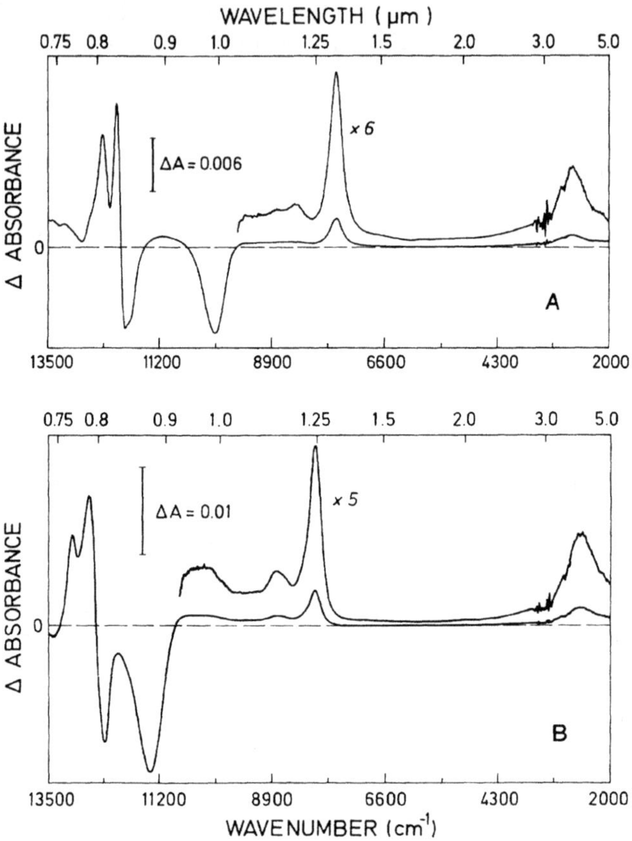

Figure 1. Absorption changes associated with the formation of P^+ at 100 K in RCs of (*A*) *Rp. viridis* and (*B*) *Rb. sphaeroides*. The spectra were obtained with RC films as described previously.[6]

recently described a broad absorption band that appears to represent the lowest-energy transition between these states.[6] The band is in the mid-IR, centered near 2750 cm^{-1} in *Rhodopseudomonas viridis* and near 2600 cm^{-1} in *Rhodobacter sphaeroides* (Fig. 1). Its dipole strength is about 85 $debye^2$ in the former species and about 25 $debye^2$ in the latter. The band is not seen with the radical cation of monomeric BChl *in vitro*, and both we[6] and Davis et al.[7] have shown that it is missing or greatly attenuated in "heterodimer" mutants, in which BChl P_L or P_M is replaced by bacteriopheophytin. It is not affected significantly by replacing the solvent H_20 by D_20.

In the previous paper,[6] we developed a simple quantum mechanical model with two basis states in which the positive charge of P^+ is localized on either P_L or P_M. Resonance interactions between the BChls mix these nonstationary states to give two supermolecular eigenstates. If the mid-IR transition is viewed as an excitation from the lower of these eigenstates to the higher, its measured energy and dipole strength can be related simply to the energy difference (δ) between the basis states and to the interaction matrix element (U). However, the values of U and δ obtained in this way are not necessarily very reliable, in part because the two-state model neglects configuration interactions with other transitions. Here we present a more refined model that includes configuration interactions and also considers possible effects of hydrogen bonding of the acetyl groups of the BChls.

To define a general set of basis states for P^+, let

$$\Phi_i = |\phi_i \prod_{j \neq i} \phi_j \bar{\phi}_j| , \qquad (1)$$

where ϕ_i and ϕ_j are molecular π-orbitals of P_L and P_M and j runs over all the filled orbitals of both molecules. Φ_i represents a radical with an unpaired electron of spin α in orbital ϕ_i (Fig. 2). We can restrict our attention to states with a fixed spin because changes of spin are formally forbidden in the absence of a magnetic field. Because the highest-occupied molecular orbital (HOMO) of BChl-b lies on the order of 7500 cm^{-1} above the next deeper orbital (the HOMO-1),[8] the two basis states with a vacancy in the HOMO of P_L or P_M (Φ_1 or Φ_2 in Fig. 2) will make the major contributions to the two lowest eigenstates of P^+. States with a vacancy in the HOMO-1 of one of the molecules (Φ_3 or Φ_4) also need to be considered, but states with a vacancy in a deeper orbital probably can be neglected for our present purposes.

The interaction matrix element $U_{k,i}$ that mixes basis states Φ_i and Φ_k can be evaluated by expanding the individual π-orbitals in terms of atomic orbitals:

$$\phi_i = \sum_s v_s^i \chi_s , \qquad (2)$$

where χ_s represents a $2p_z$ orbital on atom s. The off-diagonal elements of the interaction matrix **U** then take the form

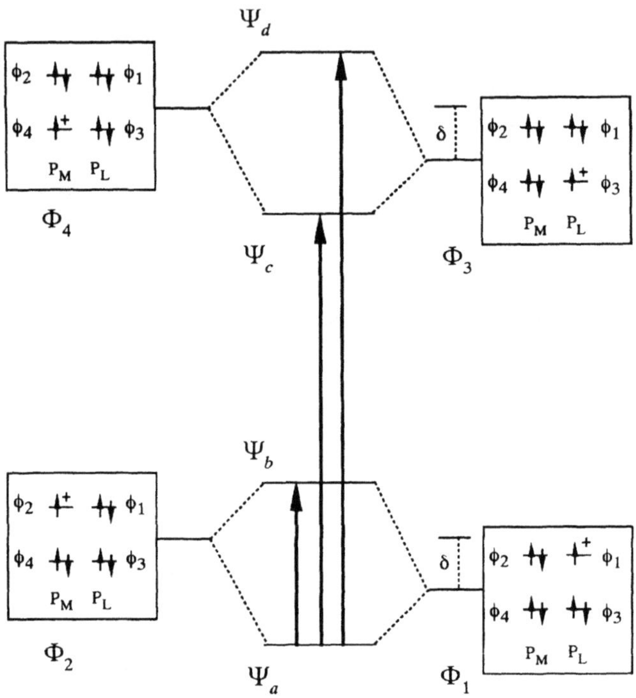

Figure 2. Basis states and excitations of P^+. The HOMO and HOMO-1 π-orbitals of the individual BChls (P_L and P_M) are labeled ϕ_1 to ϕ_4. Basis states for P^+ (Φ_1 to Φ_4) are constructed from these orbitals and combined to give the eigenstates Ψ_a to Ψ_d. The energy difference δ between Φ_2 and Φ_1 is defined to be positive if Φ_2 ($P_M{}^+$) lies above Φ_1 ($P_L{}^+$).

$$U_{k,i} \equiv \langle \Phi_k | \mathcal{H} | \Phi_i \rangle \approx -\sum_s \sum_t v_s^i v_t^k \beta_{s,t} \tag{3}$$

for ϕ_i and ϕ_k on different molecules, and $U_{k,i} \approx 0$ for ϕ_i and ϕ_k on the same molecule. $\beta_{s,t}$ is the atomic resonance integral between atom s of one BChl and atom t of the other. By expressing the resonance integrals as semi-empirical functions of the interatomic distances and the orientations of the two BChls, it thus is possible to calculate approximate values for the $U_{k,i}$ on the basis of the crystal structure.[5,6,8,9]

The energies of the basis states, which form the diagonal elements of **U**, cannot be calculated very reliably because they are sensitive to long-range electrostatic interactions and small distortions of the structure.[5] We therefore have used an adjustable parameter, δ, for the energy difference between the two basis states in which the vacancy is in the HOMO of P_L or P_M (Φ_1 or Φ_2). For simplicity, we assume that Φ_3 and Φ_4 are separated by the same energy difference.

Diagonalizing the 4x4 matrix **U** gives four eigenstates, Ψ, with expansion coefficients **d**:

$$\Psi_p = \sum_{i=1}^{4} d_p^i \Phi_i . \tag{4}$$

The two lowest eigenstates consist largely of combinations of Φ_1 and Φ_2, as in the two-state model;[6] the third and fourth states are composed largely of combinations of Φ_3 and Φ_4.

Transitions from the lowest state (Ψ_a) to the higher levels can give three absorption bands (Fig. 2). The transition dipole for excitation to state Ψ_p is

$$\vec{\mu}_{ap} = \langle \Psi_p | \tilde{\mu} | \Psi_a \rangle = \sum_i \sum_j d_a^i d_p^j \langle \Phi_j | \tilde{\mu} | \Phi_i \rangle \tag{5a}$$

$$\approx \sum_i \sum_j d_a^i d_p^j \vec{w}_{ij} \tag{5b}$$

with

$$\vec{w}_{ij} = e \sum_s v_s^i v_s^j \vec{r}_s \qquad \text{for } j \neq i \tag{5c}$$

and

$$\vec{w}_{ii} = e \sum_s \{(v_s^i)^2 + \sum_{k \neq i} 2(v_s^k)^2\} \vec{r}_s . \tag{5d}$$

Here $\tilde{\mu}$ is the dipole operator; $\vec{r}_s$, the position of atom s; and e, the electronic charge. Equation 5c can be modified straightforwardly to use the gradient operator[8] in place of $\tilde{\mu}$ for the intramolecular excitations $\Phi_1 \rightarrow \Phi_3$ and

$\Phi_2 \rightarrow \Phi_4$. We used this modification for the calculations described below, but it had only minor effects on the results.

Configuration interactions can mix the three IR excitations with each other and with other intramolecular excitations of P_L and P_M. If we expand the basis set to include the first four excited singlet states of each of the individual BChls, the configuration-interaction (c.i.) matrix **A** becomes an 11x11 matrix. Three of its diagonal elements consist of the differences between the eigenvalues of **U** ($A_{p,p} = E_p - E_a$); the others are the energies of the BChl Q_y, Q_x, B_x and B_y transitions. The off-diagonal matrix element that connects two IR excitations, $\Psi_a \rightarrow \Psi_p$ and $\Psi_a \rightarrow \Psi_q$, is

$$A_{p,q} \approx \sum_i \sum_{j \neq i} \sum_{k \neq j} \sum_{l \neq k} d_p^i d_a^j d_q^k d_a^l \sum_s \sum_t \{(\delta_{i,k} v_s^j v_t^l - \delta_{j,l} v_s^i v_t^k)\beta_{s,t} + (2v_s^i v_s^j v_t^k v_t^l - v_s^j v_s^l v_t^i v_t^k)\gamma_{s,t}\}, \quad (6)$$

where $\delta_{i,k}$ is a delta function (1 if $i = k$, zero otherwise), s and t again refer to atoms of different BChls, and $\gamma_{s,t}$ is the electron-electron repulsion integral.[8] The first four sums are taken over the HOMOs and HOMO-1s of both BChls.

The off-diagonal matrix element of **A** that couples an IR excitation, $\Psi_a \rightarrow \Psi_p$, with a Q_y, Q_x, B_x or B_y transition of the unoxidized BChl can be evaluated by representing the latter transition as

$$^1\Psi_m = \sum_N c_m^N \, {}^1\Psi_M^N . \quad (7)$$

Here $^1\Psi_M^N$ is a singlet wavefunction created by an excitation from ϕ_{n1} (the HOMO or HOMO-1) of the unoxidized molecule (BChl M) to ϕ_{n2} (the LUMO or LUMO+1) of the same BChl, and the c_m^N are c.i. coefficients for these excitations.[8] The relevant interaction matrix element is

$$A_{m,p} \approx \sum_i \sum_{j \neq i} \sum_N d_p^i d_a^j c_m^N \sum_s \sum_t \{\delta_{i,n1} v_s^j v_t^{n2} \beta_{s,t} + 2v_s^i v_s^j v_t^{n1} v_t^{n2} \gamma_{s,t}\}, \quad (8)$$

where $n1$, $n2$ and t refer to BChl M, and s refers to the oxidized BChl.

The eigenvalues of **A** give the energies of the mixed transitions; the eigenvectors provide the coefficients needed to express the overall transition dipoles in terms of the three $\vec{\mu}_{ap}$ obtained by eq. 5 and the transition dipoles of the intramolecular transitions. To implement this treatment, we optimized the c_m^N for monomeric BChl-a or -b in solution, using the gradient operator to calculate all the intramolecular transition dipoles,[8] and assuming as a first approximation that oxidation of one of the BChls does not alter the intramolecular c.i. coefficients of the other BChl. (A more complete theory

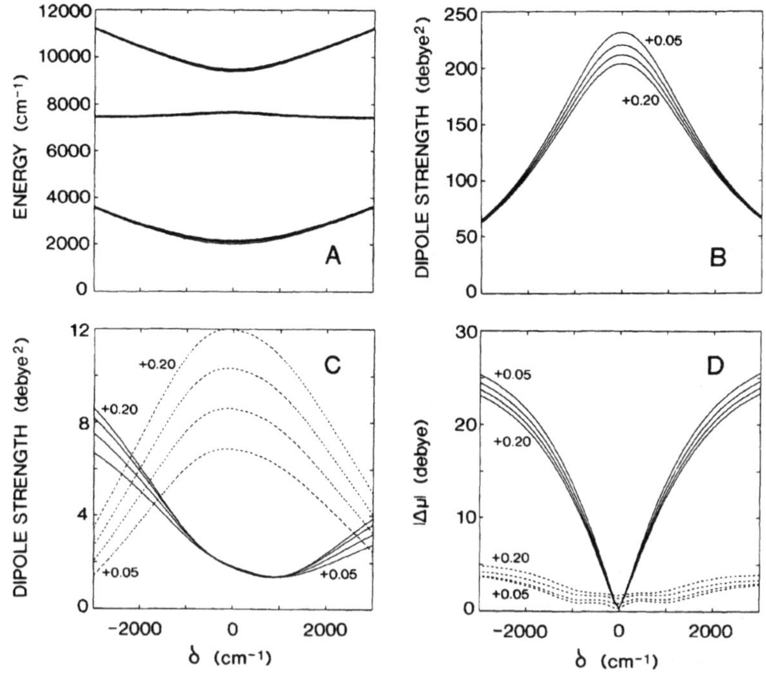

Figure 3. Calculated energies, dipole strengths, and changes in permanent dipole moment for the first three mixed transitions of P^+ in *Rp. viridis* RCs, as functions of the separation between the P_M^+ and P_L^+ basis states (δ). Panel *A* shows the energies; *B*, the dipole strength of the band in the 2700-cm^{-1} region; *C*, the dipole strengths of the bands near 7590 cm^{-1} (—) and 10000-cm^{-1} (---); and *D*, the changes in permanent dipole moment associated with the 2700-cm^{-1} (—) and the 7590-cm^{-1} (---) transitions. The families of curves in each panel were obtained by changing the sigma charge on the acetyl oxygen atoms by +0.05, +0.10, +0.15 or +0.20, relative to the standard charge used previously[6,8] in the program QCFF-PI to generate the π-orbitals for the BChls. Adding a greater positive charge models the formation of an increasingly strong H-bond to the acetyl group. In panels *B*-*D*, the curves obtained with the smallest and largest adjustments of the sigma charge are indicated; in *A*, adding a larger positive charge shifts all three transitions to slightly higher energies. (Changing the sigma charge over this range increases the calculated resonance interaction matrix element that connects Φ_1 and Φ_2 from -1010 to -1090 cm^{-1}. With the standard sigma charge on the oxygens, this matrix element is calculated to be -910 cm^{-1}.)

might consider effects of the local electrical field on the energies and c.i. coefficients of the intramolecular transitions.)

Figure 3A shows the energies of the first three transitions calculated for the *Rp. viridis* RC, as a function of the energy difference between the P_L^+ and P_M^+ basis states (δ). The lowest-energy transition probably accounts for the absorption band found near 2700 cm^{-1}. The energy observed experimentally is obtained when the basis states are separated by about $\pm 2000\ cm^{-1}$. Taking the experimental energy to coincide with the shoulder on the long-wavelength side of the band instead of with the peak of the band would reduce this estimate of $|\delta|$ by about $500\ cm^{-1}$. The next transition has approximately the right energy to match the absorption band seen at 1250 nm ($8000\ cm^{-1}$) in *Rb. sphaeroides* and other species that contain BChl-a, and at 1320 nm ($7590\ cm^{-1}$) in *Rp. viridis* (Fig. 1). The calculated energy of this band is not very sensitive to the value of δ. The third transition has a calculated energy of about $10{,}000\ cm^{-1}$, again depending on the value of δ. A broad, poorly resolved feature that could reflect this transition appears near 1 μm in the spectra for both *Rb. sphaeroides* and *Rp. viridis* (Fig. 1). The 8 higher transitions that are not represented in Fig. 3A are largely intramolecular in composition and are shifted only slightly by interactions with the IR transitions. The small blue shift of the Qy transition of the unoxidized BChl could contribute to the near-IR absorption changes that accompany the formation of P^+.

As shown in Fig. 3B, the calculated dipole strength of the mid-IR absorption band depends strongly on δ. It also depends on the distance between the centers of π-electron density of the HOMOs of P_L and P_M. If $\delta \approx 0$, the transition dipole is approximately half the change in dipole moment associated with moving an electron from one of these centers to the other.[6] Hydrogen-bonding to the ring-I acetyl groups can alter this distance by increasing the HOMO electron densities in the region where P_L and P_M are closest together. This tends to enhance the resonance interactions between the BChls, but decreases the calculated dipole strength as shown in Fig. 3B. The acetyl groups of both P_L and P_M and the ring-V keto group of P_L probably are H-bonded in *Rp. viridis*,[10] and the acetyl group of P_L may have a similar bond in *Rb. sphaeroides*.[11] The dipole strength becomes less sensitive to this effect if $|\delta| >> 0$. Again, setting $|\delta|$ in the region of $2000\ cm^{-1}$ gives results that agree with the experimental value of $85 \pm 28\ debye^2$. In *Rb. sphaeroides*, the calculated resonance interactions between P_L and P_M are weaker and both the measured and the calculated dipole strength of the mid-IR absorption band are smaller.[6]

The absorption bands in the regions of 7590 and $10000\ cm^{-1}$ are expected to be much weaker than the band in the $2700\text{-}cm^{-1}$ region (Fig. 3C). This is because the two near-IR transitions depend largely on excitation of an electron from the HOMO-1 to the HOMO within one or the other BChl. Such excitations are forbidden because both orbitals have approximately the same inversion symmetry.[12] Structural distortions or intermolecular interactions that decrease the orbital symmetry thus could strengthen these transitions. Hydrogen-bonding of the acetyl groups evidently can have such an effect (Fig. 3C).

Increasing the absolute magnitude of δ increases the calculated dipole strength of the $7590\text{-}cm^{-1}$ band and decreases that of the $10000\text{-}cm^{-1}$ band (Fig. 3C). Negative values of δ appear to be more effective than positive values in this regard, but this may depend on the bacterial species. (In a less extensive set of calculations on *Rb. sphaeroides* RCs, negative values of δ favored the band in the 10000-cm^{-1} region.) In the measured spectra for both species the absorption band in the 7590- or 8000-cm^{-1} region is considerably stronger than the diffuse feature seen at higher energies (Fig. 1). This suggests that the correct value of $|\delta|$ is relatively large, in accord with the calculations on the 2700-cm^{-1} band. However, any conclusions based on the higher-energy bands must be viewed cautiously, in the light of the sensitivity of the calculations to distortions of the orbital symmetry. Even with $\delta \approx -2000\ cm^{-1}$, the calculated dipole strength of the 7590-cm^{-1} band falls well short of the experimental value (25 ± 8 $debye^2$).

With $\delta \approx \pm 1900\ cm^{-1}$, the ground state of P^+ in *Rp. viridis* (Ψ_a) is calculated to have about 83% of its positive charge localized on one or the other of the two BChls. If δ is negative, as suggested tentatively by the calculations on the near-IR bands, it is P_M that holds the larger share of the charge. Long-range electrostatic interactions with the protein appear to favor such a localization, although the structural distortions seen in the crystal structure would tend to concentrate the charge on P_L.[5,13] From an analysis of the ENDOR spectrum of P^+, Lendzian et al.[13] have concluded that about 70% of the charge resides on one of the BChls, which they suggest to be P_L. Recent studies of the ENDOR spectra of P^+ in single crystals of wild-type and mutant *Rb. sphaeroides* RCs[14] and FTIR studies of heterodimer mutants of *Rb. capsulatus* (E. Nabedryk, S. J. Robles, E. Goldman, D. C. Youvan and J. Breton, *submitted*) indicate that the charge concentrates preferentially on P_L in wild-type RCs of these two species, but no such studies have been described yet for *Rp. viridis.* In the case of *Rb. sphaeroides*, the ENDOR spectra are fit well by assuming that 67% of the charge is on P_L,[14] which is a somewhat more symmetrical distribution than our calculations would suggest. Another observation favoring a more symmetrical dimer is that the difference between the E_m values for oxidation of P in the two heterodimer mutants of *Rb. sphaeroides* is only about 30 mV.[7]

One way to examine the magnitude of δ experimentally would be to measure the Stark effect on the 2700-cm^{-1} absorption band. Figure 3D shows the calculated changes in permanent dipole moment ($|\Delta\mu|$) associated with the 2700- and 7590-cm^{-1} transitions. If $|\delta|$ is on the order of 1800 to 2000 cm^{-1}, as suggested by Figs. 3A, B and C, the 2700-cm^{-1} transition has enough charge-transfer character to result in a substantial change in dipole moment. The absorption band thus would be expected to exhibit a large Stark effect. The predicted effect depends somewhat on the strength of the H-bonding to the acetyl groups, but is not very sensitive to the sign of δ. To our knowledge, no Stark measurements have been made in the mid-IR region of the spectrum. However, the 8000-cm^{-1} absorption band of *Rb. sphaeroides* RCs apparently does not exhibit a significant Stark effect (S. Boxer, personal communication). This is consistent with the smaller value of $|\Delta\mu|$ calculated for this band (Fig. 3D).

The expected sensitivity of the IR absorption bands to the resonance interactions between P_L and P_M and to the electrostatic interactions that contribute to δ should make these bands useful for probing structural changes in the RC. We found that cooling RCs to 100 K in the dark has little or no effect on the energy or dipole strength of the 2700-cm^{-1} band.[6] However, if RCs are cooled under illumination, and the reversible absorbance changes associated with formation and decay of $P^+Q_A^-$ then are measured at 100 K, the band appears to be shifted to higher energies by about 100 cm^{-1}. This fits with earlier observations that cooling during illumination can trap RCs in a metastable state with altered electron-transfer kinetics.[15]

Acknowledgements: This work was supported by a Human Frontier Science Program grant to J.B. and NSF Grant DMB-9111599 to W.W.P. We thank D. Bocian, S. Boxer, S. Creighton, W. Mäntele, V. Nagarajan, C. Schenck and A. Warshel for helpful discussions.

REFERENCES

1. L. K. Hanson, *Photochem. Photobiol.* 47: 903-921 (1988).
2. R. A. Friesner and Y. Won, *Biochim. Biophys. Acta* 977: 99-122 (1989).
3. A. Warshel and W. W. Parson, *Ann. Rev. Phys. Chem.* 42: 279-309 (1991).
4. J. J. Katz, L. L. Shipman, and J. R. Norris, *Ciba Foundation Symp. (new series)* 61: 1-34 (1979).
5. W. W. Parson, Z.-T. Chu, and A. Warshel, *Biochim. Biophys. Acta* 1017: 251-272 (1990).
6. J. Breton, E. Nabedryk, and W. W. Parson, *Biochem.* in press (1992).
7. D. Davis, A. Dong, W. S. Caughey, and C. C. Schenck, *These Proceedings.*
8. A. Warshel, and W. W. Parson, *J. Am. Chem. Soc.* 109: 6143-6152 (1987).
9. W. W. Parson, and A. Warshel, *J. Am. Chem. Soc.* 109: 6152-6163 (1987).
10. J. Deisenhofer, and H. Michel, *Science* 245: 1463-1473 (1989).
11. O. El-Kabbani, C. -H. Chang, D. Tiede, J. Norris, and M. Schiffer, *Biochem.* 30: 5361-5369 (1991).
12. C. Weiss, *J. Mol. Spectrosc.* 44: 37-80 (1977).
13. F. Lendzian, W. Lubitz, H. Scheer, A. J. Hoff, M. Plato, E. Tränkle, and K. Möbius, K. *Chem. Phys. Lett.* 148: 377-385 (1988).
14. M. Huber, R. A. Isaacson, E. C. Abresch, G. Feher, D. Gaul, and C. C. Schenck, *These Proceedings.*
15. D. Kleinfeld, M. Y. Okamura, and G. Feher, *Biochem.* 23: 5780-5786 (1984).

^{15}N ENDOR EXPERIMENTS ON THE PRIMARY DONOR CATION RADICAL $D^{+\cdot}$ IN BACTERIAL REACTION CENTER SINGLE CRYSTALS OF *RB. SHAEROIDES* R-26

F. Lendzian[1,2], B. Bönigk[2], M. Plato[1], K. Möbius[1] and W. Lubitz[2]

[1]Institut für Molekülphysik,
Freie Universität Berlin,
D-1000 Berlin 33, Germany

[2]Max-Volmer-Institut für Biophysikalische
und Physikalische Chemie
Technische Universität Berlin
D-1000 Berlin 12, Germany

INTRODUCTION

The rates of the electron transfer processes in reaction centers (RC's) of photosynthetic bacteria are controlled both by the spatial and the electronic structure of the involved donor and acceptor molecules. The spatial structure of bacterial RC's has been determined by X-ray diffraction for *Rhodopseudomonas (Rp.) viridis*[1] and for *Rhodobacter (Rb.) sphaeroides*.[2] The electronic structure of the transient radical species formed in the charge separation process can be elucidated by EPR and ENDOR techniques.[3] The information is contained in the electron-nuclear hyperfine couplings (hfc's) which, after assignment to specific nuclei, yield a detailed picture of the valence electron spin density distribution in the respective molecules.[3,4]

Already in the early seventies, EPR and ENDOR investigations of the cation radical of the primary donor, $D^{+\cdot}$, have lead to the proposal of a bacteriochlorophyll a (BChl a) dimer for this species, based on a narrowing of the EPR spectrum[5] (factor $1/\sqrt{2}$) and a reduction of hfc's (factor 1/2) as compared with monomeric BChl $a^{+\cdot}$ *in vitro*.[6,7] Later, based on ENDOR experiments in liquid RC solutions, this interpretation became a matter of controversy.[8,9] For $D^{+\cdot}$ in *Rb. sphaeroides* even a monomeric BChl a structure has been discussed.[10] The structure of BChl a is shown in figure 1.

A major problem with all proton ENDOR spectra of $D^{+\cdot}$ in frozen[6,7] and liquid[8,9] RC solutions was the poor spectral resolution caused by the large number of hydrogen nuclei interacting with the unpaired electron spin, and the lack of an unambiguous experimental assignment of hfc's to specific nuclei in the dimer halves D_L and D_M (L and M denote the protein subunits to which the dimer halves are bound).

A promising approach towards determination of the symmetry of the spin density distribution in $D^{+\cdot}$ is to measure the hfc's of the nitrogen nuclei by ENDOR, because there are only four of them in each BChl a molecule and an asymmetry in the spin density distribution should show up immediately in the number of nitrogen ENDOR lines observed. However, several difficulties are expected for such experiments: In frozen disordered RC solutions the expected large anisotropy of the nitrogen hf interaction leads to a significant broadening of the observed lines. An additional broadening arises from the quadrupolar coupling of ^{14}N. No individual hfc's can be resolved under these

The Photosynthetic Bacterial Reaction Center II
Edited by J. Breton and A. Verméglio, Plenum Press, New York, 1992

conditions. In liquid RC solutions the quadrupolar interaction of ^{14}N provides an additional source of relaxation, which makes it difficult to observe ^{14}N ENDOR lines especially for small hfc's, where no large hyperfine enhancement factors of the applied radio frequency field can be expected.[12] The isotope ^{15}N (natural abundance 0.37%) is much more favorable, since it has no electric quadrupolar moment (nuclear spin 1/2), and its magnetic moment is 40% larger than that of ^{14}N.

Consequently, experiments have been performed on $D^{+\cdot}$ in liquid solutions of ^{15}N-enriched RC's of *Rb. sphaeroides* in a previous investigation.[11] The results have been interpreted in terms of a symmetric dimer. Further studies, using electron spin echo techniques applied to frozen RC solutions, came to a similar conclusion.[13,14]

Here, we report on ^{15}N ENDOR experiments on $D^{+\cdot}$ in reaction center single crystals of *Rb. sphaeroides* R-26, which, for the first time, clearly show that there are at least six different ^{15}N hfc's in $D^{+\cdot}$ and that the spin density is asymmetrically distributed over the dimer halves D_L and D_M. This result is in agreement with recent proton ENDOR/TRIPLE experiments performed on $D^{+\cdot}$ in RC's of single crystals of *Rb. sphaeroides* R-26 with natural isotopic composition.[15]

MATERIALS AND METHODS

Rb. sphaeroides R-26 was grown, using [^{15}N] ammonium chloride (95%)as the sole nitrogen source in a yeast-free minimal medium, similar to that used in ref. 11. The ^{15}N-labeled RC's (>90%) were purified and crystallized according to the procedure given in ref. 16. The resulting crystals (space group $P2_12_12_1$) reached a size of ca. 0.4x0.8x4.0 mm^3 and exhibited a rhombic cross section. They were mounted together with a small amount of buffer solution inside a sealed quartz tube in a two-axes goniometer. The EPR-, ENDOR- and TRIPLE-spectrometer has been described elsewhere.[17] The crystals were illuminated inside the EPR- and ENDOR cavity in the range from 830 to 950 nm, using a 100 W tungsten/halogen lamp and appropriate filters. Details of the crystal mounting and alignment have been published elsewhere.[15] Calculation of spin densities and dipolar hf-tensors of $D^{+\cdot}$, using the RHF-INDO/SP method[4] and the structural data from X-ray diffraction studies,[23] have been described in refs. 15,18.

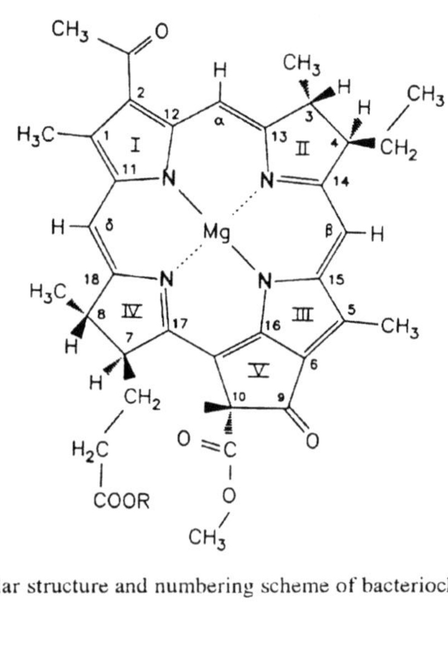

Figure 1. Molecular structure and numbering scheme of bacteriochlorophyll a (BChl a)

RESULTS AND DISCUSSION

In general the maximum ENDOR signal is observed when electronic and nuclear relaxation rates are of comparable magnitude.[17] In the solid state, this condition is mostly met at low temperatures. However, an irreversible loss of order of the crystals and concomitant changes in the ENDOR spectra have been observed, when RC crystals were cooled to low temperatures in proton ENDOR experiments (unpublished results). Therefore, the ^{15}N ENDOR experiments were performed near room temperature. The unfavorable relaxation rates found in RC single crystals at room temperature, were circumvented for the protons in $D^{+\cdot}$ by using the electron-nuclear-nuclear Special TRIPLE resonance technique.[15,17] In this experiment, both high- and low-frequency NMR transitions are induced simultaneously. Consequently, the observed signal intensities are less dependent on the nuclear relaxation rates.[12,17] However, Special TRIPLE resonance can be applied only if high- and low-frequency ENDOR lines are displaced symmetrically about the nuclear Larmor frequency. For ^{15}N nuclei in $D^{+\cdot}$ this condition is not met, because the magnetic moment is much smaller than that of ^{1}H; whereas, the hfc's are comparable in magnitude with those of the protons. As a consequence, we were restricted to the normal ENDOR experiment. The observed ^{15}N ENDOR signal intensities were less than 10% of the ^{1}H Special TRIPLE intensities of $D^{+\cdot}$ recorded for the same crystals.

Figure 2a shows the observed ^{15}N-ENDOR spectra for the magnetic field B_0 parallel to the crystal symmetry axes $\vec{a}$, $\vec{b}$, and $\vec{c}$. For these three orientations all four RC's in the crystal unit cell are magnetically equivalent. The axes were aligned (estimated error $\pm 2°$) by recording the angular dependent ^{1}H Special TRIPLE spectra, which are shown for the same three canonical orientations in figure 2b. The ^{1}H Special TRIPLE spectra agree well with those obtained for $D^{+\cdot}$ in RC single crystals with natural isotopic composition.[15] An accurate alignment of the symmetry axes $\vec{a}$, $\vec{b}$ and $\vec{c}$ with respect to $\vec{B}_0$ was essential for the ^{15}N-ENDOR experiments, because the ^{15}N hf-tensors exhibit large anisotropic contributions which, even for small misalignments, could lead to significant line broadening due to the magnetic inequivalence of the four RC sites. Because of the poor signal intensity, so far ^{15}N ENDOR spectra have been recorded only along the three crystal symmetry axes, where all four RC sites contribute to the observed intensity of each line.

According to our MO (RHF-INDO/SP) calculations,[15,18] the ^{15}N nuclei carry significant p_z spin densities and, therefore, the largest dipolar contribution to the hf couplings is expected, when $\vec{B}_0$ is oriented parallel to the normal vector $\vec{n}$ of the plane of the respective BChl $\underline{a}$ macrocycle. The largest principal value of the calculated full ^{15}N hf-tensor (isotropic and anisotropic parts) is also obtained for this orientation. Figure 3 shows the four BChl $\underline{a}$ dimers in the unit cell projected on the crystallographic b-c plane. This corresponds to an approximate "edge on view" of the dimers; the indicated normal vectors $\vec{n}$ on the average planes of the two dimer halves, D_L and D_M, are all approximately in the b-c plane and form angles of $\pm 35°$ with the b axis.

The angles between the normal vectors $\vec{n}$, obtained from the X-ray data, and the respective crystal axes $\vec{a}$, $\vec{b}$ or $\vec{c}$, are indicated in figure 2a. As expected, the largest hf splittings resulting in the largest ^{15}N-ENDOR frequencies are observed for $\vec{B}_0$ parallel to $\vec{b}$, since this crystal axis forms the smallest angle (35°) with $\vec{n}$.

In general, the frequencies of ENDOR lines in a single crystal are given to first order by[20]

$$\nu^{\pm} = [\nu_N^2 + M_S^2 \vec{l}' \frac{\underline{G}\,\underline{A}^2\,\underline{G}}{|\underline{G}|^2} \vec{l} - 2 M_S \nu_N \vec{l}' \frac{\underline{G}\,\underline{A}}{|\underline{G}|} \vec{l}]^{1/2} \qquad (1)$$

where $\underline{G}$ is the g-tensor, $\underline{A}$ the hf-tensor (including the isotropic part) and $M_s = \pm 1/2$ the electronic magnetic quantum number. $\vec{l}$ and $\vec{l}'$ are the column and row unit vectors of the direction cosines of $\vec{B}_0$ in the chosen axes system. Assuming $\underline{G}$ to be isotropic, which is a good approximation for $D^{+\cdot}$, reduces eq. (1) for the case $\vec{B}_0$ parallel to the crystal axis $\vec{a}$ to

$$\nu_a^{\pm} = [\nu_N^2 + \tfrac{1}{4}(\underline{A}^2)_{aa} \pm \nu_N A_{aa}]^{1/2} \qquad (2a)$$

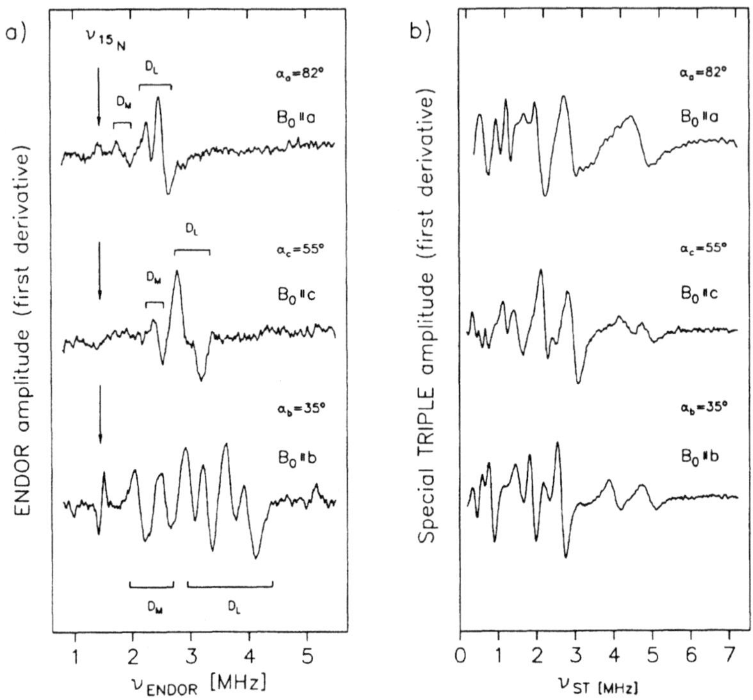

Figure 2. ^{15}N ENDOR (a) and ^{1}H Special TRIPLE spectra (b) of $D^{+\cdot}$ in single crystals of ^{15}N-labeled RC's. For the shown orientations, the field B_0 is parallel to one of the three symmetry axes of the crystal, and all four RC's in the unit cell are magnetically equivalent. Line positions in (a) correspond to the high-frequency ^{15}N ENDOR transition (see eq. (2b)); ν_{15_N} = 1.46 MHz; the indicated assignment to D_L and D_M is based on the ordering in magnitude of calculated ^{15}N hfc's (see text). In (b) line positions correspond to one half of the respective proton hf-tensor element A_{ii} (i = a,b,c); the frequency axis gives the deviation from the proton Larmor frequency, ν_H. Experimental conditions: MW: 20 mW; RF: 300 W, corrsponding to 30 Gauss rotating frame in (a), and 2x100 W corresponding to 2x10 Gauss rotating frame in (b); FM: 10 kHz, ±50 kHz deviation. Accumulation time: approximately 3 h for each spectrum of (a) and 20 min for each spectrum of (b); T = 17±1 °C. α_i(i = a,b,c) is the angle between the normal vector on the average plane of the two macrocycles and the respective crystal axis a, b or c (see fig. 3).

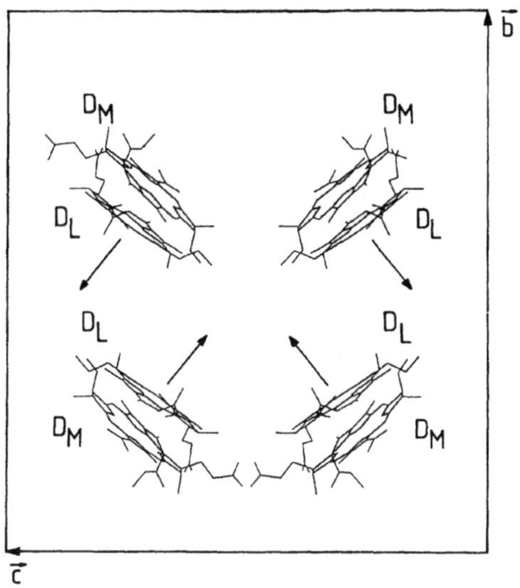

Figure 3. Projection of the unit cell of the RC crystal in the b-c plane, showing only the primary donors D (dimer halves D_L and D_M are indicated) of the four RC sites (enlarged scale). The normal vectors $\vec{n}$ on the average plane spanned by the two macrocycles D_L and D_M are indicated by arrows. They are oriented approximately in the b-c plane and form angles of approximately ±35° with the b-axis and ±55° with the c-axis.

which is equal to

$$\nu_a^{\pm} = [(\nu_N \pm ½A_{aa})^2 + ¼(A_{ab}^2 + A_{ac}^2)]^{½} \tag{2b}$$

In our case A_{aa} is the diagonal element and A_{ab} and A_{ac} are the off-diagonal elements of the ^{15}N hf-tensor in the crystal axis system. Analogous equations hold for the ENDOR frequencies for the other two crystal axes, relating $\nu_b^{\pm}$ and $\nu_c^{\pm}$ to the respective diagonal and off-diagonal tensor elements.

If the second term in eq. (2b), containing the off-diagonal elements, is small compared with the first quadratic term, eq. (2b) reduces to

$$\nu_i^{\pm} = |\nu_N \pm 1/2\, A_{ii}| \tag{3}$$

for i = a, b or c. Due to the lack of experimental values for the off-diagonal elements in eq. (2b), a first analysis of the spectra shown in figure 2a is performed using eq. (3).

All ^{15}N ENDOR lines observed in figure 2a are the respective high-frequency (ν^+) NMR transitions; the corresponding low-frequency transitions (ν^-), obtained from eq. (3) with $\nu_{^{15}N}$ = 1.46 MHz are all below 1 MHz and are, therefore, outside the useful range of the spectrometer. The spectra of figure 2a were analyzed using a deconvolution program;[19] the obtained line frequencies ν_i^+ are given in table 1.

For the direction with the largest hf splittings (B_0 parallel to b) at least six lines are clearly resolved in figure 2a; the line at the lowest frequency can be split into two components using spectral deconvolution. This result immediately shows that the spin density distribution in $D^{+\cdot}$ must be asymmetric, since there are only four ^{15}N nuclei in each BChl a molecule constituting $D^{+\cdot}$.

An experimental assignment of the lines in the three spectra of figure 2a to individual ^{15}N nuclei in D_L and D_M is not possible, because no full rotation patterns have been obtained so far for the three planes of the crystal. Moreover, since the planes of the two macrocycles of D_L and D_M are nearly coplanar, and the largest hf splitting is expected for all ^{15}N nuclei for B_0 perpendicular to this plane, no large differences between the directions of the principal axes (corresponding to the largest principal values) of all ^{15}N-hf-tensors are expected. The assignments given in table 1 are, therefore, based on the ordering of magnitude of the ^{15}N ENDOR frequencies, ν_i^+, (i = a,b,c) obtained from eq. (2b) using the calculated ^{15}N hf-tensors.

A first estimate of the isotropic ^{15}N hfc's, a_{iso}(est), can be deduced from the trace of the approximate values for the diagonal elements, A_{ii}, obtained from eq. (3). Thereby, the contribution of the off-diagonal tensor elements to the observed ^{15}N ENDOR frequencies ν^+ (eq. 2b) are neglected. The values a_{iso}(est) are listed in table 1.

In the case of the proton hfc's of $D^{+\cdot}$, eq. (3) is a good approximation for calculating the diagonal elements of the hf tensors from the observed ENDOR frequencies.[15] However, for ^{15}N nuclei, this is not the case because of the much smaller nuclear Larmor frequency ν_N (see eq. 2b). Improved values for the diagonal elements of the ^{15}N hf tensors are obtained from eq. (2b) using the ^{15}N-ENDOR frequencies ν_i^+ from the experiment and the off-diagonal elements A_{ij} ($i \neq j$) from the RHF-INDO/SP calculation[4,15,18] (see table 1). The resulting ^{15}N hf-tensor elements A_{aa}, A_{bb} and A_{cc} as well as the deduced isotropic hfc's are given in table 1. The obtained isotropic hfc's 1/3 Tr{A} are, on the average, 10% smaller than a_{iso}(est).

Table 1. ^{15}N ENDOR frequencies and hf-tensor elements [MHz] in the crystal axis system

	D_L a)				D_M a)			
	II	III	IV	I	II	IV	III	I
b) ν_a^+	2.56	2.56	2.56	2.31	1.93	1.93	1.82	1.82
b) ν_b^+	3.98	3.68	3.28	3.01	2.57	2.13	2.13	2.13
b) ν_c^+	3.10	2.84	2.95	2.84	2.46	2.46	2.46	2.46
c) a_{iso}(est)	3.52	3.13	2.94	2.52	1.72	1.43	1.35	1.35
d) A_{aa}	2.18	2.13	2.17	1.70	0.91	0.94	0.69	0.71
d) A_{bb}	4.70	4.19	3.28	2.92	2.05	1.14	1.25	1.25
d) A_{cc}	2.84	2.48	2.58	2.59	1.81	1.83	1.91	1.92
e) A_{ab}	-.34	-.71	-.47	-.17	-.37	-.22	-.28	-.10
e) A_{ac}	-.23	-.46	-.34	-.17	-.33	-.20	-.30	-.07
e) A_{bc}	2.29	1.72	2.09	1.39	1.30	1.26	0.84	0.87
f) 1/3 Tr{A}	3.24	2.93	2.68	2.40	1.59	1.30	1.28	1.29

a) Assignment of ^{15}N nuclei to pyrrol rings I to IV in the BChl a dimer halves D_L and D_M
b) High-frequency ENDOR line positions obtained from fig. 2a.
c) Isotropic ^{15}N hfc's obtained from the trace of the estimated diagonal tensor elements from eq. (3), see text.
d) Diagonal elements of the ^{15}N hf-tensors in the crystal axis system obtained from eq. (2b), using the experimental values ν_{ii}^+ and calculated values for A_{ij} ($i \neq j$); (i,j = a,b,c).
e) Calculated off-diagonal elements of the ^{15}N hf tensors obtained from RHF-INDO/SP spin densities based on the X-ray coordinates.[2,18,23] These values have been scaled by a constant factor of 0.8 to obtain agreement between the calculated ^{15}N-ENDOR frequencies in eq. (2b) and the experimental ones. This scaling also gives a fair agreement between experimental and calculated sums of all isotropic ^{15}N hfc's (16.7 MHz and15.5 MHz, respectively).
f) Isotropic ^{15}N hfc's (1/3 (A_{aa} + A_{bb} + A_{cc})); the diagonal elements of the hf tensors are obtained from eq. (2b) using the calculated elements A_{ij} ($i \neq j$).

The ^{15}N hf-tensors in the crystal axis system given in table 1 have been diagonalized, the resulting principal values of the purely dipolar tensor A_{ii} are collected in table 2. In addition, the angles, ϕ, between the principal axes corresponding to A'_{11} and the normal vectors $\vec{n}_L$ or $\vec{n}_M$ of the macrocycles DL and DM are given. ϕ is 6° for D_L and 12° for DM (average over the four respective ^{15}N nuclei). An experimental assignment of the ^{15}N hf-tensors to D_L and D_M on the basis of these angles is not possible, since they do not change significantly when the largest four ^{15}N hf tensors are assigned to D_M. The assignment given in tables 1 and 2 is based on the ordering of magnitude of the calculated ^{15}N hf tensors. The obtained ratio of the isotropic ^{15}N hfc's on D_L versus DM (2 : 1) agrees well with the observed asymmetry of spin densities obtained from recent ^1H Special TRIPLE resonance experiments on RC single crystals.[15] The observed EPR spectra of $D^{+\cdot}$ in the ^{15}N-labeled RC single crystals could be simulated very well for all three orientations (B_o parallel to a, b and c) using the ^{15}N hf-tensor elements and multiplicities given in table 1 and the ^1H hf-tensors from Special TRIPLE resonance (figure 2b) with the assignments given in ref. 15.

Table 2. Anisotropic ^{15}N hf-tensor principal values and isotropic ^{15}N hfc's [MHz]

	D_L a)				D_M a)			
	II	III	IV	I	II	IV	III	I
b) A'_{11}	3.04	2.54	2.48	1.78	1.74	1.54	1.29	1.23
b) A'_{22}	-1.10	-1.02	-0.62	-0.73	-0.78	-0.41	-0.57	-0.56
b) A'_{33}	-1.94	-1.52	-1.86	-1.05	-0.97	-1.13	-0.72	-0.66
c) ϕ	2°	9°	6°	6°	3°	12°	15°	17°
d) 1/3 Tr{$\underline{A}$}	3.24	2.93	2.68	2.40	1.59	1.30	1.28	1.29
e) a_{iso}			2.60	2.22	1.61		1.05	

a) Assignment of ^{15}N nuclei to pyrrol rings I to IV in the BChl a dimer halves
b) Principal values of ^{15}N hf tensors after diagonalization of the tensors in the crystal axis system given in Table 1; only dipolar components $A'_{ii} = A_{ii} - 1/3\ Tr\{\underline{A}\}$.
c) Angles between principal axes belonging to A'_{11} and the normal vectors $\vec{n}_L$ and $\vec{n}_M$ of D_L and D_M as obtained from the X-ray coordinates.[1,23]
d) Isotropic part of the ^{15}N hf-tensors obtained in this study (tensor elements given in Table 1).
e) Isotropic ^{15}N hfc's measured in liquid solution of ^{15}N-labeled RC's (ref. 11)

Most of the deduced isotropic ^{15}N hfc's 1/3 Tr{$\underline{A}$} agree reasonably well with those obtained in an earlier investigation of $D^{+\cdot}$ in liquid RC solutions (table 2).[11] However, the two largest hfc's (2.93 and 3.24 MHz) have not been clearly identified in liquid solution, probably because of the large hyperfine anisotropy (e.g., $A'_{11} - A'_{33} = 5$ MHz for position II in D_L), which, together with slow rotational tumbling, gives rise to significant line broadening.

It should be pointed out that our analysis of the ^{15}N ENDOR spectra (figure 2a) and the deduced ^{15}N hf tensors are based on the calculated values for the off-diagonal elements A_{ij} ($i \neq j$), table 1. Inaccuracies of the calculated tensor elements might explain the remaining discrepancy between the principal axes of the ^{15}N hf-tensor (for the largest principal value) and the normal vectors on the dimer halves D_L and, in particular, D_M. Experiments are in progress to obtain the full ^{15}N hf-tensors (i.e. rotation patterns in three planes of the crystal). The isotropic ^{15}N hfc's given in table 1 (1/3 Tr{$\underline{A}$}) should be only slightly affected by changes in the magnitudes of the off-diagonal elements; the maximum expected change is approximately 10%, see table 1). The anisotropic dipolar ^{15}N hf-tensor elements given in table 2 are much larger than those reported in earlier ENDOR and electron spin echo envelope modulation (ESEEM) studies of frozen solutions of RC's of *Rb. sphaeroides*[13] and of chromatophores of *Rhodospirillum rubrum.*[14] The reason for this discrepancy remains unclear. Recent ESEEM experiments performed in our laboratory[21] are in good agreement with the ^{15}N hf-tensor values obtained in this single crystal study.

CONCLUSION

^{15}N ENDOR experiments on the primary donor cation radical $D^{+\cdot}$ in RC single crystals clearly reveal the asymmetry of the spin density distribution by simply counting the number of observed lines. Assignment of these lines is based on the RHF-INDO/SP calculations: the obtained ratio of the sums of ^{15}N hfc's on D_L versus D_M is approximately 2 : 1 and agrees well with the value deduced from ^1H Special TRIPLE experiments for the carbon p_z-spin densities.[15]

Experiments to obtain values for the off-diagonal elements of the ^{15}N hf-tensors from rotation patterns in three crystal planes are in progress. From these results an experimental distinction between the ^{15}N hf-tensors of D_L and D_M may be possible.

The observed asymmetry of the spin density distribution in favor of D_L has been explained using a model that assumes an energetic difference between the dimer halves D_L and D_M of a magnitude comparable to that of the interdimer interaction energy.[18] Significant shifts of spin density between D_L and D_M have been observed, when RC's of different native bacteria and mutants were compared, indicating differences in the energetics of the primary donors in the different bacteria.[22] The orbital asymmetry of the primary donor is obviously a common feature in many bacterial photosystems[22] and may play a functional role for the unidirectional electron transfer in bacterial photosynthesis.

ACKNOWLEDGMENTS

The authors thank Drs. J.P. Allen (Arizona State University, Tempe), D.C. Rees (California Institute of Technology, Pasadena), and G. Feher (University of California, San Diego) for providing the X-ray coordinates of the RC of *Rb. sphaeroides* R-26 prior to publication. Dr. E. Tränkle is acknowledged for supplying the spectral deconvolution program. This work was supported by the Deutsche Forschungsgemeinschaft (Sfb 312 and Sfb 337) and by NATO (CRG910468).

REFERENCES

1. J. Deisenhofer and H. Michel, The photosynthetic reaction centre from the purple bacterium *Rhodopseudomonas viridis*, *EMBO J.* 8:2149 (1989).
2. G. Feher, J.P. Allen, M.Y. Okamura, and D.C. Rees, Struture and function of bacterial photosynthetic reatction centers, *Nature* 339:111 (1989).
3. W. Lubitz, EPR and ENDOR studies of chlorophyll cation and anion radicals, *in*: "Chlorophylls," H. Scheer, ed., CRC Press, Boca Raton (1991).
4. M. Plato, K. Möbius, and W. Lubitz, Molecular orbital calculations on chlorophyll radical ions, *in*: "Chlorophylls," H. Scheer, ed., CRC Press, Boca Raton (1991).
5. J.R. Norris, R.A. Uphaus, H.L. Crespi, and J.J. Katz, Electron Spin Resonance of chlorophyll and the origin of signal I in photosynthesis, *Proc. Natl. Acad. Sci. USA* 68:625 (1971).
6. G. Feher, A.J. Hoff, R.A. Isaacson, and L.C. Ackerson, ENDOR experiments on chlorophyll and bacteriochlorophyll *in vitro* and in the photosynthetic unit, *Ann. N.Y. Acad. Sci.* 244:239 (1975).
7. J.R. Norris, H. Scheer, and J.J. Katz, Models for antenna and reaction center chlorophylls, *Ann. N.Y. Acad. Sci.* 244:260 (1975).
8. W. Lubitz, F. Lendzian, M. Plato, K. Möbius, and E. Tränkle, ENDOR studies of the primary donor in bacterial reactions centers, in: "Antennas and Reaction Centers of Photosynthetic Bacteria," M.E. Michel-Beyerle, ed., Springer, Berlin (1985).
9. M. Plato, K. Möbius, W. Lubitz, J.P. Allen, and G. Feher, Magnetic resonance and molecular orbital studies of the primary donor states in bacterial reaction centers, *in*: "Perspectives in Photosynthesis," J. Jortner and B. Pullmann, eds., Kluwer Academic Publishers, The Netherlands (1990).
10. P.J.. O'Malley and G.T. Babcock, A proposed monomeric nature of antenna and reaction center chlorophylls, *in*: "Advances in Photosynthesis Research," Vol. 1, C. Sybesma, ed., Nijhoff/Junk, The Hague (1984).
11. W. Lubitz, R.A. Isaacson, E.C. Abresch, and G. Feher, ^{15}N electron nuclear double resonance of the primary donor cation radical $P_{865}^{+\cdot}$ in reaction centers of *Rhodopseudomonas sphaeroides*: Additional evidence for the dimer model, *Proc. Natl. Acad. Sci. USA* 81:7792 (1984).
12. H. Kurreck, B. Kirste, and W. Lubitz, "Electron Double Resonance Spectroscopy of Radicals in Solution," Chapter 4, VCH Publishers, Weinheim (1988).

13. A. DeGroot, A.J. Hoff, R. DeBeer, and H. Scheer, ^{14}N and ^{15}N coupling constants of the oxidized primary donor P-860 of baceterial photosynthesis obtained by electron spin echo envelope modulation spectroscopy, *Chem. Phys. Lett.* 113:286 (1985).
14. C.P. Lin and J.R. Norris, Interpretation of the nitrogen spin densities in the primary donor cation of photosynthetic reaction centers, *FEBS Lett.* 197:281 (1986).
15. F. Lendzian, M. Huber, R.A. Isaacson, B. Endeward, M. Plato, B. Bönigk, K. Möbius, W. Lubitz, and G. Feher, The electronic structure of the primary donor cation radical in *Rhodobacter sphaeroides* R-26: ENDOR and TRIPLE resonance studies in single crystals of reaction centers, *Biochim. Biophys. Acta*, submitted.
16. J.P. Allen, G. Feher, T.O. Yeates, D.C. Rees, J. Deisenhofer, H. Michel, and R. Huber, Structural homology of reaction centers from *Rhodopseudomonas sphaeroides* and *Rhodopseudomonas viridis* as determined by X-ray diffraction, *Proc. Natl. Acad. Sci. USA* 83:8589 (1986).
17. K. Möbius, W. Lubitz, and M. Plato, Liquid-state ENDOR and TRIPLE resonance, *in*: "Advanced EPR," A.J. Hoff, ed., Elsevier, Amsterdam (1989).
18. M. Plato, F. Lendzian, W. Lubitz, and K. Möbius, Molecular orbital study of electronic asymmetry in primary donors of bacterial reaction centers, these proceedings.
19. E. Tränkle and F. Lendzian, Computer analysis of spectra with strongly overlapping lines. Application to TRIPLE resonance spectra of the chlorophyll a cation radical, *J. Magn. Res.* 84:537 (1989).
20. N.M. Atherton and A.J. Horsewill, Proton ENDOR of $Ca(H_2O)_6^{2+}$ in $Mg(NH_4)_2(SO_2)_4 \cdot 6\ H_2O$, Mol. Phys. 37:1349 (1979).
21. H. Käß, P. Höfer, B. Bönigk, A. Rautter, and W. Lubitz, to be published
22. A. Rautter, Ch. Geßner, F. Lendzian, W. Lubitz, J.C. Williams, H.A. Murchison, S. Wang, N.W. Woodbury, and J.P. Allen, EPR and ENDOR studies of the primary donor cation radical in native and genetically modified bacterial reaction centers, these proceedings
23. J.P. Allen, G. Feher, T.O. Yeates, H. Komiya, and D.C. Rees, Structure of the reaction center from *Rhodobacter sphaeroides* R-26: The cofactors, *Proc. Natl. Acad. Sci. USA* 84:5730 (1987)

EPR AND ENDOR STUDIES OF THE PRIMARY DONOR CATION RADICAL IN NATIVE AND GENETICALLY MODIFIED BACTERIAL REACTION CENTERS

J. Rautter,[1] Ch. Geßner,[1] F. Lendzian,[1] W. Lubitz[1]
J.C. Williams,[2] H.A. Murchison,[2] S. Wang,[2] N.W. Woodbury,[2]
and J.P. Allen[2]

[1]Max-Volmer-Institut
für Biophysikalische und Physikalische Chemie
Technische Universität Berlin
Str. d. 17. Juni 135
D-1000 Berlin 12
Germany

[2]Department of Chemistry and Biochemistry and
the Center for the Study of Early Events
in Photosynthesis
Arizona State University
Tempe AZ 85287-1604
USA

INTRODUCTION

Electron transfer in bacterial reaction centers (RCs) is determined by the three-dimensional arrangement and by the electronic structure of the reacting pigments. The primary donor D - a bacteriochlorophyll (BChl) dimer - is of particular interest due to its specialized function in the primary charge separation step.[1] The structure of BChl a, the pigment constituting the dimer in all RCs investigated in this work, is shown in Figure 1.

The electronic structure of the cation radical $D^{+\cdot}$ can be obtained by measuring the electron-nuclear hyperfine couplings (hfc's). After assignment to specific molecular positions, a map of the valence electron spin density distribution over the molecule can be deduced.[2,3]

The experimental techniques employed to resolve the hfc's are EPR, ENDOR and TRIPLE resonance (reviewed in ref. 4). Detailed results were obtained recently for $D^{+\cdot}$ in RC single crystals of *Rhodobacter (Rb.) sphaeroides*.[5-7] Unambiguous assignments of the hf tensors to specific nuclei were achieved by a comparison with the X-ray structure data of the RC[8] and predictions from molecular orbital (MO) calculations.[7,9] The results of this investigation show that the unpaired electron is asymmetrically distributed over the dimer halves D_L and D_M with approximately 68% of the spin density on D_L (L and M subscripts denote the protein subunits binding the BChls constituting the dimer).

The Photosynthetic Bacterial Reaction Center II
Edited by J. Breton and A. Verméglio, Plenum Press, New York, 1992

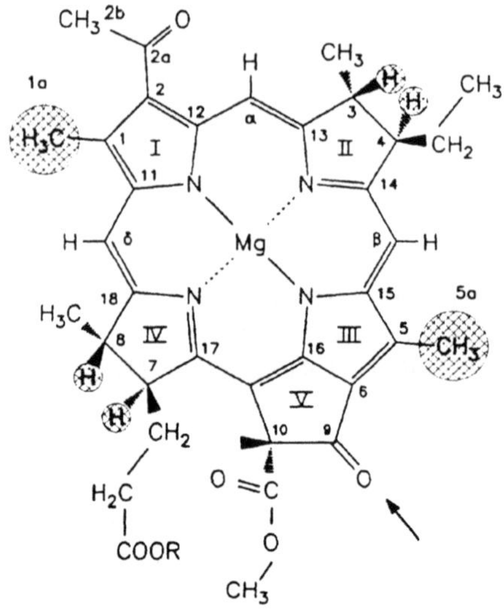

Figure 1. Structure of bacteriochlorophyll a (BChl a) including numbering scheme. Molecular positions for which large proton hfc's are observed are marked by shaded circles. The keto-group oxygen to which a hydrogen bond could be formed in the mutants is indicated by an arrow.

A possible explanation for this asymmetry is the different protein environment of the respective dimer halves.

The effects of specific amino acid residues on the spin density distribution of $D^{+\cdot}$ can be investigated by a comparison of native and genetically modified RCs.[10] In this paper we present ENDOR/TRIPLE experiments on D^{+} both in liquid and frozen RC solutions of two mutants of *Rb. sphaeroides* each having a single amino acid residue near the dimer altered.[11,12] The two mutants were designed to introduce hydrogen bonds to the 9-keto groups of the dimer. The mutations are Leu to His at L131, LH(L131), and Leu to His at M160, LH(M160) (Figure 2). No amino acid residues capable of forming such a hydrogen bond with either the L or M side of the dimer are present in the wild type.

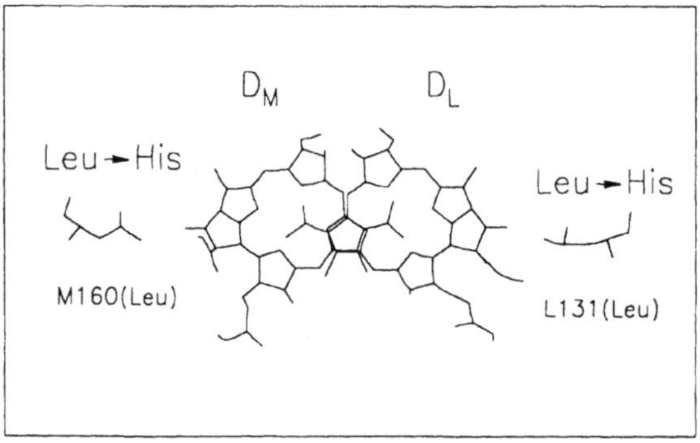

Figure 2. The primary donor BChls (D_L and D_M) with residues exchanged by site-directed mutagenesis. Residue Leu M160 or Leu L131 of the wild type reaction center structure is replaced by a histidine.

It will be shown that the different mutants exhibit drastic differences in the spin density distribution of $D^{+\cdot}$, attributed to the interaction between the dimer and the respective amino acid residue. An attempt is made to explain the observed effects by different energetics of the dimer in the framework of a simple MO model.

Furthermore, we report on a comparative investigation of $D^{+\cdot}$ in RCs of different BChl a-containing purple bacteria. The EPR and ENDOR/TRIPLE experiments reveal significant differences in the spin density distribution of $D^{+\cdot}$ in *Rhodospirillum (Rs) rubrum*,[13] *Rb. sphaeroides*,[14] and *Rb. capsulatus*.

MATERIALS AND METHODS

The mutagenesis is only briefly described as details have been published elsewhere.[11] The mutations were constructed by altering the genes encoding the L and M subunits of the reaction center. For each mutation, a plasmid containing the altered genes was introduced by conjugation into a *Rb. sphaeroides* strain in which the wild type reaction center genes on the chromosome had been deleted. The complemented *Rb. sphaeroides* strains were grown semiaerobically in rich media, and reaction centers were isolated following the protocols developed for the wild type reaction centers.

Wild type reaction centers from *Rb. sphaeroides* were isolated from the deletion strain complemented with the wild type genes. Reaction centers from *Rhodobacter capsulatus* were isolated as previously described.[15]

$D^{+\cdot}$ was generated by *in situ* illumination (830-900 nm) in the ENDOR/TRIPLE cavity at 288K. For frozen solution experiments (150K) the RCs were illuminated near room temperature for 5-10 sec followed by rapid freezing in liquid nitrogen under continuous illumination. For all experiments a 100 W tungsten/halogen lamp was used.

The EPR, ENDOR and TRIPLE spectra were recorded using either a homebuilt spectrometer[4] or a Bruker ER200D equipped with a high power ENDOR cavity of local design.[16] . The ENDOR spectra were analyzed using a spectral deconvolution program.[17]

RESULTS

Figure 3 shows the Special TRIPLE spectra of $D^{+\cdot}$ in *Rb. sphaeroides* wild type and the two mutants LH(M160), LH(L131). The hfc's of $D^{+\cdot}$ in the wild type spectrum have been unambiguously assigned by experiments performed on RC single crystals.[7] The largest coupling constants arise from the β-protons at rings II and IV and the methyl protons at positions 1a and 5a (see Figure 1). The latter are indicated in Figure 3 for the respective dimer halves D_L and D_M. The β-proton hfc's at positions 3,4,7, and 8 are strongly dependent on the geometry of the respective hydrated rings (II and IV). Therefore, they are less suited for probing the π-spin density on the macrocycle. The methyl groups are, however, better suited since the isotropic proton hfc's are directly proportional to the π-spin density on the neighboring carbon atom in the macrocycle. In contrast to other β-protons, the methyl protons exhibit relatively sharp and intense lines in frozen solution, because of the free rotation of the methyl groups even at low temperatures.[18] This effect is used here for the identification of methyl hfc's in the spectra. As an example Figure 4 shows a comparison between liquid and frozen solution spectra of LH(M160). The four line pairs (a-d) belonging to methyl protons are indicated in the figure.

All spectra shown in Figure 3 exhibit four methyl proton hfc's. Since there are only two methyl groups directly attached to the π-macrocycle of each BChl a molecule, the presence of four lines is indicative of an asymmetric distribution of the unpaired electron over the dimer halves. In contrast to the wild type, an individual assignment of all observed hfc's to the specific nuclei in $D^{+\cdot}$ could not be obtained so far; single crystal studies are planned but have not yet been started. An assignment of the methyl protons is, however, possible on the basis of the following assumptions: (i) spin density shifts occur only between D_L and D_M; there is no significant shift within each dimer half.

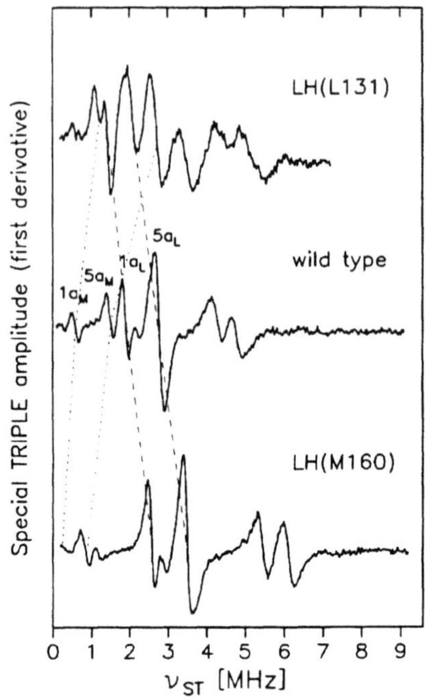

Figure 3. Special TRIPLE spectra of $D^{+\cdot}$ in the RCs of the investigated mutants in liquid solution at 288K. Line positions correspond to one half of the hfc's. The assigned methyl proton hfc's are indicated for the respective halves. Note that in the special TRIPLE spectra all line intensities decrease considerably towards smaller hfc's due to a nonselective EPR saturation for small hfc's.[19]

Table 1. Methyl proton hyperfine coupling constants (hfc's) [MHz] and ratios of hfc's in BChl a$^{+\cdot}$ and in D^+ of *Rb. sphaeroides* wild type and mutants

	BChla$^+$	WT#	LH(L131)	LH(M160)
A(5aL)	9.62	5.75	3.95	6.90
A(5aM)		3.20	5.45	1.65
A(1aL)	4.93	3.95	2.90	5.10
A(1aM)		1.40	2.30	0.70
A(5aL)/A(1aL)	1.90	1.45	1.40	1.35
A(5aM)/A(1aM)		2.30	2.40	2.35
Σ A(CH_3)§	14.5	14.3	14.6	14.4
$\frac{A(5aL)+A(1aL)}{\Sigma A(CH_3)}$	----	0.68	0.47	0.84

These values are in good agreement with those reported for *Rb. sphaeroides* 2.4.1 in ref. 20.
§ Sum over all methyl hfc's at positions 1a and 5a

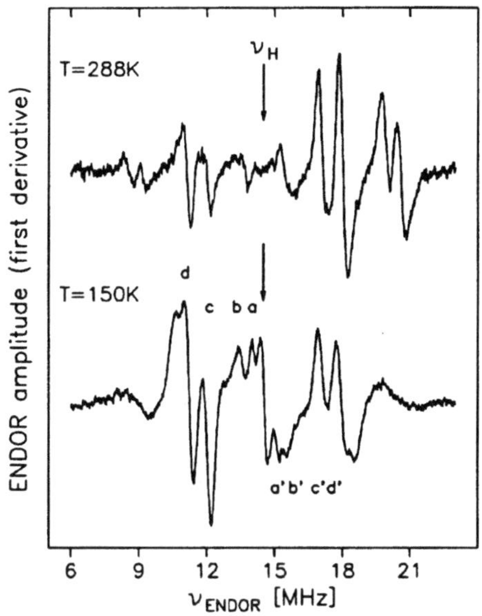

Figure 4. Comparison of the ENDOR spectra of $D^{+\cdot}$ in LH(M160) in liquid and frozen solution. The four line pairs arising from methyl proton hfc's are indicated (a,a' to d,d'). In the frozen solution spectra (bottom) the β-proton hfc's are broadened due to their larger anisotropy. The shoulders at line pair dd' are attributed to hf anisotropy and not to an additional coupling. Note that even the smallest methyl proton hfc aa' (A_{iso} = 0.7 MHz) is clearly visible in the frozen solution spectrum.

(ii) The ratio of methyl proton hfc's A(5a)/A(1a) is different in the two dimer halves and, therefore, characteristic for D_L or D_M, respectively.[21] In wild type RCs we obtain values for A(5a)/A(1a) of 1.45 (D_L) and 2.3 (D_M).* These two assumptions seem to be justified, since the sum of all four methyl proton hfc's is a constant within 3% in all RCs investigated here (see Table 1). The sum of the hfc's at positions 1a and 5a is considered as a measure of the spin density in the respective dimer halves.

Two large and two small methyl hfc's are observed in the spectrum of LH(M160) shown in Figures 3 and 4. The two large methyl hfc's are assigned to D_L and the two smaller ones to D_M based upon the calculated A(5a)/A(1a) values (see Table 1). This clearly shows that most of the spin density, 84%, resides on the L-half. This represents a considerable shift compared with the wild type that has 68% of the spin density on D_L. It should be noted that the spectrum of LH(M160) is quite similar to that of the heterodimer HL(M202).[22]

The interpretation of the spectrum of LH(L131) (Figure 3) is less obvious than that of LH(M160), because the four identified CH_3 hfc's are all of comparable magnitude. An assignment to D_L and D_M was made based on assumptions (i) and (ii). The obtained spin density is shifted towards the M-half with 53% present on D_M. This corresponds to a more symmetric case compared with the wild type. The two mutants LH(M160) and LH(L131) exhibit significant spin density shifts within $D^{+\cdot}$ in opposite directions.

All hfc's including their assignments are presented in Table 1. It is pointed out that the sum Σ A(CH_3) of all methyl proton hfc's is indeed nearly a constant for all reaction centers. The shift of spin density observed for $D^{+\cdot}$ in the two mutants, in comparison with the wild type, is not very pronounced in the Gaussian envelope linewidth of the EPR spectrum, in which only effects of extreme asymmetry can be detected.

---

* This difference in the ratios of D_L and D_M is also reflected in the heterodimers of *Rb. sphaeroides* HL(M202) and HL(L173).[22]

In LH(M160) the EPR linewidth (peak-to-peak) is 11±0.2 G. The other mutant shows a linewidth comparable to that of the wild type, 9.8±0.3 G. We have performed EPR simulations for the mutants, using the assignment of methyl proton hfc's in this work and β-proton assignments in analogy to those given in ref. 7. The simulated EPR specta agree well with the observed ones.

DISCUSSION

The results can be understood qualitatively on the basis of a simple model introduced by Plato (for details see references 7, 9). In this model a weakly coupled dimer is assumed with an energy difference $\Delta\alpha$ between the highest occupied molecular orbitals (HOMO's) of the dimer halves. *In vivo*, this energy is influenced by the protein environment. The intermolecular coupling expressed by the resonance integral, β_D, is assumed to be small compared with the intramolecular resonance integrals β_L and β_M (Figure 5a). The calculated curves in Figure 5b relate the ratios of the Hückel spin densities of the two halves, ρ_L/ρ_M or ρ_M/ρ_L, to the ratio $\Delta\alpha/\beta_D$ for a dimer consisting of a pair of two π−centers[7,9]. Assuming β_D to be constant, Figure 5b can be used to obtain the energetic difference $\Delta\alpha$ between the two halves of such a model dimer from the observed ratio of the sums of the spin densities ρ_L/ρ_M or vice versa.

Application of this model to $D^{+\cdot}$ of the wild type yields $\Delta\alpha/\beta_D \approx 0.35$ (Figure 5b). Independently, RHF-INDO/SP calculations on the supermolecule $D^{+\cdot}$ of the wild type

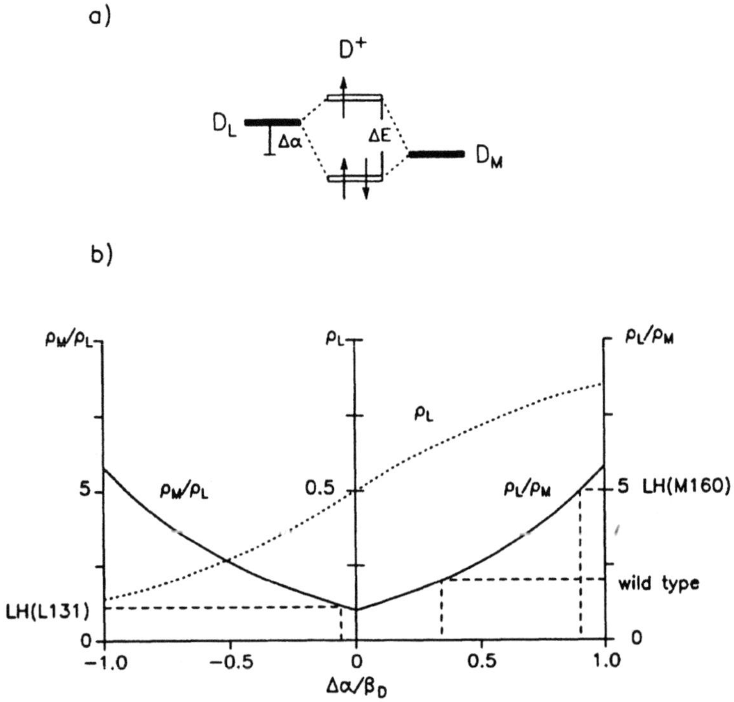

Figure 5. (a) MO scheme for the model dimer. The HOMO energy difference $\Delta\alpha$ between D_L and D_M is indicated. The splitting of the two highest occupied super-MO's of D^+ is $\Delta E = \sqrt{(\Delta\alpha)^2 + (2\beta_D)^2}$ in the approximation of $\beta_D << \beta_L, \beta_M$. (b) Relation between the spin density ratio ρ_L/ρ_M or ρ_M/ρ_L and $\Delta\alpha/\beta_D$ (solid lines). The spin density fraction on the L half, ρ_L, is given by the dashed line. Note that the left part is valid for $\rho_M > \rho_L$ and the right one for $\rho_L > \rho_M$.

give $\beta_D \approx 0.3$ eV.[9] Using this in the model results in a HOMO energy difference $\Delta\alpha = \varepsilon_L - \varepsilon_M = 0.11$ eV. This value agrees fairly well with $\Delta\alpha \approx 0.14$ eV obtained from independent RHF-INDO/SP calculations performed on the separate dimer halves of *Rb. sphaeroides*, using the X-ray structure data.[8]

For further analysis it is assumed that the dimer geometry in the four mutants remains unchanged;* i.e., the intermolecular coupling β_D is constant ($\beta_D = 0.3$ eV). However, the formation of a hydrogen bond to D_L or D_M will result in a change of $\Delta\alpha$ that can be estimated from Figure 5a, using the experimentally obtained spin density ratios.

For LH(M160) the ratio ρ_L/ρ_M is 5, which yields $\Delta\alpha/\beta_D = 0.9$, giving $\Delta\alpha = 0.27$ eV. Compared with the wild type, this corresponds to an increase of approximately 160 meV in the energetic difference of the HOMO's of the two halves. Formation of a hydrogen bond between the 9-keto group of D_M and the introduced histidine residue at position M160 is expected to lower the energy of the highest occupied π-orbital (HOMO) of D_M, thereby increasing $\Delta\alpha$. The experimentally observed spin density shift is, therefore, a strong indication for the formation of this hydrogen bond.

For LH(L131) a ratio of $\rho_M/\rho_L = 1.1$ is obtained which corresponds to $\Delta\alpha/\beta_D = -0.06$, resulting in $\Delta\alpha \approx -0.02$ eV. Compared with the wild type, D_L is lowered by approx. 130 meV. This can be explained by the formation of a hydrogen bond between the histidine residue L131 and the 9-keto oxygen of D_L, lowering its HOMO energy slightly below that of D_M, resulting in a fairly symmetric dimer.

The observed energy shifts in the two mutants are well within the range of typical hydrogen bond energies.[23] RHF-INDO/SP calculations performed on the dimer halves, using the X-ray structure and modeling the respective H-bonds by point charges, confirm the effects on the energetics concerning direction and magnitude. The formation of these hydrogen bonds in LH(M160) and LH(L131) has also been observed in recent FTIR experiments by Nabedryk et al.[24]

CONCLUSION AND OUTLOOK

We have shown that single point mutations in the vicinity of the special pair have a profound influence on the spin density distribution between the two dimer halves. The major effect observed is a shift of spin density from one half of the dimer to the other, caused by asymmetric hydrogen bonding. Experiments are planned to independently assign the hfc's by single crystal studies in order to confirm the conclusions derived in this paper.

In the light of these results it is challenging to test whether in different BChl a-containing purple bacteria the primary donor also shows differences in the symmetry of the spin density distribution. Figure 6 shows a comparison of the Special TRIPLE spectra of *Rb. sphaeroides*, *Rs. rubrum* and *Rb. capsulatus* with the BChl a^+ monomer *in vitro*. A preliminary analysis of the spectra shows for each RC the presence of a $(BChl\ a)_2$-supermolecule with different degrees of asymmetry. Whereas, *Rs. rubrum* and *Rb. sphaeroides* show significant delocalization between D_L and D_M (1.5 : 1 and 2 : 1, respectively), in *Rb. capsulatus* the spin density is mainly localized on one dimer half. This is astonishing in view of the high homology of the amino acid sequences near the dimer and the almost identical optical spectra of these bacteria. The observed differences in the distribution of the unpaired electrons in $D^{+\cdot}$ indicate differences in the energetics of the primary donors in these bacteria. A detailed analysis of the spectra will be published elsewhere.

We also performed experiments on two other mutants. In these mutants a single amino acid residue His to Phe at L168, HF(L168), or the symmetry related residue Phe to His at M197, FH(M197) was altered. The ENDOR/TRIPLE spectra of D^+ in these RCs also show changes of the spin density distribution in comparison with the wild type. A detailed analysis is still in progress and will be published elsewhere.

---

* The unaltered dimer geometry has to be confirmed by X-ray structure analysis.

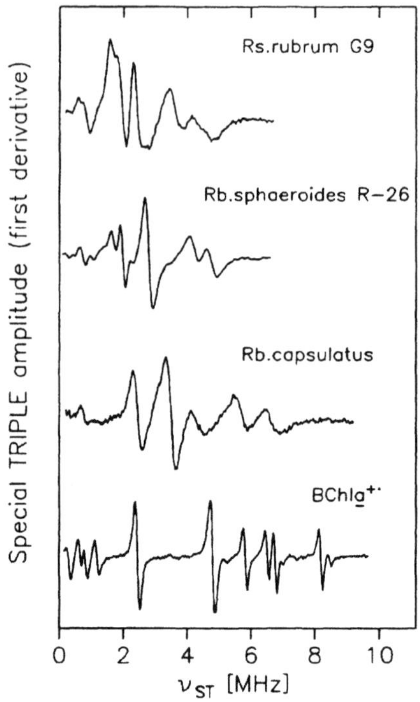

Figure 6. Comparison of the special TRIPLE spectra of $D^{+\cdot}$ in different BChl-a containing purple bacteria, with the BChl $a^{+\cdot}$ cation radical *in vitro*.

ENDOR measurements provide a means to investigate small energetic effects like the presence or absence of hydrogen bridges to the primary donor. By measuring the asymmetry of the spin density distribution one obtains information on the energy difference and interaction energies between the dimer halves. Recently, interest in the energetics of $D^{+\cdot}$ has been stimulated by the observation of a new cation-radical absorption band in the far infrared, at 2600 cm^{-1}.[25] This "new band" is also a measure of the interaction energy of the two dimer halves. In future studies we intend to correlate the observed energetic effects with other spectroscopic and kinetic data reported for these mutants.[12]

ACKNOWLEDGMENTS

We thank Dr. M. Plato for theoretical support, Dr. E. Tränkle for providing the spectral deconvolution program and Profs. Mehring and Möbius for providing experimental facilities. The work in Berlin has been supported by the Deutsche Forschungsgemeinschaft (Sfb 312, TP A4) and by NATO (CRG 910468). The work in Tempe has been supported by grants GM41300 and GM45902 from the N.I.H., grants DMB89-177729 and DMB91-58251 from N.S.F. (Publication #108 from the Arizona State University Center for the Study of Early Events in Photosynthesis; the Center is funded by D.O.E. grant DE-FG-88-ER12969.)

REFERENCES

1. G. Feher, J.P. Allen, M.Y. Okamura, and D.C. Rees, Structure and function of bacterial photosynthetic reaction centers, *Nature* 339:111 (1989).
2. W. Lubitz, EPR and ENDOR studies of chlorpohyll cation and anion radicals, *in*: "Chlorophylls," H. Scheer, ed., CRC Press, Boca Raton (1991).
3. M. Plato, K. Möbius, and W. Lubitz, Molecular orbital calculations on chlorophyll radical ions, *in*: "Chlorophylls," H. Scheer, ed., CRC Press, Boca Raton (1991).
4. K. Möbius, W. Lubitz, and M. Plato, Liquid-state ENDOR and TRIPLE resonance, *in*: "Advanced EPR", A.J. Hoff, ed., Elsevier, Amsterdam (1989).
5. F. Lendzian, B. Endeward, M. Plato, D. Bumann, W. Lubitz, and K. Möbius, ENDOR and TRIPLE resonance investigation of the primary donor cation radical P_{865}^{+} in single crystals of *Rhodobacter sphaeroides* R-26 reaction centers, *in*: "Reaction Centers of Photosynthetic Bacteria," M.-E. Michel-Beyerle, ed., Springer, Berlin (1990).
6. E.J. Lous, M. Huber, R.A. Isaacson, and G. Feher, EPR and ENDOR studies of the oxidized primary donor in single crystals of reaction centers of *Rhodobacter sphaeroides* R-26, *in*: "Reaction Centers of Photosynthetic Bacteria," M.-E. Michel-Beyerle, ed., Springer, Berlin (1990).
7. F. Lendzian, M. Huber, R.A. Isaacson, B. Endeward, M. Plato, B. Bönigk, K. Möbius, W. Lubitz, and G. Feher, The electronic structure of the primary donor cation radical in *Rhodobacter sphaeroides* R-26: ENDOR and TRIPLE resonance studies in single crystals of reaction centers, *Biochim. Biophys. Acta*, submitted.
8. T.O. Yeates, H. Komiya, A. Chirino, D.C. Rees, J.P. Allen, and G. Feher, Structure of the reaction center from *Rb. sphaeroides* R-26 and 2.4.1: Protein-cofactor (bacteriochlorophyll, bacteriopheophytin, and carotenoids) interactions, *Proc. Natl. Acad. Sci. USA* 85:7993 (1988).
9. M. Plato, F. Lendzian, W. Lubitz, and K. Möbius, Molecular orbital study of electronic asymmetry in primary donors of bacterial reaction centers, these proceedings.
10. W.J. Coleman and D.C. Youvan, Spectroscopic analysis of genetically modified photosynthetic reaction centers, *Annu. Rev. Biophys. Chem.* 19:333 (1990).
11. J.C. Williams, R.G. Alden, H.A. Murchison, J.M. Peloquin, N.W. Woodbury, and J.P. Allen, Effects of mutations near the bacteriochlorophylls in reaction centers from *Rhodobacter sphaeroides* , these proceedings
12. J.C. Williams, N.W. Woodbury, A.K.W. Taguchi, J.M. Peloquin, H.A. Murchison, R.G. Alden, and J.P. Allen, Mutations which affect the P/P^{+} redox couple in reaction centers from *Rb. sphaeroides* , these proceedings
13. W. Lubitz, F. Lendzian, H. Scheer, J. Gottstein, M. Plato, and K. Möbius, Structural studies of the primary donor cation radical P_{870}^{+} in reaction centers of *Rhodospirillum rubrum* by electron-nuclear double resonance in solution, *Proc. Natl. Acad. Sci. USA* 81:1401 (1984).
14. F. Lendzian, W. Lubitz, H. Scheer, C. Bubenzer, and K. Möbius, *In vivo* liquid solution ENDOR and TRIPLE resonance of bacterial reaction centers of *Rhodopseudomonas sphaeroides* R-26, *J. Am. Chem. Soc.* 103:4635 (1981).
15. S. Wang and J.P. Allen, Isolation and characterization of reaction centers from *Rhodobacter capsulatus*, *Biochim. Biophys. Acta*, submitted.
16. R. Thanner, W. Zweygart, H. Käß, and W. Lubitz, A high sensitivity low temperature ENDOR cavity, *J.Magn. Res.*, in preparation.
17. E. Tränkle and F. Lendzian, Computer analysis of spectra with strongly overlapping lines. Application to TRIPLE resonance spectra of the chlorophyll a cation radical, *J. Magn. Res.* 84:537 (1989).
18. J.S. Hyde, G.H. Rist, and L.E.G. Eriksson, ENDOR of methyl, matrix, and α-protons in amorphous and polycrystalline matrices, *J. Phys. Chem.* 72:4269 (1968).
19. R.D. Allendoerfer and A.G. Maki, A phenomenological description of ENDOR in solution; Example: the tri-t-butyl phenoxyl radical, *J. Magn. Res.* 3:396 (1970).
20. Ch. Geßner, F. Lendzian, B. Bönigk, M. Plato, K. Möbius, and W. Lubitz, ENDOR and Special TRIPLE investigations of $P_{865}^{+\cdot}$ in reaction center single crystals of *Rb. sphaeroides* wild type strain 2.4.1, *Appl. Magn. Res.* (1992) in press.

21. G. Feher, Bruker Lecture: Identification and characterization of the primary donor in bacterial photosynthesis: A chronological account of an EPR/ENDOR investigation, *J. Chem. Soc.* (Faraday Trans.), in press.
22. M. Huber, E.L. Lous, R.A. Isaacson, G. Feher, D. Gaul, and C.C. Schenck, EPR and ENDOR studies of the oxidized donor in reaction centers of *Rhodobacter sphaeroides* strain R-26 and two heterodimer mutants in which histidine M202 or L173 was replaced by leucine, *in*: "Reaction Centers of Photosynthetic Bacteria," M.-E. Michel-Beyerle, ed., Springer, Berlin (1990).
23. G.A. Jeffrey and W. Sänger, Hydrogen bonding in biological structures, Springer, Berlin (1991).
24. E. Nabedryk, J. Breton, J.P. Allen, H.A. Murchison, A. Taguchi, J.C. Williams, and N.W. Woodbury, FTIR characterization of Leu M160→His, Leu L131→His, and His L168→Phe mutations near the primary electron donor in *Rb. sphaeroides* reaction centers, these proceedings.
25. J. Breton, E. Nabedryk, and W.W. Parson, A new infrared electronic transition of the oxidized primary electron donor in bacterial reaction centers provides a way to access resonance interactions between the bacteriochlorophylls, *Biochemistry* (1992), in press.

MOLECULAR ORBITAL STUDY OF ELECTRONIC ASYMMETRY IN PRIMARY DONORS OF BACTERIAL REACTION CENTERS

Martin Plato[1], Friedhelm Lendzian[2],
Wolfgang Lubitz[2] and Klaus Möbius[1]

[1]Institut für Molekülphysik
Freie Universität Berlin
D-1000 Berlin 33, Germany

[2]Max-Volmer-Institut für Biophysikalische
und Physikalische Chemie
Technische Universität Berlin
D-1000 Berlin 12, Germany

INTRODUCTION

Recent Electron-Nuclear-Double and -Triple Magnetic Resonance measurements on single crystals of reaction centers (RC's) of the bacterium Rb. sphaeroides R-26 have revealed an asymmetric spin density distribution of the primary donor radical cation $P_{865}^{+\bullet}$ [1]. The observed ratio of spin densities, ρ_L/ρ_M, on the monomeric halves D_L and D_M of this dimer is approximately 2 : 1. A similar result had been found earlier for the primary donor radical cation $P_{960}^{+\bullet}$ in RC's of Rps. viridis in liquid solution [2,3].

In the case of $P_{960}^{+\bullet}$, the assignment of experimental isotropic proton hyperfine couplings (hfc's) to nuclear positions rested mainly on the comparison with theoretical isotropic hfc's. The latter were derived from hydrogen s-spin densities obtained by MO calculations on the basis of X-ray structural data [4]. The same procedure was applied later to $P_{865}^{+\bullet}$ [5] using X-ray structural data of Rb. sphaeroides R-26 [6] in an early refinement stage (March 1990).

In the recent single crystal study on $P_{865}^{+\bullet}$ [1], positional assignments were achieved by

The Photosynthetic Bacterial Reaction Center II
Edited by J. Breton and A. Verméglio, Plenum Press, New York, 1992

comparing directions of principal axes of experimentally determined dipolar proton hyperfine tensors with the respective axes of calculated tensors. The latter were derived from theoretical s- and p-spin densities on all H-, C- and N-atoms using the X-ray structure of Rb-sphaeroides R-26 RC's (refinement of December 1990) [6]. This method of assignment is superior to the method used earlier for $P_{960}^{+\bullet}$ [4] and $P_{865}^{+\bullet}$ [5] (see above) because dipolar hyperfine tensor axes are more closely related to local structural properties (e.g. bond directions) than isotropic hfc's.

It is the purpose of this paper to present the methodology and some exemplary results of these tensor calculations.

A Hückel MO treatment of a simple dimer model is introduced to rationalize the observed spin density asymmetry of primary donor radical cations in terms of differences of molecular orbital energies of the isolated monomeric moieties. Such a model can provide a semiquantitative method to trace spin density asymmetries back to structural and/or environmental asymmetries.

METHODS

Spin densities of $P_{865}^{+\bullet}$ have been calculated by the RHF-INDO/SP method described in ref. 7. The same method had been applied earlier to $P_{960}^{+\bullet}$ [3]. Whereas, in the latter case, only proton s-spin densities were evaluated to obtain theoretical isotropic proton hfc's, the complete set of (440) s- and p_x-, p_y-, p_z-spin densities were used to calculate dipolar hyperfine tensors in $P_{865}^{+\bullet}$ in the present work.

Most generally, the cartesian components of a dipolar hyperfine tensor in any molecular axis system x, y, z are given by [8]:

$$t_{ij} = -g\beta g_N \beta_N \sum_{k,l} \rho_{kl} \langle \phi_k | \hat{T}_{ij} | \phi_l \rangle \tag{1}$$

where

$$\hat{T}_{ij} = [\vec{r}_{en}^{\,2} \delta_{ij} - 3(\vec{r}_{en})_i (\vec{r}_{en})_j] / |\vec{r}_{en}|^5 \tag{2}$$

is the classical electron nuclear dipole dipole hyperfine interaction operator with the electron-nucleus distance vector $\vec{r}_{en}$. The ρ_{kl} are the elements of the spin density matrix in the valence orbital basis and ϕ_k describe the various atomic valence (Slater) orbitals. The expression $\langle\ \rangle$ represents one-, two- and three-center integrals over one-electron coordinates. Since all polycenter electron distributions are neglected in the INDO wavefunction calculation, the corresponding interaction integrals are assumed negligibly small [8]. This leaves two-center integrals with ϕ_k and ϕ_l centered on a single atom A interacting with a nucleus of atom N $\neq$ A and one-center integrals with ϕ_k on A interacting with the nucleus of A. Analytical expressions for these one- and two-center integrals $\langle \phi_k | \hat{T}_{ij} | \phi_l \rangle$ using Slater functions are given in ref. 8. Some of these expressions had to be corrected; their corrected forms are given in the appendix. Slater

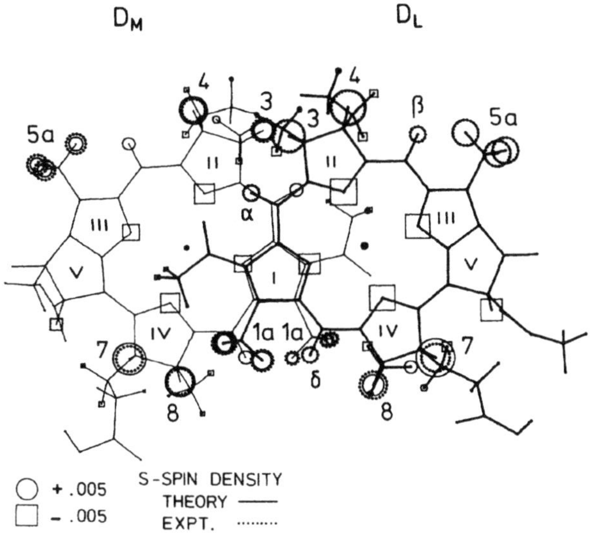

Fig. 1. Comparison of experimental (dotted) and calculated (RHF-INDO/SP) s-spin densities (solid lines) of $D^{+\bullet}$ in Rb. sphaeroides R-26. Experimental values from isotropic proton hfc's using $a_{iso} = Q_H \rho_H(1s)$ with $Q_H = 1420$ MHz [7]. Geometry from X-ray structure analysis [6] (refinement of Dec. 1990). For details of the calculation see text. S-spin densities are proportional to the areas of the respective squares ($\rho < 0$) and circles ($\rho > 0$).

orbital exponents used in this study were: $\zeta(H) = 1.00$ au^{-1}, $\zeta(C) = 1.40$ au^{-1} and $\zeta(N) = 1{,}65$ au^{-1} (au = atomic units).

Structural aspects

Hydrogen atoms were attached to the heavy atom skeleton of the dimer $P_{865} \equiv D$ by standard rules [9]. Geometry optimization by energy minimization techniques [7] was applied in ambiguous situations such as spatial overcrowding of H-atoms (e.g. from different dimer halves) and/or severe distortions of ring structures (e.g. puckering of pyrrole rings).

Bond geometries of the four methyl groups ($C\text{-}CH_3$), $1a_L$, $5a_L$, $1a_M$, $5a_M$ (subscripts refer to the dimer halves D_L and D_M, resp., see fig. 1), whose isotropic and dipolar hfc's are particularly crucial for positional assignments (see below), were all individually optimized.

Acetyl groups attached to rings I on both dimer halves were allowed to rotate around the C(acetyl)-C(ring I) bond, starting from their positions given by the X-ray structure ($O{=}C\text{-}CH_3$ planes within 10° from ring I planes) [6]. The L acetyl group settled at an angle of 45° out of the ring I_L plane with the oxygen atom coming into the nearest possible distance ($r_{ON} = 2.8$ Å) to one of the nitrogen atoms of the nearby HIS L 168. In this case additional electrostatic stabilization was achieved by inserting a positively charged H-atom

($\delta = +0.2$ e) between both atoms with $r_{NH} = 1.3$ Å, thus insinuating the existence of a hydrogen bond in this region *. This deviates from the analysis given in ref. 6 but is in accord with X-ray studies reported by El-Kabbani et al. [10]. The M acetyl group settled at an out-of-plane angle of 6° ** with the C-O bond pointing to the outside of the dimer but outside of any hydrogen bonding range with surrounding amino acid residues.

Positional corrections (in- and out-of-plane movements) were also permitted for peripheral carbon atoms on rings II and IV of both dimer halves (see fig. 1). The positions of these atoms are particularly crucial for the hfc's of non-methyl β-protons (two bonds away from the π-system) [1].

All corrections remained within ± 0.3 Å of the original X-ray coordinates except for the L acetyl group atoms which were displaced by up to 0.8 Å.

Electrostatistic effects from 92 surrounding amino acid residues (AAR's) within a distance of 10 Å of closest approach to both dimer halves (excluding phytyl chains) and from the two adjacent "accessory" bacteriochlorophyll a molecules, B_A and B_B, were included in the spin density and energy calculations. For the AAR's this was done by assigning partial atomic charges taken from CNDO calculations on individual AAR's including backbone atoms [11] to corresponding positions given by the X-ray structure data. The same procedure was used for the neutral molecules B_A and B_B on the basis of INDO calculations performed in our laboratory.

The only modification of the original INDO parametrization [9] was an increase of atomic overlap integrals $S_{\mu\nu}$ by a factor of 2 for interatomic distances $r_{\mu\nu} \geq 3$ Å. This is a better approximation to the long-distance behavior of Hartree-Fock atomic wavefunctions as compared to Slater orbitals [12] and is equivalent to a reduction of the orbital exponent ζ_C (Slater) $= 1{,}625$ au^{-1} to $= 1.325$ au^{-1} at $r = 3$ Å. This results in a stronger resonance coupling β_D (see dimer model below) of the two dimer halves in the overlap region of rings I (see fig. 1) and in a better agreement between the calculated and observed net spin density ratio ρ_L/ρ_M without, however, affecting the ordering $\rho_L > \rho_M$.

RESULTS AND DISCUSSION

The final result of the spin density calculations is shown for s-spin densities in fig. 1 in comparison with experimental values and for p_z-spin densities in fig. 2 (the z-direction is aligned perpendicular to an average plane through the four nitrogen atoms on D_L). The assignment of experimental s-spin densities, obtained from isotropic proton hfc's a_{iso} using the relation $\rho_s^H = a_{iso}^H / Q_H$ with $Q_H = 1420$ MHz [7], rests on the comparison of experimental dipolar hyperfine tensor principal axes with the calculated axes based on the

---

* This required a rotation of the HIS L 168 imidazol ring by 180°, otherwise $r_{ON} = 3.4$ Å with a reduced stabilization effect.

** This angle corresponds to the lower limit of a very shallow rotational potential reaching up to ca. 45°.

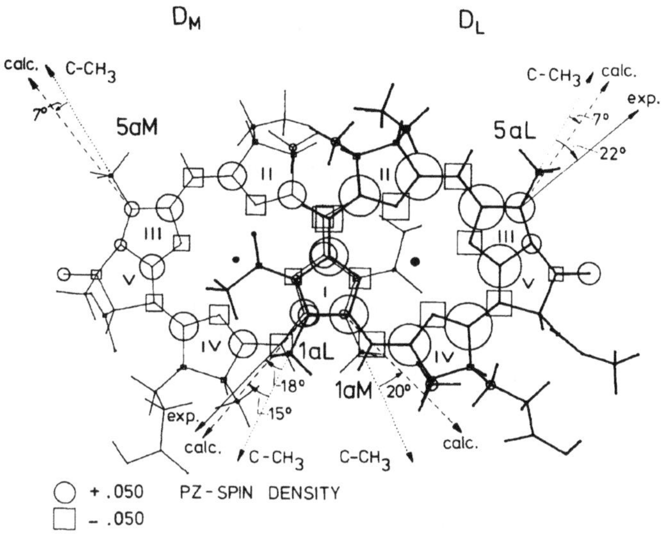

Fig. 2. Calculated p_z-spin densities of $D^{+\bullet}$ in Rb. sphaeroides R-26. Geometry as in fig. 1. The directions of the principal axes of the hf tensors of methyl protons 1a and 5a on D_L and D_M are indicated (solid arrows: experiment [1], only shown for the chosen assignmen to D_L; dashed and dotted arrows: calculations and bond directions, resp.) For details, see text.

X-ray structural data [1]. This comparison was restricted to the axes with the largest positive principal value as these are generally better related to the local bond geometry - particularly in case of near axial symmetry as in the case of rotating methyl groups. For example, for an isolated C-CH_3 fragment, theory predicts a practically axially symmetrical dipolar hyperfine tensor with its symmetry axis (largest positive principal value) parallel to the C-C bond. However, in the presence of large spin densities in the vicinity of a C-CH_3 group, quite considerable deviations between the directions of the tensor symmetry axis and the C-CH_3 bond direction can occur. Fig. 2 shows the directions of theoretical principal hyperfine tensor axes for the methyl protons $1a_L$, $5a_L$ and $1a_M$, $5a_M$. The experimental axes of the two largest methyl hfc's have been assigned to the dimer half D_L. All axes shown in fig. 2 are very close (< 9°) to the plane of the figure. (For position $1a_M$, the calculated tensor deviates significantly from axial symmetry and the axis with the largest positive principle value points towards ring I on D_L; in this case the axis closest to the C-CH_3 bond direction is depicted). The chosen assignment gives a significantly better agreement between experimental and theoretical tensor axes than the alternative assignment of the two largest methyl hfc's to the dimer half D_M. In the latter case, the smallest angles between theoretical and experimental axes are 47° ($1a_M$) and 33° ($5a_M$) * as compared to 3° ($1a_L$) and 15° ($5a_L$). The chosen assignment (to D_L) is less

* In this comparison all possible sign combinations for the off-diagonal elements t_{ij} due to site ambiguities were taken into consideration.

evident if only comparing experimental axes with C-CH_3 bond directions. A more detailed discussion of this assignment and remaining discrepancies between experimental and theoretical directions of principal axes is given in ref. 1.

The experimental ratio of net isotropic methyl proton s-spin densities $(\rho^s_{1aL} + \rho^s_{5aL})/(\rho^s_{1aM} + \rho^s_{5aM}) \equiv \rho_L{}^s/\rho_M{}^s$ is close to 2 : 1, the theoretical ratio is 3.4 : 1. The latter value practically equals the value one obtains if summing up ρ for all valence orbitals on D_L and D_M. Since most of the spin density ($\cong$ 90%) is located in the p_z-orbitals of the π-system, this number is also close to the ratio of π-spin densities $\rho_L{}^\pi/\rho_M{}^\pi$. One possible source for the remaining discrepancy between theoretical and experimental spin density ratios $\rho_L{}^\pi/\rho_M{}^\pi$ (3.4 as compared with 2.0) is the uncertainty of the value of the resonance integral β_D between the dimer halves (see dimer model below). The variation of $\rho_L{}^\pi/\rho_M{}^\pi$ with β_D is as follows: $\rho_L{}^\pi/\rho_M{}^\pi$ = 7.0; 3.4; 1.9 for $\beta_D/\beta_D{}^\circ$ = 1.0; 2.0; 3.0, respectively, where $\beta_D{}^\circ$ is the unmodified INDO resonance integral value [9]. Since $\beta_D{}^\circ$ is strongly dependent on interatomic distances, its value is also strongly affected by errors of the X-ray structural data. For example, a reduction of the interplanar distance between rings I_L and I_M from 3.8 Å (x-ray value) to 3.2 Å would result in an increase of $\beta_D{}^\circ$ and thus of β_D by ca. 30 % accompanied by an almost equally strong reduction of $\rho_L{}^\pi/\rho_M{}^\pi$.

The essential experimental result is a significant deviation of the electronic symmetry of $P^{+\bullet}_{865}$ from C_2 symmetry as was found earlier also for $P^{+\bullet}_{960}$ [4]. The theoretical tensor calculations and the ensuing assignment strongly support a ratio of $\rho_L{}^\pi/\rho_M{}^\pi \cong 2$. It must be remembered that the latter quantitative conclusion rests exclusively on the comparison of experimental and theoretical dipolar hyperfine tensor *axes*. These are only weakly affected by a common scaling of spin densities on either D_L or D_M if one excludes contributions to dipolar hyperfine tensors on one dimer half (D_L or D_M) from spin densities on the other dimer half (D_M or D_L). In other words, the present result is not based on absolute theoretical numbers for spin densities as in the case of $P^{+\bullet}_{960}$ [2,3] and in an earlier study of $P^{+\bullet}_{865}$ [5].

Recent single crystal EPR measurements of the complete g-tensor of $P^{+\bullet}_{865}$ in the W-band ($\lambda \cong$ 3 mm) in the Berlin laboratory have also revealed a breakage of the C_2 symmetry of this dimer [13]. Since the g-tensor measures a global molecular property and is difficult to calculate theoretically, no value for $\rho_L{}^\pi/\rho_M{}^\pi$ could be specified.

Deviations of the spin densities from a symmetric distribution, expressed by the deviation of the ratio ρ_L/ρ_M of net spin densities on the L- and M-half of the dimer (summed over all respective atomic valence orbitals) from unity, can be traced back to differences in the energies ϵ_L and ϵ_M of the highest filled molecular π orbitals of the isolated monomeric halves, D_L and D_M, respectively. For $\rho_L/\rho_M < 1$, as in $D^{+\cdot}$, it is necessary that $\epsilon_L > \epsilon_M$. This puts the unpaired electron in the energetically higher MO of the combined monomers, which has more L- than M-character.

Dimer model calculations

In the following we present a simple Hückel MO (HMO) description of an ethylene dimer as a model for establishing a semi-quantitative relation between ρ_L/ρ_M and the relevant energetic parameters of a dimeric molecular species.

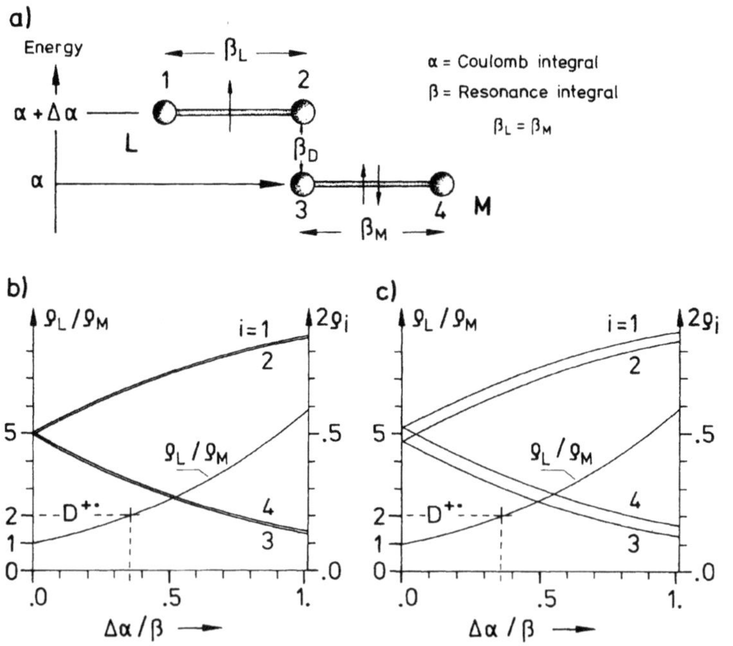

Fig. 3. Results of HMO calculations on a simple dimer model (ethylene-dimer). (a) dimer configuration and the chosen Hückel parametrization. (b) and (c): spin densities 2 ρ_i in the dimer halves L and M, and ratios of the sums of spin densites ρ_L/ρ_M for two different strengths of coupling: β_D/β_M = 0.01 (b) and β_D/β_M = 0.1 (c). For details, see text.

Fig. 3a shows the chosen Hückel parametrization for the π-electron system of an ethylene dimer consisting of four π-centers, labeled 1 to 4. Atomic Coulomb integrals on the L-half of the dimer are assumed to be more positive (i.e. higher in energy) than those on the M-half by an amount of $\Delta\alpha$. This shifts the energies of the molecular π-orbitals on the isolated L-half by the same amount $\Delta\alpha$ as compared with those on the isolated M-half. The energy shift is introduced to account for differences in structure or environmental interactions of the two monomeric halves. Bond or resonance integrals, β_L and β_M, on both monomers are assumed to be equal. The only interaction between the two monomers, described by the resonance integral β_D, is assumed to be in the overlap region between atoms 2 and 3. The molecular orbitals of the dimer follow as usual from the diagonalization of the corresponding 4 $\times$ 4 Hückel Hamiltonian.

Figs. 3a and 3c show calculated Hückel spin densities for the dimer cation radical, given by the squares of the molecular orbital coefficients on the different atoms in the half-filled molecular orbital for the two coupling cases β_D/β_M = 0.01 and 0.1, respectively. Plotted curves show the ratio of net spin densities $\rho_L/\rho_M = (\rho_1 + \rho_2)/(\rho_3 + \rho_4)$ and individual atomic spin densities ρ_1, ... ρ_4 (normalized to their values $\rho = 0.5$ in the isolated monomers) as a function of the energy ratio $\Delta\alpha/\beta_D$ (note: different vertical scales for ρ_L/ρ_M and $2\rho_i$).

The major results of the HMO-calculations are:
(i) the spin density ratio ρ_L/ρ_M increases with increasing ratio $\Delta\alpha/\beta_D$ and is practically independent of the strength of the coupling (in the given "weak coupling" range $0.01 \leq \beta_D/\beta_M \leq 0.1$).
(ii) Atomic spin densities $\rho_1 \ldots \rho_4$ show characteristic deviations from their symmetric distributions ($\rho_1 = \rho_2, \rho_3 = \rho_4$) in the isolated monomers with increasing values of β_D/β_M ("supermolecule effect"). Spin densities are "pushed" out of the overlap region (atoms 2 and 3) towards the outer atoms (1 and 4).
(iii) The spin density ratio $\rho_L/\rho_M \approx 2$ observed in $P_{865}^{+\bullet}$ would correspond to $\Delta\alpha/\beta_D \approx 0.4$, if extrapolated from this simple ethylene dimer model. This would give $\Delta\alpha \approx 0.1$ eV using $\beta_D \approx 0.3$ ev from the RHF-INDO/SP calculations on $P_{865}^{+\bullet}$, in fair agreement with the rigorous RHF-INDO/SP result $\epsilon_L - \epsilon_M = 0.14$ eV. Possible reasons for the energy difference $\epsilon_L - \epsilon_M$ in $P_{865}^{+\bullet}$ might be different positions of the acetyl groups, a difference in the puckering of the macrocycles, or an asymmetry in the protein environment (e.g. asymmetrical hydrogen bonding, see ref. 10). Thus, the ratio ρ_L/ρ_M can be regarded quite generally as a useful probe for sensing small but possibly functionally important energetic asymmetries in the dimeric primary donors of different photosynthetic bacteria.

CONCLUSION

The measurement of dipolar hyperfine tensors, in particular of their principal axes directions, was shown to be an inportant means to assign isotropic and anisotropic hfc's to nuclear positions in molecules [1], provided single crystals and sufficiently accurate X-ray structural data are available. The comparison of experimentally determined hf tensors with those calculated on the basis of MO theory was shown to be necessary in cases where hf tensor axes show large deviations from bond directions. This procedure has been applied to the primary donor radical cation $P_{865}^{+\bullet}$ of Rb. sphaeroides R-26 on the basis of methyl proton hfc's and has revealed an asymmetrical spin density distribution in favor of the L-half D_L of this dimer. This result is in accordance with the result found earlier for the primary donor radical cation $P_{960}^{+\bullet}$ of Rps. viridis. This common feature of these two bacterial donors may reflect an important functional principle underlying the primary charge separation process in bacterial reaction centers. A frequently offered hypothesis is a switching behavior of orbital charge distributions in the highest occupied (HOMO) and lowest unoccupied (LUMO) molecular orbitals of the dimer which enhances the forward reaction in the photoactive L-branch (unidirectionality) and inhibits the back reaction to the dimer ground state (high quantum efficiency); see ref. 7 and citations therein. At the present it is not clear if this purely electronic mechanism is of relevance (it may however, be one of several concurrent factors) or if other properties like asymmetrical Franck-Condon factors are of greater importance [14].

ACKNOWLEDGMENTS

The authors thank Drs. J.P. Allen (Arizona State Univ., Tempe), D.C. Rees (Caltech, Pasadena) and G. Feher (Univ. of California, San Diego) for providing the X-ray coordinates of the Rb. sphaeroides R-26 RC. The authors also thank Dr. M. Huber for helpful discussions. This work was supported by the Deutsche Forschungsgemeinschaft (SFB 337 and SFB 312).

Appendix

Corrections to formulae given in ref. 8:

Eq. 22: Replace $(1/\sqrt{3})$ by $-(1/\sqrt{3})$
Eq. 24: Replace (15/2a) by $(5\sqrt{3}/2a)$
Eq. 25: Replace $18a^2$ by $18/a^2$
Eq. 28: Replace $(9/a^2)$ by $(9/2a^2)$
Eq. 30: Replace $(5/2)^{1/2}$ by 1 (unity)
Replace [....] by
$[(2/9)a^5 + (2/3)a^4 + (4/3)a^3 + 2a^2 + 2a + 1]e^{-2a}$
Add relation $<2p_x/r^{-5}(3xy)/2p_y> = \zeta^3/5$ to the one-center integrals.

REFERENCES

1. F. Lendzian, M. Huber, R.A. Isaacson, B. Endeward, M. Plato, B. Bönigk, K. Möbius, W. Lubitz, and G. Feher, The electronic structure of the primary donor cation radical in Rhodobacter sphaeroides R-26: ENDOR and Triple resonance studies in single crystals of reaction centers, *Biochim.Biophys. Acta.* (1992), submitted.
2. F. Lendzian, W. Lubitz, H. Scheer, A.J. Hoff, M. Plato, E. Tränkle, and K. Möbius, ESR, ENDOR and TRIPLE resonance studies of the primary donor radical cation $P_{960}^{+\bullet}$ in the photosynthetic bacterium Rhodopseudomonas viridis, *Chem.Phys.Lett. 148*:377 (1988).
3. M. Plato, W. Lubitz, F. Lendzian, and K. Möbius, Magnetic resonance and molecular orbital studies of the primary donor cation radical $P_{960}^{+\bullet}$ in the photosynthetic bacterium Rhodopseudomonas viridis, *Isr.J.Chem. 28*:109 (1988).
4. H. J. Deisenhofer, O. Epp, K. Miki, R. Huber, and H. Michel, X-ray structure analysis of a membrane protein complex, *J.Mol.Biol. 180*:385 (1984).
5. M. Plato, K. Möbius, W. Lubitz, J.P. Allen, and G. Feher, Magnetic resonance and molecular orbital studies of the primary donor states in bacterial reaction centers, *in*: "Perspectives in Photosynthesis," J. Jortner and B. Pullman, Eds., Kluwer Academic Publishers (1990).
6. G. Feher, J.P. Allen, M.Y. Okamura, and D.C. Rees, Structure and function of bacterial photosynthetic reaction centers, *Nature 339* :111 (1989) and references therein.
7. M. Plato, K. Möbius, and W. Lubitz, Molecular orbital calculations on chlorophyll radical ions, *in:* "Chlorophylls," H. Scheer, Ed., CRC Press, Boca Raton (1991).
8. D.L. Beveridge and J.W. McIver, Jr., INDO molecular orbital study of hyperfine tensors: theory, methodology, and applications to CH, CH_3, and radicaloid derivatives of malonic acid, *J.Chem.Phys. 54*:4681 (1971).

9. J.A. Pople and D.L. Beveridge. "Approximate Molecular Orbital Theory," McGraw-Hill Inc., New York (1970).
10. O. El-Kabbbani, C.H. Chang, D. Tiede, J. Norris, and M. Schiffer, Comparison of reaction centers from Rhodobacter sphaeroides and Rhodopseudomonas viridis: overall architecture and protein-pigment interactions, *Biochem.* *30*:5361 (1991).
11. F.A. Momany, R.F. McGuire, A.W. Burgess, and H.A. Scheraga, Energy parameters in polypeptides. VII, *J.Phys.Chem.* *79*:2361 (1975).
12. G.J. Burns, Atomic shielding parameters, *J.Chem.Phys.* *41*:1521 (1964).
13. R. Klette, J.T. Törring, M. Plato, K. Möbius, B. Bönigk, and W. Lubitz, Determination of the g-tensor of the primary doner cation radical in single crystals of Rhodobacter sphaeroides R-26 reaction centers by 3 mm high-field EPR, to be published.
14. L.M. McDowell, D. Gaul, C. Kirmaier, D. Holten, and C.C. Schenk, Investigation into the source of electron transfer asymmetry in bacterial reaction centers, *Biochem.* *30*:8315 (1991).

NEAR-INFRARED-EXCITATION RESONANCE RAMAN STUDIES OF BACTERIAL REACTION CENTERS

Vaithianathan Palaniappan and David F. Bocian

Department of Chemistry
University of California
Riverside, CA 92521-0403

INTRODUCTION

Resonance Raman (RR) spectroscopy has proven to be an effective probe of the structure of the bacteriochlorophyll (BChl) and bacteriopheophytin (BPh) pigments in bacterial photosynthetic reaction centers (RCs) [1-3]. Most RR studies have been performed with excitation resonant with either the B-states or Q_x-states of the pigments. Interpretation of the RR scattering observed with B-state excitation is complicated by the fact that the absorption bands of all six pigments overlap. This is not the case for the Q_x region; however, these absorptions are relatively weak which leads to relatively poor quality RR spectra. In contrast, the Q_y absorption bands of the special pair (P) BChls, the accessory BChls, and the BPhs are energetically better separated and relatively intense. The combination of these spectral characteristics makes the Q_y region an attractive target for RR studies of RCs. Early Q_y-excitation RR studies of RCs were hampered by the lack of suitable detectors or laser excitation sources [4]. However, with the advent of the charge couple device (CCD) and Ti:sapphire laser, this is no longer a problem. Regardless, the most serious impediment to Q_y-excitation RR studies of RCs is the presence of strong fluorescence intrinsic to the RCs or from small amounts of exogenous pigments present in the preparations. Despite this limitation, high quality Q_y-excitation RR spectra of RCs can be obtained [5-7].

We recently initiated a comprehensive near-infrared-excitation RR study of photosynthetic RCs. Spectra are being acquired by using a large number of excitation wavelengths in the 675-925-nm region. The goal of these studies is to probe structures of the BChls and BPhs as selectively as possible and to characterize the interactions of these chromophores with the protein matrix. In this article, we report the results of our preliminary studies.

MATERIALS AND METHODS

The RCs from *Rb. sphaeroides* 2.4.1 were prepared as previously described [8]. The proteins were eluted from a DEAE anion exchange column by using 0.01 M Tris (pH 8.0),

The Photosynthetic Bacterial Reaction Center II
Edited by J. Breton and A. Verméglio, Plenum Press, New York, 1992

0.015 % Triton X-100, and 0.5 M NaCl. The RCs were chemically reduced or oxidized by adding an excess of $Na_2S_2O_4$ or $K_3Fe(CN)_6$, respectively.

The RR measurements were made on glassed samples (~1:1 in glycerol) contained in a 1 mm i.d. capillary tube. Temperature control was achieved by mounting the sample on the cold tip of a closed cycle refrigeration system via a home built holder. The RR scattering was collected in a 90° configuration by using a 0.6 M triple spectrograph. A liquid nitrogen cooled (-120° C) anti-reflection coated back-illuminated CCD served as the detector. The excitation wavelengths were provided by the discrete outputs of a Kr ion laser or a Ti:sapphire laser pumped by the visible multiline output of an Ar ion laser. The laser power at the sample was typically less than 1 mW. Typical slit widths used in the RR experiments yielded a resolution of 2-4 cm^{-1}. Other details of the experimental procedures are described elsewhere [7].

GENERAL CHARACTERISTICS OF THE RR SCATTERING

In order to extract structural information from the RR spectra of RCs, the bands must be corrrectly assigned to a particular chromophore. At certain excitation wavelengths, these vibrational assignments are relatively straightforward because specific pigments are selectively excited. For example, the RR spectra obtained with excitation wavelengths on the red-most end of the absorption spectrum (850-925 nm) are expected to exhibit bands due only to P. On the other hand, the assignment of the RR spectra obtained with excitation in the 700-825-nm region is considerably more complicated. Although the absorption spectrum in this region is dominated by the $Q_y(0,0)$ bands of the accessory BChls (~805 nm) and the BPhs (~760 nm), there are also absorptions due to vibronic satellites of both P and the accessory BChls. In particular, the $Q_y(1,0)$ vibronic satellites due to the high-frequency modes (1300-1750 cm^{-1}) of P are expected in the 785-815-nm region. This region overlaps the $Q_y(0,0)$ bands of the accessory BChls. Similarly, $Q_y(1,0)$ satellites due to the vibrations of the accessory BChls are expected throughout the 700-800-nm region. Satellites from modes whose frequencies are in the 500-1000-cm^{-1} range should absorb in the region of the $Q_y(0,0)$ bands of the BPhs whereas satellites due to the high-frequency modes are expected in the 705-730-nm region. Despite the fact that the vibronic absorption bands are relatively weak and completely obscured by the much stronger $Q_y(0,0)$ bands of the accessory BChls and BPhs, the RR scattering from these satellites is expected to be quite strong. This is a general characteristic of the RR cross section. Indeed, in the small displacement limit, the RR scattering from the (1-0) vibronic satellite is expected to be equal to that from the (0-0) band [9]. It is the scattering from these vibronic satellites that complicates the interpretation of the RR spectra of RCs obtained with excitation in the 700-825-nm region. On the other hand, the strong scattering expected from the $Q_y(1,0)$ vibronic bands of the BPhs aids in the assignments of the modes in the 1300-1750-cm^{-1} region. The $Q_y(1,0)$ bands of these latter vibrations occur in the 685-665-nm region of the absorption spectrum. There are no absorptions due to P or the accessory BChls in this region; consequently, the BPhs are the only contributors to the RR spectrum with excitation in this regime.

HIGH-FREQUENCY RR SPECTRA OF RCs

The high-frequency RR spectra of reduced and oxidized RCs obtained with several excitation wavelengths in the 675-875-nm region are shown in Figures 1 and 2, respectively.

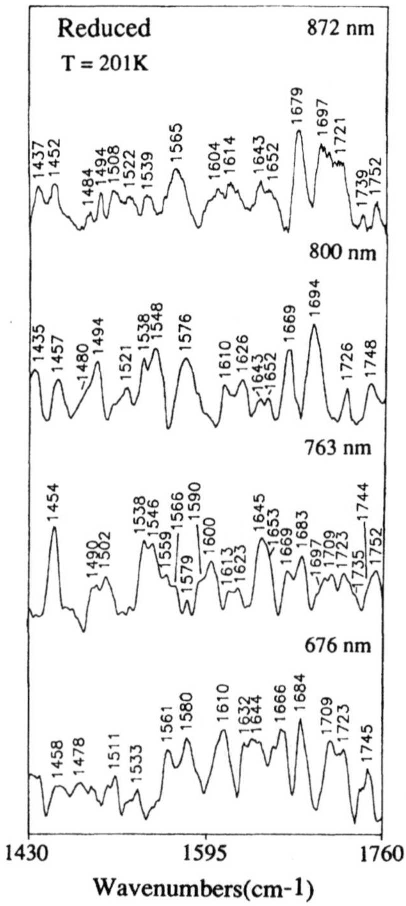

Figure 1. High-frequency Q_y-excitation RR spectra of chemically reduced RCs from *Rb. sphaeroides* 2.4.1 obtained at 201K.

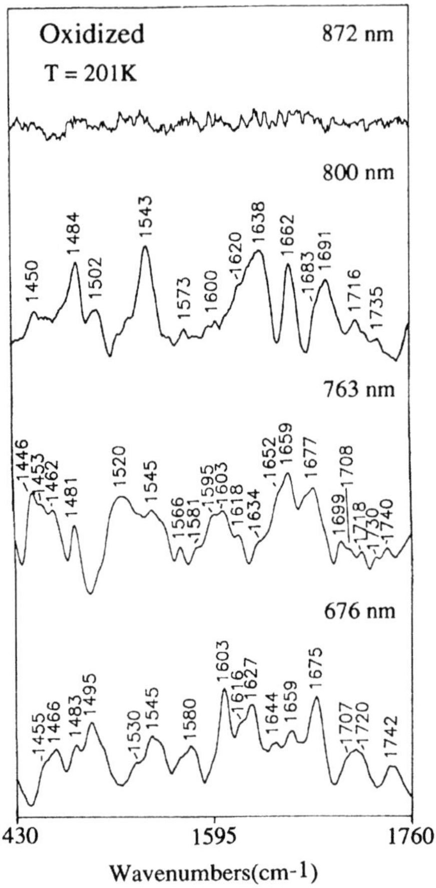

Figure 2. High-frequency Q_y-excitation RR spectra of chemically oxidized RCs from *Rb. sphaeroides* 2.4.1 obtained at 201K.

These excitation wavelengths span the Q_y absorption bands of P, the accessory BChls, and the BPhs. The RR spectra obtained at all excitation wavelengths exhibit a number of bands. The bands in the 1400-1630-cm^{-1} region are due to in-plane skeletal modes of the macrocycles whereas those above 1630 cm^{-1} are due to carbonyl stretching vibrations of various substituent groups [1-3]. The frequencies of these modes are quite sensitive to the structure of the macrocycles in the protein matrix [1-3,10]. These structure-sensitive bands will be the focus of the discussion.

Special Pair Vibrations

The RR spectra obtained with λ_{ex}=872 nm are expected to exhibit contributions only from P (*vide supra*). This is confirmed by the complete bleaching of the RR signals upon chemical (or photo) oxidation. In the carbonyl region, the RR bands observed at 1697 and 1679 cm^{-1} are due to the C_9-keto stretching modes ($\nu(C_9{=}O)$) of P_L and P_M, respectively [1-3,11]. A pair of bands is observed at 1643 and 1652 cm^{-1} which can be assigned as the C_2-acetyl carbonyl stretching vibrations ($\nu(C_2{=}O)$) of these two pigments. The 1652 cm^{-1} band has been previously identified as $\nu(C_2{=}O)$ of P_M whereas a band at 1620 cm^{-1} (not clearly observed with λ_{ex}=872 nm) has been attributed to this vibration of P_L [11]. However, the RR data reported here (*vide infra*) and normal coordinate calculations [12] suggest that bands in the 1620-cm^{-1} region are actually due to skeletal fundamental vibrations. Several bands are observed above 1700 cm^{-1} that are also due to carbonyl stretches. In particular, the 1739 and 1721-cm^{-1} can be assigned to the $\nu(C_{10}{=}O)$ vibrations of P_M and P_L, respectively. The former band occurs at a frequency typical of an unbound ester carbonyl [3] whereas the frequency of the latter is unusually low. In this regard, X-ray crystallographic studies on RCs show that the C_{10}-carbonyl group of P_M is free while that of P_L is hydrogen bonded to serine L244 [13,14]. The origin of the band observed at 1752 cm^{-1} is uncertain.

In the region of the skeletal vibrations, the Q_y-excitation RR spectra provide new insights into the structure of the individual chromophores in the dimer. In particular, there appear to be two bands in the region of the ν_{10}-like modes, one near 1614 cm^{-1} and the other near 1604 cm^{-1}. The former band occurs at a frequency typical of a five-coordinate BChl [1,2,15], whereas the latter is somewhat lower. Crystallographic studies on RCs indicate that the Mg-histidine M202 bond of P_M is typical in length and much shorter (by ~1Å) than the Mg-histidine L173 bond of P_L [13,14]. Indeed, the San Diego group has suggested that histidine L173 is not ligated to the Mg of P_L [13]. These data in conjunction with the RR results suggest that the 1614- and 1604-cm^{-1} bands are due to P_M and P_L, respectively. The low frequency observed for the ν_{10}-like band of P_L is presumably due to the weak axial ligation. This would allow the Mg atom to assume a more in-plane structure, thereby expanding the core of the macrocycle and lowering the frequency of the ν_{10}-like band [10].

Accessory BChl Vibrations

The RR spectra of reduced RCs observed with λ_{ex}=800 nm exhibit a large number of bands. Although many of these bands are due to the accessory BChls, others are clearly attributable to P. This is evident from the fewer number of bands observed in the spectra of oxidized RCs (Figures 1 and 2). For example, bands are observed at 1652 and 1643 cm^{-1} in the 800-nm excitation RR spectra of reduced RCs which bleach upon oxidation. These bands are due to scattering from the $Q_y(1,0)$ satellites of the $C_2{=}O$ vibrations of P. Bleaching also occurs in the region of the ν_{10}-like bands and other lower frequency skeletal modes.

Nominally, the comparison of the RR spectra of reduced and oxidized RCs can be used to delineate bands due to the accessory BChls; however, the spectral analysis is complicated by the fact that oxidation of P also affects the RR spectra of the accessory BChls (as well as those of BPhs, *vide infra*). Oxidation-induced changes in the RR spectra of the BChls have been previously reported based on B-state excitation data. In particular, Robert and Lutz have reported that $\nu(C_9{=}O)$ of $BChl_L$ downshifts upon oxidation of P whereas the analogous mode of $BChl_M$ is unaffected [16]. The $\nu(C_9{=}O)$ bands of these two chromophores are overlapped and occur near 1694 cm^{-1} in reduced RCs (Figure 1). Upon oxidation of P, the centroid of the band downshifts by ~3 cm^{-1} and a distinctive shoulder appears on the low-frequency side (Figure 2). Oxidation also preferentially affects the $\nu(C_2{=}O)$ vibrations of one of the two accessory BChls (presumably $BChl_L$). The $\nu(C_2{=}O)$ bands of $BChl_L$ and $BChl_M$ are also overlapped and occur near 1669 cm^{-1} in reduced RCs. Upon oxidation, one component of this band downshifts by ~7 cm^{-1} leaving a shoulder on the high-frequency side.

BPh Vibrations

The $Q_y(0,0)$ absorption bands of the BPhs lie in a congested spectral region near 750 nm whereas the $Q_y(1,0)$ vibronic satellites of the high-frequency vibrations occur in a relatively uncongested region below 700 nm. The RR spectra obtained with excitation in this latter regime can be used to selectively excite the BPhs. The RR spectra of reduced and oxidized RCs obtained with λ_{ex}=763 and 676 nm are shown in Figures 1 and 2. In the RR spectra of reduced RCs, bands assignable to the $\nu(C_9{=}O)$ modes of BPh_L and BPh_M are observed at 1684 and 1709 cm^{-1}, respectively. The $\nu(C_2{=}O)$ bands for these two pigments are overlapped and occur near 1667 cm^{-1}. It should be noted that the latter vibrations for the two BPhs have been previously assigned to the bands observed in the 1625-1640-cm^{-1} region [1-3,16]. This interpretation, which is based on B- and Q_x-excitation RR studies, leaves no logical assignment for the band observed near 1667 cm^{-1} in the Q_y-excitation spectra. Consequently, we prefer to attribute the bands in 1625-1640-cm^{-1} region to skeletal modes of the two BPhs rather than to the $\nu(C_2{=}O)$ vibrations.

Comparison of Figures 1 and 2 reveals that oxidation significantly affects the RR bands of BPh_L and not BPh_M. In particular, upon oxidation the $\nu(C_9{=}O)$ band of the former pigment downshifts from 1684 to 1675 cm^{-1} whereas the latter remains near 1709 cm^{-1}. Oxidation-induced shifts are also observed for the $\nu(C_2{=}O)$ band(s) and the skeletal modes of the macrocycle(s); however, the spectral congestion in these regions makes the interpretation of the spectra less clear. It is, however, clear that a component of the band attributable to $\nu(C_2{=}O)$ band downshifts upon oxidation (to ~1659 cm^{-1}).

FUNCTIONAL IMPLICATIONS OF THE OXIDATION-INDUCED SPECTRAL CHANGES

The RR data obtained for reduced versus oxidized RCs indicate that oxidation of P strongly perturbs $BChl_L$ and BPh_L whereas $BChl_M$ and BPh_M are mostly unaffected.

Oxidation downshifts the $\nu(C_9{=}O)$ modes of both $BChl_L$ and BPh_L. Downshifts are also observed for the $\nu(C_2{=}O)$ modes of the accessory BChls and BPhs. Although the assignments for these latter modes are less obvious than for the $\nu(C_9{=}O)$ vibrations, it appears that only one BChl and one BPh (presumably the L-branch chromophores) are affected by oxidation. The fact that both the $\nu(C_9{=}O)$ and $\nu(C_2{=}O)$ vibrators (which are on opposite sides of the macrocycle) of both $BChl_L$ and BPh_L are perturbed by oxidation of P suggests that the physical changes which elicit these frequency shifts are global in nature. Accordingly, it seems unlikely that changes in hydrogen bonding to specific carbonyl groups or structural changes in the macrocycles could be the origin of the spectral shifts. In addition, the oxidation-induced frequency shifts of both the $\nu(C_9{=}O)$ and $\nu(C_2{=}O)$ modes of BPh_L are larger than those of $BChl_L$. This observation suggests that electrostatic perturbations due to P^+ are not the principal contributor because the former pigment is significantly further from P than is the latter.

The apparent global nature of the oxidation-induced physical changes that result in the frequency shifts of the carbonyl modes of $BChl_L$ and BPh_L suggests that these changes involve a number of protein residues. One possibility is that oxidation of P results in a structural reorganization that alters the effective dielectric constant of the medium surrounding $BChl_L$ and BPh_L [17]. In this scenario, the lower frequencies of the carbonyl modes imply a larger effective dielectric constant [18] which in turn would stabilize either macrocycle with respect to anion formation. The fact that BPh_L experiences the largest perturbation is further consistent with the fact that this macrocycle forms a metastable anionic intermediate. This preferential stabilization of the chromophores on the L-branch could contribute to the path specificity of electron transfer.

In attempts to investigate further the differences in the local environment of BPh_L versus BPh_M, we have also conducted preliminary RR studies on RCs from *Rb. capsulatus* wild-type and the (L)E104L mutant. Oxidation-induced frequency shifts of the carbonyl modes are also observed for these RCs. Comparison of the RR data for RCs from the wild-type and the (L)E104L mutant indicate that hydrogen bond to the $(C_9{=}O)$ group of BPh_L is only partially responsible for the fact that the frequency of the $\nu(C_9{=}O)$ vibration of BPh_L is substantially lower than that of BPh_M. In particular, this frequency difference is ~ 25 cm^{-1} in wild-type and ~15 cm^{-1} in the mutant. The fact that the carbonyl stretching frequency of the $C_9{=}O$ group of BPh_L is much lower than that of BPh_M suggests that this pigment is intrinsically stabilized toward anion formation independent of the oxidation state of P. This could facilitate preferential electron transfer down the L-branch.

ACKNOWLEDGMENTS

We thank Drs. H. A. Frank and M. A. Aldema for providing the RCs from *Rb. sphaeroides* and Drs. D. C. Youvan and E. J. Bylina for providing the RCs from *Rb. capsulatus*. Financial support was provided by grant GM39781 from the National Institute of General Medical Sciences.

REFERENCES

1. Lutz, M. (1984) in Adv. Infrared Raman Spectrosc. 11, 211-300.
2. Lutz, M. and Robert, B. (1988) in Biological Applications of Raman Spectroscopy (Spiro, T. G., Ed.) Vol. 3, pp 347-411, Wiley, New York.

3. Lutz, M. and Mantele, W. (1991) in Chlorophylls (Scheer, H., Ed.) pp 855-902, CRC, Boca Raton, FL.
4. Bocian, D. F., Boldt, N. J., Chadwick, B. W., and Frank, H. A. (1987) FEBS. Lett. 214, 92-96.
5. Donohoe, R. J., Dyer, R. B., Swanson, B. I., Violette, C. A., Frank, H. A., and Bocian,D. F. (1990) J. Am. Chem. Soc. 112, 6716-6718.
6. Shreve, A. P., Cherepy, N. J., Franzen, S., Boxer, S. G., and Mathies, R. A. (1991) Proc. Natl. Acad. Sci. U.S.A. 77, 3105-3109.
7. Palaniappan, V., Aldema, M. A., Frank, H. A., and Bocian, D. F. (1992) Biochemistry, submitted.
8. McGann, W. J. and Frank, H. A. (1985) Biochim. Biophys. Acta 807, 101-109.
9. Myers, A. B. and Mathies, R. A. (1987 in Biological Applications of Raman Spectroscopy (Spiro, T. G., Ed.) Vol. 2, pp 1-58, Wiley, New York.
10. Schick, G. A. and Bocian, D. F. (1987) Biochim. Biophys. Acta 895, 127-154.
11. Mattioli, T. A., Hoffman, A., Robert, B., Schrader, B., and Lutz, M. (1991) Biochemistry 30, 4648-4654.
12. Donohoe, R. J., Frank, H. A., and Bocian, D. F. (1988) Photochem. Photobiol. 48, 541-548.
13. Yeates, T. O., Komiya, H., Chirino, A., Rees, D. C., Allen, J. P., Feher, G. (1988) Proc. Natl. Acad. Sci. U.S.A. 83, 7993-7997.
14. El-Kabbani, O., Chang, C.-H., Tiede, D., Norris, J., and Schiffer, M. (1991) Biochemistry, 30, 5361-5369.
15. Callahan, P. M. and Cotton, T. M. (1987) J. Am. Chem.Soc. 109, 7001-7007.
16. Robert, B. and Lutz, M. (1988) Biochemistry 27, 5108-5114.
17. Yeates, T. O., Komiya, H., Rees, D. C., Allen, J. P. and Feher,G (1987) Proc. Natl. Acad. Sci. U. S. A. 84, 6438-6442.
18. Krawczyk, S. (1989) Biochim. Biophys. Acta 976, 140-149.

ASYMMETRIC STRUCTURAL ASPECTS OF THE PRIMARY DONOR IN SEVERAL PHOTOSYNTHETIC BACTERIA: THE NEAR-IR FOURIER TRANSFORM RAMAN APPROACH

T.A. Mattioli, B. Robert, and M. Lutz

Département de Biologie Cellulaire et Moléculaire
C.E. de Saclay, 91191 Gif-sur-Yvette cedex, FRANCE

INTRODUCTION

The structural elucidation of the reaction centers (RCs) from two purple photosynthetic bacteria, *Rhodopseudomonas (Rps.) viridis* [1] and *Rhodobacter (Rb.) sphaeroides* [2,3] has revealed the specific spatial arrangement of the bacteriochlorophyll (BChl) and bacteriopheophytin (BPhe) prosthetic groups mediating electron transfer and stable charge separation which take place in the RC protein. These prosthetic groups (two BChl molecules constituting the primary donor dimer, P; two so-called accessory BChl molecules; two BPhe molecules) are arranged in pairs along a pseudo C_2 symmetry axis which runs from P (periplasmic side) to the non-heme iron (cytoplasmic side) i.e. in the direction of transmembrane electron transfer. Despite the apparent two possible pathways for the resulting charge separation, the electron transfer seems to occur predominantly via one pathway [4], the so-called L-branch. Current conventional wisdom deems that this unidirectional charge separation is controlled by the spatial arrangement of the chromophores and fine-tuned by their local protein environments.

A complete structural characterization of the primary electron donor, the origin of the transferred electron, seems to be of vital importance for a complete understanding of the asymmetric functioning of these reaction centers. To this end, resonance Raman spectroscopy seems to be a useful tool in studying the primary donor structure. As a vibrational spectroscopy, the Raman scattering technique, like its infrared absorption counterpart [5], sensitively detects small perturbations of molecular vibrations thus giving structural information concerning differences in conformations and protein interactions interactions of the chromophores. In addition to this, <u>resonance</u> Raman spectroscopy also yields <u>electronic</u> information since the vibrational modes observed are those enhanced via the resonant electronic transition of the chromophore being examined.

The Photosynthetic Bacterial Reaction Center II
Edited by J. Breton and A. Verméglio, Plenum Press, New York, 1992

We have recently been using near-infrared excited Fourier transform (FT) resonance Raman spectroscopy to study the vibrational and electronic structure of the primary electron donor in photosynthetic bacteria [6]. By using excitation at 1064 nm we have obtained, selectively, a preresonance Raman spectrum of the primary donor in its reduced state as well as a resonance Raman spectrum of the primary donor in its cation, radical state. The information obtained from these spectra shows that the primary electron donor is asymmetric in terms of its interactions with the protein as well as asymmetric in terms of the charge repartition among its BChl dimer components after oxidation. We present here FT Raman results from several BChl *a*-containing bacterial species for which crystal structures are not yet available, but whose primary sequences are strongly conserved in the P binding pocket. As compared to *Rb. sphaeroides*, the results indicate that both the asymmetric H-bonding patterns are conserved for P as well as the asymmetric + charge localization in $P^{+\cdot}$.

EXPERIMENTAL

Fourier transform Raman spectra were recorded using a Bruker IFS 66 interferometer coupled to a Bruker FRA Raman module equipped with a continuous, diode-pumped Nd:YAG laser as described in Reference 6. Approximately 200 mW of 1064 nm radiation was used to excite the Raman spectra of the reaction centers (concentration not less than 100 O.D.), contained in a sapphire cell, in the presence of ascorbate or ferricyanide. All spectra were recorded at room temperature and spectral resolution was 4 cm^{-1}.

RESULTS AND DISCUSSION

Excitation of RC Raman Spectra Using 1064 nm Radiation

The absorption spectrum of the reaction center from *Rb. sphaeroides* exhibits a broad band at ca. 860 nm arising from the two excitonically coupled BChl *a* molecules comprising the primary donor, P, in its reduced state. This band corresponds to the first/lowest excited singlet state of the primary donor, ^1P, and is the precursor state to eventual photo-induced charge separation. Excitation of the Raman spectrum of these RCs in their reduced state using 1064 nm radiation results in the preresonance enhancement of the vibrational modes of P over those of the other chromophores and of the protein [6]. This results in the selective observation of the vibrational spectrum of P; the degree of preresonance enhancement is such that difference techniques are not required. When P undergoes chemical or photochemical one-electron oxidation, the 860 nm band bleaches and a new band at ca. 1250 nm, arising from $P^{+\cdot}$, appears. In this oxidation state, exciting the Raman spectrum of the RCs results in the observation of the resonance Raman spectrum of $P^{+\cdot}$; 1064 nm is ca. 1400 cm^{-1} in energy above 1250 nm, thus a genuine resonance condition with this electronic state occurs via a vibronic satellite of the 1250 nm band. Thus, with 1064 nm excitation, Raman bands which bleach upon oxidation of P should primarily correspond to those preresonantly enhanced via the ca. 860 nm absorption band of reduced P while the appearance of new Raman bands upon oxidation should correspond to $P^{+\cdot}$.

The P Preresonance Raman Spectrum

Figure 1 shows the Fourier transform (FT) Raman spectra, excited using 1064 nm radiation, of RCs from *Rb. sphaeroides* 2.4.1 with the primary donor in its reduced and oxidized states. Inspection of these two spectra clearly indicates which bands are attributable to P and which are attributable to $P^{+\cdot}$. In the reduced P spectrum, bands in the C_2 acetyl and C_9 keto carbonyl stretching region which are bleached upon oxidation of P are the 1622, 1653, 1680, and 1691 cm^{-1} bands; these bands are attributed to P in its neutral, ground state. It is estimated that the contributions of the reduced primary donor to this spectrum is ca. 70% [6]. There is a shoulder observed at 1663 cm^{-1} which seems not to bleach upon P oxidation. The fact that there are four distinct carbonyl bands, and not two, which are (preresonantly) enhanced in this spectral region of the reduced P spectrum is reflecting the dimeric character of the electronic state of P corresponding to the 865 nm transition. The similar intensities of these four bands show that both components of P, namely P_L and P_M, are preresonantly enhanced to the same degree, indicating that the excitonic excitation associated with the 860 nm absorption is delocalized over both P_L and P_M on a time scale faster than that of the vibrational Raman effect in condensed media (i.e. ca. 10^{-13} s).

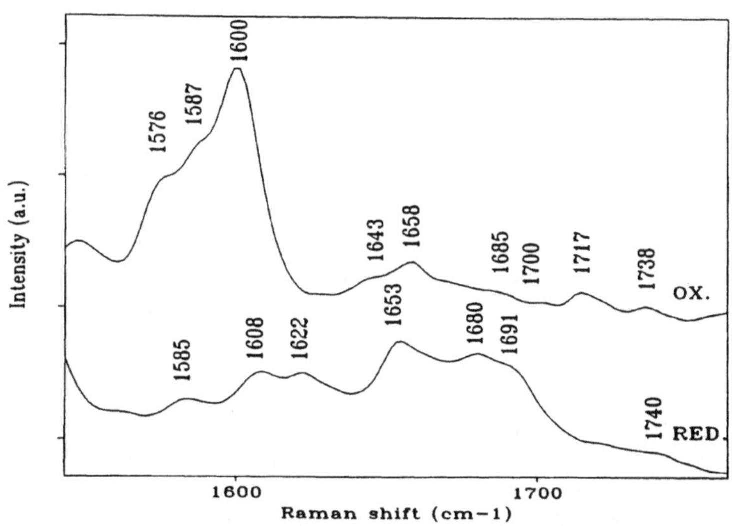

Figure 1. (Bottom) Fourier transform Raman spectrum of the RC from *Rb. sphaeroides* 2.4.1 in the presence of ascorbate (reduced RC); this spectrum is dominated by the preresonance contributions of the primary donor in its reduced state, P. (Top) Fourier transform resonance Raman spectrum of the RC in the presence of ferricyanide; this spectrum contains resonance Raman contributions of the primary donor in its oxidized state, $P^{+\cdot}$ and includes the bands at 1600, 1643, and 1717 cm^{-1}. Conditions: room temperature, 200 mW of 1064 nm excitation, coaddition of 1000 interferograms, 4 cm^{-1} resolution.

The 1680 and 1691 cm^{-1} bands are only consistent with C_9 keto carbonyl frequencies not engaged in H-bonding (the 1680 cm^{-1} band could be reflecting either a very weak H-bond interaction or a local region of high dielectric constant [7]). The 1653 and 1622 cm^{-1} bands are thus attributed to the two C_2 acetyl carbonyls of P, the latter indicating it is engaged in a strong H-bond while the former is free from such interaction. This asymmetric H-bonding pattern, as drawn from the FT Raman results here, is in agreement with the X-ray crystallographic structure of *Rb. sphaeroides* [3] where the histidine L168 residue, as in the X-ray structure of *Rps. viridis* [1], is engaged in a H-bond with the C_2 acetyl carbonyl of P_L; thus the 1622 cm^{-1} band is assigned to the C_2 acetyl carbonyl of P_L [6]. There is no candidate for H-bonding to the C_2 acetyl carbonyl of P_M in the *Rb. sphaeroides* structure.

The band at 1608 cm^{-1} is attributable to a C_aC_m methine bridge stretching mode. Its observed frequency and narrow band profile (14 cm^{-1} FWHM) fully confirm that both P_L and P_M possess one axial ligand on the central Mg atom [8].

The $P^{+}\cdot$ Resonance Raman Spectrum

Most of the new bands attributable to $P^{+}\cdot$ which appear in the FT Raman spectrum of oxidized RCs occur in the 1300-1700 cm^{-1} region (Fig. 1). Since 1064 nm is ca. 1400 cm^{-1} in the vibronic region of the 1250 nm absorption band of $P^{+}\cdot$, this resonance Raman spectrum, consistently, indicates that many in-plane skeletal BChl *a* modes are being resonantly enhanced at this wavelength. This observation constitutes strong indication that, regardless of the precise nature of the scattering mechanism(s) involved in the resonance, the electronic transition(s) that promote(s) $P^{+}\cdot$ resonant scattering at 1064 nm involve(s) significant in-plane character.

The most intense band in the FT Raman spectrum of oxidized RCs is at 1600 cm^{-1} (not including the carotenoid band). It most likely corresponds to the 1608 cm^{-1} band observed for neutral P. This C_aC_m mode thus seems to downshift by 7 cm^{-1} upon oxidation of P. A similar downshift has been observed in the one-electron oxidation of BChl *a* [9,10] and is consistent with the one-electron oxidation of metallo-porphyrins and chlorins with a_{1u}-like redox orbitals [11,12]. Thus the downshift of the C_aC_m mode upon RC oxidation is interpreted in terms of BChl *a* oxidation. Since the frequency of this mode is also senstive to the BChl core size and macrocycle conformation, its observed frequency may not accurately reflect the localization of the unpaired electron in $P^{+}\cdot$ [6].

In the carbonyl stretching region of the oxidized RC spectrum, two important bands can be unambiguously assigned to $P^{+}\cdot$, at 1643 and 1717 cm^{-1}. In this same region, other bands at 1658, 1685, and 1700 cm^{-1} cannot be exclusively assigned to $P^{+}\cdot$ because they may well have been masked in the reduced RC spectrum by the more intense P bands. The 1717 cm^{-1} band is attributed to a free C_9 keto carbonyl of $P^{+}\cdot$ which has upshifted with respect to neutral P; this shift can be either +26 or +38 cm^{-1} depending if it originates from the 1691 or the 1680 cm^{-1} band of P. Infrared studies indicated that upon one-electron oxidation of BChl *a*, the vibrational frequency of the C_9 keto carbonyl upshifts by +32 cm^{-1} [5]. Thus, the observed +26 or +38 cm^{-1} upshift of this keto carbonyl in the FT resonance Raman spectrum of $P^{+}\cdot$ suggests that the + charge in $P^{+}\cdot$ is predominantly localized on one of the two BChl components. Indeed, if the + charge were equally shared by both P_L and P_M one would expect a smaller upshift, approximately one-half the shift observed in monomeric BChl *a* upon its one-electron oxidation (i.e. ca. +16 cm^{-1}). This result indicates that the unpaired electron in $P^{+}\cdot$ for *Rb. sphaeroides* does not (fully) share a common redox orbital of P_L and P_M on the time scale of the resonance Raman effect.

The other $P^{+\cdot}$ band at 1643 cm^{-1} exhibits a vibrational frequency which, based on *in vitro* BChl *a* oxidation studies [5], is only consistent with the C_2 acetyl carbonyl of P which has upshifted from the 1622 cm^{-1}. This assignment, then, indicates that the hole is primarily localized on the P_L component to which the 1622 cm^{-1} band has been attributed [6].

In a molecular orbital picture where P_L and P_M are interacting in the $P^{+\cdot}$ state, it would appear that the electron is removed from a dimer molecular orbital comprised primarily of a P_L molecular orbital. This could arise from inequivalent protein environments and/or conformations of P_L and P_M which could result in slightly different redox orbital energies for these molecules. It is difficult to assess the relative magnitudes of the coefficients of the dimer redox orbital, but if we assume that the observed magnitude of the shift of the keto carbonyl (+26 cm^{-1} for P as compared to +32 cm^{-1} for the monomer) is proportional to the hole density, it may be estimated that the orbital of P_L contributes ca. 80% to the dimer orbital. The absence or weakness of any $P_M{}^+$ contributions may be explained by a difference in resonance Raman scattering cross sections of the $P_L{}^+$ and $P_M{}^+$ species at 1064 nm.

Structure of P and $P^{+\cdot}$ from Other Species

We have obtained the FT Raman spectra of reaction centers from other photosynthetic purple bacteria containing BChl *a* in order to see if these vibrational and electronic structural asymmetries of the primary donor are conserved. Table I summarizes these. In general, the FT Raman spectra of these RCs in their reduced and oxidized states are very similar. Although the X-ray crystal structures are not available, the amino acid sequences of *Rb. capsulatus* and *Rsp. rubrum* [13] in the vicinity of the primary donor indicate that the H-bonding patterns are not expected to be different; this is quite evident from a comparison of the C_2 and C_9 carbonyl frequencies in Table I. The primary sequence of *Rhodocyclus* (*Rc.*) *gelatinosus* (see Agalidis *et al.*, in this Volume) is not yet known, however, the high degree of similarity of the primary donor FT Raman spectra strongly indicates that its structure is very similar to that of *Rb. sphaeroides* and the others in Table I. Indeed for all the primary donors in Table I, the FT Raman data indicate that there is a strong H-bond to one of the two C_2 acetyl carbonyls of P which is most likely that of P_L originating from a histidine ligand in position L168, using the *Rb. sphaeroides* numbering scheme. The other C_2 acetyl carbonyl and the other two C_9 keto carbonyls are free from such interactions. Interestingly, the only vibrational frequency which appears to show any slight but significant variation is that of the H-bonded C_2 acetyl; this could be indicating minor changes in the geometry of the histidine residue and the carbonyl which would slightly alter the strength of the H-bond but not rupture it.

The FT Raman spectra of the oxidized primary donors are also all very similar among those in Table I. In particular, each of these spectra exhibits a band at ca. 1717 cm^{-1} which is a marker mode for $P^{+\cdot}$. The observed shift of this band upon oxidation of P is very similar for all five types of RCs and thus the estimate localization of the + charge in $P^{+\cdot}$ is also quite similar, ca. 80% and probably on the P_L component for these species.

From these results it appears clear that the charge repartition in $P^{+\cdot}$ is highly conserved among the bacterial RCs studied here. In all these RCs, the H-bonding pattern of the primary donor (at the level of the conjugated C_2 and C_9 carbonyls), as well as the kinetics and quantum yields of the primary light reactions, are very similar. In order to assess how this asymmetric charge localization relates to structural asymmetry in the P protein pocket and to the charge separation, we need to study the structure of the primary

TABLE I. Observed Carbonyl Stretching Frequencies (cm^{-1}) of P and $P^{+\cdot}$ from Several BChl *a*–Containing Purple Bacteria

	P				$P^{+\cdot}$			
					C_2	C_9	C_9 shift	$\%P_L^+$
Rb. sphaeroides R26	1620	1653	1679	1691	1641	1717	+26	80
Rb. sphaeroides 2.4.1	1622	1653	1680	1691	1643	1717	+26	80
Rb. capsulatus	1624	1654	1680	1691	1645	1718	+27	80
Rsp. rubrum G9	1618	1653	1678	1691	1638	1717	+26	80
Rc. gelatinosus	1616	1653	1682	1696	1641	1720	+24	75
carbonyl	C_2	C_2	C_9	C_9				
molecule	P_L	P_M	P_M	P_L				
H-bond donor	His L168	none	none	none				

donor in RCs where the H-bonding pattern is different (i.e. *Rps. viridis*, and site-directed mutants) and those exhibiting different electron transfer kinetics (i.e. *Chloroflexus aurantiacus*). Such FT Raman studies are now in progress.

REFERENCES

1. J. Deisenhofer and H. Michel, *EMBO J. 8*:47 (1989).
2. J.P. Allen, G. Feher, T.O. Yeates, H. Komiya, and D. Rees, *Proc. Natl. Acad. Sci. U.S.A. 84*:5730 (1987); *84*:6162 (1987).
3. O. El-Kabbani, C.-H. Chang, D. Tiede, J. Norris, and M. Schiffer, *Biochemistry 30*:5361 (1991).
4. M.E. Michel-Beyerle, M. Plato, J. Deisenhofer, H. Michel, M. Bixon, and J. Jortner, *Biochim. Biophys. Acta 932*:52 (1988).
5. W.G. Mäntele, A.M. Wollenwebber, E. Nabedryk, and J. Breton, *Proc. Natl. Acad. Sci. U.S.A. 85*:8468 (1988).
6. T.A. Mattioli, A. Hoffmann, B. Robert, B. Schrader, and M. Lutz, *Biochemistry 30*:4648 (1991).
7. S. Krawczyk, *Biochim. Biophys. Acta 976*:140 (1989).
8. B. Robert, *Biochim. Biophys. Acta 1011*:99 (1990).
9. M. Lutz and J. Kléo, *Biochim. Biophys. Acta 564*:365 (1979).
10. T.M. Cotton and R.P. Van Duyne, *J. Am. Chem. Soc. 103*:6020 (1981).
11. A. Salehi, W.A. Oertling, H.N. Fonda, G.T. Babcock, and C.K. Chang, *Photochem. Photobiol. 48*:525 (1988).
12. T.G. Spiro, R.S. Czernuszewicz, and X.-Y. Li, *Coord. Chem. Rev. 100*:541 (1990).
13. H. Komiya, T.O. Yeates, D.C. Rees, J.P. Allen, and G. Feher, *Proc. Natl. Acad. Sci. U.S.A. 85*:541 (1988).

RHODOCYCLUS GELATINOSUS REACTION CENTER: CHARACTERIZATION OF THE QUINONES AND STRUCTURE OF THE PRIMARY DONOR

Ileana Agalidis,[1] Bruno Robert,[2] Tony Mattioli,[2]
and Françoise Reiss-Husson,[1]

[1]UPR 407, CNRS, 91198 Gif-sur Yvette, France
[2]Dpt Biologie Cellulaire et Moléculaire, CE Saclay, 91191 Gif, France

INTRODUCTION

Rhodocyclus gelatinosus belongs to the many species of photosynthetic bacteria which contain a reaction center tightly associated with a cytochrome subunit. This was shown early by the observation of haem photooxidation down to 77 K[1] and confirmed more recently by studies of photoreceptor units consisting of the reaction center , a tetrahaem cytochrome and B875 antenna[2,3] . Unlike *Rhodopseudomonas viridis*, the cytochrome subunit is loosely bound to the reaction center and becomes detached during solubilization[2] (and unpublished results). Such behavior could, in principle, open the way to studies of interactions between isolated reaction center and tetrahaem cytochrome. These experiments however are hampered by the difficult isolation of *Rhodocyclus gelatinosus* reaction centers in a native form. Indeed, the H subunit was lost and only LM subunits were isolated when either LDAO[4] or octylthioglucoside and Triton[2] were used as detergent.

Recently we improved the reaction center isolation procedure[5,6]. A purified preparation still containing the H subunit, and displaying a stable photochemical activity, was obtained[6]. In this article we will compare some functional properties of this reaction center preparation with reference to *Rhodopseudomonas viridis* and *Rhodobacter sphaeroides.* Similarities are found with the former at the level of the acceptor quinone complex; the structure of the primary donor in both the reduced and oxidized states, as determined by Fourier transform resonance Raman spectroscopy, resembles the latter.

METHODS

Instability of photochemical activity in presence of a number of detergents (including lauryl-dimethyl amineoxide) compelled us to devise a more gentle isolation method[5,6].

The Photosynthetic Bacterial Reaction Center II
Edited by J. Breton and A. Verméglio, Plenum Press, New York, 1992

Photoreceptor units were dissociated by overnight incubation at 5°C with a mixture of two detergents, octylthioglucoside and decyltetraethylene oxide, followed by phase separation at 20°C. After a short centrifugation, crude reaction centers depleted in antenna and cyt c were recovered in the supernatant; further purification was done by ion-exchange chromatography. The reaction center fraction contained the three polypeptides H, M and L (apparent molecular weights resp. 38, 28 and 24 kDa), and residual impurities of higher M_{app}. Best purity index (A280nm/A800nm) was 1.30.

Kinetic measurements were done as described[6]. For Fourier transform resonance Raman experiments, reaction centers were pelleted by ultracentrifugation at 200000 g during 24 hrs. Fourier transform (FT) Raman spectra were recorded using a Bruker IFS 66 interferometer coupled to a Bruker FRA 106 Raman module, as described in Ref. 17. Approximately 200 mW of 1064 nm radiation was used to excite the Raman spectra of the RCs in their reduced (in the presence of ascorbate) or oxidized (in the presence of ferricyanide) states. All spectra were recorded at room temperature and were the result of the coaddition of 2000 interferograms. Spectral resolution was 4 cm^{-1}.

CHARACTERIZATION OF QUINONE ACCEPTORS

Flash-induced absorbance changes measured at 780nm or 865nm on a purified reaction center fraction decayed in the dark as a monoexponential with a mean lifetime of 35 ms; addition of herbicide did not modify the observed kinetic trace[6]. This observation implies that Q_B has been removed during reaction center purification and therefore the decay is exclusively due to $P^+Q_A^-$.

Previous studies of the photoreceptor units led us to conclude that the first quinone acceptor in *Rc. gelatinosus* is menaquinone 8[3,7]. This was confirmed by measuring on reaction center fraction the differential absorption spectrum (Q_A minus Q_A). In the presence of diaminodurol which is a fast electron donor to P^+, the lifetime of Q_A could be extended to several seconds; the measurements were done 35-50 ms after the flash, when P^+ was already rereduced by DAD and not interfering with the semiquinone signal (Fig.1).

Between 380 and 500 nm the complicated profile is clearly similar to that reported for semimenaquinone anion in the *Rps. viridis* reaction center[8]. The maximum is at 412 nm, which corresponds to a shift of the 395 nm semimenaquinone anion band observed for menaquinone in vitro[9]. Secondary maxima and shoulders may arise from electrochromic shifts of nearby pigments. In the near infrared, the absorbance change is similar to that ascribed in *Rb. sphaeroides* [10] to the electrochromic shift of $Bpheo_A$ Q_y band, induced by the negative charge on the quinone; this change is independent of the nature of the semiquinone.

Despite the removal of the secondary quinone the binding site of Q_B was still native. In presence of an excess of various ubiquinones, a large decrease of P^+ rereduction rate after a flash indicated that they all could function as secondary acceptors. With UQ_6, UQ_9 and UQ_{10}, the back reaction had a major exponential phase with a mean time of 1-2.3 s, without any contribution of the 35 ms fast phase; full reconstitution of Q_B thus took place. Activity of Q_B as a two electron gate was demonstrated by observation of binary oscillations of absorbance changes measured at 450 nm in a series of saturating flashes, in presence of UQ_{10} and DAD. These indicated the alternative formation of semiquinone and quinol [6].

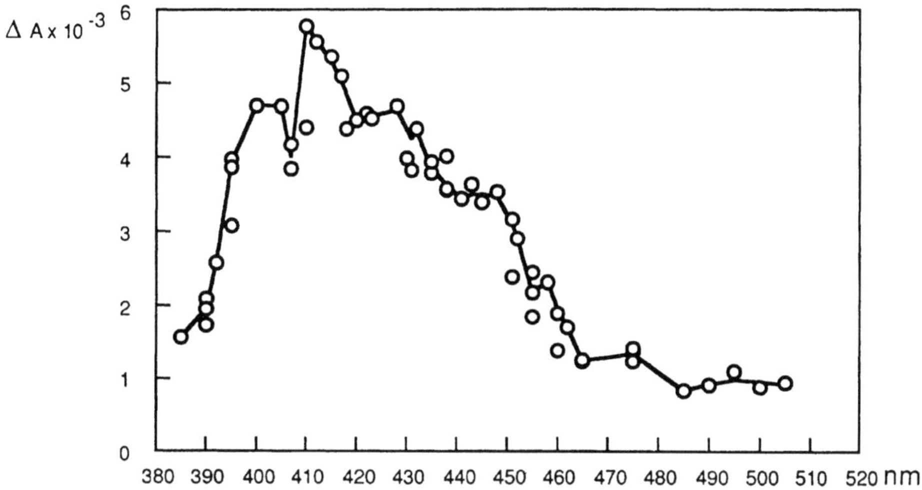

Fig.1. Flash-induced difference spectrum of Q_A^-. Signals were measured 35 ms after the flash. Spectral band pass: 3 nm. RCs (3 μM) were suspended in 10mM Tris buffer, pH 8.0, containing 1 mM EDTA, 0.2 M NaCl, 0.5 mg/ml $C_{10}E_4$, 7.8% glycerol and 500 μM DAD.

In contrast, menaquinone MK8 did not function as a secondary acceptor, the rate of recombination reaction after a flash being, in its presence, as fast as that measured in Q_B-less reaction center. Thus, as in *Rps. viridis*[8], the Q_B site binds preferentially ubiquinones when the Q_A site is occupied by the native menaquinone species. This probably is due more to the low redox potential of menaquinones rather than to steric factors. In *Rb. sphaeroides*, the functional occupancy of the Q_B site is governed by the free energy difference between the quinones respectively bound at the Q_A and Q_B sites[11]. In *Rc. gelatinosus*, if MK were bound to the Q_B site, this free energy difference between Q_A and Q_B (both MK8) would probably be too small.

The value of the apparent equilibrium constant K_{2app} of the electron transfer between Q_AQ_B and Q_AQ_B may be deduced from the P^+ rereduction rate after a flash by Q_A and Q_B respectively[12]. When the Q_B site is occupied by UQ10, this value is about 60, and may be compared with those measured in purified reaction centers of *Rb. sphaeroides* (K_{2app}=16 [13]) and *Rps. viridis* (K_{2app}=100, pH 9[8]). The redox energy gap between primary and secondary quinones in *Rc. gelatinosus* is thus of the same order of magnitude as in *Rps. viridis*, and significantly higher than in *Rb. sphaeroides*. Assuming that for the primary quinone the E_m value is -140 mv as in *Rc. gelatinosus* chromatophores[14], and taking for Q_B (UQ10) E_m=40 mv at pH 8 as in *Rb.sphaeroides* [15], the redox potential difference may be estimated to 200 mv.

STRUCTURE OF THE DIMER IN THE REDUCED AND OXIDIZED STATES

A typical absorption spectrum of purified reaction center is given in Fig. 2. It is similar to that of *Rb. sphaeroides* reaction center, indicating the same Bchl and Bpheo contents. The

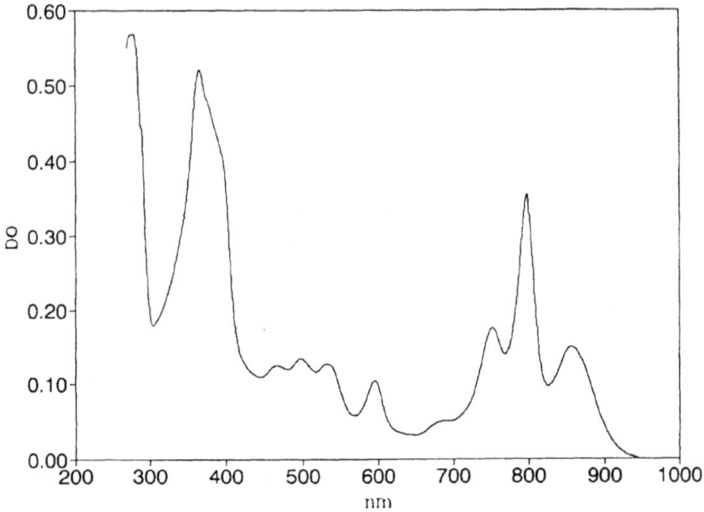

Figure 2. Absorption spectrum of purified reaction centers, suspended in the same buffer as in Fig.2, except for DAD.

Soret region is dominated by the strong contributions of these pigments, as expected from the very low residual cyt c content (< 0.1 heme per reaction center). The only difference concerns the bound carotenoid, which visible absorption bands are blue-shifted by about 7 nm as compared to 15-15' cis-spheroidene in *Rb. sphaeroides*. In addition, oxidized reaction centers (in presence of ferricyanide) displayed a weak absorption band centered at 1245nm (not shown) attributed as in *Rb. sphaeroides* [16] to a P^{+} transition.

We have obtained the near-infrared Fourier transform Raman spectra of the primary donor of *Rc. gelatinosus* in its reduced and oxidized states and have compared them with those of *Rb. sphaeroides*. Using 1064 nm excitation, i) the reresonance Raman spectrum of the primary donor in its reduced state (P) via its 865 nm absorption band, and ii) the resonance Raman spectrum of P in its cation, radical state ($P^{+\cdot}$) via its 1245 nm absorption band, are selectively observed. The details of these resonance enhancements and the selective observation of the primary donor in these two redox states are described for the specific case of *Rb. sphaeroides* elsewhere[17].

Table I. Observed Carbonyl Stretching Frequencies (cm^{-1}) of P in *Rc. gelatinosus* Compared to Those for *Rb. sphaeroides*

gelatinosus[a]	*sphaeroides*[b]	carbonyl	molecule[c]	H-bond donor[d]
1616	1620	acetyl	P_L	His L168
1654	1653	acetyl	P_M	none
1680	1679	keto	P_M	none
1696	1691	keto	P_L	none

[a] This work. [b] From Ref. 17. [c] Assignments from Ref. 17. [d] H-bond donor candidates from Ref. 18.* Estimated from Fourier deconvolution.

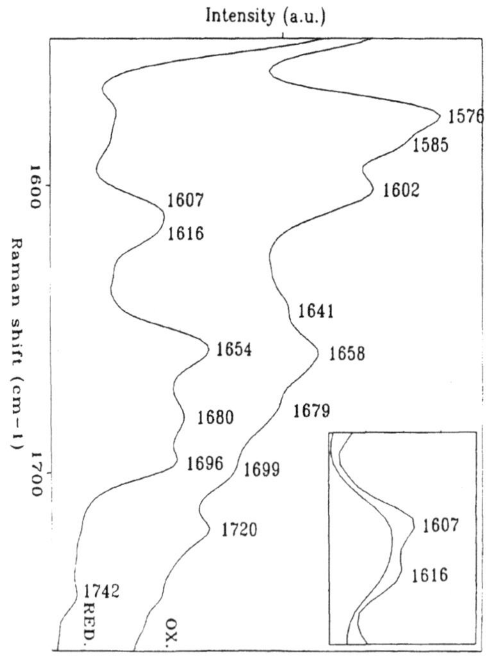

Figure 3. (Bottom) Fourier transform Raman spectrum of the RC from *Rc. gelatinosus* in the presence of ascorbate (reduced RC); this spectrum is dominated by the preresonance Raman contributions of the primary donor in its reduced state, P. (Top) Fourier transform resonance Raman spectrum of the RC in the presence of ferricyanide; this spectrum is dominated by the resonance Raman contributions of the primary donor in its oxidized state, $P^{+\cdot}$. Obvious new bands in this spectrum include those at 1602, 1641, and 1720 cm^{-1} which are all atributable to $P^{+\cdot}$ (see text for discussion). Spectral resolution is 4 cm^{-1}. Conditions: room temperature, 200 mW of 1064 nm excitation, coaddition of 2000 scans. Inset: Fourier deconvolution of the complex band at ca. 1610 cm^{-1} in the reduced P spectrum showing components at 1607 and 1616 cm^{-1}.

In general, the P and $P^{+\cdot}$ FT Raman spectra of *Rc. gelatinosus* and *Rb. sphaeroides* are quite similar indicating that in these species the main structural features of the dimer are similar; the results are summarized in Figure 3 and Table I.

For *Rb. sphaeroides*, the FT Raman spectrum of reduced P exhibits a narrow (ca. 14 cm^{-1} FWHM) band at 1607 cm^{-1} corresponding to the C_aC_m methine bridge stretching mode. This mode is sensitive to the number of axial ligands on the central Mg atom of the BChl molecules comprising P. The width and frequency of this band indicates that both BChl *a* components of the P dimer possess only one axial ligand each. As well for *Rb. sphaeroides*[17], there is a 1620 cm^{-1} band observed in the reduced P spectrum which arises from a H-bonded acetyl carbonyl group. From the X-ray crystal structure of the *Rb. sphaeroides* reaction center[18], the histidine L168 residue is the candidate for this H-bonding to the acetyl carbonyl of the P_L component of P. For the *Rc. gelatinosus* P spectrum, these corresponding bands are not spectrally resolved and resulted in a complex band at ca. 1610 cm^{-1}. Fourier deconvolution in this spectral region clearly indicated a 1607 cm^{-1} component and a 1616 cm^{-1} component. Thus, as in the case of *Rb. sphaeroides*, both

BChl *a* components of the primary donor of *Rc. gelatinosus* also possesses one axial ligand each, and furthermore, an acetyl carbonyl of P is also engaged in a H-bond, as indicated by the 1616 cm^{-1} component; the fact that this frequency is estimated to be 4 cm^{-1} lower in frequency than that of the corresponding band for *Rb. sphaeroides* could be indicating a stronger H-bond for the case of *Rc. gelatinosus*. In the absence of amino acid sequence data, presently it can only be speculated that the axial ligands to the primary donor are histidine residues. Similarily, it is not yet known if the histidine L168 residue in *Rb. sphaeroides* is conserved in *Rc. gelatinosus* ; the conservation of this residue would place it as the prime candidate for a H-bond donor of to the acetyl carbonyl of P_L.
The observed frequencies of the other acetyl carbonyl and two other keto carbonyl modes in the *Rc. gelatinosus* P spectrum indicate that no other H-bonds are associated with these carbonyls (see Table I). Thus, the assymetric H-bonding pattern of the C_2 acetyl and C_9 keto carbonyls of P in *Rc. gelatinosus* is the same as that of P in *Rb. sphaeroides*.
The $P^{+\cdot}$ spectra of *Rc. gelatinosus* and *Rb. sphaeroides* are also very similar as seen in Table II.

Table II. Observed frequencies (cm^{-1}) of $P^{+\cdot}$ in *Rc. gelatinosus* compared to those for *Rb. sphaeroides*

gelatinosus	*sphaeroides*	group	molecule	shift from P	
1602	1600	C_aC_m		-5	-7
1641	1641	acetyl	P_L	+25	+21
1720	1717	keto	P_L	+24	+26

The upshift of the C_9 keto carbonyl of P upon oxidation to $P^{+\cdot}$ indicates that for *Rc. gelatinosus*, as in the case of *Rb. sphaeroides*[17], the +1 charge appears to be primarily localized on one of the two BChl *a* components of P. For *Rb. sphaeroides* the +1 charge was estimated to be ca. 80% localized on the P_L component. Although it is not possible at present to conclude on which component of P in *Rc. gelatinosus* the +1 charge is localized, the FT Raman data indicate that the charge is primarily localized on one BChl *a* component (probably P_L) to an extent similar as that in *Rb. sphaeroides*.

In conclusion, the strong similarites of the P and $P^{+\cdot}$ FT Raman spectra of *Rc. gelatinosus* compared to those of *Rb. sphaeroides* indicate that in these species the main structural features of the primary donor are the same. More thorough interpretation of the minor differences in the FT Raman spectra of these two reaction centers will await a more detailed characterization of the *Rc. gelatinosus* reaction center (i.e. primary sequence data).

REFERENCES

1. T. Kihara and B. Chance, Biochim. Biophys. Acta 189:116 (1969)
2. A. Fukushima, K. Matsuura, K. Shimada and T. Satoh, Biochim. Biophys. Acta 933:399 (1988).
3. I. Agalidis, E. Rivas and F. Reiss-Husson, Photosynth. Res. 23:249 (1990).
4. R.G. Prince, P.L. Dutton, B.J. Clayton and R.K. Clayton, Biochim. Biophys. Acta 502:354 (1978).
5. I. Agalidis and F. Reiss-Husson, Biochem. Biophys. Res. Commun. 177:1107 (1991).

6. I. Agalidis and F. Reiss-Husson, Biochim. Biophys. Acta 1098:201 (1992).
7. I. Agalidis, E. Rivas and F. Reiss-Husson, Z. Naturforsch. 46c:99 (1991).
8. R.J. Shopes and C. A. Wraight, Biochim. Biophys. Acta 806:458 (1985).
9. P.S. Rao and E. Hayon, Biochim. Biophys. Acta 191:516 (1983).
10. A. Vermeglio and R.K. Clayton, Biochim. Biophys. Acta 461:159 (1977).
11. K.M. Giangiacomo and P.L. Dutton, Proc. Natl. Acad. USA 86:2658 (1989).
12. C.A. Wraight and C.A. Stein, in "The Oxigen Evolving System of Photosynthesis" Y. Inoue et al.,eds. Academic Press, New York (1983).
13. L.J. Mancino, D.P. Dean and R.E. Blankenship, Biochim. Biophys. Acta, 764:46 (1984).
14. P.P. Dutton, Biochim. Biophys. Acta, 226:63 (1971).
15. A.W. Rutherford and M.C.W. Evans, FEBS Lett 110:257 (1979).
16. P.L.Dutton, K.J. Kaufmann, B. Chance and P.M. Rentzepis, FEBS Lett. 60:275 (1975).
17 T.A. Mattioli, A. Hoffmann, B. Robert, B. Schrader, and M. Lutz, Biochemistry 30:4648 (1991).
18. O. El-Kabbani, C.-H. Chang, D. Tiede, J. Norris, and M. Schiffer, Biochemistry 30:5361 (1991).

FTIR CHARACTERIZATION OF LEU M160→HIS, LEU L131→HIS AND HIS L168→PHE MUTATIONS NEAR THE PRIMARY ELECTRON DONOR IN *RB. SPHAEROIDES* REACTION CENTERS

E. Nabedryk and J. Breton
SBE/DBCM, CEN Saclay, 91191 Gif-sur-Yvette, France

J. Allen, H. Murchison, A. Taguchi, J. Williams and N. Woodbury
Dept of Chemistry and Biochemistry and
Center for the Study of Early Events in Photosynthesis
Arizona State University, Tempe, USA 85287-1604

Light-induced FTIR difference spectroscopy has been used to monitor structural changes in *Rb. sphaeroides* mutant reaction centers (RCs) associated with the substitution of amino acids near the primary electron donor (P). In the wild type *Rb. sphaeroides* RC, the 9keto carbonyls for both BChls constituting P (P_L and P_M) have no specific interactions with the protein. A hydrogen bond probably exists between the 2a acetyl C=O of P_L and His L168[1,2] but no such bond is possible with the symmetry related amino acid on the M side, a Phe residue (M197). The mutations Leu L131→His and Leu M160→His[3] (see also Williams et al., these proceedings) were designed to introduce a proton donating residue that could form a hydrogen bond with the keto C=O of ring V of each BChl of the dimer. In addition, the mutation His L168→Phe was designed to break a hydrogen bond between the 2a C=O of P_L and His L168.

The construction and initial characterization of the mutants are described elsewhere (ref. 3 and Williams et al., these proceedings). Light-induced FTIR difference spectra of purified chromatophores are performed as described previously[4].

The Photosynthetic Bacterial Reaction Center II
Edited by J. Breton and A. Verméglio, Plenum Press, New York, 1992

Fig. 1 shows the $P^+Q_A^-/PQ_A$ FTIR difference spectra at 100K for *Rb. sphaeroides* chromatophores of wild type (Wt) (1a), Leu M160→His (1b), Leu L131→His (1c), and His L168→Phe (1d). The strong bands evident at 1752/1740 cm^{-1} and 1705/1683 cm^{-1} in the $P^+Q_A^-/PQ_A$ spectrum of Wt chromatophores have been previously correlated to a predominant contribution of the 10a ester and 9keto carbonyls, respectively, of the P_L moiety[5]. For the Leu M160→His and Leu L131→His mutants, dramatic changes are observed in the 1660 to 1720 cm^{-1} region in both the amplitudes and the frequencies of the bands. Two large differential signals at 1718/1696 cm^{-1} and 1678/1664 cm^{-1} are found in Leu M160→His while in Leu L131→His, a complex structured positive signal is observed between 1678 cm^{-1} and 1710 cm^{-1}. These changes indicate a very large perturbation of the environment of the keto C=O groups of the BChl dimer or, at least, a perturbation of the charge density on ring V in P and P^+. Positive bands at ≈1290 cm^{-1}, 1500–1430 cm^{-1} and 1580–1530 cm^{-1}, characteristic of the dimeric BChl state of P^+ in Wt RCs[5], are still present in the spectra of the mutants. However, frequency shifts and/or amplitude changes are detectable, for example the ≈1290 cm^{-1} band is shifted to 1309 cm^{-1} in Leu M160→His and a band is seen at ≈1560 cm^{-1} in Leu L131→His. In addition, while the Leu L131→His mutant exhibits the broad absorption band at ≈2600 cm^{-1} that is typical for a dimer of BChl in the P^+ state of Wt *Rb. sphaeroides* RCs[6], the corresponding band in the Leu M160→His spectrum is centered at ≈2800 cm^{-1} with a reduced amplitude (data not shown). This is attributed to differences in the structure of the dimer and/or changes in interactions induced by the protein environment. These data suggest a redistribution of the unpaired electron on P_L and P_M in the Leu M160→His mutant. Significant distribution changes are evident from ENDOR measurements that show 84% of the electron spin density residing on P_L compared to 68% for the Wt (Rautter et al., these proceedings).

Hydrogen bonding results in specific shifts of carbonyls bands. Based upon *in vitro* studies of the electrochemically generated BChla radical cation[7], it is likely that the 1718/1696 cm^{-1} is due to a non–bonded 9keto C=O and the 1678/1664 cm^{-1} signal arises from a strongly hydrogen–bonded 9keto C=O of P in the Leu M160→His spectrum. Small signals at 1677 cm^{-1} and 1665 cm^{-1} are always present in the Wt spectrum. On the basis of polarized light–induced spectroscopy on oriented films of *Rb. sphaeroides* RCs[8], it has been established that the 1665 cm^{-1} signal cannot be identified as a keto C=O mode and is more probably related to a peptide C=O in the environment of P. Thus, the 1678/1664 cm^{-1} signal in the mutant could also be interpreted in terms of the superposition of the contributions from peptide and keto carbonyls. Using this hypothesis, part of the 1678/1664 cm^{-1} signal is assigned to the keto C=O of P_M hydrogen bonded to His M160 while the keto of P_L absorbs at a value of 1718/1696 cm^{-1} which is shifted to higher frequencies compared to Wt. Such an upshift for the neutral keto of P_L has also been observed at 1696 cm^{-1} in the His M200→Leu heterodimer[5]. We cannot exclude the alternate possibility that the whole 1678/1664 cm^{-1} signal is due to a light–induced protein amide I change.

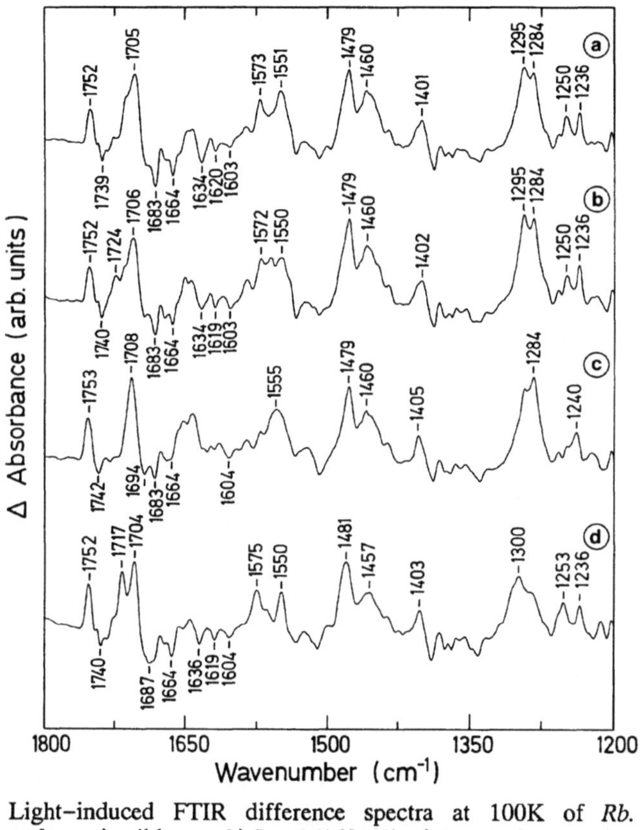

FIGURE 1. Light–induced FTIR difference spectra at 100K of *Rb. sphaeroides* chromatophores from a) wild type, b) Leu M160→His c) Leu L131→His d) His L168→Phe. 4 cm^{-1} resolution.

In the Leu L131→His spectrum (Fig. 1c), the keto C=O region is also drastically perturbed. A small but broad positive peak is observed at 1710 cm^{-1} with shoulders at 1692 cm^{-1} and 1678 cm^{-1}. The negative signal at 1665 cm^{-1} is still present, as well as a very small signal at 1684 cm^{-1}. The much reduced amplitude of this signal compared to the large negative signal occurring at 1683 cm^{-1} in Wt is presumably due to the formation of a strong hydrogen bond between His L131 and the 9keto of P_L. In addition, the 10a ester C=O signal at 1752/1740 cm^{-1} in Wt is upshifted to 1753/1744 cm^{-1} in Leu L131→His (it is unchanged in Leu M160→His), indicating a perturbation of the whole ring V of P_L in both P and P^+ states. A hydrogen bond between the Ser L244 hydroxyl side chain and the 10 ester C=O of P_L has been proposed from the X-ray structure of *Rb. sphaeroides* RC[2]. The present IR data on the Leu L131→His mutant suggest a breaking of this (weak) hydrogen bond.

The light-induced $P^+Q_A^-/PQ_A$ FTIR difference spectra at 100K for *Rb. sphaeroides* chromatophores of the His L168→Phe mutant is displayed on Fig. 1d. Although the spectra of this mutant and Wt are very similar, some differences are observed. In the ≈1290 cm^{-1} region, a 1283 cm^{-1} component is the major peak in the mutant. The differential 9keto signal is found at 1705/1681 cm^{-1}, and the positive shoulder observed at 1713 cm^{-1} in Wt is reduced in this mutant. In addition, the 10a ester signal is downshifted to 1751/1738 cm^{-1}. It therefore appears that in the His L168→Phe mutant, changing the 2a acetyl environment of P_L perturbs the vibrational properties of at least the 9keto carbonyl. It is worth noting that the $P^+Q_A^-/PQ_A$ FTIR difference spectrum at 100K for *Rb. sphaeroides* chromatophores containing the Phe M197→Tyr mutation (see Nabedryk et al., these proceedings) also indicates a strong perturbation of the keto groups with a large splitting of the signal in the P^+ state. In addition, while a small negative band at 1620 cm^{-1} is present in Wt (Fig. 1a), it is absent in His L168→Phe (Fig. 1d). In Wt, this band could correspond to the 2a acetyl C=O of P_L hydrogen-bonded to His L168. This hydrogen bond would be lost in His L168→Phe.

CONCLUSION

Large changes in the FTIR spectra have been observed that are associated with changes of amino acids at single sites near the primary donor. These changes support the conclusion that a hydrogen bond has been introduced to the dimer at the 9keto group of P_L and P_M due to the mutations Leu L131 to His and Leu M160 to His, respectively. Small changes in the C=O frequency region are observed in the spectrum of RCs with the mutation His L168 to Phe, which should cause the loss of a hydrogen bond to the 2a acetyl of P_L. In addition, the broad absorption band at 2600 cm^{-1} in wild type is present in the spectrum of Leu L131→His mutant but shifted to 2800 cm^{-1} for Leu M160→His mutant. One interpretation of this shift is a redistribution of the unpaired electron as has been shown by ENDOR measurements (Rautter et al., these proceedings).

ACKNOWLEDGMENTS. This work was in part supported by grants GM41300 and GM45902 from the N.I.H., grants DMB89–177729 and DMB91–58251 from the N.S.F. This is publication #109 of the Arizona State University Center for the Study of Early Events in Photosynthesis. The Center is funded by D.O.E. grant DE–FG–88–ER13969.

REFERENCES

1. J.P. Allen, G. Feher, T.O. Yeates, H. Komiya, and D.C. Reeves, *Proc. Natl. Acad. Sci. USA* 84:5730 (1987).
2. O. El–Kabbani, C.–H. Chang, D. Tiede, J. Norris, and M. Schiffer, *Biochemistry* 30:5361 (1991)
3. J.C. Williams, R.G. Alden, H.A. Murchison, J.M. Peloquin, N.W. Woodbury, and J.P. Allen, *Biochemistry,* submitted.
4. E. Nabedryk, K.A. Bagley, D.L. Thibodeau, M. Bauscher, W. Mäntele, and J. Breton, *FEBS Lett.* 266:59 (1990).
5. E. Nabedryk, S.J. Robles, E. Goldman, D.C. Youvan, and J. Breton, *Biochemistry*, in press.
6. J. Breton, E. Nabedryk, and W.W. Parson, *Biochemistry*, in press (1992).
7. W.G. Mäntele, A.M. Wollenweber, E. Nabedryk, and J. Breton, *Proc. Natl. Acad. Sci. USA* 85:8468 (1988).
8. D.L. Thibodeau, E. Nabedryk, and J. Breton in: Spectroscopy of Biological Molecules, pp. 69–70, R.E. Hester and R.B. Girling, eds., The Royal Society of Chemistry, Cambridge (1991).

FTIR SPECTROSCOPY OF THE $P^+Q_A^-/PQ_A$ STATE IN MET L248→THR, SER L244→GLY, PHE M197→TYR, TYR M210→PHE, TYR M210→LEU, PHE L181→TYR AND PHE L181-TYR M210→TYR L181-PHE M210 MUTANTS OF *RB. SHAEROIDES*

E. Nabedryk and J. Breton
SBE/DBCM, CEN Saclay, 91191 Gif-sur-Yvette, France

J. Wachtveitl, K. A. Gray and D. Oesterhelt
Max-Planck-Institut für Biochemie, 8033 Martinsried, FRG

There are several differences concerning the amino acids environment surrounding the primary electron donor (P) in the bacterial reaction center (RC) of *Rb. sphaeroides and Rps. viridis.* In *Rps. viridis* RC[1], the keto carbonyl group of ring V of P_L forms a hydrogen-bond to Thr L248. The equivalent residue in *Rb. sphaeroides*[2] is Met that cannot hydrogen-bond. In both RCs, the corresponding residue on the M side is the conserved Ile M282. Thus, in *Rb. sphaeroides* RC, the 9keto C=O for both P_L and P_M are not interacting with the protein[3,4]. In contrast, the 10a ester C=O of ring V of P_L in *Rb. sphaeroides* is hydrogen-bonded to Ser L244[4] which is replaced by a Gly residue in *Rps. viridis*. In *Rps. viridis*, the ring I acetyl carbonyls of P_L and P_M are bound to His L168 and Tyr M195 side chains, respectively[1]. This symmetry in hydrogen-bonding is not preserved in *Rb. sphaeroides* as the residue at the equivalent M197 position is Phe. However, His L168 is retained in the *Rb. sphaeroides* sequence and a hydrogen bond between the His L168 side chain and the 2a acetyl C=O of P_L has also been proposed[4].

In the present work, light-induced FTIR difference spectroscopy is used to investigate structural changes in *Rb. sphaeroides* RC (at the level of both the chromophore and the protein) resulting from site-directed mutagenesis of amino acids in the vicinity of

The Photosynthetic Bacterial Reaction Center II
Edited by J. Breton and A. Verméglio, Plenum Press, New York, 1992

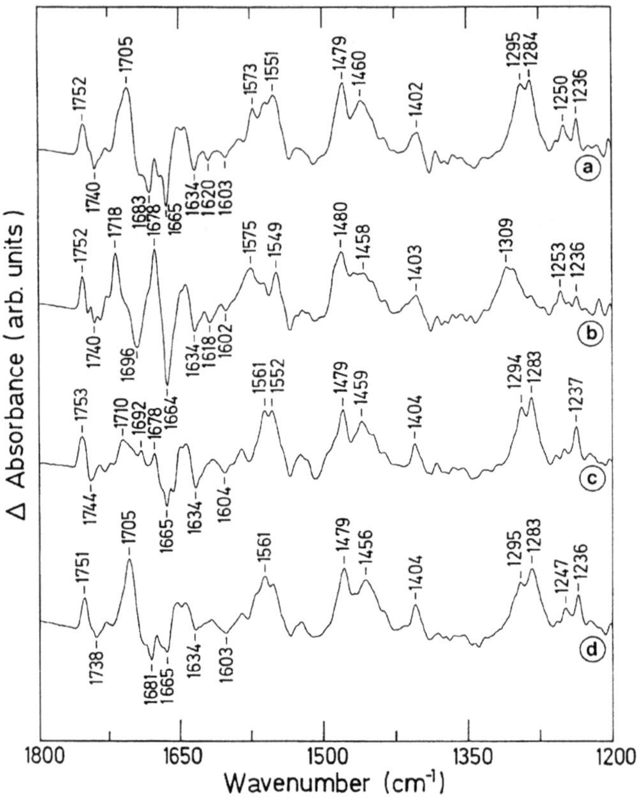

FIGURE 1. Light-induced FTIR difference spectra at 100K of *Rb. sphaeroides* chromatophores from a) wild type, b) Met L248→Thr, c) Ser L244→Gly, d) Phe M197→Tyr. 4 cm^{-1} resolution.

P. The following mutations, Met L248→Thr, Ser L244→Gly and Phe M197→Tyr were designed to make amino acid residues that surround the special pair in *Rb. sphaeroides* identical to those found in *Rps. viridis*. In addition, the role of a Tyr residue which is close to P, B_L and H_L in the crystal structures of the two RCs (Tyr M210 in *Rb. sphaeroides*, Tyr M208 in *Rps. viridis*), has been investigated with the Tyr M210→Phe and Tyr M210→Leu mutants. In these mutants, the initial electron transfer rate constants are considerably decreased, compared to the WT RCs[5,6]. Furthermore, the possibility of a hydrogen bond between the Tyr M210 hydroxyl side chain and the 2a acetyl of P_M has been discussed[7]. In both *Rb. sphaeroides* and *Rps. viridis* RCs, the symmetry related residue on the M side is a Phe (L181) which is close to P, B_M and H_M. The FTIR spectroscopic properties of the single-site Phe L181→Tyr mutant and of the double-site Phe L181-Tyr M210→Tyr L181-Phe M210 are also described here.

The mutants were constructed and their chromatophores were prepared as described elsewhere[8]. Purified chromatophores were deposited on CaF_2 windows for FTIR measurements performed as in[9].

Light-induced FTIR difference spectra at 100K between the charge-separated state $P^+Q_A^-$ and the relaxed state PQ_A, designated as $P^+Q_A^-/PQ_A$ spectra are displayed in Fig. 1 for *Rb. sphaeroides* chromatophores from wild type (WT) (1a), Met L248→Thr (1b), Ser L244→Gly (1c) and Phe M197→Tyr (1d). The $P^+Q_A^-/PQ_A$ spectra of the mutants display several significant differences compared to the WT spectrum.

The photooxidation of P as well as the photoreduction of Q_A have been previously characterized in RCs of *Rb. sphaeroides* and *Rps. viridis*[9–13]. In the C=O frequency region, the $P^+Q_A^-/PQ_A$ spectrum of *Rb. sphaeroides* chromatophores from WT (Fig. 1a) shows a negative peak at 1683 cm^{-1} with a shoulder at 1693 cm^{-1}, and a positive peak at 1705 cm^{-1} with a shoulder at 1713 cm^{-1}. The largest differential signal at 1705/1683 cm^{-1} has been assigned to a shift of the 9keto C=O of P_L upon photooxidation, favoring a predominant localization of the positive charge on the P_L side in the P^+ state[14]. The smaller 1713/1693 cm^{-1} signal is tentatively related to P_M. The vibrational frequency range observed for the keto carbonyls of P indicates that these groups are not hydrogen-bonded to the protein, in agreement with the X-ray model[3,4].

The positive signal at 1752 cm^{-1} (1751 cm^{-1} at 275K, see ref. 14) in the $P^+Q_A^-/PQ_A$ spectrum of WT (Fig. 1a) has been attributed[14] to a predominant contribution of the 10a ester C=O from P_L^+. A negative signal at ≈1740 cm^{-1} is clearly observed at 100K while at 275K, weakly negative structured features are observed[14]. A sharp negative signal at 1740 cm^{-1} has been also observed at 100K for RCs containing only Q_A[10] while at 250K, the same RCs show a structured signal without a clear negative signal. It therefore appears that the sharpening of the 1740 cm^{-1} signal at 100K can be related to a temperature effect.

No clear assignment for 2a acetyl C=O groups of P is possible in the $P^+Q_A^-/PQ_A$ spectrum of WT, due to superimposed contribution in the 1620–1665 cm^{-1} region

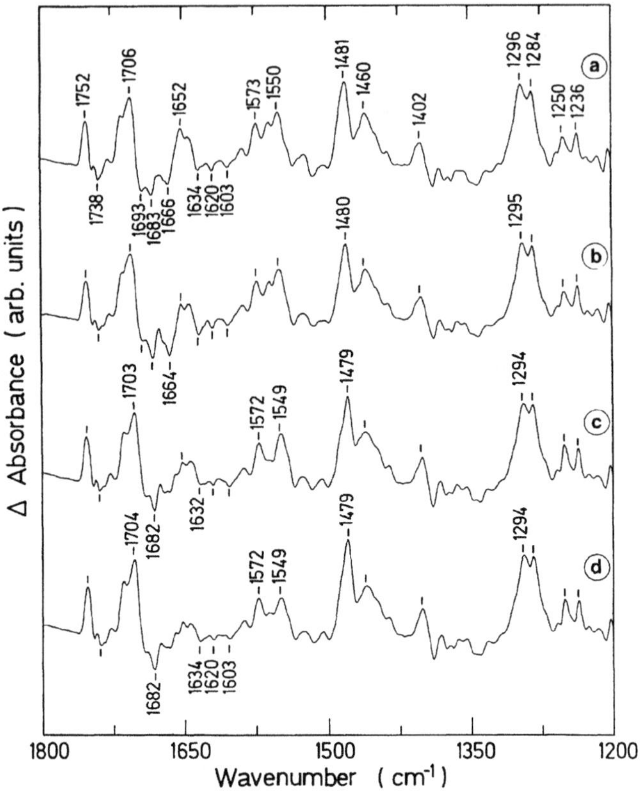

FIGURE 2. Light-induced FTIR difference spectra at 100K of *Rb. sphaeroides* chromatophores from a) Tyr M210→Leu, b) Tyr M210→Phe, c) Phe L181→Tyr, and d) Phe L181-Tyr M210→Tyr L181-Phe M210 mutants.

from peptide and quinone carbonyls as well as from amino acid side chains. In this spectrum, several bands, especially the ones at 1664 cm^{-1}, 1634 cm^{-1} and 1619 cm^{-1} are candidates for acetyl carbonyls.

In WT chromatophores and RCs, the three large positive bands at 1580–1530 cm^{-1}, 1500–1430 cm^{-1} and ≈1290 cm^{-1}, appear characteristic of a BChl dimer state[14] of P^+. However, other contributions are also expected such as quinone anion modes[15] at ≈ 1470 cm^{-1} and BChla cation C–C modes[16] at ≈1550–1500 cm^{-1}. These bands are either small or altogether absent in the FTIR difference spectrum of the radical cation of the monomeric BChla *in vitro*[16]. In the mutants spectra, the three bands are retained as well as a broad positive band at ≈2600 cm^{-1} (data not shown) that has been recently correlated to the dimeric cation state of the photooxidized P in bacterial RCs[17].

The main differences between the spectra of WT and mutants are found in the keto C=O frequency range where in WT the positive band at 1705 cm^{-1} with its shoulder at 1713 cm^{-1} has been assigned to the 9keto C=O of P^+. A sharper P^+ keto band is observed at 1708 cm^{-1} in Ser L244→Gly (Fig. 1c) with no distinct shoulder. In Met L248→Thr (Fig. 1b), the P^+ keto band appears at 1706 cm^{-1} with structures at

1714 cm^{-1} and 1724 cm^{-1}. In Phe M197→Tyr (Fig. 1d), a large splitting of the 9keto C=O P+ band is observed at 1717 cm^{-1} and 1704 cm^{-1}. The 10a ester signal at 1752/1739 cm^{-1} in WT is upshifted by 1 to 3 cm^{-1} in the spectra of the mutants. All these observations suggest that these mutations influence the repartition of the charge density on the ring V of P_L and/or P_M. The keto carbonyls of P+ appear to be very sensitive to the local amino acid environment not only of these groups but also of the ester and acetyl groups.

The Met L248→Thr mutation was designed to introduce a potential proton donating residue to the 9keto of P_L. In this mutant, the keto C=O of P absorb at 1683 cm^{-1} and 1693 cm^{-1} (Fig. 1b) as it is also observed in WT and there is no evidence for an additional band at lower frequency that would be expected from the formation of a hydrogen bond between the Thr side chain and the keto carbonyl of P_L. However, the observation of a structured positive band at 1706 cm^{-1} in P+ with pronounced shoulders at 1714 cm^{-1} and 1724 cm^{-1} suggests that different structures or environments for the 9keto C=O could be involved in the P+ state of this mutant.

Surprisingly, the main effect resulting from the expected perturbation of the environment of the 10a ester C=O in the Ser L244→Gly mutant is observed on the P+ band of the 9keto C=O (Fig. 1c) which becomes more symmetric than in the WT spectrum. The keto signals for the neutral P state are still found at 1683 cm^{-1} and 1694 cm^{-1}, the amplitude of both signals being now comparable in the mutant in contrast to the observation of a peak at 1683 cm^{-1} with a shoulder at 1693 cm^{-1} in WT. The single positive band at 1708 cm^{-1} is interpreted in terms of the superposition at the same frequency of the absorptions from both keto of P_L and P_M in the photooxidized special pair. These observations suggest that the positive charge in the P+ state of the Ser L244→Gly mutant is less asymmetrically localized on P_L than it is in WT. In the ester C=O region, the disappearance of the small negative shoulder at 1744 cm^{-1}, the 3 cm^{-1} upshift from 1739 cm^{-1} to 1742 cm^{-1} of the negative signal, the reproducible 1 cm^{-1} upshift from 1752 cm^{-1} to 1753 cm^{-1} of the positive signal indicate a small perturbation of the ester C=O group. From the X-ray RC structure, a hydrogen bond between Ser L244 and the 10a ester C=O of P_L has been proposed[4]. The frequencies upshifts detected in the C=O ester region of the Ser L244→Gly mutant are not inconsistent with the breaking of a weak hydrogen bond.

The mutation Phe M197→Tyr was designed to introduce a proton donating residue to the 2a acetyl C=O of P_M in *Rb. sphaeroides*. When light-induced $P^+Q_A^-/PQ_A$ FTIR difference spectra for WT chromatophores and the Phe M197→Tyr mutant (Fig. 1d) are compared, marked spectral changes are observed at ≈1300 cm^{-1}, at ≈1580 cm^{-1} where a band is lacking in the mutant and at 1717 cm^{-1} and 1704 cm^{-1} where a large splitting of the 9keto C=O P+ band appears. This large splitting indicates a strong perturbation of at least one keto C=O of P. The signal at 1704 cm^{-1} is assigned to the keto C=O of P_L^+ (and is only downshifted by 1 cm^{-1} with respect to WT). The 1718 cm^{-1} band is assigned to the 9keto C=O of P_M^+ whose intensity is increased and frequency is upshifted compared to the 1713 cm^{-1} shoulder observed in WT. Although the spectrum of the Phe M197→Tyr mutant does not show evidence for an additional band in the 1620- to 1640-cm^{-1} C=O region, resonance Raman spectra of isolated RCs from the same mutant[18] exhibit a new shoulder at 1625 cm^{-1} which has been assigned to hydrogen bonding formation between the 2a acetyl group of P_M and Tyr. Quantum chemical calculations have shown that a slightly different geometry of the ring I acetyl C=O group can induce selective perturbation on the P optical spectrum[19]. The strong perturbation of the 9keto C=O band in the Phe M197→Tyr mutant could thus reflect the different environment of the 2a acetyl of P_M compared to WT, and possibly its involvement in hydrogen bond formation with Tyr M197.

Light-induced $P^+Q_A^-/PQ_A$ FTIR difference spectra of chromatophores from Tyr M210→Leu, Tyr M210→Phe, Phe L181→Tyr and the double site-directed mutant Phe L181-Tyr M210→Tyr L181-Phe M210 are displayed on Fig. 2a,b,c,d, respectively. The general similarity of the FTIR difference spectra of these mutants as compared to the WT spectrum indicates that no significant alteration in the P chromophore-protein interactions and/or the protein itself occurs in these mutants. In particular, there is no evidence for hydrogen-bonding of Tyr M210 to any carbonyl group of P. These FTIR data are quite in agreement with resonance Raman spectra[20] showing that the local environment of the two BChls constituting P is not altered in both Tyr M210→Phe and Tyr M210→Leu mutant RCs. Low temperature absorption and linear dichroism spectra of RCs of these two mutants also show basically no difference between WT and mutants[8,20] with however, a 3-nm red shift of the Q_y absorption band of the monomer BChls. In the FTIR difference spectra of the Tyr M210→Leu and Tyr M210→Phe mutants (Fig. 2a,b), the keto C=O P^+ band at 1706 cm^{-1} appears to be reproducibly upshifted by 1 cm^{-1} as compared to WT, while in the Phe L181→Tyr mutant, a downshift of 2 cm^{-1} is observed for the equivalent band. For this last mutant, the frequency of the keto in the neutral state is also downshifted (to 1682 cm^{-1}). These frequency shifts can best be evidenced by calculating the difference between $P^+Q_A^-/PQ_A$ FTIR spectra of mutants and WT (data not shown). In the double-site mutant (Phe L181-Tyr M210→Tyr L181-Phe M210), the keto C=O of P_L^+ peaks at 1704 cm^{-1}, in agreement with the spectra of the corresponding single-site mutants. It is interesting to note that in *Rb. capsulatus* RC, the single-site mutant exhibits an initial electron transfer rate faster than that observed in WT, while the double-site mutant shows the same decay time compared to WT[21]. Nevertheless, the present IR data indicate that the Phe L181→Tyr mutation leads to more detectable but however very small changes on the $P^+Q_A^-/PQ_A$ FTIR difference spectrum than the Tyr M210→Phe mutation.

REFERENCES

1. H. Michel, O. Epp, and J. Deisenhofer, EMBO J. 5:2445 (1986).
2. H. Komiya, T.O. Yeates, D.C. Reeves, J.P. Allen, and G. Feher, Proc. Natl. Acad. Sci. USA 85:9012 (1988).

3. J.P. Allen, G. Feher, T.O. Yeates, H. Komiya, and D.C. Reeves, Proc. Natl. Acad. Sci. USA 84:5730 (1987).
4. O. El–Kabbani, C.–H. Chang, D. Tiede, J. Norris, and M. Schiffer, Biochemistry 30:5361 (1991).
5. U. Finkele, C. Lauterwasser, W. Zinth, K.A. Gray, and D. Oesterhelt, Biochemistry 29:8517 (1990).
6. V. Nagarajan, W.W. Parson, D. Gaul, and C. Schenck, Proc. Natl. Acad. Sci. USA 87:7888 (1990).
7. D.M. Tiede, D.E. Budil, J. Tang, O. El–Kabbani, J.R., Norris, and M. Schiffer in: "The Photosynthetic Bacterial Reaction Center, Structure and Dynamics", J. Breton and A. Verméglio, eds., pp. 13–20, Plenum, New York, (1988).
8. K.A. Gray, J.W. Farchaus, J. Wachtveitl, J. Breton, and D. Oesterhelt, EMBO J. 9: 2061 (1990).
9. E. Nabedryk, K.A. Bagley, D.L. Thibodeau, M. Bauscher, W. Mäntele, and J. Breton FEBS Lett. 266:59.
10. K.A. Bagley, E. Abresch, M.Y. Okamura, G. Feher, M. Bauscher, W. Mäntele, E. Nabedryk, and J. Breton, in: "Current Research in Photosynthesis, M. Baltscheffsky, ed., Vol.I, pp. 77–80, Kluwer Academic Publishers, Dordrecht (1990).
11. D.L. Thibodeau, E. Nabedryk, R. Hienerwadel, F. Lenz, W. Mäntele, and J. Breton, Biochim. Biophys. Acta 1020:253 (1990).
12. S. Buchanan, H. Michel, and K. Gerwert, in: "Reaction Centers of Photosynthetic Bacteria", Springer Series in Biophysics, Vol. 6, pp. 75–85, M.–E. Michel–Beyerle, ed., Springer–Verlag, Berlin, (1990).
13. E.H. Morita, H. Hayashi, and M. Tasumi, The Chemical Society of Japan 1583 (1991).
14. E. Nabedryk, S.J. Robles, E. Goldman, D.C. Youvan, and J. Breton, Biochemistry, in press.
15. J. Breton, D.L. Thibodeau, C. Berthomieu, W. Mäntele, A. Verméglio, and E. Nabedryk, FEBS Lett. 278:257 (1991).
16. W.G. Mäntele, A.M. Wollenweber, E. Nabedryk, and J. Breton, Proc. Natl. Acad. Sci. USA 85:8468 (1988).
17. J. Breton, E. Nabedryk, and W.W. Parson, Biochemistry, in press (1992).
18. T.A. Mattioli, K.A. Gray, J. Wachtveitl, J.W. Farchaus, M. Lutz, D. Oesterhelt, and B. Robert, in: "Spectroscopy of Biological Molecules", pp. 71–72, R.E. Hester and R.B. Girling, eds., The Royal Society of Chemistry, Cambridge (1991).
19. M.A. Thompson, M.C. Zerner, and J. Fajer, J. Phys. Chem. 95:5693 (1991).
20. T.A. Mattioli, K.A. Gray, M. Lutz, D. Oesterhelt, and B. Robert, Biochemistry 30:1715 (1991).
21. C.–K. Chan, L.X.–Q. Chen, T.J. DiMagno, D.K. Hanson, S.L. Nance, M. Schiffer, J.R. Norris, and G.R. Fleming, Chem. Phys. Lett. 176:366 (1991).

LIGHT-INDUCED CHARGE SEPARATION IN PHOTOSYNTHETIC BACTERIAL REACTION CENTERS MONITORED BY FTIR DIFFERENCE SPECTROSCOPY: THE Q_A VIBRATIONS

J. Breton, J.-R. Burie, C. Berthomieu,
D.L. Thibodeau, S. Andrianambinintsoa,
D. Dejonghe, G. Berger and E. Nabedryk
SBE/DBCM, CEN Saclay, 91191 Gif-sur-Yvette, France

The crystal structure of the photosynthetic bacterial reaction center (RC) suggests that the localization and the conformation of the cofactors involved in the electron transport is optimized by the protein environment which would then be responsible, at least in part, for the efficiency and quasi-irreversibility of the charge separation. For example, the difference in the nature and organization of the amino acids lining the binding pocket of the primary (Q_A) and secondary (Q_B) quinones, which are both a ubiquinone (UQ_{10}) in *Rb. sphaeroides*, might explain the differences in the redox properties of the two quinones as well as their very distinct roles in the electron transfer and proton transport mechanisms. However, X-ray studies yield an essentially static picture of the RC in the neutral state and do not provide information on the light-induced structural changes accompanying the charge separation and stabilization processes.

Infrared (IR) spectroscopy, which is highly sensitive to even minute alterations in bond lengths and energies, thus constitutes an attractive method for probing the structural changes in the RC at the level of individual bonds of both the cofactors and the protein. The vibrational spectrum of the quinones in their different states of ionization and/or protonation *in vivo* is expected to provide information on the geometrical and energetic

The Photosynthetic Bacterial Reaction Center II
Edited by J. Breton and A. Verméglio, Plenum Press, New York, 1992

factors (H–bonding, distortion of the ring and substituents...) involved in the charge stabilization processes and the protonation events taking place within the RC. With this goal in mind, we have implemented light–induced FTIR difference spectroscopy of various primary photosynthetic reactions involving the photoreduction of the quinones. At first, the $P^+Q_A^-/PQ_A$ and $P^+Q_B^-/PQ_B$ of RCs of both *Rb. sphaeroides* and *Rp. viridis* have been compared (Bagley et al.,1990; Nabedryk et al.,1990; Thibodeau et al.,1990; see also Buchanan et al.,1990). It was soon realized however, notably through the use of chemically modified quinones (with duroquinone replacing UQ_{10} as Q_A) or isotopically labelled (^{13}C or ^{18}O) UQ_{10} (Bagley et al.,1990), that the dominating contribution from P^+/P in these spectra tends to swamp out the vibrations associated with quinone reduction. In order to overcome this problem, a time–resolved FTIR technique (rapid–scan) has been developed and has led to the first $Q_A^-Q_B/Q_AQ_B^-$ double difference spectra of *Rb. sphaeroides* (Thibodeau et al.,1990) and *Rp. viridis* (Thibodeau et al.,1992). A double difference spectrum calculated between a light–induced $P^+Q_A^-/PQ_A$ spectrum and an electrochemically generated P^+/P spectrum has provided the first Q_A^-/Q_A spectrum in *Rb. sphaeroides* RCs (Mäntele et al.,1990).

Using a different approach, we have recently demonstrated that pure Q^-/Q difference spectra free from contribution of P^+/P (or Cyt^+/Cyt in *Rp. viridis*) could be directly obtained with a very high signal to noise ratio by illuminating RCs or chromatophores in the presence of a reductant (ascorbate) and a mediator (diaminodurene, DAD). These compounds rapidly rereduce P^+ so that the reduced quinone becomes the **only** detectable species which photoaccumulates. Using this method, the Q_A^-/Q_A (Breton et al.,1991a; Nabedryk et al.,1991) and Q_B^-/Q_B (Breton et al.,1991b) spectra could be precisely characterized for both *Rb. sphaeroides* and *Rp. viridis*.

However, these spectra cannot be directly interpreted in terms of the quinone vibrations inasmuch as any bond affected by the photoinduced change of state of the quinone (such as protein backbone or side chains, water, other cofactors...) will also contribute to the difference spectrum. It is thus necessary to reconstitute RCs with **chemically modified** or **isotopically labelled** quinones in order to separate the contributions of the quinones from those of the protein. The results of such an approach for the Q_A^-/Q_A vibrations of *Rb. sphaeroides* reconstituted with a series of 1,4–naphtoquinones (NQ) are presented in this contribution.

In order to demonstrate the validity of the approach that consists in removing and reconstituting Q_A, the Q_A^-/Q_A spectrum of Q_A–depleted *Rb. sphaeroides* RCs reconstituted with UQ_{10} (Fig.1a) is compared to that of native RCs (Fig.1b). The reconstitution was achieved under a flow of argon by adding a large molar excess (about 5–10 times) of a solution of the quinone in n–hexane to a 10–20 μl droplet of RCs deposited on a CaF_2 disc. The mediators used were either DAD or tetramethyl–p–phenylenediamine (TMPD). Within the noise level, the two spectra (Fig.1a,b) are practically identical. This observation demonstrates not only that the added UQ_{10} does

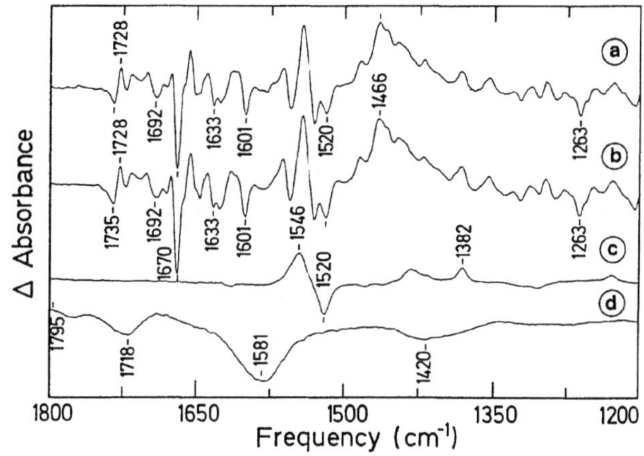

Figure 1. Q_A^-/Q_A spectra of *Rb. sphaeroides* RCs generated in the presence of TMPD of (a) Q_A-depleted RCs reconstituted with UQ_{10} and (b) native RCs. Electrochemically generated spectra of (c) $TMPD^+$/TMPD and (d) $ascorbate^+$/ascorbate.

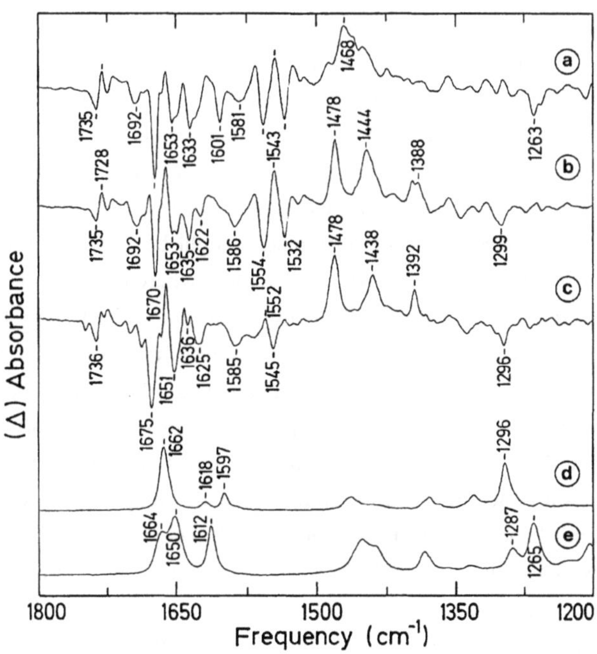

Figure 2. Q_A^-/Q_A spectra in the presence of DAD of (a) native RCs of *Rb. sphaeroides*, (b) Q_A-depleted *Rb. sphaeroides* RCs reconstituted with vitamin K_1 and (c) RCs of *Rp. viridis*. Absorption spectra of (d) a film of vitamin K_1 and (e) UQ_{10} in CCl_4.

indeed reconstitute the Q_A function but, at least as important, that the binding site of Q_A appears not to be perturbed by the rather drastic treatment required to remove the primary quinone (Okamura and Feher, 1975). Furthermore these spectra (Fig.1a,b) are very close to the Q_A^-/Q_A spectrum reported previously (Breton et al.,1991a; see also Fig.2a) except for a contribution from the couple $TMPD^+/TMPD$ (negative at 1520 cm^{-1}, positive at 1546 cm^{-1}). This contribution could be easily corrected for by subtracting a $TMPD^+/TMPD$ spectrum generated in a spectroelectrochemical cell (Fig.1c). As discussed previously (Breton et al.,1991a), no contribution from ascorbate could be detected in these Q_A^-/Q_A spectra. This is best shown by the absence of the 1795-cm^{-1} positive band characteristic of oxidized ascorbate (Fig.1d).

The Q_A^-/Q_A spectrum of Q_A-depleted *Rb. sphaeroides* RCs reconstituted with vitamin K_1 (2-Methyl-3-phytyl-NQ) is shown (Fig.2b) in comparison with the Q_A^-/Q_A spectra of RCs of *Rb. sphaeroides* containing the native UQ_{10} (Fig.2a) and of *Rp. viridis* (Fig.2c) containing the native menaquinone 9 (2-Methyl-3-isoprenyl-NQ). The three positive bands at 1478 cm^{-1}, 1438 cm^{-1} and 1392 cm^{-1} in Fig.2c have been previously assigned (Buchanan et al.,1990; Breton et al.,1991a,c) to vibrations (presumably C$\underline{\cdots}$O and C$\underline{\cdots}$C) of the menaquinone anion. The close analogy between the spectra 2b, with the three positive bands peaking at 1478 cm^{-1}, 1444 cm^{-1} and 1388 cm^{-1}, and 2c strongly supports this assignment. The negative band around 1300 cm^{-1}, common to both Q_A^-/Q_A spectra of menaquinone-containing RCs (Fig.2b,c), and which corresponds to an intense band at 1296 cm^{-1} in the absorption spectrum of vitamin K_1 (Fig.2d), can be ascribed to a quinone vibration. This band, absent in fig. 2a, is also missing in the absorption spectrum of UQ_{10} (Fig.2e), where a band at 1265 cm^{-1} appears instead.

The comparison of the two Q_A^-/Q_A spectra from *Rb. sphaeroides* RCs (Fig.2a,b) is also instructive as the common features should correspond to the non quinonic contributions while the most noticeable differences might be ascribed to the vibrations of the quinones. Besides the 1500- to 1370-cm^{-1} region, where the differences between the spectra 2a and 2b are essentially due to the nature of the anion bands, only the region from 1650 cm^{-1} to 1570 cm^{-1} shows prominent differences. Comparison of the absorption spectra of the isolated quinones (Fig.2d,e) reveals a much larger amplitude of the C=C bands (around 1590-1620 cm^{-1}) relative to the C=O bands (around 1640-1670 cm^{-1}) in UQ_{10} than in vitamin K_1. The presence of a large negative band at 1601 cm^{-1} in the spectrum 2a and its absence in spectrum 2b thus provide compelling evidence in favor of an assignment of this band to the C=C vibration of UQ_{10} *in vivo*. This observation strongly supports the previous assignment of a band at 1604 cm^{-1} in $P^+Q_A^-/PQ_A$ spectra of *Rb. sphaeroides* RCs containing native UQ_{10}, and which disappears in spectra of RCs containing ^{13}C-labelled UQ_{10}, to the C=C vibration of UQ_{10} (Bagley et al.,1990). On the other hand, the comparison of the absorption spectra of the isolated quinones (Fig.2d,e) clearly suggests that the band observed around 1585 cm^{-1} in spectra 2b and 2c can be attributed to the C=C vibration of menaquinone *in vivo*.

Although the differences observed in the 1650– to 1620–cm^{-1} region of spectra 2a and 2b are most probably assignable to the C=O vibrations of the quinones, the present data do not make it possible to unambiguously ascribe a specific negative band to one of these modes. As previously noticed (Breton et al.,1991c), these vibrations are likely to overlap with positive and negative absorption changes from protein modes so that the position of the observed minima may give only little information on the precise location of the pure quinone carbonyl vibrations. Furthermore, the C=O vibrations *in vivo* could appear split in view of the probable unequivalence of the bonding interactions between the two C=O groups and their respective proteic binding sites.

A similar problem arises when comparing spectra 2b and 2c in an attempt to locate the C=O and C=C vibrations of the neutral quinone in *Rp. viridis*, although the close analogy in the frequencies of the three anion bands in spectra 2b and 2c provides the additional information that the bands of the neutral species should also arise at almost the same position in the two spectra. While the C=C vibration of Q_A in *Rp. viridis* can be safely assigned to the band located at 1585 cm^{-1}, it remains uncertain to which of the three negative bands common to spectra 2b and 2c at about 1650 cm^{-1}, 1635 cm^{-1} or 1625 cm^{-1} one should assign the C=O vibration(s) *in vivo*.

It is worth noting that the signals observed in the 1560– to 1530–cm^{-1} and 1690– to 1740–cm^{-1} regions of Fig.2b are much closer to those observed in Fig.2a than in Fig.2c. For the former region, which corresponds to the Amide II absorption of the protein, this observation shows that the protein conformational change concomitant with the charge stabilization on Q_A is different in *Rb. sphaeroides* and *Rp. viridis* RCs and, at least in *Rb. sphaeroides*, is not markedly affected by the chemical nature of the quinone acting in the Q_A site. This makes it unlikely that a conserved residue like Try M250 (M252 in *Rb. sphaeroides*), which is strategically located in between H_A and Q_A, is primarily responsible for the differential features observed in this frequency region.

For the region extending between 1740 cm^{-1} and 1690 cm^{-1}, which displays strikingly similar spectral features in the two spectra (Fig.2a,b), protein side chains, notably those of carboxylic amino acids, could contribute. However, it is also possible that the signals in this frequency region involve changes at the level of other cofactors such as H_A, which is known to experience an electrochromic shift upon Q_A^- formation. In this scheme, the differential feature at 1735/1728 cm^{-1} could be assigned to the 10a C=O ester vibration of H_A (Mäntele et al.,1990), while the negative signal at 1692 cm^{-1} could correspond to its 9keto C=0 carbonyl.

The Q_A^-/Q_A spectra of Q_A–depleted *Rb. sphaeroides* RCs reconstituted with vitamin K_1 (Fig.3a), 2,3–Dimethyl–1,4–naphtoquinone (DMNQ, Fig.3b), 2–Methyl–1,4–naphtoquinone (MNQ, menadione, Fig.3c) and NQ (Fig.3d) can be compared to the absorption spectra of the respective quinones *in vitro* (Fig.4a–d). The *in vivo* spectra all show a number of highly comparable features, notably in the 1560– to 1530–cm^{-1} and 1660– to 1740–cm^{-1} regions. They also differ in the position of the C$\cdots$O and C$\cdots$C anion

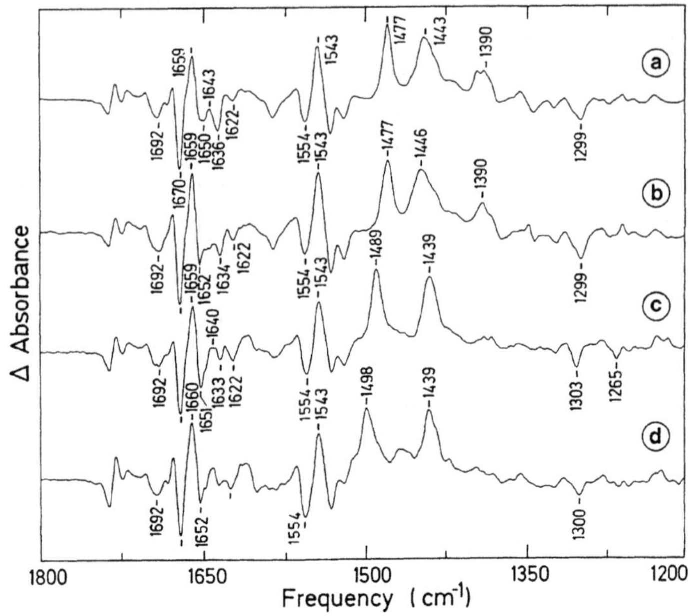

Figure 3. Q_A^-/Q_A spectra in the presence of TMPD of Q_A-depleted *Rb. sphaeroides* RCs reconstituted with (a) vitamin K_1, (b) 2,3-Dimethyl-1,4-naphtoquinone, (c) 2-Methyl-1,4-naphtoquinone and (d) 1,4-naphtoquinone.

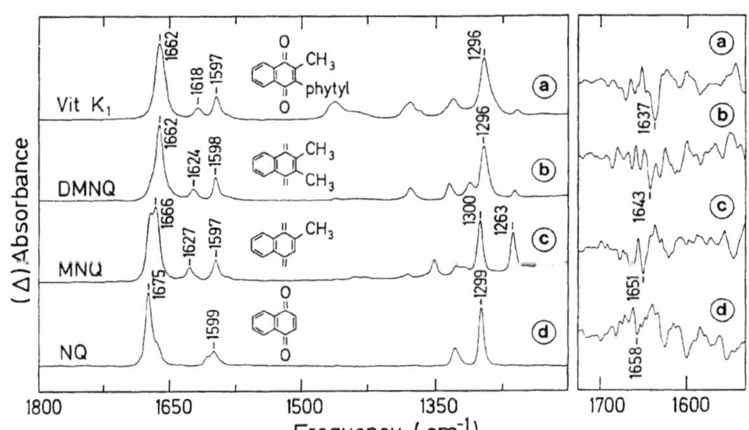

Figure 4 (Left). Absorption spectra of (a) vitamin K_1, (b) 2,3-Dimethyl-1,4-naphtoquinone (DMNQ), (c) 2-Methyl-1,4-naphtoquinone (MNQ) and (d) 1,4-naphtoquinone (NQ) in CCl_4.
Figure 5 (Right). *In vivo* double difference spectra calculated, as explained in the text, for Q_A-depleted *Rb. sphaeroides* RCs reconstituted with (a) vitamin K_1, (b) 2,3-Dimethyl-1,4-naphtoquinone, (c) 2-Methyl-1,4-naphtoquinone and (d) 1,4-naphtoquinone.

bands which appears highly specific of the nature of the substituents at positions 2 and 3 of the naphtoquinone ring. The spectra of RCs containing NQ (Fig.3d) and MNQ (Fig.3c) exhibit two main anion bands, as does the spectrum of an electrochemically generated anion of MNQ in acetonitrile (Breton et al.,1991c), while those of RCs reconstituted with DMNQ (Fig.3b) or vitamin K_1 (Fig.3a) show an additional peak around 1390 cm^{-1}.

The spectra in figure 3 show small but significant differences in the 1630– to 1660–cm^{-1} frequency region where the main C=O vibration of the neutral quinones is expected to absorb. Although it is difficult to extract directly from the spectra the frequency of this mode, it is nevertheless quite clear that a large negative component is observed at 1651 cm^{-1} in the spectrum of MNQ *in vivo* (Fig.3c) while, for vitamin K_1 (Fig.3a), a large component is present at 1636 cm^{-1}. On the other hand, the trough observed around 1643 cm^{-1} in the two previous spectra is much reduced for DMNQ (Fig.3b).

The position of the main C=O vibration for each of the four quinones can be estimated by constructing a first set of double difference spectra (not shown) in which the specific protein contributions observed in the 1560– to 1530–cm^{-1} and 1690– to 1660–cm^{-1} regions of the individual spectra are minimized. The bands in the two latter spectral regions appearing remarkably insensitive to the nature of the quinone used to reconstitute the RCs, a second approach has also been used. Briefly, the individual spectra in figure 3 are first normalized on both the 1554/1543–cm^{-1} and the 1670/1659–cm^{-1} differential signals and an average of the four spectra is calculated. This average spectrum (not shown) is then subtracted from each of the four normalized spectra. In the 1675– to 1625–cm^{-1} range, this second set of double difference spectra (Fig.5a–d) should show a **main** negative band which is expected to arise from the C=O contribution of the corresponding quinone. It should also be noted that these spectra are bound to be quite complex with noise originating from poor cancellation of the largest absorption changes and positive bands of reduced amplitude arising from the C=O of the three other quinones. It is gratifying to find that both sets of double difference spectra give almost the same value for the peak position of each of the four quinones *in vivo*: 1658 cm^{-1} for NQ, 1651 cm^{-1} for MNQ, 1643 cm^{-1} for DMNQ and 1637 cm^{-1} for vitamin K_1. These shifts of the *in vivo* frequencies generally follow the trend observed for the shift in frequency of the carbonyl vibration of the isolated quinones: 1675 cm^{-1} for NQ, 1666 cm^{-1} for MNQ, 1662 cm^{-1} for both DMNQ and vitamin K_1. Furthermore, when considering the *in vitro* absorption spectra of the neutral quinones (Fig.4) and of the anion of MNQ (Breton et al.,1991c), the amplitude observed in the *in vivo* spectra for the C=O vibration (Fig.5a–d) is qualitatively consistent with that of both the anion modes around 1400–1500 cm^{-1} and the neutral quinone mode at about 1300 cm^{-1} (Fig.3a–d).

The frequency found for the main C=O mode of vitamin K_1 in *Rb. sphaeroides* corresponds to a negative band at 1635 cm^{-1} (Fig.2b) which has also a counterpart at 1636 cm^{-1} in *Rp. viridis* (Fig.2c). Considering the close correspondence between the position of the three anion bands in these two spectra as a strong evidence for a similar binding of the menaquinone in the RCs of the two species, we thus provisionally assign the band at 1636 cm^{-1} in spectrum 2c to at least one C=O mode of Q_A in *Rp. viridis*.

The present study demonstrates that reconstitution of quinone-depleted RCs with chemically modified quinones can be used to tentatively assign the C=O and C=C vibrations of the neutral quinones *in vivo*. This approach clearly sets the stage for the more detailed investigations using isotopically labelled quinones which are now in progress in our laboratory.

REFERENCES

Bagley, K., Abresch, E., Okamura, M.Y., Feher, G., Bauscher, M., Mäntele, W., Nabedryk, E. and Breton, J., 1990, in: "Current Research in Photosynthesis", Baltscheffsky, M. ed., Kluwer, Dordrecht, pp.77–80

Breton, J., Thibodeau, D.L., Berthomieu, C., Mäntele, W., Verméglio, A. and Nabedryk, E., 1991a, FEBS Lett.278, 257–260

Breton, J., Berthomieu, C., Thibodeau, D.L. and Nabedryk, E., 1991b, FEBS Lett. 288, 109–113

Breton, J., Bauscher, M., Berthomieu, C., Thibodeau, D.L., Andrianambinintsoa, S., Dejonghe, D., Mäntele, W. and Nabedryk, E., 1991c, in "Spectroscopy of Biological Molecules", Hester, R.E. & Girling, R.B., eds., The Royal Society of Chemistry, Cambridge, pp. 43–46

Buchanan, S., Michel, H. and Gerwert, K., 1990, in "Reaction Centers of Photosynthetic Bacteria", Michel-Beyerle, M.-E., ed., Springer, Berlin, pp.75–85

Mäntele, W., Leonhard, M., Bauscher, M., Nabedryk, E., Breton, J. and Moss, D., 1990, in "Reaction Centers of Photosynthetic Bacteria", Michel-Beyerle, M.-E., ed., Springer, Berlin, pp.31–44

Nabedryk, E., Bagley, K., Thibodeau, D.L., Bauscher, M., Mäntele, W. and Breton, J., 1990, FEBS Lett. 266, 59–62

Nabedryk, E., Berthomieu, C., Verméglio, A. and Breton, J., 1991, FEBS Lett. 293, 53–58

Okamura, M.Y., Isaacson, R.A. and Feher, G., 1975, Proc. Natl. Acad. Sci. USA, 72, 3491–3495

Thibodeau, D.L., Nabedryk, E., Hienerwadel, R., Lenz, F., Mäntele, W. and Breton, J., 1990, Biochim. Biophys. Acta, 1020, 253–259

Thibodeau, D.L., Nabedryk, E., Hienerwadel, R., Mäntele, W. and Breton, J., 1992, in "Time-resolved Vibrational Spectroscopy V", Takahashi ed., Springer, Berlin, pp.79–82

TIME-RESOLVED INFRARED AND STATIC FTIR STUDIES OF $Q_A \rightarrow Q_B$ ELECTRON TRANSFER IN *RHODOPSEUDOMONAS VIRIDIS* REACTION CENTERS

Rainer Hienerwadel, Eliane Nabedryk*, Jacques Breton*, Werner Kreutz, and Werner Mäntele

Institut für Biophysik und Strahlenbiologie, Universität Freiburg
Albertstrasse 23, 7800 Freiburg, Germany

*DBCM/SBE, CEN Saclay, 91191 Gif-sur-Yvette Cédex, France

Introduction

Light-induced electron transport in bacterial photosynthetic reaction centers leads to the creation of a charge-separated state stable for milliseconds to seconds. The structures provided by X-ray crystallography (Michel et al., 1986; Allen et al., 1988; Deisenhofer & Michel, 1989; El-Kabbani et al., 1991) constitute a unique guideline to address questions on how the function may be related to the arrangement of the cofactors and of specific amino acid residues in their vicinity. The sequence of electron transfer reactions, the identity of the reaction partners, and the reaction mechanisms have been characterized from static and time-resolved absorbance measurements (for a review, see Parson & Ke, 1982). Transfer of the first electron to the primary (Q_A) and secondary (Q_B) quinone electron acceptors has received considerable attention, since it is associated with intraprotein protolytic reactions (for a recent review, see Okamura & Feher, 1992), which have a potential role in electrostatic charge stabilization.

While electronic transitions of the pigments and the quinones can be used to trace the immediate involvement of these cofactors in the reactions, only very limited information is available on the dynamic nature of the cofactor-protein interactions, on protolytic reactions in the quinone environment, and on local protein conformational changes, all possibly contributing to the stabilization of the charged radical states in the RC.

Infrared spectroscopy can be used to address these questions, since it is not selective for pigment or quinone vibrational modes. Using the intrinsic photochemical reactions, IR difference spectra of primary charge separation P^+Q^-/PQ, and, in combination with exogeneous electron donors, H^-/H or Q^-/Q difference spectra

The Photosynthetic Bacterial Reaction Center II
Edited by J. Breton and A. Verméglio, Plenum Press, New York, 1992

were obtained. Furthermore, electrochemical reactions in an IR spectroelectrochemical cell have been used to obtain IR difference spectra of individual cofactors (P^+/P, Q_A^-/Q_A, Q_B^-/Q_B). The current state of FTIR studies on the RC is summarized in a recent review (Mäntele, 1992). The FTIR difference spectra show highly detailed band structures in the 1800 cm^{-1} to 1000 cm^{-1} spectral region. In the case of P^+Q^-/PQ and P^+/P difference spectra, the changes of bond order and interaction with the protein environment upon formation of the π-cation radical of the primary donor dominate the spectrum, and the main band patterns in the 1750-1620 cm^{-1} region could be interpreted upon comparison with suitable model compound IR spectra. Only few but distinct protein modes respond or contribute to P^+ formation and stabilization (Leonhard, 1992).

The situation is different for quinone reduction. Even in the case of *Rb. sphaeroides*, where the same ubiquinone species (UQ-10) occupies the Q_A and the Q_B binding site, the Q_A^-/Q_A and Q_B^-/Q_B spectra differ markedly. Furthermore, the Q^-/Q difference spectra of RC are significantly more complex than Q^-/Q model spectra (Bauscher et al., 1990; Bauscher 1991) and exhibit numerous bands which cannot be associated with genuine quinone modes.

The band patterns of the Q_A^-/Q_A and Q_B^-/Q_B spectra (Mäntele et al., 1990; Breton et al., 1991a,b; Bauscher, 1991; see also Breton et al., these proceedings) indicate contributions from the protein backbone (as evidenced by differential bands in the amide I and amide II region), possibly from aromatic side chain residues (in the 1520-1580 cm^{-1} region) and possibly from protolytic reactions at ASP and GLU side chain residues (mainly in the 1700-1750 cm^{-1} range), in addition to the quinone C=O and C=C and the semiquinone anion C-O and C-C modes. Questions arise on the dynamic nature of these protein conformational changes, and whether stationary Q^-/Q difference spectra of a relaxed Q and Q^- state represent the situation of forward electron transfer.

Recently, we have analyzed the molecular processes upon $Q_A \rightarrow Q_B$ electron transfer in *Rb. sphaeroides* RC by time-resolved IR spectroscopy using a kinetic photometer with tunable IR diode lasers (Hienerwadel et al., 1992) . We were able to identify rapid (<0.5 μsec) IR absorbance changes arising from the $P \rightarrow P^+$ or the $Q_A \rightarrow Q_A^-$ transition, correlating in frequency and amplitude with P^+/P and Q_A^-/Q_A stationary spectra. In addition, transient $\approx$120 μsec signals could be characterized which arise either from the $Q_A^- \rightarrow Q_A$ or the $Q_B \rightarrow Q_B^-$ transitions. These signals largely correlate with $Q_A^-Q_B/Q_AQ_B^-$ double difference spectra obtained by time-resolved FTIR spectroscopy (Thibodeau et al., 1990). A third class of signals was detected, however, which are triggered by $Q_A \rightarrow Q_B$ electron transfer, but are kinetically uncoupled and exhibit half times around 1 msec and slower. We have attributed these signals to changes of protonation in the Q_B (and probably also the Q_A) environment. A preliminary investigation of RC from mutants of *Rb. sphaeroides* (Glu L212 → Gln or Asp L213 → Asn) has indicated that, upon transfer of the first electron, both residues contribute to the IR signals in the 1750 cm^{-1} to 1700 cm^{-1} frequency region (Hienerwadel et al., 1993). We presume that these changes of protonation play a dominant role in the stabilization of the negative charge on Q_B, and follow further IR signals in the 1750 cm^{-1} to 1700 cm^{-1} range in order to trace the pathway of protons into the RC to the Q_B binding site.

The structural homology of the *Rb. sphaeroides* RC and the *Rps. viridis* RC (Allen et al., 1986) indicates that a similar protein involvement upon $Q_A \rightarrow Q_B$ electron transfer can be expected. Unlike in *Rb. sphaeroides* RC, the acceptor side in *Rps. viridis* is occupied by different quinones, a menaquinone 9 for Q_A and an ubiquinone 9 for Q_B. In the case of *Rps. viridis* RC, the light-induced $P^+Q_A^-/PQ_A$ and $P^+Q_B^-/PQ_B$ (Nabedryk et al., 1990; Buchanan 1990), the Q_A^-/Q_A and Q_B^-/Q_B (Breton et al., 1991a, b) as well as the electrochemically generated P^+/P (Leonhard et al., 1991; Leonhard, 1992) difference spectra have been previously reported. In the work presented here, we use time-resolved IR spectroscopy in the 1760 cm^{-1} to 1450 cm^{-1} range to study in *Rps. viridis* RC the vibrational modes of the quinones and of their protein pocket in order to determine the dynamics of quinone binding and interaction, of protein conformational changes, and of protolytic reactions.

Materials and Methods

Rps. viridis RC were prepared according to standard procedures or by chromatofocusing (Welte et al., 1983). The RC were kept in a buffer containing 20 mM Tris pH 8 and 0.1% LDAO. For time-resolved IR spectroscopy, samples were prepared as described in Hienerwadel et al. (1992), except that UQ-9 was added according to the procedure described in Breton et al. (1991b) and 15 mM ferricyanide was added to oxidize the hemes. For steady-state Q^-/Q FTIR spectra, samples with sodium ascorbate and diaminodurene were prepared as described in Breton et al. (1991a,b).

Time-resolved IR measurements and control measurements in the near-IR were performed as described in Hienerwadel et al. (1992). FTIR difference spectra were performed as described in Breton et al. (1991a,b).

Results

RC charge recombination pathways in *Rps. viridis* are more complex than in *Rb. sphaeroides*, since electron donation from the cytochrome subunit can lead to multiple kinetic components. In order to avoid contributions from the cytochrome, ferricyanide was added to oxidize the hemes. Under these conditions, repetitive excitation with approx. one flash/sec leads to the recombination reactions $P^+Q_A^- \rightarrow PQ_A$ and $P^+Q_B^- \rightarrow PQ_B$, with decay times in the 1 msec and 100 msec time range, respectively (Shopes & Wraight, 1985). Figure 1 shows recovery of the primary electron donor at 960 nm of an IR sample of *Rps. viridis* RC after charge separation by a laser flash. The time course of recovery can be approximated by a sum of two exponentials yielding half-times of 30 msec and 90 msec, respectively. The initial amplitude of the 30 msec component was around 65 % of the total amplitude.

These values are in qualitative agreement with those found by Baciou et al. (1990) for RC solubilized in LDAO. They were attributed to two different populations exhibiting different $P^+Q_B^-$ charge recombination. Measurements at higher time-resolution neither exhibited time constants in the msec range (characteristic for $P^+Q_A^-$ recombination (Shopes & Wraight, 1985)), nor a μsec component, which would be indicative for the fast $C_{559} \rightarrow P$ electron transfer, which occurs in 0.2 - 0.3 μsec (Shopes et al., 1987; Dracheva et al., 1988).

Additional evidence for the presence of simple $P^+Q_B^- \rightarrow PQ_B$ charge recombination is derived from time-resolved IR measurements at a number of frequencies between 1760 cm^{-1} and 1450 cm^{-1}. The decay of these signals only exhibit the 30 msec and 90 msec phases with no evidence for longer phases. We thus conclude that more than 90% of the observed signal at 960 nm (figure 1) arises from $P^+Q_B^- \rightarrow PQ_B$ charge recombination.

Figure 2 shows a time-resolved IR trace at 1479 cm^{-1} on two different time scales. At this frequency, the signal is a superposition of contributions from the $P \rightarrow P^+$ transition, as evidenced from electrochemically-induced P^+/P difference spectra (Leonhard, 1992), and from the Q_A^-/Q_A and Q_B^-/Q_B transitions, as evidenced from steady-state FTIR difference spectra (Breton et al., 1991 a,b). A good fit of the signal (corresponding to the inset of figure 2) was obtained with two fast decaying components of 35 μsec and 180 μsec of comparable amplitudes (further on called transient components), a fixed 30 msec decay and an offset (to account for the 90 msec decay). These two transient components we attribute to contributions from Q_A^- which decay upon electron transfer to Q_B. For the electron transfer from Q_A to Q_B, a 35 μsec component was found by Mathis et al. (1992) for the bacteriopheophytin band shift. Although biphasic $Q_A \rightarrow Q_B$ electron transfer was not reported there, the heterogeneity observed for the $P^+Q_B^- \rightarrow PQ_B$ recombination might be also present for the $Q_A \rightarrow Q_B$ forward electron transfer. Baciou et al. (1990) have proposed different interactions between Q_A, Q_B and the Fe^{2+} located between the two quinones as an explanation for the heterogeneity of charge recombination. Such interactions could also influence the forward electron transfer. In this view, the 180 μsec component would reflect a heterogeneity in forward electron transfer in the preparation used here, as it has been monitored for charge recombination at 960 nm (see above).

Shopes & Wraight (1986) have described the action of ferricyanide on the acceptor side of the RC, which could be an alternative explanation for the 180 μsec component, although the ability of ferricyanide to reoxidize Q_A^- is much lower in the presence of Q_B than in its absence. However, since the 180 μsec component was also

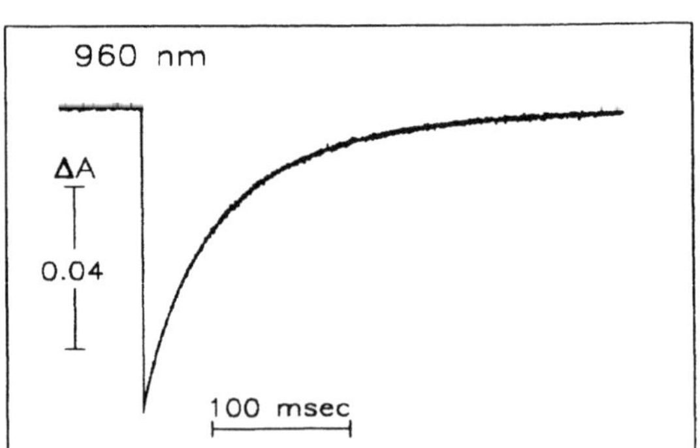

Figure 1. Flash-induced signal at 960 nm from *Rps. viridis* RC.

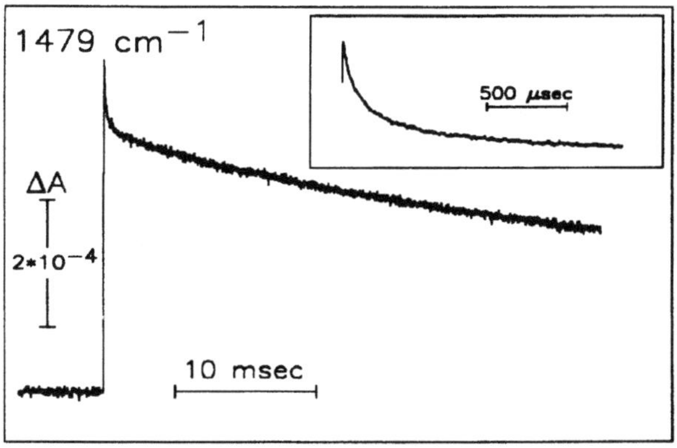

Figure 2. Flash-induced IR signal at 1479 cm^{-1} (average of 50 flashes). Inset: Same signal on a faster time scale (average of 500 flashes).

found in RC not treated with an oxidant (data not shown), such an explanation seems rather unlikely. Because the flashes were not saturating, double flash experiments could not rule out the possibility that the 180 μsec-component arises from transfer of a second electron to Q_B, in case ferrocyanide would act as a donor to P^+.

We have analyzed the transient components at various frequencies in the regions from 1760 to 1700 cm^{-1} and 1650 cm^{1} to 1450 cm^{-1}. The problem that a decreasing transient component can arise from Q_A^- or from Q_B, and an increasing transient component from Q_A or Q_B^- (and from their respective host sites) has been extensively discussed in Hienerwadel et al. (1992). Most of the time-resolved signals correlate well in frequency and amplitude with the constructed $Q_A^-Q_B/Q_B^-Q_A$ double difference spectrum shown in figure 3, which was obtained by subtraction of normalized Q_A^-/Q_A and Q_B^-/Q_B spectra (Breton et al., 1991a, b). Furthermore, they correlate with the double difference spectrum of $Q_A^-Q_B/Q_AQ_B^-$ obtained by time resolved FTIR measurements taking the difference in charge recombination time of the states $P^+Q_A^-$ and $P^+Q_B^-$ (Thibodeau et al., 1992). An analysis (analogous to that of the 1479 cm^{-1} signal) of the time course of transient components found in the frequency region from 1650 cm^{-1} to 1450 cm^{-1} revealed two components for most of the signals, with halftimes in the range of 30-70 μsec and 150-250 μsec. The amplitude ratios of the fast (30-70 μsec) to slow (150-250 μsec) transient component are not the same at all frequencies. However, the slow and fast transient signals at a given frequency always have the same sign. Whether additional slow components are present for some frequencies cannot be determined with the present signal to noise ratio. Comparison of the sign of the transient components in the frequency range from 1650 cm^{-1} to 1450 cm^{-1} with the double difference spectrum in figure 3 shows good agreement for all frequencies with the exception of the signal found at 1570 cm^{1}, where no significant feature is observed, while the kinetic measurements revealed a negative difference band in the region from 1590 cm^{-1} to 1560 cm^{-1}, with a maximum around 1570 cm^{-1}. The signal obtained at 1570 cm^{-1} is shown in figure 4.

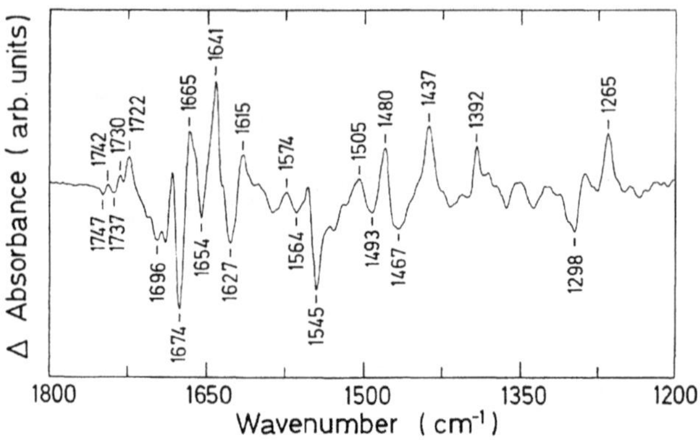

Figure 3. $Q_A^-Q_B/Q_AQ_B^-$ double difference spectrum from RC of *Rps. viridis* obtained by subtracting light-induced FTIR difference spectra of Q_A^-/Q_A and Q_B^-/Q_B. The amplitude (peak to peak) of the Q_B^-/Q_B spectrum was adjusted to 0.75 of the amplitude of the Q_A^-/Q_A spectrum.

The double difference spectrum of figure 3 shows small differential features in the frequency region between 1760 cm^{-1} and 1700 cm^{-1}. These bands were investigated by the time resolved IR method, some transient signals are shown in figure 5. Data fits analogous to the one described above again revealed two components with the same sign for all signals (with exception of the one found at 1707 cm^{-1}). Half times and relative amplitudes of the transient components are summarized in Table 1. As an example for a signal exhibiting only weak transient components, the one at 1758 cm^{-1} is shown in figure 5.

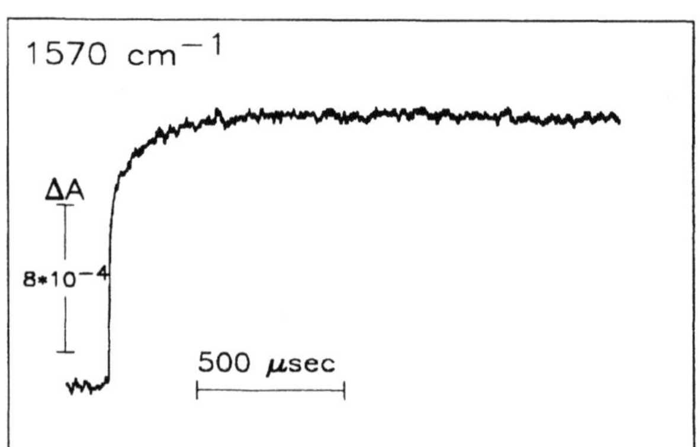

Figure 4. Flash-induced IR signal at 1570 cm^{-1} (average of 500 flashes).

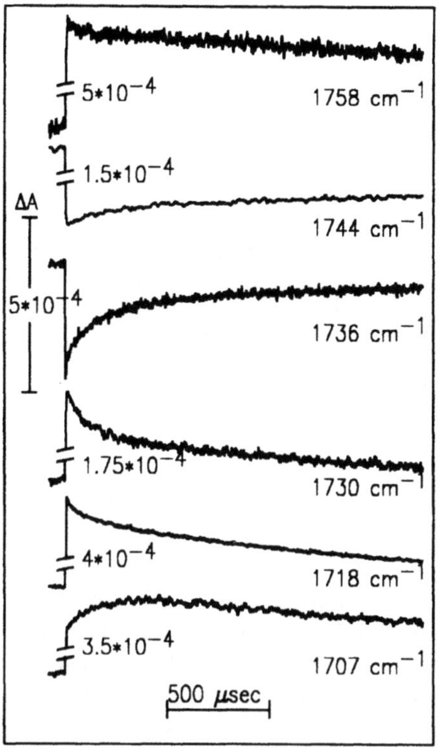

Figure 5. Flash-induced IR signals in the frequency region from 1700-1760 cm^{-1}. For some signals the initial amplitude is shown on a reduced scale (each signal is the average of 500 flashes).

Table 1. Half times and relative amplitudes (in parentheses) of the data fits for the transient components shown in figure 5. For fitting procedure see text.

wave-number (cm^{-1})	$t_{1/2}$ Fast	$t_{1/2}$ Slow
1744	100 μsec (-0.3)	1.2 msec (-0.7)
1736	60 μsec (-0.5)	340 μsec (-0.5)
1730	60 μsec (0.45)	1.2 msec (0.55)
1718	50 μsec (0.25)	1.1 msec (0.75)
1707	100 μsec (-0.35)	2.2 msec (0.65)

Discussion

In the case of *Rps. viridis*, the most appropriate spectral region to follow quinone modes with minimum interference from the protein is in the major semiquinone anion bands whose maxima are located at 1478 cm^{-1} for Q_A^- and 1475 cm^{-1} for Q_B^- (Breton et al., 1991a, b; Bauscher 1991). Thus, we can assume that the transient signal observed at 1479 cm^{-1} (figure 2) indeed represents the recovery of Q_A upon electron transfer from Q_A^- to Q_B. This interpretation is consistent with the observation that the relative amplitudes for most of the transient components in the frequency range from 1760 cm^{-1} to 1450 cm^{-1} correlate well with the $Q_A^-Q_B/Q_AQ_B^-$ double difference spectrum constructed from the steady state light-induced Q_A^-/Q_A and Q_B^-/Q_B spectra. This observation also shows that the majority of the processes of protein relaxation which accompany the formation of Q_B^- are occuring within the 30- to 250-μsec kinetics of Q_B reduction. In the frequency region above 1700 cm^1, transient components are observed which are comparable or slightly slower for the fast component but are significantly slower for the slow component. We thus note that in this spectral range the IR signals do not only monitor events which follow the electron transfer without delay, but also other processes which are initiated by electron transfer but are kinetically uncoupled. In the spectral region above 1700 cm^{-1} , the quinones both in their neutral and anion form do not absorb (Bauscher, 1991). Contributions from other cofactors, and notably of the 10a C=O ester group(s) of P^+/P around 1740-1760 cm^{-1}, have been proposed (Mäntele et al. 1985, 1988). This agrees with the kinetics observed at 1758 cm^{-1} (figure 5) which we assign to a mode of P^+, since no notable transient component is observed.

The most obvious canditates for the slow transient components observed above 1700 cm^{-1} are the C=O modes of protonated ASP and GLU side chains. IR signals in this range could either correspond to protolytic reactions or to a change of the COOH absorbance in response to the electrostatic field created upon electron transfer to Q_B (see Gunner, M.R. and Honig, B., these proceedings). A change in the local field or dielectric constant in the vicinity for one of these groups will cause a shift of the absorption band resulting in a differential signal, while a shift of pK will result in a variation in the intensity of this band. In a recent analysis of RC from *Rb. sphaeroides* mutants, signals from ASP and GLU side chains could be identified in this frequency range (Hienerwadel et al, 1993).

A less probable possibility would be the involvement of the other cofactors, such as the bacteriopheophytins and monomeric bacteriochlorophylls in the RC, which are known to exhibit electrochromic shifts upon formation of Q_A^- or Q_B^- (Verméglio and Clayton, 1977; Shopes and Wraight, 1985). In this respect, it has been suggested that a differential feature around 1730 cm^{-1} in the Q_A^-/Q_A difference spectrum of *Rb. sphaeroides* could originate from the 10a C=O ester of H_A (Mäntele et al., 1990). It has been recently reported (Tiede, D. M. and Hanson, D.K., these prooceedings) that, upon formation of Q_B^-, these cofactors experience time-dependent electrostatic effects. These relaxation processes occur in the 400 μsec range and are probably related to a redistribution of protons in the RC.

Conclusions

The IR signals detected here represent a unique possibility to follow the involvement of the protein in primary charge separation and stabilization in RC. Time-resolved IR spectroscopy offers the possibility of monitoring the processes of protein relaxation and protolytic reactions upon "forward" electron transfer. Up to now, only a small number of bands in the steady-state difference spectra has been analyzed in its time evolution. We emphasize, however, that notably the signals in the 1760-1700 cm^{-1} range are considerably slower than $Q_A^- \rightarrow Q_B$ electron transfer in *Rps. viridis* RC. Although a conclusive assignment of the modes involved in this region must await further experiments, such as their pH dependence, isotopic effects, and mutants, we favour their interpretation in terms of protolytic reactions upon $Q_A \rightarrow Q_B$ electron transfer, which might be an important step for the stabilization of the separated charges.

Acknowledgements

The authors would like F. Fritz and Dr. T. Wacker for providing *Rps. viridis* RC. Part of this work was supported by grants from the Deutsche Forschungsgemeinschaft to W.M. and from the European Communities (SC100335) to W.M. and J.B. W.M. gratefully acknowledges a Heisenberg fellowship of the Deutsche Forschungsgemeinschaft.

References

Allen, J. P., Feher, G., Yeates, T. O., Rees, D. C., Deisenhofer, J., Michel, H., and Huber, R. (1986) Proc. Natl. Acad. Sci. USA **83**, 8589-8593
Allen, J. P., Feher, G., Yeates, T. O., Komiya, H., & Rees, D. C. (1988) Proc. Natl. Acad. Sci. USA **85**, 8487-8491
Baciou, L., Rivas, E., and Sebban, P.(1990) Biochemistry **29**, 2966-1976
Bauscher, M. (1991) PhD Thesis, Faculty of Chemistry, Universität Freiburg
Bauscher, M., Nabedryk, E., Bagley, K. A., Breton, J., & Mäntele, W. (1990) FEBS Lett. **261**, 191-195
Breton, J., Thibodeau, D. L., Berthomieu, C., Mäntele, W., Verméglio, A., and Nabedryk, E. (1991a) FEBS Lett. **278**, 257-260
Breton, J., Berthomieu, C., Thibodeau, D. L., & Nabedryk, E. (1991b) FEBS Lett. **288**, 109-113
Buchanan, S., Michel, H., and Gerwert, K. (1990) in "*Reaction Centers of Photosynthetic Bacteria*" (M. E. Michel-Beyerle, ed.) Springer Series in Biophysics **6**, 75-85
Deisenhofer, J., and Michel, H. (1989) The EMBO Journal **8**, 2149-2170
Dracheva, S.M., Drachev, L.A., Konstantinov, A.A., Semenov, A.Y., Skulachev, V.P., Arutjunian, A.M., Shuvalov, V.A., and Zaberezhnaya, S.M. (1988) Eur. J. Biochem. **171**, 253-264
El-Kabbani, O., Chang, C. -H., Tiede, D., Norris, J. & Schiffer, M. (1991) Biochemistry **30**, 5361-5369
Hienerwadel, R., Thibodeau, D., Lenz, F., Nabedryk, E., Breton, J., Kreutz, W., and Mäntele, W. (1992) Biochemistry **31**, 5799-5808
Hienerwadel, R., Nabedryk, E., Paddock, M.L., Rongey, S., Feher, G., Okamura, M.Y., Mäntele, W., and Breton, J. (1993) Proceedings of the IX International Congress on Photosynthesis (Nagoya, Japan)

Leonhard,M., Moss,D., Bauscher, M., Nabedryk, E., Breton, J., and Mäntele, W. (1991) in *Spectroscopy of Biological Molecules* (Hester, R.E. and Girling, R.B., ed.) The Royal Society of Chemistry, 75-76
Leonhard, M. (1992) PhD Thesis, Faculty of Chemistry, Universität Freiburg
Mäntele, W., Nabedryk, E., Tavitian, B.A., Kreutz, W., and Breton, J. (1985) FEBS Lett **187**, 227-232
Mäntele, W., Wollenweber, A.M., Nabedryk, E., Breton, J. (1988) Proc. Natl. Acad. Sci. **85**, 8468-8472
Mäntele, W. (1992) in *The Photosynthetic Bacterial Reaction Center* (Deisenhofer, J. and Norris, J., eds.) Vol II, Chapter 14, Academic Press, N.Y., in press
Mäntele, W., Leonhard, M., Bauscher, M., Nabedryk, E., Breton, J., and Moss, D. A. (1990b) in "*Reaction Centers of Photosynthetic Bacteria*" (M. E. Michel-Beyerle, ed.) Springer Series in Biophysics **6**, 31-44
Mathis, P., Sinning, I., and Michel, H. (1992) Biochim. Biophys. Acta **1098**, 151-158
Michel, H., Epp, O., & Deisenhofer, J. (1986) The EMBO Journal **5**, 2445-2451.
Nabedryk, E., Bagley, K. A., Thibodeau, D. L., Bauscher, M., Mäntele, W., and Breton, J. (1990) FEBS Lett **266**, 59-62
Okamura, M., & Feher, G. (1992) Ann. Rev. Biochemistry, **61**, 861-896
Parson, W. W., & Ke , B. (1982) in "*Photosynthesis: Energy conversion by Plants and Bacteria*" (Govindjee, ed.) Academic Press, p.331-385
Shopes, R.J., and Wraight, C.A. (1985) Biochim. Biophys. Acta **806**, 348-356
Shopes, R.J., and Wraight, C.A. (1986) Biochim. Biophys. Acta **848**, 364-371
Shopes, R.J., Levine, L.M.A., Holten,D. and Wraight, C.A. (1987) Photosynth. Res. **12**, 165-180
Takahashi, E., & Wraight, C. A. (1990) Biochim. Biophys. Acta **1020**, 107-111
Thibodeau, D. L., Nabedryk, E., Hienerwadel, R., Lenz, F., Mäntele, W., and Breton, J. (1990), Biochim. Biophys. Acta **1020**, 253-259
Thibodeau, D. L., Nabedryk, E., Hienerwadel, R., Lenz, F., Mäntele, W., and Breton, J. (1992), Proceedings of the V Int. Conf. on Time-Resolved Vibrational Spectroscopy; Springer Verlag, in press
Verméglio, A., and Clayton, R.K. (1977) Biochim. Biopyhs. Acta **461**, 159-165
Welte, W., Hüdig, H., Wacker, T., and Kreutz, W. (1983) J. Chromatog. **259**, 341-346

IS DISPERSIVE KINETICS FROM STRUCTURAL HETEROGENEITY RESPONSIBLE FOR THE NONEXPONENTIAL DECAY OF P870* IN BACTERIAL REACTION CENTERS?

Stephen V. Kolaczkowski, Paul A. Lyle and Gerald J. Small

Ames Laboratory-USDOE and Department of Chemistry
Iowa State University
Ames, Iowa 50011, USA

INTRODUCTION

Spectral hole burning has revealed the structure underlying the broad (~ 100-500 cm^{-1}) Q_y-absorption band of chlorophylls and other cofactors in a wide variety of antenna and reaction center (RC) complexes (for reviews see refs. 1 and 2). It is now known that the contribution from site inhomogeneous broadening (Γ_I) to this band is ~ 50-200 cm^{-1}, depending on the complex. Although the magnitude of Γ_I for a complex is influenced by the solvent and detergent or isolation procedure, there is an intrinsic contribution that stems from what we refer to as normal glass-like structural heterogeneity of the protein.[3,4] That is, the existence of a large number of conformational substates of the protein,[5] which are unlikely to be discernable by X-ray diffraction, give rise, nevertheless, to a broad distribution of transition frequencies. It is not accidental that the above Γ_I -values are comparable to those observed for chromophores imbedded in glasses and polymers. Hole burning has also yielded the *homogeneous* contributions to the absorption profile (bandwidth) from linear electron-phonon coupling (Γ_{ep}) and exciton level structure/ scattering (Γ_{ex}, important in the case of certain antenna complexes).

Since proteins are glass-like and the kinetics of many processes (e.g., dielectric relaxation, photoconductivity and spectral diffusion) in amorphous solids are highly non-exponential (dispersive), it is natural to ask whether the structural heterogeneity of the RC complex is sufficient to yield measurable dispersion in the kinetics of primary charge separation. The existence of Γ_I for the Q_y-band proves that the adiabatic electronic energy gap(s) and the pure electronic coupling matrix element(s) relevant to charge separation have distributions of values. It is only whether the widths of the distributions are broad enough to yield non-exponential decay for P^* (primary donor state of special pair). The question is timely since very recent ultra-fast experiments on bacterial RC have shown that P^* decay is non-exponential[6-9] at room temperature. For example, in ref. 6 the stimulated emission of $P870^*$ of *Rhodobacter sphaeroides* (R26 mutant) was fit with a biexponential decay with

The Photosynthetic Bacterial Reaction Center II
Edited by J. Breton and A. Verméglio, Plenum Press, New York, 1992

components of 2.9 ps (65%) and 12 ps (35%). The biexponential decay could, however, be a manifestation of something other than glass-like structural heterogeneity.

In this paper we report the results of our initial theoretical and experimental investigations into the question of dispersive kinetics for P^* decay in bacterial RC arising from glass-like structural heterogeneity. The decay of P^* is due to electron-transfer and it is generally taken that[10] P^+B^-H figures importantly in the formation of P^+BH^-, where B and H denote the active bacteriochlorophyll and bacteriopheophytin monomers. Discussion continues on the relative importance of the one-step (superexchange) and two-step mechanisms for the production of P^+BH^- from P^*BH.[6,11-16] In the latter mechanism P^+B^-H serves as a real intermediate state.

EXPERIMENTAL

Sample preparation for hole burning spectroscopy generally involves dilution of the RCs (>10 μM RCs 0.05 % (wt/vol) Triton X-100 20 mM Tris/HCl pH 8.0) into glycerol to form a 1:2 aqueous to glycerol glass with an absorbance at 870 of 0.25-0.4. The RCs were reconstituted with UQ_2 (2,3-dimethoxy-5-decyl-benzoquinone) at 40 μM to refill any of the Q_a sites left vacant by the isolation procedure. The room temperature reversible bleaching of the P870 band is used to indicate the extent of Q_a reconstitution.

Laser excitation is provided with a Coherent 899 Ti:Sapphire laser pumped with a Coherent Innova 200 Ar^+ laser using visible wavelength multiline optics. The laser burn intensity is 10 mW/cm^2. The quality of the spectra being obtained has greatly improved primarily due to the use of a Bruker IFS 120 high resolution Fourier transform spectrometer yielding a factor of 20 in S/N as compared to our earlier data. The instrument resolution is $< 0.01\ cm^{-1}$, although $2.0\ cm^{-1}$ resolution is typically used to probe the region from 24000-8750 cm^{-1}.

RESULTS AND DISCUSSION

From the standard quantum mechanical expression for the rate of nonadiabatic electron-transfer,[17] it is apparent that the pure electronic coupling matrix element (V) and adiabatic electronic energy gap (Ω) are important factors for consideration of dispersive kinetics of the $DA \rightarrow D^+A^-$ electron-transfer process, where D ≡ donor and A ≡ acceptor. By necessity, we assume that these two variables are uncorrelated. We begin by considering dispersive kinetics from a distribution of Ω-values. Next we present some of our experimental data that speak to the question of dispersive kinetics from a distribution of V-values.

A. Dispersive Kinetics from the Energy Gap (Ω)-Distribution

Spectral hole burning results[2] for P870 and P960 of *Rb. sphaeroides* and *Rps. viridis* and the temperature dependence of P^* decay[17] indicate that for the problem at hand the low frequency modes (protein phonons, special pair marker mode(s)) are most relevant. It was recently reported[18] that for strong electron-phonon coupling the electron transfer rate constant has the form

$$k_{DA} = 2\pi V^2[2\pi S(\sigma^2 + \overline{\omega}^2)]^{-1/2} \exp[-(\Omega - S_0\overline{\omega})^2/2S(\sigma^2 + \overline{\omega}^2)] \, , \tag{1}$$

where $\Omega = \omega_D - \omega_A > 0$, $S = S_0 \operatorname{ctnh}(\hbar\overline{\omega}/2kT)$, S_0 is the 0 K Huang-Rhys factor for the protein phonons of mean frequency $\overline{\omega}$ and 2σ is the width of the one-phonon profile of the

TABLE I. Parameter values for dispersive kinetics calculations. Values for $\overline{\omega}$, ω_{sp}, σ, σ_{sp} and Γ were determined from hole burning data. See text for apportionment of reorganization energy for charge separation between the phonons ($\overline{\omega}$) and marker mode (ω_{sp}).

	$\overline{\omega}$ (cm^{-1})	S (0 K)	ω_{sp_1} (cm^{-1})	S_{sp} (0 K)	σ (cm^{-1})	σ_{sp_1} (cm^{-1})	Γ (cm^{-1})	Ω_{0_1} (cm^{-1})
P960	30	7.0	135	1.1	20	25	85	300
P870 (R26)	30	6.0	120	1.5	20	25	120	300

nuclear factor associated with the usual Fermi-Golden rule expression for the rate. Equation 1 is trivially generalized to include the special pair marker mode (ω_{sp}, $S_{sp} = S_{sp}$ (0 K) ctnh($\hbar\omega_{sp}/2kT$)). For a Gaussian distribution (f_Ω) of Ω-values centered at Ω_o, the average value for the rate is

$$\langle k_{DA}\rangle = 2\pi V^2[2\pi(\Gamma^2 + S(\sigma^2 + \overline{\omega}^2))]^{-1/2} \exp[-(\Omega_o - S_o\overline{\omega})^2/2(\Gamma^2 + S(\sigma^2 + \overline{\omega}^2))] \, , \quad (2)$$

where 2Γ is the width of the normal distribution for Ω. The time dependence of the P^* population is

$$P^*(t) = P^*(t{=}0) \int d\Omega \, f_\Omega \exp(-k_{DA}(\Omega)t) \quad (3)$$

From Eq. 2 (generalized to include the marker mode), it is apparent that when $\Gamma^2 < [S(\sigma^2 + \overline{\omega}^2) + S_{sp}(\sigma_{sp}^2 + \omega_{sp}^2)] \equiv \gamma_{ep}^2$, dispersive kinetics will not be observed, i.e., one is in the limit where the width 2Γ of the Ω-distribution is smaller than the *homogeneous width* $2\gamma_{ep}$ of the nuclear factor. Furthermore, the latter width has its minimum value in the $T \rightarrow 0$ K limit. The parameter values of Table I, many of which are obtained from the hole burning data[2], were used to investigate dispersive kinetics for the $P^*BH \rightarrow P^+B^-H$ (first step of two-step mechanism) and P^+BH^- (superexchange)[19] processes. It was found that the non-exponentiality of P^* decay from the Ω-distribution is too slight to be observable even in the low temperature limit.[18] In consideration of this result we note, for example, that for *Rb. sphaeroides* $2\gamma_{ep}$ = 350 and 870 cm^{-1} in the low T limit and at 300 K, respectively, whereas 2Γ = 240 cm^{-1}.

Our purpose here is to more critically assess the assumptions made in ref. 18 that led to certain of the parameter values given in Table I. It was taken that 2Γ, the width of the Ω-distribution, equals $2^{1/2}\Gamma_I$ with Γ_I the contribution from RC-inhomogeneity to the full width of the $P^* \leftarrow P$ absorption band. This relationship assumes that the Γ_I-value for the absorption band corresponding to the ground state to the charge-transfer (CT) state (P^+B^-H or P^+BH^-) transition is comparable to that for the P-band and, furthermore, that there is an absence of correlation between the inhomogeneous distributions of the absorption transition frequencies of the P- and CT-bands. The latter assumption receives some support from hole burning data on the RC of photosystem II.[20] The former finds support from studies of the 4 K absorption and fluorescence spectra of the 1:1 π-molecular CT complex anthracene:sym-trinitrobenzene.[21]

The anthracene:sym-trinitrobenzene complex was studied as a dilute solute in glasses and host crystals. The CT absorption and fluorescence spectra in both types of media were found to be essentially identical (e.g., vibronic bandwidths ~ 1000 cm^{-1}). Furthermore, the CT fluorescence origin band in the glass was *invariant* to the location of the laser excitation frequency within the origin band of the CT absorption spectrum. The results of ref. 21 lead to $\Gamma_I \lesssim 300$ cm^{-1} for this ~ 100% CT transition. It is ,therefore, germane to note that our calculations of $P^*(t)$ for *Rb. sphaeroides* show that 2Γ must be ~ 600 cm^{-1} (other parameter values as in Table I) rather than 240 cm^{-1} before the dispersion in P^* decay would be detectable in data obtained in the *low T limit* with typical signal to noise levels (results not shown). However, dispersive kinetics would still *not* be observable at room temperature.

The electron-phonon coupling parameters in Table I correspond to a reorganization energy of 360 cm^{-1} which corresponds to the accurately determined value for the CT absorption of the anthracene:pyromellitic acid dianhydride π-molecular complex.[22] It might be argued that this value is too small for the problem at hand. However, increasing this value would only strengthen the conclusions of ref. 18, *vide supra.*

The apportionment of the reorganization energy between the phonons ($\bar{\omega}$) and marker mode (ω_{sp}), Table I, was based on the requirement that Eq. 2 account for the T-dependence of P^* decay. Improved fits of the temperature dependence of the decays for P870* of *Rb. sphaeroides* and P960* of *Rps. viridis* are given in ref. 18.

Finally we make a few remarks on the mean energy gap value $\Omega_0 = 300$ cm^{-1}, Table I. First, it is the *effective* gap for the low frequency modes to accommodate. Second, the fact that the homogeneous width of the nuclear factor, $2\gamma_{ep}$, is ~ 1000 cm^{-1} at room temperature means, cf. Eq. 2, that the rate is quite insensitive to significant variations in Ω_0. It follows that the conclusion that[18] dispersive kinetics for P^* decay at room temperature is most improbable is not altered by such variations. We see no reason why this conclusion should not apply to the second step ($P^+B^-H \rightarrow P^+BH^-$) of the two-step mechanism although the special pair marker mode is likely to be "silent" in this step.

To conclude this subsection we observe that the utilization of the standard Fermi-Golden rule rate expression, which leads to Eq. 1, is subject to a number of constraints. In particular, when the adiabatic electronic energy gap between P^*BH and P^+B^-H is too small (~ kT at room T) its use would be questionable. Nevertheless, even in this case an exact theoretical treatment of the problem is unlikely to reveal dispersive kinetics at room T from the Ω-distribution.

B. Burn Wavelenth Dependent Studies of P870 of *Rb. Sphaeroides*

In previous work[1,2] it has been shown that the width of the zero-phonon hole (ZPH) in the photochemical hole burned spectrum of the P-band can be used to determine the lifetime of P^* from its *total* zero-point level which, incidentally, is a determination not possible by ultra-fast spectroscopy.

With an improved experimental setup, we have obtained hole spectra for P870 (R26 mutant) with a high enough S/N ratio to allow us to determine whether there is a variation in the P870* lifetime as the burn frequency (ω_B) is tuned across the inhomogeneous distribution of zero-phonon transition frequencies. Reference 18 did not address the question of dispersive kinetics for P^* decay from a distribution of values for the electronic coupling matrix element V, Eq. 1. If the width of V-distribution is significant, one might expect to observe a variation in lifetime.

The 4.2 K absorption spectrum of the sample used is shown in Fig. 1 and exhibits the sharpest P870 (R26) band measured to date, 437 cm^{-1}. Hole burned spectra were obtained for 10 ω_B-values (10902 -11049 cm^{-1}) that lie in the low energy side of P870 and span the aforementioned distribution of the zero-phonon transition frequency. Four of the hole spectra are shown in Fig. 2. The ZPH profiles were *systematically* analyzed using a fitting procedure[23] which was successfully tested using theoretically calculated P870 hole spectra. Experimental ZPHs and fits are shown in Fig. 3. For the above ω_B-values, which cover a range of 147 cm^{-1}, no variation in the ZPH width was observed with ω_B. The ZPHs yielded a width of 11.3 ± 0.5 cm^{-1} which corresponds to P870* lifetime of 0.94 ± 0.05 ps

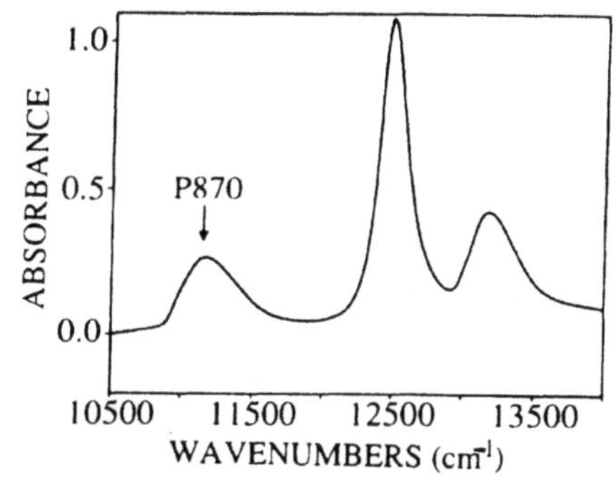

Figure 1. 4.2 K absorption spectrum of the Q_y-region of *Rb. sphaeroides* reaction centers. The P870, accessory BChl and BPheo bands are at 11120 cm^{-1} (899.2 nm), 12500 cm^{-1} (800.0 nm), and 13225 cm^{-1} (756.1 nm), respectively.

at 4.2 K for a P870 bandwidth of 437 cm^{-1} (the underscoring of this width is mandated by Eq. 2). As detailed a set of studies have been performed on P870 of deuterated wild type RC for which the P870 bandwidth was 434 cm^{-1}.[23] Again, no discernable variation in the ZPH width with ω_B was observed. The values of ω_B used yielded a ZPH width of 11.5 ± 1.0 cm^{-1}, which yields a P870* lifetime of 0.93 ± 0.1 ps.

Although the theoretical simulations (as reviewed in ref. 2) of the above ω_B-dependent P870 hole spectra are still being refined, they are satisfactory enough at this time to warrantshowing the results in Fig. 4. The simulation of the P870 (R26) hole burned spectrum obtained with ω_B = 10921 cm^{-1} was obtained with Γ_I = 150 cm^{-1} and a frequency

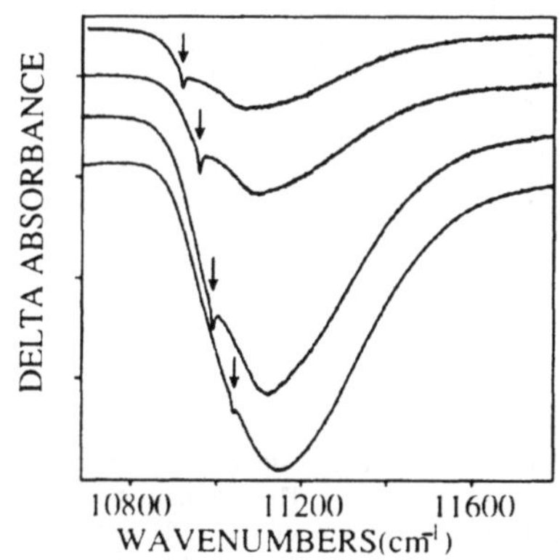

Figure 2. Photochemical hole burned spectra of P870, T= 4.2 K. From top to bottom the burn frequencies are 10921, 10957, 10992 and 11039 cm^{-1}. The arrows locate the zero-phonon holes, each of which is coincident with ω_B. The burn intensity is the same for all spectra, 10 mW/cm^2. From top to bottom the %-absorbance changes (as measured at the maximum of the broad hole) are 6.1%, 8.7%, 20.6 and 22.6%.

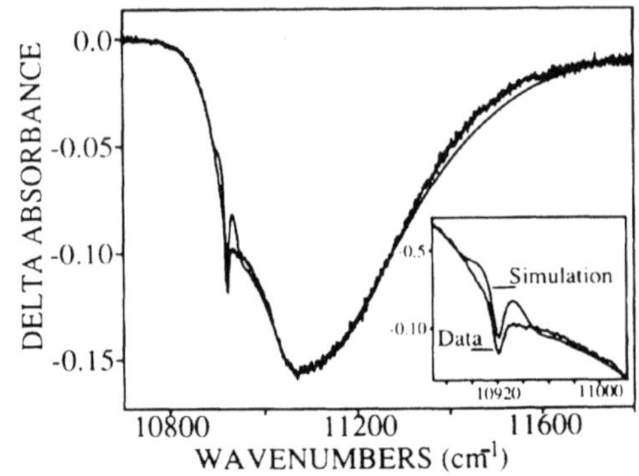

Figure 3. Lorentzian fits (dashed profiles) to the zero-phonon holes of Figure 2. The curves are normalized to the largest delta absorbance.

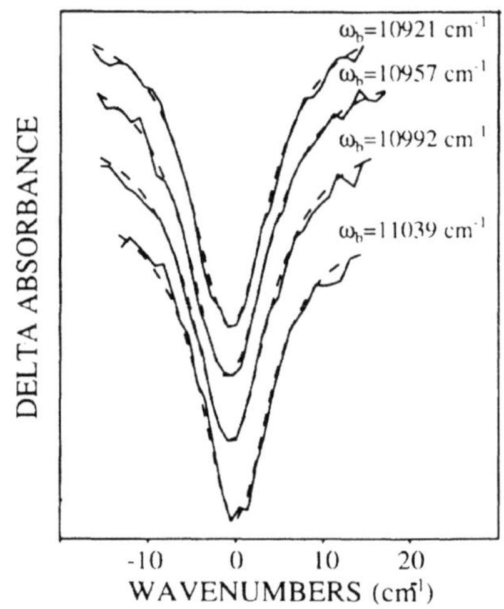

Figure 4. Experimental and simulated P870 hole spectra for ω_B = 10921 cm^{-1}. The simulation parameters are: Γ_I = 150 cm^{-1}; maximum of the ZPL distribution = 10995 cm^{-1}; homogeneous width of ZPL = 5.5 cm^{-1}; (ω_{sp}, S_{sp}) = (120 cm^{-1}, 1.65);$(\overline{\omega}, S)$ = (30 cm^{-1},1.8). Width of the one-phonon profile = 32.5 cm^{-1} with low energy half-width (gaussian) = 10 cm^{-1} and high energy half-width (lorentzian) = 22.5 cm^{-1}. The homogeneous width of the fundamental marker mode level (ω_{sp}) is 60 cm^{-1}. This width is allowed to increase with increasing quantum numbers as described in ref. 2. Inset: expanded view of region around the zero-phonon hole.

of 10993 cm^{-1} for the center of the inhomogeneous distribution of the zero-phonon transition frequency (total zero-point level). Other theoretical parameter values are given in the figure caption. The most difficult region to simulate is around the ZPH, see inset. We have determined that this is due primarily to the use of a gaussian for the low energy side of the one phonon profile. Utilization of a more slowly decaying function should lead to a significantly improved fit, while at the same time, not appreciably changing the values of the other parameters given in the caption.

The above results indicate that in the low T limit the electron-transfer time of P870* (from total zero-point) is *invariant to excitation frequency* within the inhomogeneous distribution of zero-phonon line frequencies. Unfortunately, this negative result cannot be interpreted as meaning that nonexponential decay kinetics for P* could not arise from a distribution of values for the coupling matrix element V. The reason is that the inhomogeneous ZPL-distribution depends on the distribution functions for the ground and excited state energies (E, E*) of the special pair, $f_P(E)$ and $f_P{*}(E^*)$, and the extent/type of correlation between the two distributions (for a detailed discussion see ref. 24). Noting that at any given value of E of P there is a high degree of accidental degeneracy, an absence of correlation means that Γ_I of P870 is the width of $f_P{*}(E^*)$ and, furthermore, that at any excitation frequency one essentially samples the entire ensemble of RCs, i.e., all possible V-values. At the other extreme there is perfect positive or negative correlation. In this case the width of $f_P{*}(E^*) < \Gamma_I$ of P870 and, also, different excitation frequencies would sample different subsets of the RC ensemble and, therefore, also different parts of the V-value distribution. The actual situation is likely to like between these two extremes.

The fact that the width of $f_P{*}(E^*)$ for P870 (R26) $< \Gamma_I \sim 150$ cm^{-1} (for the sample studied) is interesting in itself. Given that the excitonic interaction of the special pair is so strong, ≈ 800 cm^{-1}, the relative smallness of the width of $f_P{*}(E^*)$ suggests that the special pair geometries of the RC ensemble are not greatly different. We hasten to add that Γ_I must be due, in part, to structural variations of the protein itself.

CONCLUDING REMARKS

We have further[18] addressed the question of whether the nonexponential decay of P870* at room temperature could be due to dispersive kinetics arising from the glass-like structural heterogeneity of the protein.

On the basis of a nonadiabatic electron-transfer theory, which exposes the homogeneous width of the nuclear factor from low frequency modes (phonons), and hole burning data we conclude that this nonexponentiality is not due to a distribution of values, f_Ω, for the relevant adiabatic electronic energy gap(s) Ω. Dispersive kinetics from f_Ω in the low temperature limit are judged to be unlikely. Nevertheless, the expression (Eq. 2) for the average electron-transfer rate constant suggests that samples which exhibit sufficiently different Γ_I-values for the P-band should have measurably different values for $\langle k_{DA} \rangle$ in the low temperature limit. For example, if we increase the half-width of the f_Ω distribution for *Rb. sphaeroides* from 120 cm^{-1} (Table I) to 260 cm^{-1}, Eq. 2 predicts that the P870* lifetime would increase by ~ 30%. The former increase would correspond to an increase in the width of the P-band from ~ 440 cm^{-1} to ~ 620 cm^{-1}, as measured in the low temperature limit. Such a variation is attainable. Hole burning studies to test this prediction of Eq. 2 are planned.

The hole burning data obtained for P870 establish that the P870* lifetime (from total zero-point) is invariant to excitation frequency within the heterogeneity-induced zero-phonon line frequency distribution of P870. This is consistent with, but does not prove, an absence of dispersive kinetics from a distribution of values for the pure electronic coupling matrix element V. The proposed studies mentioned above may allow for a definite conclusion to be reached on this point.

Finally we remark that ultra-fast time domain measurements of the temperature dependence of P870*'s nonexponential decay loom as very important for its eventual understanding. Accurate measurements for $T \leq 200$ K are particularly important since the effects of thermal compressibility are likely to be negligible. If the nonexponentiality were due to a distribution of V-values, one would not expect it to carry a strong T-dependence for temperatures at which the ensemble of RC conformations is frozen in.

ACKNOWLEDGEMENTS

Research at Ames Laboratory was supported by the Division of Chemical Science, Office of Basic Energy Science, U.S. Department of Energy. Ames Laboratory is operated for the U.S. Department of Energy by Iowa State University under contract no. W-7405-Eng-82. P. L. was supported by a W. E. Catron Fellowship.

REFERENCES

1. S. G. Johnson, I.-J. Lee and G. J. Small, Solid state line narrowing spectroscopies, *in:* "Chlorophylls," H. Scheer ed., CRC Press Inc., New York (1991).
2. N. R. S. Reddy, P. A. Lyle, and G. J. Small, Applications of spectral hole burning spectroscopies to antenna and reaction center complexes, *Photosyn. Res.* 31:167 (1992).
3. G. P. Singh, H. J. Schink, H. Lohneysen, F. Parak and S. Hunklinger, Excitations in metmyoglobin crystals at low temperature, *Z. Phys.* B55:23 (1984).
4. I.-S. Yang and A. C. Anderson, Specific heat of melanin at temperatures below 3 K, *Phys. Rev. B* 34:2942 (1986).
5. H. Frauenfelder, S. G. Sligar and P. G. Wolynes, The energy landscape and motions of proteins, *Science* 254:1598 (1991).
6. M. H. Vos, J C. Lambry, S. C. Robles, D. C. Youvan, J. Breton and J.-L. Martin, Direct observation of vibrational coherence in bacterial reaction centers using femtosecond absorption spectroscopy, *Proc. Natl. Acad. Sci.* USA 88:8885 (1991).
7. M. Du, S. J. Rosenthal, X. Xie, T. J. DiMagno, M. Schmidt, J. R. Norris and G. R. Fleming, Femtosecond spontaneous emission studies of reaction centers from photosynthetic bacteria, *Proc. Natl. Acad. Sci.* USA, accepted.
8. M. G. Muller, K. Griebenow and A. R. Holtzworth, Picosecond fluorescence kinetics of the primary processes in isolated photosynthetic reaction centers from the purple bacterium *Rb. sphaeroides*, *in*: "Digest of ultrafast phenomena VIII (E.N.S.T.A.)," Paris (1992).
9. P. Hamm and W. Zinth, Ultrafast fluorescence spectroscopy of reaction centers of photosynthetic bacteria, *in*: "Digest of ultrafast phenomena VIII (E.N.S.T.A.)," Paris, (1992).
10. R. A. Friesner and Y. Won, Spectroscopy and electron transfer dynamics of the bacterial photosynthetic reaction center, *Biochim. Biophys. Acta* 977:99 (1989).
11. W. Holzapfel, U. Finkele, W. Kaiser, D. Oesterhelt, H. Scheer, H. U. Stitz and W. Zinth, Initial electron-transfer in the reaction center from *Rhodobacter sphaeroides*, *Proc. Natl. Acad. Sci.* USA 87:5168 (1990).
12. W. Holzapfel, U. Finkele, W. Kaiser, D. Oesterhelt, H. Scheer, H. U. Stitz and W. Zinth, Observation of a bacteriochlorophyll anion radical during the

primary charge separation in a reaction center, *Chem. Phys. Lett.* 160:1 (1989).
13. K. Dressler, E. Umlaut, S. Schmidt, P. Hamm W. Zinth, S. Buchanan and H. Michl, Detailed studies of the subpicosecond kinetics in the primary electron transfer of reaction centers of *Rhodopseudomonas viridis*, *Chem. Phys. Lett.* 183:270 (1991).
14. C. Lauterwasser, U. Finkele, H. Scheer, and W. Zinth, Temperature dependence of the primary electron transfer in photosynthetic reaction centers from *Rhodobacter sphaeroides*, *Chem Phys. Lett.* 183:471 (1991).
15. C.-K. Chan, T. J. DiMagno, L. X.-Q. Chen, J. R. Norris and G. R. Fleming, Mechanism of the initial charge separation in bacterial photosynthetic reaction centers, *Proc. Natl. Acad. Sci.* USA 88:11202 (1991).
16. D. Holten and C. Kirmaier, Evidence that a distribution of bacterial reaction centers underlies the temperature and detection wavelength dependence of rates of the primary electron-transfer reactions, *Proc. Natl. Acad. Sci.* USA 87:3552 (1990).
17. M. Bixon, and J. Jortner, Activationless and pseudoactivationless primary electron transfer in photosynthetic bacterial reaction centers, *Chem Phys. Lett.* 159:17 (1989).
18. G. J. Small, J. M. Hayes and R. J. Silbey, On the question of dispersive kinetics for the initial phase of charge separation in bacterial reaction centers, *J. Phys. Chem.*, submitted.
19. The adiabatic ω_D-ω_A energy gap for this process is ~ 2000 cm^{-1}. With the linear electron-phonon coupling parameter values of Table I it would be necessary to take into account the high frequency intramolecular modes of B and H to achieve a maximum rate. This would reduce the effective gap for phonons to a few hundred cm^{-1}.
20. D. Tang, R. Jankowiak, M. Siebert, C. F. Yocum and G. J. Small, Excited-state structure and energy transfer dynamics of two different preparations of the reaction center of photosystem II: a hole burning study, *J. Phys. Chem.* 94:6519 (1990).
21. J. M. Hayes and G. J. Small, Neat and mixed crystal and glass optical spectra of the anthracene-trinitrobenzene complex, *Chem. Phys.* 21:135 (1977).
22. D. Haarer and M. Philpott, Excitons and polarons in organic charge transfer crystals, *in*: "Spectroscopy and excitation dynamics of condensed molecular systems", V. M. Agranovich and R. M. Hochstrasser, ed., North-Holland Publ. Co. (1983).
23. P. A. Lyle, S. Kolaczkowski, J. R. Norris and G. J. Small, to be published.
24. H. M. Sevian, and J. L. Skinner, A molecular theory of inhomogeneous broadening, including the correlation between different transitions in liquids and glasses, *Theor. Chim. Acta* 82:29 (1992).

EFFECT OF CHARGE TRANSFER STATES ON THE ZERO PHONON LINE OF THE SPECIAL PAIR IN THE BACTERIAL REACTION CENTER

Elizabeth J.P. Lathrop and Richard A. Friesner

Department of Chemistry
Columbia University
New York, NY 10027

I. INTRODUCTION

The nature of the excited states of the special pair (SP) bacteriochlorophylls (BChl) in the bacterial photosynthetic reaction center (RC) has been a subject of considerable controversy over the past decade. The crystal structure of the RC,[1-3] in conjunction with numerous spectroscopic experiments,[4-7] has provided a substantial basis for investigating this problem. However, the existence of complicated interactions between multiple electronic surfaces and the difficulty of computing the underlying electronic wavefunctions from a reliable quantum chemical method have rendered a definitive solution difficult to achieve.

In this paper, we review the basic hypothesis for vibronic coupling we had advanced previously[8] in light of the recent structured hole-burning results.[9-13] The essential physical picture of the electronic manifold, and its consequences for the mechanism of charge separation, still appear to be consistent with both spectroscopic[4-7] and kinetic data.[14-22] In our previous work, we invoked a quasiresonant interaction between the Q_y exciton state and a nearby charge transfer(CT) state, in an attempt to explain the structureless hole-burning spectra. However, our new understanding of the vibronic coupling model leads to a much simpler and more satisfying picture of the nature of the SP absorption bands, and at the same time resolves some apparent conflicts with other workers.[9,10,23,24]

A major development over the past several years has been the acquisition of resonance Raman (RR) data obtained via excitation directly into the low energy exciton component (P^* band) of the SP.[25] While there is not yet quantitative agreement between the data of the two experimental efforts (Bocian and coworkers[26] (personal communication) vs. Boxer, Mathies, and coworkers[25]), both cases give a similar qualitative result. In particular, experiments show a substantial number of low frequency modes that are strongly coupled to the P^* band. This is unlike the BChl monomers, where these couplings appear to be weak. Here we provide a straightforward interpretation of this observation which remains consistent with other experimental data (hole-burning[9–13,27–29] and Stark spectra[30–33]) as well.

The paper is divided into five sections. In section II, we review our basic electronic model of the RC and its qualitative experimental consequences. The new interpretation of vibronic coupling in the P^* band is discussed in section III, with special attention to the hole-burning and resonance Raman experiments. Subsequently, in section IV we present results based on a simplified model system to support our picture of the vibronic coupling in the P^* band. In the conclusion, section V, a plan for comprehensive future calculations for model refinement is presented.

II. BASIC MODEL OF THE SP ELECTRONIC STRUCTURE

Our model of the SP electronic structure uses diabatic states of zeroth order coupled by configuration mixing terms. While there have been numerous attempts to calculate the electronic eigenstates of the SP using semiempirical molecular orbital theory (where a comprehensive review is available[34]), it is our view that an accurate *a priori* determination of the quantities of interest e.g. energy levels and off-diagonal couplings of the intramolecular CT states of the SP, is not yet feasible. Where possible, our model parameters are derived from constraints inherent in the experimental data rather than from postulation and ad hoc hypotheses.

As suggested by the following experimental facts, the Q_y exciton state underlying P^* band appears to be coupled to one (or more) diabatic CT states of the SP. The P^* band has an anomalously large Stark effect, where the magnitude of the change in dipole moment associated with the Q_y transition is approximately 3.4 greater than that for the monomeric BChls.[31–33] A substantially enhanced Huang-Rhys S value (~3.5 vs. ~1.0 for the monomers) is also reported, based on an analysis of the hole-burning results on the assumption that the P^* band is composed of a single electronic transition.[11] Finally, an unusually large width and temperature dependence[7] of the absorption spectra of P^* is observed.

Several years ago, Youvan and coworkers produced an important mutant RC in the bacterium *Rb.capsulatus*, in which one BChl of the SP was changed to bacteriopheophytin (BPh) by the removal of the Mg atom. This "heterodimer" mutant can successfully carry out primary charge separation, albeit at a reduced rate[36,37] and with a substantially reduced quantum yield.[37] The femtosecond transient absorption experiments of Holten, Kirmaier and coworkers revealed that a charge transfer intermediate in this system is formed very rapidly after the initial flash; the species created has been identified as the internal CT state of the heterodimer.[36,37,38] Furthermore, Stark measurements on the heterodimer reveal two features, which are hypothesized to be admixtures of the exciton and CT states.[39] Both spectroscopic and kinetic experiments were also performed on a similar heterodimer mutant for the *Rb.sphaeroides* system,[11,40–42] producing similar qualitative results.

In conjunction with *in vitro* redox measurements on BChl and BPh monomers, the measurement of the free energy gap of a mutant allows us to attempt an extrapolation of the location of the CT state in the wild type RC. Solution measurements[43] indicate that the difference in redox potential between a BChl and a BPh is on the order of 0.2-

0.3 eV. One could argue that this differential will be altered in a protein environment; however, a second mutant in which the BPh acceptor on the L side of the RC (the charge separation active branch) is replaced by the BChl provides a second estimate of this redox change. In this mutant, the free energy gap for charge separation (as measured by delayed fluorescence) changes from 0.26 eV[44] in the wild type to 0.06 eV (Schenck, personal communication), suggesting that the estimate of 0.2 eV from the solution measurements is reasonable.

If the CT intermediate observed in the heterodimer kinetics and Stark measurement is on the order of 0.1 eV below the Q_y exciton state, the wild type location can be restricted to a value no more than 0.2 eV above the exciton state. This estimate is consistent with another observation. For large energy gap between the exciton and CT states (5000 to 10000 cm^{-1}, as has been proposed by other workers), the configuration mixing would have to be correspondingly large (at least several thousand wavenumbers) to explain the large anomalous Stark effect in the P^* band. This model implies a large Stark effect for the CT state at high energy, due to the acquired transition dipole intensity from mixing with the exciton state. Indeed, such a transition is predicted to occur in the Q_x region with an oscillator strength greater than that of the Q_x band.[45] The failure to observe any such high energy state in this system favors our estimate of a lower CT state position.

The zeroth order energies we refer to here are those at the equilibrium geometry of the CT state. Semiempirical electronic structure calculations to date have been carried out for all states using the ground electronic state equilibrium geometry. As we shall argue, there is good reason to believe that the geometrical distortions in the CT state are significant. In this case, the vertical (adiabatic) energy gap between the Q_y and CT states will be substantially larger than the difference between the zeroth order energies at equilibrium. This is one possible explanation for the failure of the electronic structure calculations to yield a sufficiently low energy for the CT state. There are of course other alternative explanations, such as the numerous approximations involved in both the theoretical methods and the treatment of the protein environment, or uncertainties in the geometry of the porphyrin rings in the RC.

In this paper, we will take the above experimental constraints as working hypothesis. With the position of the intradimer CT state limited in between 0.0 and 0.2 eV above the Q_y exciton state, we hypothesize that the coupling to this state is primarily responsible for the observed Stark enhancement in the P^* band.

III. VIBRONIC COUPLING IN THE P^* BAND

A central anomalous feature of the P^* band is the apparent large amplitude of its Franck-Condon factors as compared to BChl monomers *in vitro* and the spectator BChl in the RC. Hole-burning experiments[11] yield an estimate of 3.5 for the total Huang-Rhys S factor for P^*, as compared to values of 1.0-2.0 for the monomeric and spectator BChl. This strong vibronic coupling is further manifested in the large bandwidth and strong temperature dependence of P^*.

In principle, resonance Raman spectroscopy allows the direct determination of the S value of each vibrational mode; in practice the analysis is difficult because reliable excitation profiles and, preferably, absolute intensity measurements are required. Whether the P^* Raman data permit this kind of modeling remains to be seen.

However, some qualitative conclusions can be drawn from the existing resonance Raman data. In data sets of both Bocian and coworkers[26] (and by personal communication) and Boxer, Mathies and coworkers,[25] there is a substantially greater aggregate intensity in the low frequency part of the RR spectrum than in the high frequency modes. While the enhancement patterns observed by the two groups are very different, both are compatible with the argument we will be making (Bocian and coworkers observed enhancement of virtually all of the porphyrin low frequency modes, while Boxer, Mathies, and coworkers observed a smaller subset of these low frequency modes). Present theory is incapable of determining which (if either) data set is correct; indeed, the RR results are essential for fixing vibronic parameters in model calculations.

The basic argument we propose here is as follows. Coupling of the Q_y state to the CT state leads to enhancement of the effective S values in the low frequency region. This is responsible for the small amplitude of the ZPL in hole-burning experiments on the P^* band. This interpretation is significantly different from the one we have proposed previously, where the reduction or elimination of the ZPL amplitude was proposed to be caused by quasiresonant interaction alone. This new interpretation agrees with ideas previously put forth by Small and coworkers.[9,10,23,24]

The above discussion need not imply that our previous calculations are inaccurate. Indeed, these simulations strengthen the argument described above: mixing with the CT state leads to reduced values of S and hence small amplitude ZPLs. Quasiresonant mixing can also be effective in removing sharp structure from the side holes, although it is not clear whether this is needed to explain the experimental data, given the significant number of low frequency modes observed in the Raman spectrum (the dispersion of which can also wash out side hole structure). However, a correct basic understanding of how the CT state reduces the ZPL is critical to constructing reasonable models.

The enhancement of the total S value by the CT state coupling can be understood using the simplest possible model of the system: two excited electronic states and a single vibrational mode. The first state (Q_y exciton state) is assumed to carry all the oscillator strength while the second state (CT state) is dark. In the diabatic representation, the effective Hamiltonian in the linear vibronic coupling model of the excited state surface can be written as:[46,47]

$$\begin{aligned} H &= \Delta E + J(|1><2| + |2><1|) + w b^\dagger b \\ &\quad + (\frac{g_1}{\sqrt{2}}|1><1| + \frac{g_2}{\sqrt{2}}|2><2|)(b + b^\dagger). \end{aligned} \tag{1}$$

Here, $|1>$ and $|2>$ are the diabatic electronic states (exciton and CT states, respectively), J their electronic coupling parameter, ω the vibrational frequency, $b^\dagger$ and b the boson creation and annihilation operators, and g_1 and g_2 the equilibrium position shifts in the excited states $|1>$ and $|2>$, and ΔE the zeroth order splitting between the two electronic states. The zero energy is set to ($\frac{1}{2}\Delta$E - E_g + $\frac{1}{2}\omega$) where $\hbar$ = 1. and E_g is the ground state electronic energy of a Born-Oppenheimer model.

Friesner and Silbey have previously analyzed the qualitative features of the solutions to this Hamiltonian in various regions of the parameter space.[46,47] The most important demarcation for our purposes is that between strong and weak electronic coupling, as determined by the dimensionless ratio γ = J/ω . If $\gamma \ll 1$, perturbation theory in J is valid, and we are in the weak electronic coupling regime. In this case, the vibronic eigenstates of the system are well approximated by linear combinations of diabatic eigenstates of the form:[46]

$$\Psi^{\pm}(Q) = \frac{1}{\sqrt{2}}(|1> \phi^{g_1}_{m_1}(Q + g_1) \pm |2> \phi^{g_2}_{m_2}(Q + g_2)). \tag{2}$$

Here $\phi^{g_i}_{m_i}(Q)$ is the m_ith harmonic oscillator wave function with equilibrium position

g_i, relative to ground state normal coordinate Q, and m_1 and m_2 are chosen such that the two closest vibronic configurations are nearly degenerate in energy.

In this limit, the dark CT state does not contribute to the effective value of S for the absorption band. The vibronic coupling leads to a shift in the transition energies of the vibronic eigenstates and to a small amount of increased intensity in the vicinity of the zeroth order CT state electronic energy (this is what is usually understood by vibronic borrowing). However, the relative amplitude of each vibronic line is entirely determined by the Q_y potential energy surface, as the Franck-Condon progressions appearing in both the original Q_y and the previously dark regions are minimally altered from the Q_y progression.

In the strong coupling case, $\gamma \gg 1.$, a different situation occurs. The wavefunction depends on a second ratio, S/ω. When S $\gg \omega$, the adiabatic limit is in effect. For S $\ll \omega$, the vibronic coupling can be treated as a perturbation in g_i. For this latter situation, the eigenstates are approximately of the form,[46]

$$\Psi_n^{\pm}(Q) \quad \sim \quad |\pm > \phi_n^0(Q), \tag{3}$$

where $|\pm >$ are the adiabatic states. For large J, the new vibrational wavefunction is a harmonic oscillator form with the displacement set to the average of those for the two excited state vibrational surfaces. While the adiabatic limit is more complex to analyze, we can again anticipate the vibrational wavefunction for the bright electronic component to be substantially modified from its original form.

Here in the strong coupling region, it is apparent that the magnitude of the displacement in the CT state can significantly affect the observed value of S. We note that in the adiabatic limit, the vibrational wavefunctions are anharmonic. However, for small vibrational quantum numbers, the wavefunctions can be approximated by linearly displaced oscillators with renormalized temperature dependent frequencies and displacements. A large S in the CT state can now appear to substantially increase the observed S value in the P^* band.

A final parameter of is ΔE, the zeroth order splitting between the two electronic states, analyzed previously by Friesner and Silbey.[47] Large ΔE tends to shift the wavefunction towards weak electronic coupling; however the effects can be complicated and depend upon the interplay of all three relevant parameters ΔE, J, and g_i.

IV. MODEL SYSTEM CALCULATIONS

To illustrate these effects, we present calculations based on the simple 2-state, 1-mode model system described above. The amplitude of the ZPL is determined as a function of the electronic coupling J, vibrational frequency ω, displacement of the two surfaces g_1 and g_2, and the zeroth order electronic splitting ΔE. The analysis of these results is complicated by the overlap of the two electronic bands in the intermediate coupling regime. Formally, there exist two ZPLs, one associated with each electronic state. The ZPL originating from the dark CT state may be difficult to observe if it is buried under the vibronic progression of the lower component or mixed with the higher lying component of the Q_y exciton state. For the purpose here, we have defined the effective Huang-Rhys factor S_{eff} as the ratio of the ZPL amplitude of the allowed low energy transition to the total intensity in this band.

The effective S values were determined from the homogeneous absorption lineshapes calculated using a thermally averaged Green's function approximation method

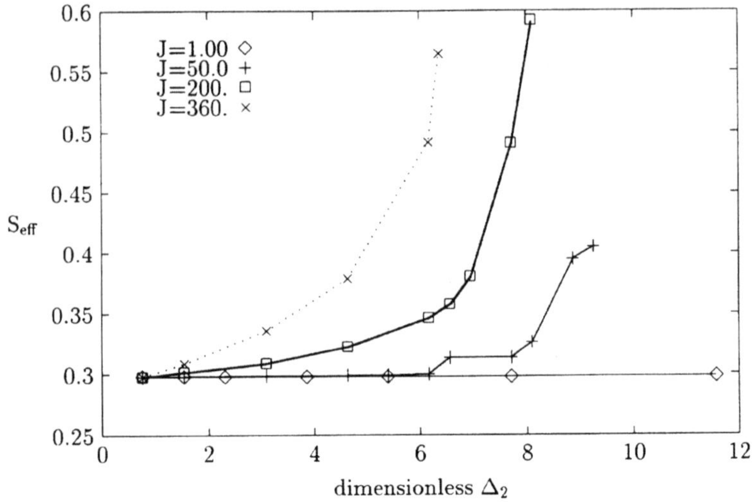

Figure 1. Effective S for increasing electronic coupling parameter J between the Q_y exciton and CT states, as a function of dimensionless displacement of CT state Δ_2. The energy gap is fixed at 2835 cm^{-1} and the S value for the Q_y exciton state at 0.297.

previously developed by Lagos and Friesner.[48] This approximate method was extensively tested against convergent basis set results[49] and was shown to reliably reproduce single-photon optical lineshapes at various temperatures for the RC system.[50] Although a direct numerical solution to the linear vibronic coupling of this 2-level, 1-vibrational mode system is possible,[51,52] we employed the Green's function method with extension to the multilevel, multimode system with future investigations in mind. In our simple one mode model, the vibrational frequency ω is set to 94 cm^{-1}, and the dimensionless displacement of this mode in the excited Q_y state as compared to the ground electronic state is set at 0.771 (Δ_1), giving an S value of 0.297 for this exciton state. These are reasonable values for a low frequency porphyrin mode coupled to the excitonic transition.

Figure 1 shows the effective value of S verses the dimensionless displacement of the CT state for various values of J. The basic point of this paper is evident here. For $J \ll \omega$ ($\gamma \ll 1$), increasing the displacement of the CT surface has little effect on the effective S value. This is to be expected in the weak coupling limit where a perturbative treatment applies. As J is increased past ω , perturbation theory in J fails and the CT surface exerts a significant influence on the effective S value.

Figure 2 shows results for the large J, zero ΔE limit, in which perturbation theory in the vibronic mixing of the (+) and (-) states becomes valid. Here, assuming that the displacement in the (+) state is the average of those in the exciton and CT states, the effective S value approaches the limiting value,

$$g^{+} = \frac{1}{2}(g_1 + g_2). \tag{4}$$

Here g_1 is from the Q_y exciton state and g_2 is from the CT state. In the case where $\Delta_2 = 1.543$, a calculated value of 1.157 for the dimensionless displacement for the (+) state yields an effective S of 0.669 in this limit. Indeed, the corresponding curve for large J reaches this value.

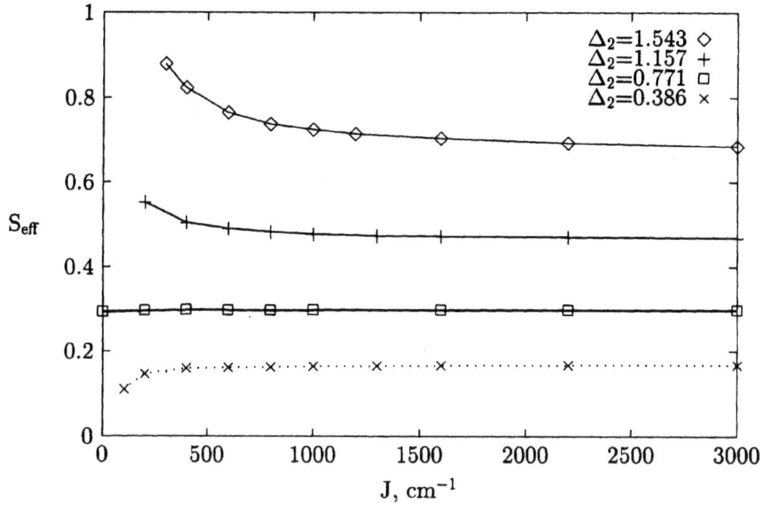

Figure 2. Effective S in the large J, zero ΔE limit, for various values of CT state displacement. Here Δ_1 = 0.771.

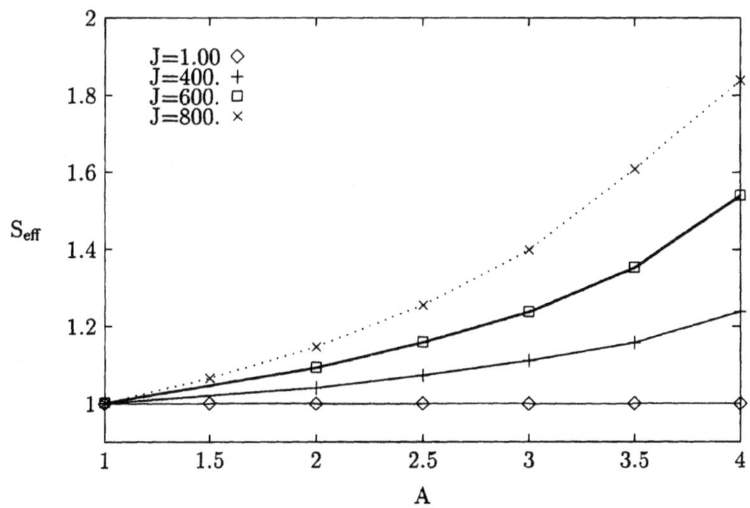

Figure 3. The multimode model. The Effective S is plotted against the scaling factor A (where g_2 = Ag_1 for all modes) for changing values of J. The energy gap is fixed at 3000 cm^{-1} and the total S value for the exciton state is 1.0.

We repeat the calculations for a multimode Hamiltonian containing 6 vibrational modes and show the results in figure 3. Here the frequencies and coupling constants for the exciton state are estimated from the resonance Raman data of Schreve et al.[25] (assuming scattering was in the short-time limit). Although these estimates of the coupling constants are not quantitatively accurate, we wish to demonstrate the effect observed for the single mode case also exists in the complex multimode systems. In this calculation, we scale the coupling constants of all 6 modes in the CT state uniformly by a scaling factor (A). As can be seen, the qualitative behavior shown in Figure 1 is again reproduced.

V. CONCLUSION

In this paper we propose a simple explanation of how the apparent Huang-Rhys factor in the P^* band of the bacterial reaction center can be enhanced by coupling to a charge transfer state. The argument requires that the electronic coupling strength to the CT state be comparable to or greater than the vibrational frequencies of the enhanced modes. As these modes were experimentally found to be in the range of 200 cm^{-1} or smaller, this requirement is compatible with previous estimates of the coupling strength based on electronic structure calculations or phenomenological estimates.

A more ambitious goal is to develop a simple quantitative relation between the observed S values, the unperturbed monomeric parameters, and the CT state parameters. Our future objective is to find the subset of the parameter space defined by parameters J, ΔE, and S_{CT}, that will lead to the appropriate values of the total S for the P^* band obtained from hole-burning experiments. This will greatly facilitate the quantitative modeling of the electronic surfaces of the reaction center, in which other spectroscopic data (e.g. Stark measurements) must also be matched.

ACKNOWLEDGEMENTS

Support from the NIH is gratefully acknowledged. RAF is a recipient of a Research Career Development Award and a Camille and Henry Dreyfus Teacher-Scholar Award.

References

[1] Michel, H., Epp, O. and Deisenhofer, J. (1986) EMBO J. 5, 2445-2451.

[2] Chang, C.H., Tiede, D.M., Tang, J., Smith, U., Norris, J.R. and Schiffer, M. (1986) FEBS Lett. 205, 82-86.

[3] Allen, J.P., Feher, G., Yeates, T.O., Komiya, H. and Rees, D.C. (1987) Proc. Natl, Acad. Sci. USA 84, 5730-5734.

[4] Phillipson, K.D. and Sauer, K. (1973) Biochemistry 12, 535-539.

[5] Shuvalov, V.A. and Asadov, A.A. (1979) Biochim. Biophys. Acta 545, 296-308.

[6] Vermeglio, A. and Paillotin, G. (1982) Biochim. Biophys. Acta 681, 32-40.

[7] Kirmaier, C. and Holten, D. (1988) in Photosynthetic Bacterial Reaction Center: Structure and Dynamics, Vol. 149 (Breton, J. and Vermegio, A., eds.), PP. 219-228, Plenum, New York.

[8] Won, Y. and Friesner, R.A. (1988) J. Phys. Chem. 92, 2214-2219.

[9] Tang, D., Jankowiak, R., Gillie, J.K., Small, G.J. and Tiede, D.M. (1988) J. Phys. Chem. 92, 4012-4015.

[10] Tang, D., Jankowiak, R., Small, G.J. (1989) Chemi. Phys. 131, 99-113.

[11] Middendorf, T.J., Mazzola, L.T., Gaul, D.F., Schenck, C.C., Boxer, S.G. (1991) J. Phys. Chem. 95, 10142-10151.

[12] Johnson, S.G., Tang, D., Jankowiak, R., Hayes, J.M. and Small, G.J. and Tiede, D.M. (1989) J. Phys. Chem. 93, 5953-5957.

[13] Johnson, S.G., Tang, D., Jankowiak, R., Hayes, J.M. and Small, G.J. and Tiede, D.M. (1990) J. Phys. Chem. 94, 5849-5855.

[14] Woodbury, N., Becker, M., Middendorf, D., Parson, W.W. (1985) Biochemistry 24, 7516-7521.

[15] Martin, J.L., Breton, J., Hoff, A.J., Migus, A. and Antonetti, A. (1986) Proc. Natl. Acad. Sci. USA 83, 957-961.

[16] Breton, J., Martin, J.L., Migus, A., Antonetti, A. and Orszag, A. (1986) Proc. Natl. Acad. Sci. USA 83, 5121-5125.

[17] Wasliewski, M. and Tiede, D. (1986) FEBS Lett. 204, 368-372.

[18] Kirmaier, C. and Holten, D. (1988) FEBS Lett. 239, 211-218.

[19] Breton, J., Martin, J.L., Fleming, G.R., Lambry, J.C. (1988) Biochemistry 27,8276.

[20] Fleming, G.R., Martin, J.L. and Breton, J. (1988) Nature 333, 190-192.

[21] Kirmaier, C., Holten, D. (1991) Biochemistry 30, 609.

[22] Chan, C., DiMagno, T.J., Chen, L.X., Norris, J.R. and Fleming, G.R. (1991) Proc. Natl. Acad. Sci. USA 88, 11202-11206.

[23] Hayes, J.M. and Small, G.J. (1986) J. Phys. Chem. 90, 4928-4930.

[24] Hayes, J.M., Gillie, J.K., Tang, D. and Small, G.J. (1988) Biochim. Biophys. Acta 932, 287-305.

[25] Schreve, A.P., Cherepy, N.J., Franzen, S., Boxer, S.G. and Mathies, R.A. (1991) Proc. Natl. Acad. Sci. USA 88, 11207-11211.

[26] Donohoe, R.J., Dyer, r.B., Swanson, B.I., Violette, C.A., Frank, H.A. and Bocian, D.F. (1990) J. Am. Chem. Soc. 112, 6716.

[27] Meech, S.R., Hoff, A.J. and Wiersma, D.A. (1985) Chem. Phys. Lett. 121, 287-292.

[28] Boxer, S.G., Lockhart, D.J. and Middendorf, T.R. (1986) Chem. Phys. Lett. 123, 476-482.

[29] Boxer, S.G., Middendorf, T.R. and Lockhart, D.J. (1986) FEBS Lett. 200, 237-241.

[30] Losche, M., Feher, G. and Okamura, M.Y. (1987) Proc. Natl. Acad. Sci. USA 84, 7537-7341.

[31] Lockhart, D.J. and Boxer, S.G. (1987) Biochemistry 26, 664-668.

[32] Lockhart, D. and Boxer, S.G. (1988) Proc. Natl. Acad. Sci. USA 85, 107-111.

[33] Braun, H.P., Michel-Beyerle, M.E., Breton, J., Buchanan, S. and Michel, H. (1987) FEBS Lett. 221, 221-225.

[34] Hanson, L.K. (1988) Photochem. Photobiol. 47, 903-921.

[35] Bylina, E.J. and Youvan, D.C. (1988) Proc. Natl. Acad. Sci. USA 85, 7226-7230.

[36] Kirmaier, C., Holten, D., Bylina, E.J. and Youvan, D.C. (1988) Proc. Natl. Acad. Sci. USA 85, 7563-7566.

[37] Kirmaier, C., Gaul, D., DeBey, R., Holten, D., Schenck, C.C. (1991) Science 251, 922.

[38] Chan, C., Chen, L.X., DiMagno, T.J., Hanson, D.K., Nance, S.L., Schiffer, M., Norris, J.R. and Fleming, G.R. (1991) Chem. Phys. Lett. 176, 366-372.

[39] DiMagno, T.J., Bylina, E.J., Angerhofer, A. (1990) Biochemistry 29, 899-907.

[40] Nagarajan, V., Parson, W.W., Gaul, D., Schenck, C. (1990) Proc. Natl. Acad. Sci. USA 87, 7888-7892.

[41] Hammes, S.L., Mazzola, L., Boxer, S.G., Gaul, D.F. and Schenck, C.C. (1990) Proc. natl. Acad. Sci. USA 87, 5682-5686.

[42] Mattioli, T.A., Gray, K.A., Lutz, M., Oesterhelt, D. and Robert, B. (1991) Biochemistry 30, 1715-1722.

[43] Davis, M.S., Forman, A., Hanson, L.K., Thornber, J.P. and Fajer, J. (1979) J. Phys. Chem. 83, 3325-3332.

[44] Goldstein, R.A., Takiff, L. and Boxer, S.G. (1988) Biochim. Biophys. Acta 934, 253-263.

[45] Parson, W.W. and Warshel, A. (1987) J. Am. Chem. Soc. 109, 6143-6152.

[46] Friesner, R. and Silbey, R. (1981) J. Chem. Phys. 74, 1166-1174.

[47] Friesner, R. and Silbey, R. (1981) J. Chem. Phys. 75, 3925-3936.

[48] Friesner, R.A. and Lagos, R.E. (1984) J. Chem. Phys. 81, 5899-5905.

[49] Won, Y., Lagos, R. and Friesner, R. (1986) J. Chem. Phys. 84, 6567-6574.

[50] Won, Y. and Friesner, R.A. (1988) J. Phys. Chem. 92, 2208-2214.

[51] Merrifield, R.E. (1963) Radiat. Res. 20, 154.

[52] Fulton, R.L. and Gouterman, M. (1960) J. Chem. Phys. 33, 872.

THEORETICAL STUDIES ON THE ELECTRONICAL STRUCTURE OF THE SPECIAL PAIR DIMER AND THE CHARGE SEPARATION PROCESS FOR THE REACTION CENTER *RHODOPSEUDOMONAS VIRIDIS*

P.O.J. Scherer and Sighart F. Fischer

Physik-Department T38
Technische Universität München
8046 Garching, Germany

INTRODUCTION

In a recent article[1] we have shown, that the electronic spectra for the absorbance, the linear dichroism and the electrochromicity can be well simulated for the prosthetic groups of the reaction centers Rp. viridis, Rb. sphaeroidis, Chloroflexus aurantiacus, Rb. capsulatus and a HIS M280 → LEU M280 mutant on the basis of an extended exciton model. The model included the Qy-transitions of the six pigments, their excitonic coupling and two internal charge transfer states of the dimer. The estimate of the coupling strength was based on quantum calculations of the INDO-S type for the structure of Rp. viridis and proper adjustments for the other reaction centers. In order to arrive at the proper strength of the electrochromicity observed for the special pair the two charge transfer states had to be separated by about 0.5 eV. This separation did not come out of the calculation of the isolated dimer and was interpreted as a field induced separation due to an asymmetric charge distribution of the surrounding protein environment.

In this article we want to explore the electronic structure of the dimer in more detail. We include double and triple substituted configurations for the configuration interaction (CI). This enables us also to study transient absorption spectra such as the excitations built on the lowest excited singlet state of the dimer P^* and a charge transfer state P^+X^-,

The Photosynthetic Bacterial Reaction Center II
Edited by J. Breton and A. Verméglio, Plenum Press, New York, 1992

where X^- could be B_L^-, H_L^- or Q_A^-.These calculations provide a consistent description of these absorption spectra, if sufficient configurations (up to 6000) are included. However the electrochromicity remains too small. Among the double excited states we find as lowest singlet a state, which consists of the proper singlet configuration from two lowest interacting triplets, each localized on one dimer half. This state interacts strongly with the charge transfer states and could provide an alternative explanation for the strong electrochromicity in the absence of strong symmetry reducing local fields of the surrounding protein. Even though there might be some question if this state is sufficiently low in energy as indicated by our calculations, we analyse this model in detail since it might be of general interest. Furthermore we investigate possible Rydberg states on the basis of an RINDO-model which includes 2s and 2p functions for the hydrogens and 3s, 3p for the C and N atoms. We find here a low lying Rydberg state, which could also induce the strong electrochromicity. It extends considerably into the pheophytine directly and could therefore provide a coupling route for the charge separation. We conclude the study on the unusual properties of the dimer by the estimate of the intermolecular potential for the ground and the lowest excited states, which lead to a consistent description of an oscillatory behaviour in the stimulated emission after coherent excitation[2].

TRANSIENT EXCITATIONS OF THE SPECIAL PAIR

As transient excitations we consider the excitations of P^* and P^+X^-, where the counter ion X^- may be also eliminated, so that the spectrum of the cation results. While the spectrum of P^* is measured[3] on the time scale of the lifetime of P^*, stationary spectra of P^+ can be observed[4]. The method of our calculations has been described elsewhere[5]. We used Zerner's parameterization[6] and included 6000 configurations, some of which have been treated pertubatively. Fig. 1 shows in the lower part the ground state spectrum with the dominating P^* transition followed in energy by the transition to the so called upper band. The P^*-state contains considerable admixture of the anti symmetric combination of the lowest charge transfer states. This is not so important for the upper dimer band, which could on the basis of symmetry interact with the symmetric combination. The reason for this is found in cancellation of local interactions not present for the lower dimer state P^*. This result is consistent with the observation of a larger inhomogeneity in the lower dimer band P^* as compared to the upper[5], since only the charge transfer component is sensitive to local inhomogeneities in the charge distribution.

Below the dimer transition P^* a state with low intensity is predicted. We denote it as the P^{**} state and will discuss it in detail in conjunction with the electrochromicity.

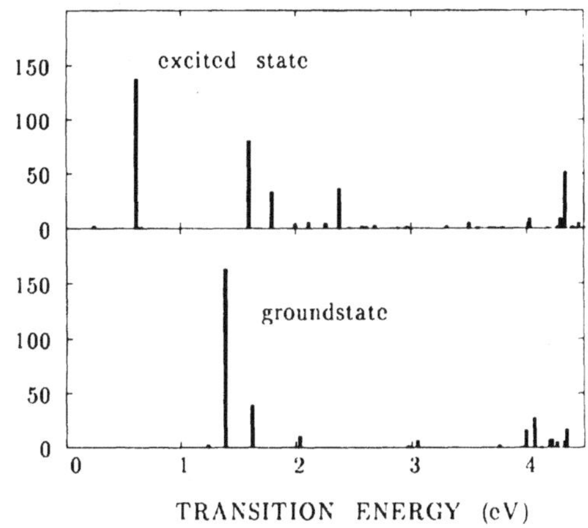

Figure 1. Absorption spectrum of the dimer. The lower and upper part show the transition from the ground state P and the transient absorption of P* rsp. .

It should be noted that the spectrum gives about the proper splitting of the lower and upper dimer band but both are blue shifted as compared to the experiments[8]. From a study of the supermolecule including all six pigments we found[5] that the interaction with the other state via double and triple configurations large accounts for the missing shift predicted for the isolated dimer. Finally we like to point out that the charge transfer states located at 3 and 4 eV carry little intensity, which is also on outcome of the multiple configuration interaction and could not be explained with single CI or within the extended exciton model.

The spectrum of P* on the upper part of Fig. 1 shows a very strong transition at about 0.6 eV. This has not been observed yet, since subpicosecond time resolution at this wavelength is difficult to achieve. We like to predict that an equivalent transition should be found for the mutant which lacks the pheophytine H_L and has a longer lifetime of 150 ps for P* [10]. The two transitions in the energy regime where the monomers absorb correlate well with measured transient absorption spectra[3]. The higher transitions are out of the region which is usually analysed experimentally. The very low intensity transition at about 0.2 eV corresponds to the transition between the lower and the upper dimer band.

In Fig. 2 the excitation spectrum of the cation P^+ is shown. The calculated spectra a) and b) and c) differ in the number of configurations included in the calculation, which are 69, 300 and 3000 respectively. Only the one with the 3000 configurations is in good agreement with experiments. This makes the analysis in terms of dominating configurations difficult but it is instructive. The ground state of the cation P^+ can be represented in terms of the dominating configurations of the orbitals from the dimer P as follows. We denote the occupied orbitals of P in order of decreasing energy with the indices 0, -1, -2,... and those of the empty orbitals with increasing energy as 1, 2, 3,... . Using further creation and destruction operators $a^*_{v,\uparrow}$, $a^*_{v,\downarrow}$ and $a_{v,\uparrow}$, $a_{v,\downarrow}$ respectively with the spin indices $\uparrow$, $\downarrow$ we can represent a cation state $a_{0,\uparrow}|P\rangle$ or $a_{0,\downarrow}|P\rangle$. This is the dominating component of $|P^+\rangle$.

There are other components admixed, which correspond to excited configurations. The dominant are

$$|P^+_\uparrow\rangle = 0.85\, a_{0,\uparrow}\, |P\rangle + 0.25\, (a^*_{1,\uparrow}\, a_{-1,\downarrow} + a^*_{1,\downarrow}\, a_{-1,\uparrow})\, a_{0,\uparrow}\, |P\rangle$$
$$+ 0.25\, a^*_{1,\uparrow}\, a^*_{1,\downarrow}\, a_{-1,\uparrow}\, a_{-1,\downarrow}\, a_{0,\uparrow} |P\rangle + \ldots$$

and the lowest excited states of P^+ are given by

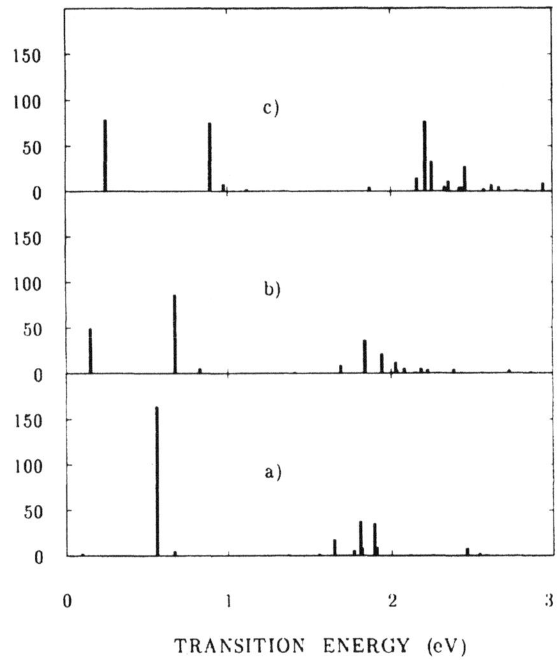

Figure 2. Absorption spectrum of P^+ calculated with different configuration interaction, case (a) 69 configurations, case (b) 300 configurations and (c) 3000 configurations, in case (c) a counter ion was placed in the B_L position.

$$|P^{+}\rangle^{*}_{1} = 0.7\, a_{-1,\uparrow}\, |P\rangle - 0.5\, a_{0,\uparrow}\, a_{0,\downarrow}\, a^{*}_{1,\uparrow}\, |P\rangle + 0.3\, a_{-1,\uparrow}\, a_{-1,\downarrow}\, a^{*}_{2,\uparrow}\, |P\rangle$$
$$-0.2\, a_{0,\uparrow}\, a_{0,\downarrow}\, a_{-1,\uparrow}\, a^{*}_{1,\uparrow}\, a^{*}_{1,\downarrow}\, |P\rangle + \ldots$$

and

$$|P^{+}\rangle^{*}_{2} = -0.5\, a_{-1,\uparrow}|P\rangle \quad -0.4\, a_{0,\uparrow}\, a_{0,\downarrow}\, a^{*}_{1,\uparrow}|P\rangle \quad +0.5\, a_{-1,\uparrow}\, a_{-1,\downarrow}\, a^{*}_{2,\uparrow}|P\rangle$$
$$-0.2\, ((a_{-1,\uparrow}\, a_{0,\downarrow} - a_{0,\uparrow}\, a_{-1,\downarrow})\, a^{*}_{2,\uparrow} + 2\, a^{*}_{2,\downarrow}\, a_{-1,\uparrow}\, a_{0,\uparrow})|P\rangle + \ldots$$

The energy location of the lowest state at 0.26 eV is close to the observed one and it has also almost exactly the correct intensity of 76 $Debye^2$. The second transition of Fig 2c corresponds to the well known 13000 nm - transition which has been used as marker for the appearance of the P^+ state after charge separation. Experimentally the low energy transition at 0.26 eV is unusually broad. We believe that this is largely due to its high polarizability and consequently its high sensitivity against environmental inhomogeneous broadening. In addition it can interact with the infrared transition from an H-bonded network as it is formed around solvated protons. Due to its strong dipole moment it can transfer intensity into such solvent dipoles. If these form a continuum around the $(P^+)^*$ band a Fano line shape can be produced. The direction of the lowest excitation of P^+ is similar to the $P \rightarrow P^*$ transition perpendicular to the symmetry axis.

CHARACTERIZATION OF THE STATE P^{**}

The nature of the state which we call P^{**} is most easily understood if we expand the dimer SCF orbitals in terms of orbitals of the isolated dimer halves PL and PM

$$\varphi_n^{P} = \Sigma\, C_{n,\nu}^{L}\, \varphi_\nu^{L} + \Sigma\, C_{n,\nu}^{M}\, \varphi_\nu^{M}$$

and transform the CI basis states by expanding the operator creating an electron with spin σ in the dimer orbital φ_n^{P} as

$$a^{*}_{P,n,\sigma} = \Sigma\, C_{n,\nu}^{L}\, a^{*}_{L,\nu,\sigma} + \Sigma\, C_{n,\nu}^{M}\, a^{*}_{M,\nu,\sigma}$$

The single excited configurations for example are this way described as mixtures of single excitations of the dimer halves plus charge transfer contributions. In the basis of the supermolecule states P^{**} is a mixture of several contributions. Using the transformed basis, however, only one configuration dominates. It is created by the operator

$$A^*_{TT} =$$

$$3^{-1/2}\,[\,a^*_{L,1,\uparrow}a_{L,0,\downarrow}a^*_{M,1,\downarrow}a_{M,0,\uparrow}+a^*_{M,1,\uparrow}a_{M,0,\downarrow}a^*_{L,1,\downarrow}a_{L,0,\uparrow}$$
$$-\,(a^*_{L,1,\uparrow}a_{L,0,\uparrow}-a^*_{L,1,\downarrow}a_{L,0,\downarrow})(a^*_{M,1,\uparrow}a_{M,0,\uparrow}-a^*_{M,1,\downarrow}a_{M,0,\downarrow})/2\,]$$

where 1 (0) denotes the lowest unoccupied (highest occupied) orbital of the dimer half L or M.

The operator A^*_{TT} creates a singlet excitation of four electrons occupying the half filled orbitals (open shells) $\varphi_{L,0}$, $\varphi_{L,1}$, $\varphi_{M,0}$, and $\varphi_{M,1}$. and can be easily interpreted as the singlet combination of two triplet excitations located on the dimer halves as follows. The two electrons in the orbitals $\varphi_{L,0}$ and $\varphi_{L,1}$ are combined to give the three components of a triplet state :

$$A^*_{L,1,-1} = a^*_{L,1,\downarrow}a_{L,0,\uparrow}$$
$$A^*_{L,1,0} = (a^*_{L,1,\uparrow}a_{L,0,\uparrow}-a^*_{L,1,\downarrow}a_{L,0,\downarrow})/\sqrt{2}$$
$$A^*_{L,1,1} = a^*_{L,1,\uparrow}a_{L,0,\downarrow}$$

and corresponding operators create the triplet states on the other dimer half. The singlet combination of the two triplets now is given by

$$3^{-1/2}\,(A^*_{L,1,-1}A^*_{M,1,1}+A^*_{M,1,-1}A^*_{M,1,1}-A^*_{L,1,0}A^*_{M,1,0})$$

which is just A^*_{TT} as defined above.

Besides the double triplet excitation A^*_{TT} an independent singlet excitation of the four electrons exists which is the coupling of two singlet excitations on the dimer halves

$$A^*_{SS} = (a^*_{L,1,\uparrow}a_{L,0,\uparrow}+a^*_{L,1,\downarrow}a_{L,0,\downarrow})(a^*_{M,1,\uparrow}a_{M,0,\uparrow}+a^*_{M,1,\downarrow}a_{M,0,\downarrow})/2$$

The coupling matrix element between SS and TT

$$V(SS,TT) = (\,\langle\varphi_{L,1}\varphi_{M,0}|\varphi_{L,1}\varphi_{M,0}\rangle+\langle\varphi_{M,1}\varphi_{L,0}|\varphi_{M,1}\varphi_{L,0}\rangle$$
$$-\,\langle\varphi_{L,1}\varphi_{M,1}|\varphi_{L,1}\varphi_{M,1}\rangle-\langle\varphi_{L,0}\varphi_{M,0}|\varphi_{L,0}\varphi_{M,0}\rangle\,)\,\tfrac{1}{2}\sqrt{3}$$

involves the exchange of two electrons between the dimer halves. It vanishes at large distance of the dimer halves and is still very small for the Rps.viridis structure as it is proportional to the squared overlap of orbitals on L and M. The diagonal energy of the double triplet state

$$E_{TT} = E_L^T + E_M^T + E_{el} + E_{exch}$$

is the sum of three contributions: the energy of the independent triplets

$$E_L^T = \varepsilon_{L,1} - \varepsilon_{L,0} - \langle \varphi_{L,0}\varphi_{L,0} | \varphi_{L,1}\varphi_{L,1} \rangle$$

$$E_M^T \quad \varepsilon_{M,1} - \varepsilon_{M,0} - \langle \varphi_{M,0}\varphi_{M,0} | \varphi_{M,1}\varphi_{M,1} \rangle,$$

the electrostatic interaction

$$E_{el} = \langle \varphi_{L,1}\varphi_{L,1} - \varphi_{L,0}\varphi_{L,0} | \varphi_{M,1}\varphi_{M,1} - \varphi_{M,0}\varphi_{M,0} \rangle = \langle \Delta\rho_L | \Delta\rho_M \rangle$$

plus an exchange contribution

$$E_{exch} = \frac{1}{2}(\langle \varphi_{L,1}\varphi_{M,1} | \varphi_{L,1}\varphi_{M,1} \rangle + \langle \varphi_{L,0}\varphi_{M,0} | \varphi_{L,0}\varphi_{M,0} \rangle$$

$$+ 3\langle \varphi_{L,1}\varphi_{M,0} | \varphi_{L,1}\varphi_{M,0} \rangle + 3\langle \varphi_{L,0}\varphi_{M,1} | \varphi_{L,0}\varphi_{M,1} \rangle)$$

which is again proportional to squared overlaps.

At large distances of L and M we have $E_{TT} \simeq E_L^T + E_M^T$. The energy difference

$$E_{SS} - E_{TT} = (E_L^S - E_L^T) + (E_M^S - E_M^T)$$

$$- \langle \varphi_{L,1}\varphi_{M,1} | \varphi_{L,1}\varphi_{M,1} \rangle - \langle \varphi_{L,0}\varphi_{M,0} | \varphi_{L,0}\varphi_{M,0} \rangle$$

$$- \langle \varphi_{L,1}\varphi_{M,0} | \varphi_{L,1}\varphi_{M,0} \rangle - \langle \varphi_{L,0}\varphi_{M,1} | \varphi_{L,0}\varphi_{M,1} \rangle$$

is mainly determined by the singlet-triplet splitting of the dimer halves and the double singlet SS is at approximately the double of the lowest singlet excitation energy whereas the double triplet TT can be dropped down by the singlet-triplet splitting and may fall into the region of the lowest singlet excitations.

In the single CI approach the lowest dimer singlet excitation is a mixture of the monomer Q_y transitions plus appreciable contribution from intradimer charge transfer states. The most important charge transfer configurations

$$A^*_{L^+M^-} = (a^*_{M,1,\uparrow} a_{L,0,\uparrow} + a^*_{M,1,\downarrow} a_{L,0,\downarrow}) / \sqrt{2} \text{ and}$$

$$A^*_{M^+L^-} = (a^*_{L,1,\uparrow} a_{M,0,\uparrow} + a^*_{L,1,\downarrow} a_{M,0,\downarrow}) / \sqrt{2}$$

are coupled to the double triplet excitation via

$$V(TT,L^+M^-) = -\sqrt{3/2}\; U(\varphi_{M,0}\, , \varphi_{L,1})$$

$$V(TT,M^+L^-) = -\sqrt{3/2}\; U(\varphi_{L,0}\, , \varphi_{M,1})$$

where the one electron interactions U between the HOMO of one monomer and the LUMO of the other are of the order of 0.2 eV [1] which is rather strong. The matrix element between TT and the HOMO-LUMO transitions of the monomers which are the dominant contributions to the Q_y transitions of the isolated monomers vanishes as long as singlet-triplet coupling is neglected.

A SIMPLIFIED MODEL FOR THE LOWEST DIMER EXCITATIONS

We consider the coupling of the states L^*,M^*,L^+M^-,M^+L^- and P^{**} which for a symmetric dimer is represented by the matrix

$$\begin{bmatrix} E_* & V_{exz} & U_u & U_o & 0 \\ V_{exz} & E_* & U_o & U_u & 0 \\ U_u & U_o & E_{CT} & 0 & U_{uo} \\ U_o & U_u & 0 & E_{CT} & U_{uo} \\ 0 & 0 & U_{uo} & U_{uo} & E_{P**} \end{bmatrix}$$

where V_{exz} is the direct coupling of the two Q_y transitions at energy E_*, U_u,U_o and U_{uo} are one electron couplings between the two monomers and interactions proportional to overlap squares are neglected.This matrix becomes block diagonal by forming symmetric and anti symmetric combinations of the local Q_y transitions and of the CT states. An orthogonal transformation gives the coupling matrix for the even states

$$|(+)\rangle = 2^{-1/2}\,|(L^* + M^*)\rangle\ ,\quad |CT(+)\rangle = 2^{-1/2}\,|(L^+M^- + M^+L^-)\rangle\ ,\quad P^{**}$$

$$H(+)= \begin{bmatrix} E_* + V_{exz} & U_u + U_o & 0 \\ U_u + U_o & E_{CT} & \sqrt{2}\,U_{uo} \\ 0 & \sqrt{2}\,U_{uo} & E_{P^{**}} \end{bmatrix}$$

and for the odd states

$$|(-)\rangle = 2^{-1/2}|(L^* - M^*)\rangle,\quad |CT(-)\rangle=2^{-1/2}|(L^+M^- - M^+L^-)\rangle$$

$$H(-) = \begin{bmatrix} E_* - V_{exz} & U_u - U_o \\ U_u - U_o & E_{CT} \end{bmatrix} .$$

In the presence of an electric field $\vec{F}$ states of different symmetry are mixed by the operator

$$W(\vec{F}) = |L^+M^-> \vec{p}\vec{F} < L^+M^-| - |M^+L^-> \vec{p}\vec{F} < M^+L^-| =$$
$$= \vec{p}\vec{F} \, (\, |CT(+)><CT(-)| \; + \; |CT(-)><CT(+)| \,) .$$

To obtain simple analytical expressions we study a special case where $U_u \approx -U_o$ and the CT states are above $|(-)>$ and P^{**}. For a discussion of the lowest dimer band we consider the zero order states

$$|1> = \cos(\alpha)|(-)> + \sin(\alpha)|CT(-)> \text{ and } |2> = \cos(\beta)\, |P^{**}> + \sin(\beta)|CT(+)>$$

which are coupled by the perturbation $V = <1|W|2> = \sin(\alpha)\sin(\beta)\, \vec{p}\vec{F}$. We note that the zero order states show no change of permanent dipole but a strong polarizability due to the coupling W. If the two states are both within the dimer band the resulting Stark effect may look very similar to a second derivative as will be discussed in the following. The two state system is described by the hamiltonian matrix

$$H = \begin{bmatrix} 0 & V \\ V & \Delta \end{bmatrix}$$

which is diagonalized by an orthogonal transformation $\tilde{H} = R^T \, H \, R$

$$\text{with } R = \begin{bmatrix} c & -s \\ s & c \end{bmatrix} \quad \text{and } \; cs\Delta + (c^2 - s^2)V = 0.$$

The eigenvalues are $E_1 = 2csV + s^2\Delta$ and $E_2 = -2csV + c^2\Delta$ and the intensities of the two transitions are $I_1 = I_o c^2$ and $I_2 = I_o s^2$ where $I_o = (\vec{\mu}_o \vec{e})^2$. $\vec{\mu}_o$ is the transition dipole moment of the state $|1>$ and $\vec{e}$ is the polarization of the optical field.

The two transitions are dressed with a line profile function $f(\varepsilon)$. If we assume that both transitions are within the dimer band we may expand the line profile functions according to $f(\varepsilon - E) = f(\varepsilon) - Ef'(\varepsilon) + f''(\varepsilon)\, E/2 + \ldots$

The electrochromicity spectrum then reads

$$\Delta I = I_0 c^2 \,(f(\varepsilon) - (2csV + s^2\Delta)\, f'(\varepsilon) + \tfrac{1}{2}(2csV + s^2\Delta)\, f''(\varepsilon) + ...$$
$$+ I_0 s^2 \,(f(\varepsilon) - (-2csV + c^2\Delta)\, f'(\varepsilon) + \tfrac{1}{2}(-2csV + c^2\Delta)\, f''(\varepsilon) + ...$$
$$- I_0 f(\varepsilon)$$

The contributions proportional to $f(\varepsilon)$ and $f'(\varepsilon)$ are zero as $c^2+s^2 = 1$ and $-(c^2-s^2)\, 2csV - 2c^2s^2\Delta = 2c^2s^2\Delta - 2c^2s^2\Delta = 0$.

The second derivative contribution is

$$\tfrac{1}{2} f''(\varepsilon)\,(4c^2s^2V^2 + c^2s^2\Delta^2) = \tfrac{1}{2} f''(\varepsilon)\,(4c^2s^2V^2 + (c^2-s^2)^2V^2) = \tfrac{1}{2} f''(\varepsilon)\, V^2$$

and the third derivative contribution is

$$- \tfrac{1}{6} f'''(\varepsilon)\; c^2\,(2csV + s^2\Delta)^3 \;- \tfrac{1}{6} f'''(\varepsilon)\; s^2\,(-2csV + c^2\Delta)^3$$
$$= - \tfrac{1}{6} f'''(\varepsilon)\,(8c^5s^3V^3 - 8c^3s^5V^3 + 12c^4s^4V^2\Delta + 12c^4s^4V^2\Delta + 6c^3s^5V\Delta^2 - 6c^5s^3V\Delta^2$$
$$+ c^2s^6\Delta^3 + c^6s^2\Delta^3)$$
$$= - \tfrac{1}{6} f'''(\varepsilon)\,(- 8c^4s^4\Delta V^2 + 24c^4s^4\Delta V^2 + 6c^4s^4\Delta^3 + c^2s^2(s^4+c^4)\Delta^3)$$
$$= - \tfrac{1}{6} f'''(\varepsilon)\,(16c^4s^4\Delta V^2 + c^2s^2(4c^2s^2 + 1)\Delta^3)$$
$$= - \tfrac{1}{6} f'''(\varepsilon)\,\Delta V^2\,(16c^4s^4 + (c^2-s^2)^2(4c^2s^2 + 1))$$
$$= - \tfrac{1}{6} f'''(\varepsilon)\,\Delta V^2\,((2cs)^4 + (c^2-s^2)^2\,((2cs)^2 + 1))$$
$$= - \tfrac{1}{6} f'''(\varepsilon)\,\Delta V^2\,((2cs)^4 + (1 - (2cs)^2)^2\,((1 + (2cs)^2))$$
$$= - \frac{\Delta V^2}{6}\; f'''(\varepsilon).$$

COMPARISON WITH THE CONVENTIONAL MODEL

In the conventional model (without P^{**}) the Stark effect of the dimer band is due to its coupling to the charge transfer states which must have different energies. We start with the states $|(+)>$, $|(-)>$, $|L^+M^->$ and $|M^+L^->$. The model hamiltonian including the electric field is

$$\begin{bmatrix} E_*+V_{exz} & 0 & (U_u+U_o)/\sqrt{2} & (U_u+U_o)/\sqrt{2} \\ 0 & E_*-V_{exz} & (U_u-U_o)/\sqrt{2} & -(U_u-U_o)/\sqrt{2} \\ (U_u+U_o)/\sqrt{2} & (U_u-U_o)/\sqrt{2} & E_{CT}+\delta+\vec{p}\vec{F} & 0 \\ (U_u+U_o)/\sqrt{2} & -(U_u-U_o)/\sqrt{2} & 0 & E_{CT}-\delta-\vec{p}\vec{F} \end{bmatrix}$$

For $U_u \approx -U_o$ perturbation theory gives the state

$|1\rangle = |(-)\rangle + s_1\, |L^+M^-\rangle + s_2 |M^+L^-\rangle$

with

$$s_1 = \frac{U_u - U_o}{\sqrt{2}\,(E_* - V_{exz} - E_{CT} - \delta - pF\,)} \approx \frac{U_-}{\Delta - \delta}\left(1 + \frac{\vec{p}\vec{F}}{\Delta - \delta} + \left(\frac{\vec{p}\vec{F}}{\Delta - \delta}\right)^2 + \dots\right)$$

$$s_2 = \frac{U_u - U_o}{\sqrt{2}\,(E_* - V_{exz} - E_{CT} + \delta + pF)} \approx \frac{U_-}{\Delta + \delta}\left(1 - \frac{\vec{p}\vec{F}}{\Delta + \delta} + \left(\frac{\vec{p}\vec{F}}{\Delta + \delta}\right)^2 + \dots\right)$$

$$\Delta = E_* - V_{exz} - E_{CT}\ ,\quad U_- = (U_u - U_o)/\sqrt{2}$$

The transition energy is

$$E = E_* - V_{exz} + s_1(F)\,\frac{U_u - U_o}{\sqrt{2}} + s_2(F)\,\frac{U_u - U_o}{\sqrt{2}}$$

$$\approx E_* - V_{exz} + \frac{U_-^2}{\Delta - \delta}\left(1 + \frac{\vec{p}\vec{F}}{\Delta - \delta} + \left(\frac{\vec{p}\vec{F}}{\Delta - \delta}\right)^2\right) + \frac{U_-^2}{\Delta + \delta}\left(1 - \frac{\vec{p}\vec{F}}{\Delta + \delta} + \left(\frac{\vec{p}\vec{F}}{\Delta + \delta}\right)^2\right)$$

$$\approx E(F{=}0) + \vec{p}\vec{F}\, U_-^2\left((\Delta-\delta)^{-2} - (\Delta+\delta)^{-2}\right) + (\vec{p}\vec{F})^2\, U_-^2\left((\Delta-\delta)^{-3} + (\Delta+\delta)^{-3}\right)$$

and the intensity of the transition is

$$I(F) = (\vec{\mu}_o\vec{e})^2 / (1 + s_1^2 + s_2^2)$$

$$\approx (\vec{\mu}_o\vec{e})^2/\left(1 + \left(\frac{U_-}{\Delta - \delta}\right)^2\left(1 + 2\frac{\vec{p}\vec{F}}{\Delta - \delta} + 3\left(\frac{\vec{p}\vec{F}}{\Delta - \delta}\right)^2\right) + \left(\frac{U_-}{\Delta+\delta}\right)^2\left(1 - 2\frac{\vec{p}\vec{F}}{\Delta + \delta} + 3\left(\frac{\vec{p}\vec{F}}{\Delta + \delta}\right)^2\right)\right)$$

$$\approx (\vec{\mu}_o\vec{e})^2/\Big(1 + \left(\frac{U_-}{\Delta - \delta}\right)^2 + \left(\frac{U_-}{\Delta+\delta}\right)^2 + \vec{p}\vec{F}\, 2U_-^2\left((\Delta-\delta)^{-3} - (\Delta+\delta)^{-3}\right)$$
$$+ (\vec{p}\vec{F})^2\left(3U_-^2\left((\Delta-\delta)^{-4} + (\Delta+\delta)^{-4}\right)\right)\Big)$$

$$\approx \frac{(\vec{\mu}_o\vec{e})^2}{1 + U_-^2\left((\Delta-\delta)^{-2} + (\Delta+\delta)^{-2}\right)}\left[1 - \vec{p}\vec{F}\,\frac{2\,U_-^2\left((\Delta-\delta)^{-3} - (\Delta+\delta)^{-3}\right)}{1 + U_-^2\left((\Delta-\delta)^{-2} + (\Delta+\delta)^{-2}\right)}\right.$$

$$\left. - (\vec{p}\vec{F})^2\,\frac{3U_-^2\left((\Delta-\delta)^{-4} + (\Delta+\delta)^{-4}\right)}{1 + U_-^2\left((\Delta-\delta)^{-2} + (\Delta+\delta)^{-2}\right)} + (\vec{p}\vec{F})^2\,\frac{4\,U_-^4\left((\Delta-\delta)^{-3} - (\Delta+\delta)^{-3}\right)^2}{\left(1 + U_-^2\left((\Delta-\delta)^{-2} + (\Delta+\delta)^{-2}\right)\right)^2}\right]$$

and finally the electrochromicity is

$$\Delta I = I_o \Big(- \vec{p}\vec{F} \frac{2\ U_-^2 ((\Delta-\delta)^{-3} - (\Delta+\delta)^{-3})}{1 + U_-^2((\Delta-\delta)^{-2}+(\Delta+\delta)^{-2})}$$

$$+ (\vec{p}\vec{F})^2 \Big(- \frac{3U_-^2((\Delta-\delta)^{-4} + (\Delta+\delta)^{-4})}{1 + U_-^2((\Delta-\delta)^{-2}+(\Delta+\delta)^{-2})} + \frac{4\ U_-^4 ((\Delta-\delta)^{-3} - (\Delta+\delta)^{-3})^2}{(1 + U_-^2((\Delta-\delta)^{-2}+(\Delta+\delta)^{-2}))^2} \Big) \mathbf{f}(\varepsilon - E_o)$$

$$- I_o \Big(\vec{p}\vec{F}\, U_-^2 ((\Delta-\delta)^{-2} - (\Delta+\delta)^{-2}) + (\vec{p}\vec{F})^2 U_-^2 ((\Delta-\delta)^{-3} + (\Delta+\delta)^{-3}$$

$$- (\vec{p}\vec{F})^2 U_-^2 ((\Delta-\delta)^{-2} - (\Delta+\delta)^{-2}) \frac{2\ U_-^2 ((\Delta-\delta)^{-3} - (\Delta+\delta)^{-3})}{1 + U_-^2((\Delta-\delta)^{-2}+(\Delta+\delta)^{-2})} \Big) \mathbf{f}'(\varepsilon - E_o)$$

$$+ I_o (\vec{p}\vec{F})^2 \frac{1}{2} U_-^4 ((\Delta-\delta)^{-2} - (\Delta+\delta)^{-2})^2\ \mathbf{f}''(\varepsilon - E_o)$$

This equation simplifies if we assume an isotropic probe, i.e. terms linear in F average to zero and that $\delta \ll \Delta$. Then we find

$$\Delta I = \Delta I = I_o\ (\vec{p}\vec{F})^2 \Big(- \frac{6U_-^2\ \Delta^{-4}}{1 + 2U_-^2\ \Delta^{-2}} + \frac{144U_-^4\ \Delta^{-8}\ \delta^2}{(1 + 2U_-^2\ \Delta^{-2})^2} \Big) \mathbf{f}(\varepsilon - E_o)$$

$$- I_o \Big((\vec{p}\vec{F})^2 (U_-^2\ 2\Delta^{-3} - U_-^2\ \Delta^{-3}\ 4\ \delta \frac{12\ U_-^2\ \Delta^{-4}\ \delta}{1 + 2U_-^2 \Delta^{-2}} \Big) \mathbf{f}'(\varepsilon - E_o)$$

$$+ I_o (\vec{p}\vec{F})^2\ 2U_-^4\ \Delta^{-6}\ \delta^2\ \ \mathbf{f}''(\varepsilon - E_o)$$

The second derivative appears only if $\delta \neq 0$. It may become dominant if the width of the profile is small.

THE LOWEST RYDBERG STATE OF THE REACTION CENTER

The electron transfer from the dimer P to the pheophytine H_L is a tunneling process via a potential barrier or two barriers if $P^+B_L^-$ is considered as real intermediate. The electronic coupling matrix elements between the state P^* and $P^+B_L^-$ and the next to $P^+H_L^-$ are rather small. We calculated 10 cm^{-1} and 7 cm^{-1} rsp. .

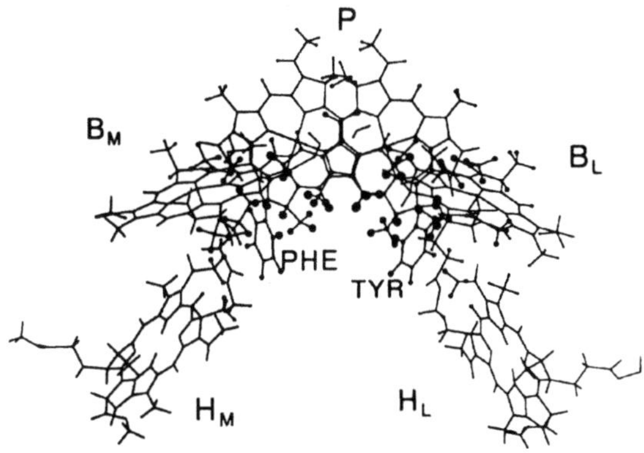

Figure 3. Electron density of the lowest Rydberg orbital. The partial densities on the molecules are: HM 0.01, BM 0.04, PL 0.39, PM 0.35, BL 0.09, HL 0.01, PHE 0.04, TYR 0.07.

Somewhat larger values could arise if the effective potential barrier could be lowered by other low-lying states, which extend from the dimer to the pheophytine. Such states could be of the Rydberg type.

We investigated such Rydberg states by extending the basis set for the INDO-program to incorporate 2s and 2p functions for the hydrogen atoms and 3s and 3p for the C and N atoms. We found that the dimer can form one low-lying state at 1.7 eV, the orbital of it is shown in Fig. 3. It extends with a small percentage directly into H_L. The next higher Rydberg states are higher by more than 1 eV so that our parameterization, which reduces the resonance integral by a factor of 0.05 as compared to the valence orbital parameters, is consistent with predictions on smaller molecules like benzene, where more is known about the energy location. The Rydberg state has a strong dipole of 20 Debyes in the direction of the symmetry axis and could induce a proper Stark effect of P* if it falls into the dimer band.

AN INTERMOLECULAR POTENTIAL FOR THE DIMER

We evaluated the potential energy for the ground state with a force field approximation (Charmm) and for the lowest excited states of the dimer within our extended exciton model. Fig. 4 shows the dependence of the potential energy upon the distance of the two pyrrol rings I of the two monomers. It can be seen that the distance between the molecules for the excited state P* is shortened relative to the ground state.

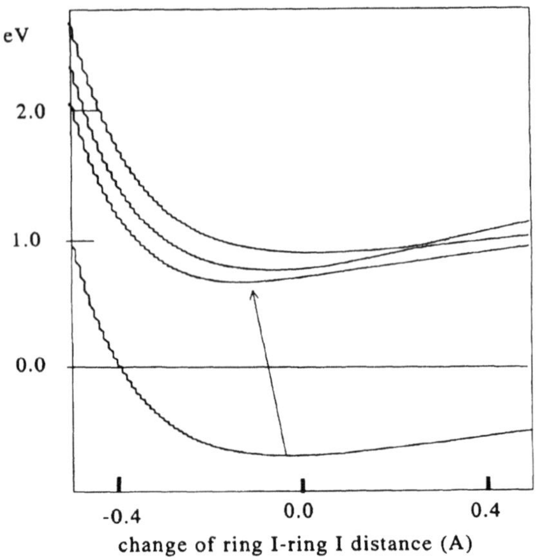

Figure 4. Potential diagram for the ground state P and three excited states including P*, the upper dimer band and the lowest charge transfer state.

The vibrational states along this coordinate may be identified with those found around 100 cm^{-1} by hole burning. Due to the displacement the two lowest states could be excited coherently and the resulting stimulated emission would show an oscillation of the form

$$A(t)= A_0 \exp(2\lambda^2(\cos(\omega t)-1)-t/\tau)$$

Here λ is the dimensionless displacement with $\lambda^2=2.8$. The lifetime τ is the lifetime of the excited state and $\hbar\omega \approx 100$ cm^{-1}. This calculation is consistent with the observation of vibrational coherence [8].

CONCLUSION

We have shown that the quantum calculations with multiple configuration interaction can predict the electronic spectrum as well as transient spectra quite well. Due to the strong interaction of the two dimer halves some new exotic states arise as low lying excitations which might be of interest in connection with the strong electrochromicity. We presented a detailed analysis of a "double triplet" and a Rydberg state and their possible implication on the Stark spectrum. However, we did not find strong evidence that these states are essential for the charge separation. In a model study of the influence of the protein we could show that the protein motion can become unimportant for the charge separation, if the process proceeds in the inverted regime, that means the charge transfer state must be below the excited state P^* not necessarily at equal energy at the initial configuration.

REFERENCES

1. P.O.J. Scherer and S.F. Fischer, Interpretation of optical reaction center spectra, in: Chlorophylls, H. Scheer, ed. , CRC Press Boca Raton (1991)
2. M.H. Vos, J.L. Lambry, S.J. Robles, D.C. Youvan, J. Breton and J.L. Martin,Excited state dynamics of bacterial reaction centers,Proc. Natl. Acad. Sci. USA 88: 8885 (1991)
3. U. Finkele, K. Dressler and W. Zinth,Analysis of transient absorption data from reaction centers of purple bacteria, in: Reaction centers of photosynthetic bacteria, ed. M.E. Michel-Beyerle, Springer Series in Biophysics Vol. 6 (1990)
4. W.W. Parson, E. Nabedryk and J. Breton, A new I.R. electronic transition of P^+, in this book
5. P.O.J. Scherer, Quantum chemical investigation of electronically excited states of molecular systems, Journal of Luminescence 53:133 (1992)
6. W.D. Edwards, J.D. Head and M.C. Zerner, On the electronic excited states of model chlorophyll, J. Am. Chem. Soc. 104:5833 (1982)
7. S.G. Johnson, I-J. Lee and G.J. Small Solid State Spectral line-narrowing spectroscopies, in: Chlorophylls, ed. H. Scheer, CRC Press Boca Raton (1991)
8. J. Breton, Low temperature linear dichroism stdy of the orientation of the pigments in reduced and oxidized reaction centers of Rps. viridis and Rb. sphaeroides,in: The photosynthetic bacterial reaction center structure and dynamics, eds. J. Breton and A. Vermeglio Nato ASI Series A 149 (1987)
9. M.A. Thompson, M.C. Zerner and J. Fajer,A theoretical examination of the electronic structure and excited states of the bacterochlorophyll-b dimer from Rhodospeudomonas viridis, J. Phys. Chem. (1991).
10. J. Breton, J.L. Martin, J.C. Lambry, S.J. Robles and D.C. Youvan, Ground state and femtosecond transient absorption spectroscopy of a mutant of Rb. capsulatus which lacks the initial electron acceptor bacterio pheophytin, in: Reaction centers of photosynthetic bacteria, ed. M.E. Michel-Beyerle, Springer Series in Biophysics Vol. 6 (1990)

RECENT EXPERIMENTAL RESULTS FOR THE INITIAL STEP OF BACTERIAL PHOTOSYNTHESIS

Theodore J. DiMagno[1], Sandra J. Rosenthal[1], Xiaoling Xie[2], Mei Du[1], Chi-Kin Chan[1], Deborah Hanson[3], Marianne Schiffer[3] James R. Norris[4,1] and Graham R. Fleming[1]

[1]Department of Chemistry, The University of Chicago
Chicago, IL, 60637

[2]Battelle Research Center, Richland, WA, 99352

[3]Biological and Medical Research Division and [4]Chemistry Division
Argonne National Laboratory, Argonne, IL, 60439

INTRODUCTION

This paper discusses experiments pertaining to two major issues of the primary events of photosynthesis: Why is only a single side (A-side) of the highly C2 symmetrical reaction center active? How does the electron get from the special pair donor (P) to the bacteriopheophytin acceptor (H_A)? The investigation presented here highlights the correlation between the reduction potential of the primary donor and the initial charge separation rate measured by femtosecond spectroscopy in a series of mutated reaction centers from *Rb. capsulatus*[1].

A standard procedure for determining the rate of photoinduced charge separation is to measure the lifetime of the excited, singlet state of the primary donor, P^*. The vast majority of the determinations of the lifetime of P^* in the bacterial reaction center are based on stimulated emission. Prior to approximately a year ago, most stimulated emission results were interpreted in terms of a single exponential decay. In this paper the lifetime of P^* is measured by monitoring either stimulated emission or spontaneous emission. Importantly, we have found that the disappearance of P^* when followed by spontaneous emission is never characterized by a single exponential process. Moreover, within the last year studies employing either spontaneous or stimulated emission are in quantitative agreement that the decay of P^* in *Rb. sphaeroides* is definitely not a single exponential process [1,2,3,4,5].

This recent consensus has been achieved by improved signal-to-noise and a better definition of the baseline zero. The analysis of stimulated emission data is complicated by absorption from excited states, including intermediates, and by ground state bleaching. Consequently, establishing the correct zero level for measuring the disappearance of P^*

The Photosynthetic Bacterial Reaction Center II
Edited by J. Breton and A. Verméglio, Plenum Press, New York, 1992

is a non-trivial task. Additionally, an accurate baseline requires high signal-to-noise. Unfortunately, obtaining a sufficient signal-to-noise ratio with stimulated emission is very demanding. In contrast, spontaneous emission measurements are not bothered by ground state bleaching and excited state absorption such that defining the baseline does not appear to present a serious problem in these fluorescence experiments. Also an excellent signal-to-noise is possible in the spontaneous emission studies compared to stimulated emission in part because of the high laser-repetition rate of the fluorescence technique employed here.

That P^* does not decay as a single exponential adds a significant complication to the interpretation of the primary events of bacterial photosynthesis. First of all, how to treat the data becomes a major issue. Is the data bi-exponential, multi-exponential, stretched-exponential or another possibility? In this paper we will describe the decay of P^* with a bi-exponential process, keeping in mind that distinguishing the bi-exponential situation from the other cases is difficult.

Second of all, the possible origins of the non-single exponential decay must be considered. Here we list four possibilities. (1) Incomplete vibrational relaxation after excitation is a distinct prospect, especially since oscillations in the stimulated emission decay of P^* have been observed at very low temperatures in certain reaction-center preparations[2,3]. (2) Obviously sample heterogeneity is a possible origin of the non-single exponential behavior. Moreover, several types of heterogeneity appear plausible. (a) The normal chemistry of the reaction center may serve as an intrinsic source of sample heterogeneity. For example, a mixture of the protonation states of the quinones may produce structural variations of the reaction center which then lead to variations in reaction rates. (b) Possible irregularities in sample preparation may also contribute to the sample heterogeneity. (c) Finally, the sample heterogeneity may arise purely from the inherent nature of the protein environment of the donor-acceptor complex. In other words, even if the quality of the sample preparation is perfect and the state of the redox chemistry is perfectly homogeneous, significant heterogeneity in the electron transfer rate arises because of the intrinsic nature of the protein environment. As a result the decay of P^* can not be accurately characterized by a single exponential function. 3) The reaction centers are reasonably homogeneous with respect to electron transfer rates such that the complicated, non-single exponential decay is intrinsic to the primary events. For example, a single intermediate state accompanying the formation of $P^+B_AH_A^-$, such as $P^+B_A^-H_A$, results in kinetic equations for which the decay of P^* is bi-phasic[6]. (Here B_A is the accessary bacteriochlorophyll and H_A is the bacteriopheophytin on the active A-side of the reaction center protein.) Of course more complicated kinetic schemes could also be proposed and would result in multi-exponential decay. However, in this paper more complex possibilities will be neglected and emphasis will be placed on a single intermediate state accompanying the formation of H_A^-. 4) Finally, and undoubtedly to some extent, the three previous classifications may coexist as contributors to the observed complex decay. Hopefully, one of the three will be the dominant source of the non-single exponential decay.

At this point, distinguishing among the various possibilities is impossible. Instead, one purpose of this paper is to discuss some of the consequences of assuming that the observed decay of P* is an intrinsic bi-exponential process arising from homogeneous reaction centers, i.e., case (3) above. Another purpose of this report is to explore the correlation between reduction potential of the primary donor and the reaction rate of the primary event.

EXPERIMENTAL APPROACH

In order to examine in more detail the non-single exponential decay, we have studied a series of symmetry related mutants of *Rb. capsulatus*. Since the amino acid sequence of the reaction-center protein of *Rb. capsulatus* is so similar to *Rb. sphaeroides* R-26, we have assumed that the wild type *Rb. sphaeroides* R-26 reaction-center structure is an accurate model for the mutated, reaction-center proteins of *Rb. capsulatus*. The manipulations of the amino acid sites L181 and M208, related by C2 symmetry as shown in Figure 1, result in modified chemical kinetics for the initial electron transfer event. Some amino acid groups speed up the chemistry while others slow it down. Since the L181 group, located primarily on the inactive B-side, has almost the same effect in altering the chemistry as does the M208 moiety on the active A-side, we assume that the dominant role of these two sites is via the special pair. Finally, we are also assuming that the mutations induce no structural changes large enough to alter the electron transfer rates[7].

In order to understand the origin of the kinetic differences induced by the mutations at sites L181 and M208, we have attempted to correlate the electron transfer rate with the free energy of the initial chemistry. We have assumed that the electron transfer chemistry is dominated by a single, intermediate state of charge separation such as $P^+B_A^-H_A$. To perform this correlation we have measured the reduction potential of the primary donor **in the dark**. In wild-type and mutant reaction centers, the free energy gaps G(WT) and G(Mutant), respectively, between P^* and the initial product, perhaps $P^+B_A^-H_A$, can be approximated by the following equations:

$$G(WT) = -E_D(WT) + E_A(WT) + E_C(WT) + G_{P^*}(WT) \tag{1a}$$

$$G(Mutant) = -E_D(Mutant) + E_A(Mutant) + E_C(Mutant) + G_{P^*}(Mutant) \tag{1b}$$

where in Equation 1a $E_D(WT)$ is the reduction potential of the donor, $E_A(WT)$ is the reduction potential of the relevant acceptor, $E_C(WT)$ is a coulombic correction factor taking into account such things as short distance charge separation, change in dielectric constant of solvent, etc. and $G_{P^*}(WT)$ is the free energy offset from P^*. Since the optical spectrum of P is essentially unchanged by the mutations we assume that the relative

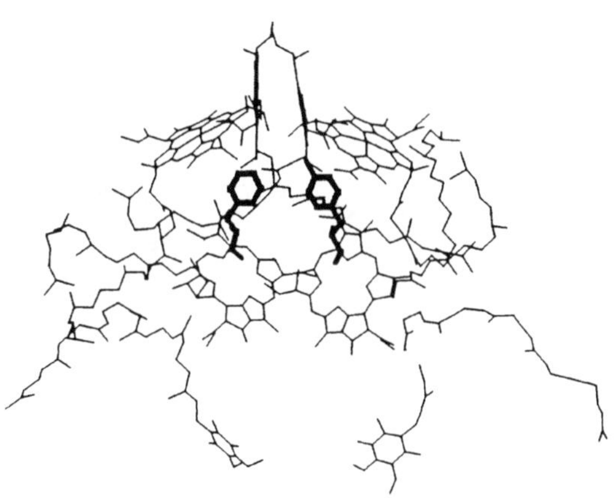

Figure 1. *Rb. sphaeroides* R26 reaction center structure with the C_2 related residues phenylalanine L181 (left) and tyrosine M208 (right) along with the other chromophores. The active A-side is on the right.

energy of P^* is not changing with the mutation. Equation 1b is the same as Equation 1a except that it is written explicitly for mutated reaction centers.

Previously we found that the major effect of the M208 and L181 amino acids is on the special pair[8]. Thus, in this paper the effect of both the M208 and the L181 sites on B_A and H_A will be neglected even if the chemistry involves the intervening B_A which is close to the M208 site. In other words, we are assuming that the mutations studied here affect only the reduction potential of the donor (i.e., $E_A(WT) \equiv E_A(Mutant)$). Additionally the coulombic correction factor E_C will be assumed to be independent of the genetic manipulation (i.e., $E_C(WT) \equiv E_C(Mutant)$). Upon subtracting Equation 1b from 1a the free energy is expressed purely in terms of the reduction potential of the donor and the energy offset from P^*, G_{P*}, as described in Equation 2b,

$$\Delta G(Mutant) \equiv G(WT) - G(Mutant) =$$

$$- E_D(WT) + E_D(Mutant) + G_{P^*}(Mutant) - G_{P^*}(WT) \tag{2a}$$

or

$$\Delta G(Mutant) = - E_D(WT) + E_D(Mutant) + G_0 \tag{2b}$$

where

$$G_0 \equiv G_{P^*}(Mutant) - G_{P^*}(WT)$$

is a reference free energy. Thus, the absolute free energy for either wild-type or a particular mutant reaction center, or of a particular mutant is not needed, only the difference in the reduction potential between the donor in wild-type and the donor in mutated reaction centers. And finally, by assuming that the reorganization energy λ and the electronic matrix element for electron transfer, V, is not changed by these mutations, a simple Marcus parabola can be used to correlate reduction potential to the log of the rate based on Equation 3. That the experimental data fits a parabola so accurately appears to justify the assumptions.

$$k = \frac{2\pi}{\hbar(4\pi k_B T)^{1/2}} V^2 \exp\left[\frac{\{\Delta G(WT\ or\ Mutant) - \lambda\}^2}{4 k_B T \lambda}\right] \tag{3}$$

Since ten different reaction centers have been constructed and measured, we have ten equations to determine only four parameters (G(WT), G_0, λ, and V). Given these four parameters and the measured difference in reduction potential the rate of any reaction-center mutant can be calculated. In this fitting process, G_0 is initially assumed to be based on the standard hydrogen reaction. After fitting the rates as a function of free energy for our series of mutants, G_0 will represent the offset in free energy between the putative intermediate state and P^* in wild type reaction centers.

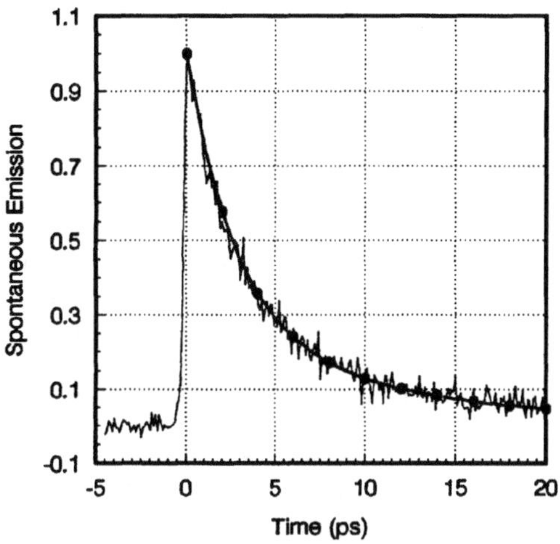

Figure 2. The spontaneous emission and bi-exponential fit (smooth line with dots) for wild type *Rb. capsulatus* with time constants τ_1 = 2.7 ps and τ_2 = 11.1 ps.

EXPERIMENTAL METHODS

Two rate constants for nine reaction-center mutants as well as wild type *Rb. capsulatus* are presented in Table I. However, for five reaction centers stimulated emission was used such that only a single rate constant was extracted for four of these mutants. Details of the fluorescence measurements[1] and the stimulated emission measurements[8,9] have been presented in separate publications. In most of the cases that have been studied by both methods, the fast component of the spontaneous emission agrees with the single rate constant reported in the past by stimulated emission. Moreover we anticipate that future, improved stimulated emission studies that previously exhibited a single time constant will also yield bi-exponential data. For comparison to other work Table I also includes spontaneous emission results for both ordinary *Rb. sphaeroides* as well as 99.9% deuterated *Rb. sphaeroides*.

The redox potentials were determined by a chemical titration of the reaction center in tris-LDAO buffer at pH 7.8 with a potassium ferricyanide/ferrocyanide redox couple. The state of oxidation of the special pair was monitored by optical absorption spectroscopy. At most 70% reversibility to the original reduced state of the special pair was achieved with potassium ferrocyanide. At that point the solutions were so dilute that the concentration of the special pair could not be measured with sufficient accuracy.

EXPERIMENTAL RESULTS

A typical spontaneous emission curve and its bi-exponential fit are shown in Figure 2. The bi-exponential fit to the data statistically appears slightly better than a stretched exponential[1]. Table 1 also presents the observed reduction potentials for nine mutants of *Rb. capsulatus* plus wild type. The reference potential for Table I is the standard hydrogen reaction. Typically, reproducibility was better than ± 5 mV and the value determined for wild type is in reasonably good agreement with other laboratories. Similar mutants in *Rb. sphaeroides* also show similar reduction potential changes[10]. Experiments are in progress to establish the accuracy of the reduction potentials reported in Table I.

DISCUSSION OF RESULTS

Figure 3 shows the parabolic relationship observed between the logarithm of the fast rate constant and the reduction potential of the special pair in this series of ten reaction centers using Equation 3. Basing the parabolic fit on the average rate constant, the average lifetime, or the slow rate constant will not change the shape of the parabola sufficiently to alter our main conclusions. Thus, for the present data set the logarithm of the rate versus the measured reduction potential of the "dark" donor special pair can be described rather accurately by a parabola. A variety of theoretical approaches will result in such a parabolic relationship. For simplicity of interpretation we will use the simple non-adiabatic electron transfer theory of Marcus[11,12].

The correlation between data and theory shown in Figure 3 gives a G(WT) for wild type reaction centers of *Rb. capsulatus* of around 50 cm^{-1}. This value is well within k_BT at room temperature. Not as evident, but perhaps more important, the apparent reorganization energy, λ, is always ~100 cm^{-1} or less and is extremely small. The very small reorganization energy is required by simple Marcus theory to produce the steep dependence of rate on reduction potential. Use of a modified Marcus theory, such as Jortner theory[13,14], does not appear to alter this observation. Rates and redox potentials for similar mutants of *Rb. sphaeroides* from other laboratories[15,10] agrees sufficiently well with our data presented in Figure 2 and Table 1, thus verifying the steepness of the parabola. Within the confines of our assumptions the small reorganization energy suggests that an explanation other than
simple Marcus theory, or Jortner theory[13,14], may be needed. As a possible alternative explanation the electron transfer may proceed via a nearby intermediate state that is essentially degenerate with P^*. For example, perhaps the intermediate is operating partially via the superexchange mechanism. Then, if the relevant superexchange energy gap is close to zero, the maximum of the parabolic relationship shown in Figure 3 would be enhanced by a resonance phenomenon associated with the three level degeneracy. In this case the steep curvature of the log of the rate versus reduction potential arises from the resonance condition and not from a small reorganization energy. This possible explanation might alleviate the necessity of a small reorganization energy as obtained from the standard rate vs. reduction potential fit.

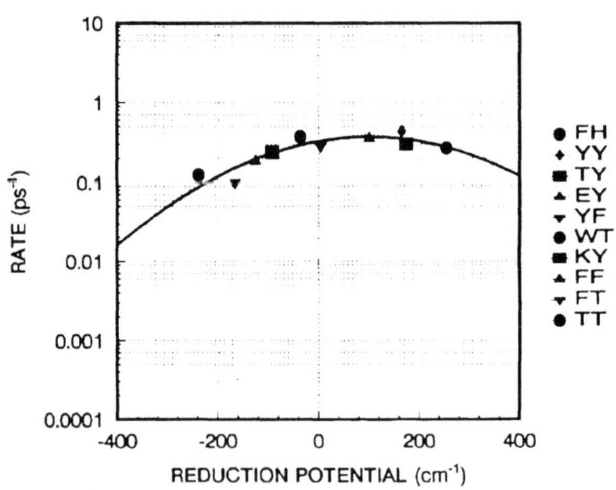

Figure 3. Experimental rate versus reduction potential for reaction centers from *Rb. capsulatus*. Left letter = L181. Right letter = M208. WT = wild type.

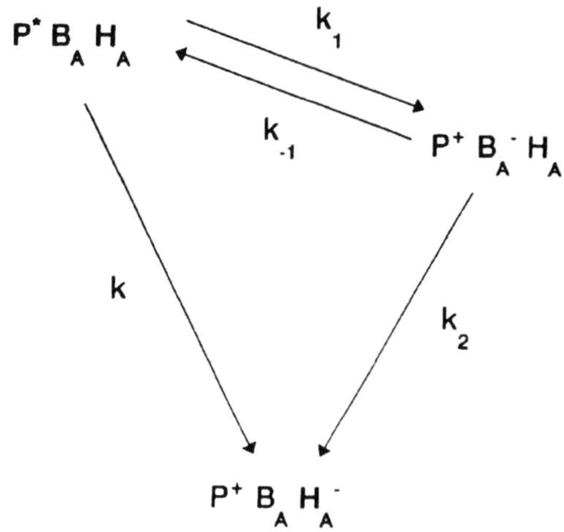

Figure 4. Kinetic scheme incorporating an intermediate state to produce the complex bi-exponential kinetics observed in the fluorescence spectra.

The homogeneous kinetic scheme typically invoked for the primary events (see Figure 4) also results in bi-phasic kinetics[6] where we assume that $k_1/k_{-1} = \exp\{-G(WT)/(k_BT)\}$. Most previous interpretations of the decay of P^* were based on a dominant, single exponential process where $k_2 > k_1$ and the energy level of the intermediate state $P^+B_A^-H_A$ is more than k_BT away from the energy level of P^* [9,16]. However, if the intermediate state energy of $P^+B_A^-H_A$ is within k_BT of the energy of P^*, then the amplitude of the second component becomes significant and should be experimentally measurable. For the relative amplitudes of the two components presented for wild type reaction centers in Table 1, then k_2 is required to be smaller than k_1 (see reference 1), i.e., $k_2 < k_1$. In this situation the reaction becomes dominated by the superexchange process where $P^+B_A^-H_A$ serves as a "parking state". The one step superexchange mechanism has been supported by measurements of Kirmaier and Holten[17,18]. Ogrodnik has suggested such a parking state to explain the loss of quantum yield in the presence of a high external electric field applied to the sample[19]. Also, we note that a superexchange parking state based on $k_2 < k_1$ leads to a complicated temperature dependence for delayed fluorescence.

Finally we emphasize that other possibilities can not be ruled out. For example, one way to maintain the more conventional view with $k_2 > k_1$ is to assume that the solution of the above homogeneous kinetic scheme results in a dominant single exponential process but that the observed bi-phasic or multi-phasic decay of P^* is the result of reaction-center heterogeneity and/or incomplete vibrational relaxation. Such a view would be consistent with past interpretations based on stimulated emission data[16,9].

CONCLUSIONS

Assuming homogeneous reaction centers and complete vibrational relaxation the spontaneous emission data require the formation of a "parking state". However, we have not ruled out other factors such as sample heterogeneity. In addition, we have presented a correlation of redox potential of the primary donor with electron transfer rate and have discussed the two following interpretations: 1) the free energy of an intermediate state, $\Delta G(WT)$, and the reorganization energy, λ, for electron transfer of the primary events in the reaction center are extremely small; 2) superexchange degeneracy may be occurring. New experiments presently underway involving mutants on each end of the parabola of Figure 3 (i.e., both small and large reduction potentials for the special pair) should help

Table 1

Mutation Inactive-Active	Lifetime ps (fast)	%	Lifetime ps (long)	%	Redox Potential (mV)
Phe-His	3.7*				466
Tyr-Tyr	2.3†	84.4	10.5	15.6	477
Thr-Tyr	3.2*				476
Glu-Tyr	2.8*				485
Tyr-Phe	3.5†	59.5	24.2	40.5	497
Phe-Tyr (WT)	2.7†	72.3	11.1	27.7	502
^{1}H sphaeroides	2.7†	80.8	12.1	19.2	530
^{2}H sphaeroides	3.2†	82.6	16.5	17.4	---
Lys-Tyr	4.2*				509
Phe-Phe	5.4†	53.3	40.0	46.7	513
Phe-Thr	10.0†	38.0	70.0	62.0	518
Thr-Thr	8.0*	51.8	56.3	48.2	527

*Measured by stimulated emission.
†Measured by spontaneous emission.

distinguish between the possibilities of sample heterogeneity versus a parking state and superexchange resonance versus low reorganization energy.

Acknowledgments

We gratefully acknowledge the efforts of Maxim Popov for calibrating the reduction potential of wild type *Rb. capsulatus* presented here. This work was supported by the U.S. Department of Energy, Office of Basic Energy Sciences, Division of Chemical Sciences (JRN), and Office of Health and Environmental Research (MS, DH) under contract W-31-109-Eng-38. MS was also supported by Public Health Service Grant GM36598.

References

1. M. Du, S. J. Rosenthal, X. Xie, T. J. DiMagno, M. Schmidt, J. R. Norris and G. R. Fleming, *Proc. Natl. Acad. Sci. USA*, in press.
2. M. H. Vos, J. C. Lambry, S. J. Robles, D. C. Youvan, J. Breton and J. -L. Martin, *Proc. Natl. Acad. Sci. USA*, 88:8885 (1991).
3. M. H. Vos, J. C. Lambry, S. J. Robles, D. C. Youvan, J. Breton and J. -L. Martin, *Proc. Natl. Acad. Sci. USA*, 89:613 (1992).
4. P. Hamm and W. Zinth, *Ultrafast VIII*, Springer-Verlag, Eds. J.L. Martin, H. Migus, A. Zewail and G.A. Mourou, in press.
5. W. Holzapfel, V. Finkele, W. Kaiser, D. Oesterhelt, H. Scheer, H. U. Stilz and W. Zinth, *Proc. Natl. Acad. Sci. USA*, 87:5168-5172 (1989).

6. M. Bixon, J. Jortner and M. E. Michel-Beyerle, *Biochim. Biophys. Acta* 1056:301-312 (1991).
7. A. J. Chirino, E. J. Lous, M. Huber, J. P. Allen, G. Feher, C. C. Schenck and D. C. Rees, *Cadarche II*, this volume.
8. C. -K. Chan, L. X.-Q. Chen, T. J. DiMagno, D. K. Hanson, S. L. Nance, M. Schiffer, J. R. Norris and G. R. Fleming, *Chem. Phys. Lett.*, 176:366-372 (1991).
9. C. -K. Chan, T. J. DiMagno, L. X.-Q. Chen, J. R. Norris and G. R. Fleming, *Proc. Natl. Acad. Sci. USA*, 88:11202-11206 (1992).
10. W. W. Parson, *Cadarche II*, this volume.
11. R. A. Marcus, R. A. *J. Chem. Phys.*, 24:966-978 (1956).
12. R. A. Marcus, *J. Chem. Phys.*, 24:979-989 (1956).
13. J. Jortner, *J. Chem. Phys.* 64:4860 (1976).
14. J. Jortner, *J. Am. Chem. Soc.*, 102:6676 (1980).
15. V. Nagarajan, W. W. Parson, D. Gaul and C. Schenk, *Proc. Natl. Acad. Sci. USA* 87:7888 (1990).
16. W. Holzapfel, U. Finkele, W. Kaiser, D. Oesterhelt, H. Scheer, H. U. Stilz and W. Zinth, *Chem. Phys. Lett.*, 160:1-7 (1989).
17. C. Kirmaier and D. Holten, *Israel J. Chem.* 28:79-85 (1988).
18. C. Kirmaier and D. Holten, *Biochemistry* 30:609-613 (1991).
19. A. Ogrodnik, *Cadarche II*, this volume.

MODEL CALCULATIONS ON THE FLUORESCENCE KINETICS OF ISOLATED BACTERIAL REACTION CENTERS FROM *RHODOBACTER SPHAEROIDES*

Alfred R. Holzwarth, Marc G. Müller, and Kai Griebenow

Max-Planck-Institut für Strahlenchemie
Stiftstraße 34-36, W-4330 Mülheim a.d. Ruhr, Germany

INTRODUCTION

Purple bacterial reaction centers (RCs) contain the special pair P860 (primary electron donor) and two cofactor pigment/protein branches denoted L and M which are formed by one accessory bacteriochlorophyll (B) and one bacteriopheophytin (H) each[1,2,3]. The L-branch contains also a quinone Q_A which acts as the final electron acceptor in isolated RCs[4]. The sequence and kinetics of the primary electron transfer processes in isolated purple bacterial reaction centers (RCs) are still discussed controversially. The model which seems to have the strongest experimental support at present is the so-called sequential two-step model[5]. It involves a reduced accessory bacteriochlorophyll (B_L^-) in the L-branch as a short-lived intermediate which is formed in 3.5 ps and is reoxidized by H_L in 0.9 ps. Alternatively, a direct reduction of H_L in a single step from P860* involving superexchange has been proposed[6,7] possibly coupled with a distribution in the rates of electron transfer[8]. Basically all kinetic data of RC processes so far are based on transient absorption techniques only[9]. The interpretation of these data presents problems both in view of the complex kinetics and the overlapping absorption difference spectra of the various intermediates. Thus an additional time-resolved method which can give complementary information to transient absorption is needed. Fluorescence kinetics is the method of choice in view of its inherently high S/N ratio and its modified relative weighting of the various kinetic components, as compared to transient absorption. Also fluorescence may be able to detect kinetic processes, like e.g. conformational protein/radical pair relaxation steps[10] that may be associated with minor absorbance changes only. Several groups have studied fluorescence kinetics of bacterial RCs in the past[11,12]. However, the time-resolution was generally not sufficient in those experiments to resolve the kinetics of the primary process(es).

We have recently reported the first measurements of fluorescence kinetics for bacterial as well as photosystem II RCs with high time resolution in the picosecond range[10,13]. The fluorescence kinetics of RCs from *Rb. sphaeroides* was presented in that work with-

The Photosynthetic Bacterial Reaction Center II
Edited by J. Breton and A. Verméglio, Plenum Press, New York, 1992

out detailed discussion[13]. For RCs from *Rb. sphaeroides* new kinetic features, not observed in transient absorption measurements so far, were a 12 ps component of substantial amplitude, in addition to the expected main 3 ps decay component[9] and also a 100 ps component. It is the purpose of this contribution to present a detailed kinetic analysis of the fluorescence data from *Rb. sphaeroides* RCs by testing various kinetic models suitable to explain these data.

MATERIALS AND METHODS

RCs were isolated from *Rb. sphaeroides* w.t. by anion exchange chromatography. The measurements were carried out by the time-correlated single-photon-counting technique as described[13]. The full width at half maximum of the system temporal response function was about 30 ps[13]. All measurements were carried out at room temperature (22°C) accumulating up to about 20000 counts in the peak channel with a time resolution per channel of 2.0 ps. From test measurements we verified that we are able to easily resolve lifetimes down to about 2 to 3 ps. The sample was circulated through a flow cell from a dark reservoir in order to keep the RCs in the open state. Excitation occurred at a wavelength of 800 nm. The ΔG-values for the various reaction steps are also calculated based on the relationship between the forward and reverse rate constants: $\Delta G = -k_B T \cdot ln(k^{forw}/k^{rev})$.

RESULTS AND DISCUSSION

Fluorescence decays were recorded at various emission wavelengths across the whole emission range (840 to 940 nm in intervals of 20 nm). The kinetic data were analyzed both by global lifetime and global target analysis procedures[14,15,16]. The global lifetime analysis requires a minimum of four decay components for good fits, as judged by the χ^2-values ($\chi^2 = 1.117$) and plots of the weighted residuals (short and long analysis window). The results, presented as decay-associated spectra (DAS), are shown in Fig. 1. The two shortest-lived components ($\tau_1 = 3$ ps and $\tau_2 = 12$ ps) show DAS with a maximum wavelength of about 900 nm, indicating that their origin is fluorescence from the special pair (P*). Also the component of 100 ps has a substantial contribution in the emission range of the special pair but also extends to somewhat shorter wavelengths. In contrast the DAS of the 930 ps component is substantially blue-shifted. It agrees with the fluorescence emission of decoupled BChl with an emission maximum at about 860 nm. It represents apparently a small amount of impurity in the samples which is excited preferentially at the excitation wavelength of 800 nm.

The 3 ps component has a dominant amplitude in the DAS. This component is expected in the fluorescence kinetics and represents the kinetics of the primary charge separation process, as characterized previously in transient absorption measurements[5,6]. The 12 ps component shows about one third of the amplitude of the 3 ps component. It seems also to be associated with a charge separation process in the reaction center, as judged by its DAS (Fig. 1). This component has not been observed before in transient absorption measurements[9]. Furthermore, a 100 ps component is clearly present which to its largest part also represents fluorescence form P*. We can clearly exclude an "impurity" as the origin of these components, since standard checks on our RC preparations show properties (absorption, bleaching spectrum and difference spectrum, polyacrylamide gel electrophoresis etc.) that agree very well with generally accepted standards. In addition analytical HPLC checks have proven a homogeneous band.

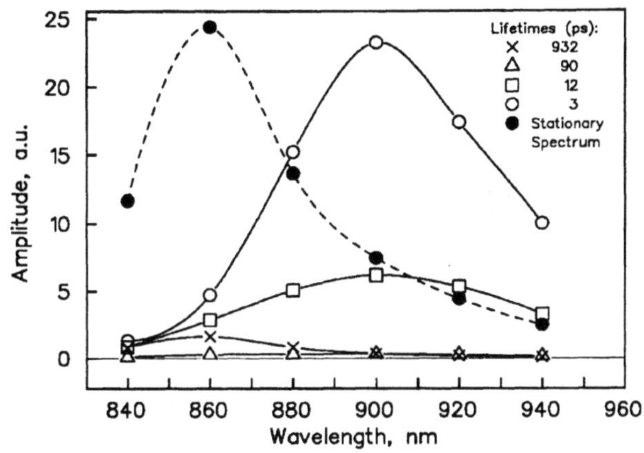

Figure 1. Decay-associated fluorescence spectra of isolated RCs from *Rb. sphaeroides* as calculated by the global lifetime analysis. λ_{exc} = 800 nm.

Kinetic models: In order to explain the appearance of a 12 ps and possibly also the 100 ps components in the fluorescence kinetics we tested a large variety of kinetic models by global target analysis. The rate constants and species-associated spectra (SAS) of a model were directly fitted to the original data. Of those models tested only three are presented here which formally explain the data about equally well. A common feature of all these models is that they assume P* as the only radiative excited state and X^- as an a priori unknown intermediate. This means that all the lifetimes (except for the BChl "impurity fluorescence" at 860 nm) have the same emission spectrum (Fig. 3) and that there is no spectral development of the fluorescence at times longer than about 2 ps. Several models that allowed for two emitting states were also analyzed (not shown here) but seem to be quite unlikely on the basis of the fitted SAS and rate constants. The three "best-fit" models with only one emitting state are shown in Fig. 2 along with the results of the kinetic analysis:

i) A sequential model where a secondary charge transfer or conformational relaxation step following the much faster primary charge separation process is assumed (The reverse order can be excluded).
ii) A branching model which describes the 12 ps component in terms of a parallel electron transfer to another chromophore, located in the so far considered "inactive" M-branch in addition to electron transfer in the normally active M-branch.
iii) An inherent heterogeneity in the RCs giving rise to a heterogeneity in the rate of primary and possibly also secondary electron transfer for different RCs.

The sequential and the branching models describe the fluorescence kinetics about equally well as judged by their χ^2-values and the residuals while the heterogeneity model appears to have a slightly lower χ^2-value. The improvement in the fit parameters for the latter model is not sufficient, however, to exclude the other models. From these data at the present level of S/N ratio (which is quite good) it is not possible to distinguish between these models. We will thus discuss in the following the relative merits of each of these models.

Sequential model: The a priori most reasonable interpretation of X^- in this model appears to be B_L^-. In this case the rate of formation of X^- would be consistent with transient absorption data of Zinth and coworkers[5]. However, the rate of reoxidation would be by more than an order of magnitude slower than in that model. Furthermore, the decay rate of P^+H^- is predicted to be substantially faster (11 ns^{-1}) than by most transient absorption data (about 5 ns^{-1}, see however[8]. Thus an alternative explanation could be invoked in which the intermediate state P^+X^- in our model is interpreted as a conformationally unrelaxed radical pair state $(P^+H_L^-)_{unrelax}$ possibly following a state P^+B^- similar to photosystem II reaction centers[10]. The latter state would not be resolved in our data if the rates in the model of Holzapfel et al.[5] are assumed. In this interpretation the fluorescence data could possibly be consistent with transient absorption data of Holzapfel et al.[5].

Branching model: This model predicts that about 1/3 of the primary charge separation occurs in an equilibration reaction into the M-branch. X^- could either be B_M^- or H_M^-. The state P^+X^- is predicted to be slightly higher in energy ($\Delta G = +7$ meV) than P*. The new element in this model consists in the suggestion that unidirectionality would be much less pronounced in the primary step than assumed so far[9,17,18]. Instead, most of the selectivity would be brought about by a) a slow decay of P^+X^- in a secondary step and b) the unidirectionality enforced by the electron transfer in the L-branch to Q_A. Again the ΔG-value in the L-branch for P^+H^- is only -127 meV, somewhat too positive, which might support an unrelaxed $P^+H_L^-$ state (vide supra).

Heterogeneity model: An important difference of this model to the ones discussed above is the fact that it involves one decay component more with a lifetime of about 2.3 ns. This model would allow the main lifetime (3 ps) component to be explained within a kinetic scheme and with rates that are fully consistent with existing transient absorption data both with respect to the primary charge separation process as well as with respect to the charge stabilization (4 ns^{-1})[5,6]. (We again ignore here the possible involvement of an intermediate B_L^- state[5]. Neglecting this state does not influence our kinetic parameters for the other states.) The only argument against this scheme might be the value of $\Delta G = -142$ meV which may be expected to be somewhat more negative. For about 22% of the RCs as determined by the global target analysis the heterogeneity model then would

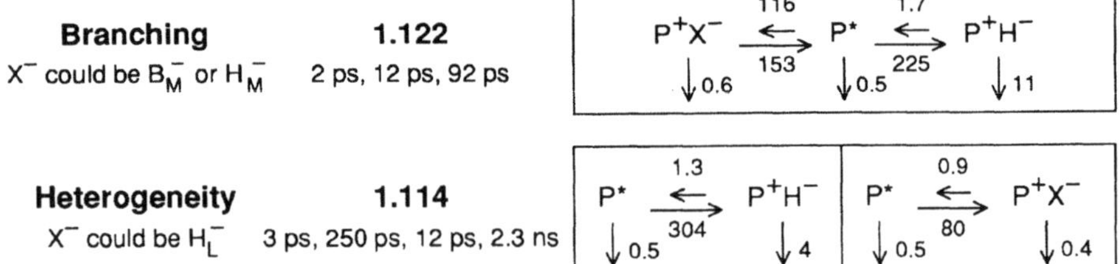

Figure 2. Rate constants, lifetimes and χ^2-values for three different kinetic models as calculated by the global target analysis. Possible assignments for X^- are given.

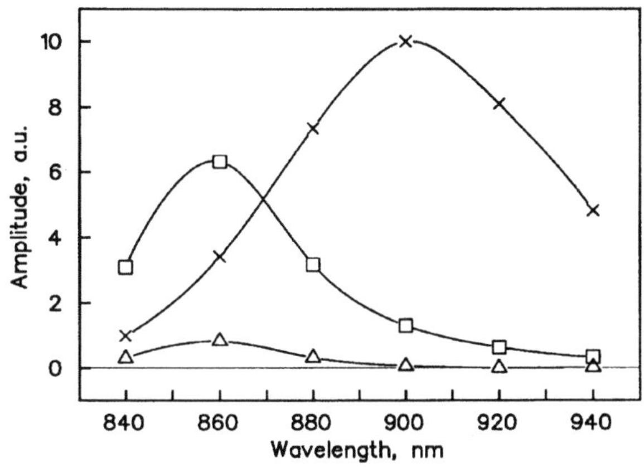

Figure 3. Species-associated fluorescence spectra of RCs from *Rb. sphaeroides* as calculated for the three kinetic models discussed in this paper. The short-wavelength components are due to the free BChl while the longer-wavelength component represents the SAS of P860*.

predict a primary charge separation which is reduced by about a factor of nearly 4. If one interprets X^- as H_L^-, also the charge stabilization would be slowed down by a factor of about 10 in these centers (possibly to 0.4 ns^{-1}) as compared to the main proportion of RCs, if the 2.3 ns component is taken as a genuine RC component which is actually suggested by its spectrum. Such a slowed down rate would most probably have been noticed in transient absorption data. However Kirmaier and Holten reported quite a large range of rates for this process[8]. As an alternative explanation one might assume that the second set represents closed RCs (Q_A reduced) which would explain at least the drastically reduced rate of secondary electron transfer. However two arguments speak against this possibility: a) The primary charge separation step has been measured with reduced Q_A and was found to be slowed down by less than 50%[19,20] thus inconsistent with the rate of 80 ns^{-1} in this model. b) It is unlikely that we would accumulate PQ_A^- under our measurement conditions. Rather it would be much more likely that we accumulate some $P^+Q_A^-$ state because of insufficient sample flow speed to exchange the sample volume between flashes. This state would be non-fluorescent, however, and could not explain the 12 ps component. Thus, on the basis of this model, the data would be consistent with a large heterogeneity in both the primary and secondary electron transfer rates in open RCs, as has been proposed, albeit to a somewhat lesser extent, in[8].

We have discussed here the relative merits of each of the three models. None of them can be definitely excluded at present on the basis of our data and those in the literature. In any case it is obvious that the electron transfer process in RCs of *Rb. sphaeroides* are more complex than assumed so far on the basis of transient absorption data alone. The 12 ps component has a substantial relative amplitude and can not be ignored. Most likely the 100 ps component must also be included in a model. The observation of these new components poses important new questions as to the nature of the primary processes in purple bacterial RCs. It will be necessary to include these components in a coherent mechanistic framework which is consistent with both fluorescence and transient absorption data. We suspect that these components will also be present in transient absorption data but have possibly not been resolved so far in most data sets in view of the generally smaller S/N ratio of transient absorption data. However, reports on "heterogeneity" in the rates seen in transient absorption[8,21] may be taken as evidence for the presence of one or both fluorescence components.

ACKNOWLEDGEMENTS

We acknowledge Mr. M. Reus for valuable technical support. We also should like to thank Prof. K. Schaffner for his interest and support of this work. Partial financial support was provided by the Deutsche Forschungsgemeinschaft (SFB 189).

REFERENCES

1. J. Deisenhofer, O. Epp, K. Miki, R. Huber, and H. Michel, X-ray structure analysis of a membrane protein complex. Electron density map at 3 Å resolution and a model of the chromophores of the photosynthetic reaction center from *Rhodopseudomonas viridis*, *J. Mol. Biol.* 180:385 (1984).
2. C.-H. Chang, O. El-Kabbani, D. Tiede, J. Norris, and M. Schiffer, Structure of the membrane-bound protein photosynthetic reaction center from *Rhodobacter sphaeroides*. *Biochemistry* 30:5352 (1991).
3. O. El-Kabbani, C.-H. Chang, D. Tiede, J. Norris, and M. Schiffer, Comparison of reaction centers from *Rhodobacter sphaeroides* and *Rhodopseudomonas viridis*: Overall architecture and protein-pigment interactions. *Biochemistry* 30:5361 (1991).
4. M.R. Gunner, D.E. Robertson, and P.L. Dutton, Kinetic studies on the reaction center protein from *Rhodopseudomonas sphaeroides*: The temperature and free energy dependence of electron transfer between various quinones in the Q_A site and the oxidized bacteriochlorophyll dimer. *J. Phys. Chem.* 90:3783 (1986).
5. W. Holzapfel, U. Finkele, W. Kaiser, D. Oesterhelt, H. Scheer, H.U. Stilz, and W. Zinth, Observation of a bacteriochlorophyll anion radical during the primary charge separation in a reaction center. *Chem. Phys. Lett.* 160:1 (1989).
6. J. Breton, J.-L. Martin, A. Migus, A. Antonetti, and A. Orszag, Femtosecond spectroscopy of excitation energy transfer and initial charge separation in the reaction center of the photosynthetic bacterium *Rhodopseudomonas viridis*. *Proc. Natl. Acad. Sci. USA* 83:5121 (1986).
7. M. Bixon, J. Jortner, M.E. Michel-Beyerle, A. Ogrodnik, and W. Lersch, The role of the accessory bacteriochlorophyll in reaction centers of photosynthetic bacteria: Intermediate acceptor in the primary electron transfer? *Chem. Phys. Lett.* 140:626 (1987).
8. C. Kirmaier and D. Holten, Evidence that a distribution of bacterial reaction centers underlies the temperature and detection-wavelength dependence of the rates of the primary electron-transfer reactions. *Proc. Natl. Acad. Sci. USA* 87:3552 (1990).
9. W.W. Parson, Reaction centers, *in*: Chlorophylls, H. Scheer, ed., CRC Press, Boca Raton (1991).
10. T.A. Roelofs, M. Gilbert, V.A. Shuvalov, and A.R. Holzwarth, Picosecond fluorescence kinetics of the D1-D2-cyt-b559 photosystem II reaction center complex. Energy transfer and primary charge separation processes. *Biochim. Biophys. Acta* 1060:237 (1991).
11. N.W.T Woodbury and W.W. Parson, Nanosecond fluorescence from isolated photosynthetic reaction centers of *Rhodopseudomonas sphaeroides*. *Biochim. Biophys. Acta* 767:345 (1984).
12. A.M. Freiberg, V.I. Godik, S.G. Kharchenko, K.E. Timpmann, A.Y. Borisov, and K.K. Rebane, Picosecond fluorescence of reaction centres from *Rhodospirillum rubrum*. *FEBS Lett.* 189:341 (1985).

13. M.G. Müller, K. Griebenow, and A.R. Holzwarth, Primary processes in isolated photosynthetic bacterial reaction centers from *Chloroflexus aurantiacus* studied by picosecond fluorescence spectroscopy. *Biochim. Biophys. Acta* 1098:1 (1991).
14. A.R. Holzwarth, J. Wendler, and G.W. Suter, Studies on chromophore coupling in isolated phycobiliproteins. II. Picosecond energy transfer kinetics and time-resolved fluorescence spectra of C-phycocyanin from *Synechococcus 6301* as a function of the aggregation state. *Biophys. J.* 51:1 (1987).
15. J.M. Beechem, M. Ameloot, and L. Brand, Global and target analysis of complex decay phenomena. *Anal. Instrum.* 14:379 (1985).
16. M.G. Müller, Picosekundenuntersuchungen an Photosyntheseantennen, Dissertation, Schriftenreihe des MPI für Strahlenchemie No. 65, ISSN 0932-5131 (1992).
17. M.E. Michel-Beyerle, M. Plato, J. Deisenhofer, H. Michel, M. Bixon, and J. Jortner, Unidirectionality of charge separation in reaction centers of photosynthetic bacteria. *Biochim. Biophys. Acta* 932:52 (1988).
18. W. Aumeier, U. Eberl, A. Ogrodnik, M. Volk, G. Scheidel, R. Feick, M. Plato, and M.E. Michel-Beyerle, Unidirectionality of charge separation in reaction centers of *Rb. sphaeroides* and *Chloroflexus aurantiacus*, *in*: Current Research in Photosynthesis I, M. Baltscheffsky, ed., Kluwer Academic Publishers, Dordrecht (1990).
19. J. Breton, J.-L. Martin, A. Migus, A. Antonetti, and A. Orszag, Femtosecond spectroscopy of excitation energy transfer and initial charge separation in the reaction center of the photosynthetic bacterium *Rhodopseudomonas sphaeroides*, *in*: Ultrafast Phenomena V. G.R. Fleming and A.E. Siegman, eds., Springer, Berlin (1986).
20. N.W. Woodbury, M. Becker, D. Middendorf, and W.W. Parson, Picosecond kinetics of the initial photochemical electron-transfer reaction in bacterial photosynthetic reaction centers. *Biochemistry* 24:7516 (1985).
21. C. Kirmaier and D. Holten, An assessment of the mechanism of inital electron transfer in bacterial reaction centers. *Biochemistry* 30:609 (1991).

FEMTOSECOND SPECTROSCOPY OF THE PRIMARY ELECTRON TRANSFER IN PHOTOSYNTHETIC REACTION CENTERS

Wolfgang Zinth[1], Peter Hamm[1], Karl Dressler[2], Ulrich Finkele[2], Christoph Lauterwasser[1]

[1]Institut für Medizinische Optik der Universität München
Barbarastr 16, 8000 München 40, Germany

[2]Physik Department E11 der Technischen Universität München
James-Franck-Straße, 8046 Garching, Germany

INTRODUCTION

In the primary photosynthetic process of bacterial reaction centers (RC) light energy is stored by a rapid electron transfer (ET). The structural arrangement of the reaction centers with the six chromophores kept in two symmetry related branches A and B predetermines the ET path. The branches begin at two strongly interacting bacteriochlorophyll molecules forming the special pair P which acts as a primary electron donor. Subsequently each branch contains a monomeric bacteriochlorophyll (BChl) molecule B_A and B_B, a bacteriopheophytin (BPhe) (H_A, H_B) and a quinone (Q_A, Q_B) [1, 2]. Spectroscopy on reaction centers has revealed that the two pigment branches are spectroscopically non-equivalent and that electron transfer uses predominantly the A branch. It is generally accepted that the ET starts after excitation by light from the special pair P and that the primary reaction is finished by the ET from the bacteriopheophytin H_A to the quinone Q_A which occurs with a time constant of 200 ps. However, there exist different opinions on the first part of the ET reaction: In the super-exchange ET model the electron is transferred directly from the special pair P to the bacteriopheophytin H_A on the A branch. The monomeric bacteriochlorophyll is only used as a virtual

The Photosynthetic Bacterial Reaction Center II
Edited by J. Breton and A. Verméglio, Plenum Press, New York, 1992

electron carrier [3-7]. In the stepwise ET model the monomeric bacteriochlorophyll B_A is a real electron carrier and the electron undergoes two reaction steps before it reaches the bacteriopheophytin. This model is supported by recent experimental results on RC from Rhodobacter (Rb.) sphaeroides which indicate that the electron transfer to B_A occurs in approximately 3.5 ps while the second transfer step to the bacteriopheophytin H_A should be faster taking less than one picosecond (0.9 ps) [8, 9].

In this paper we give additional information on the primary ET reaction obtained by transient absorption and emission experiments. In a detailed discussion we will use these results in order to obtain new insight into the nature of the primary electron transfer.

MATERIAL AND METHODS

RC from Rps. viridis and Rb. sphaeroides R26.1 were prepared as described in Ref. 8, 10. Room temperature experiments were performed in cuvettes with a pathlength of 1 mm under stirring. Measurements on low temperature RC were performed on quinone depleted RC from Rb. sphaeroides strain R26.1 desolved in glycerol [12]. The time resolved absorption experiments used the excite-and probe technique. Details of the experimental set-up are described elsewhere [8, 10]. The main features of the experiments are: Excitation beam: short pulses of a duration of about 200 fs at a repetition rate of 10 Hz, excitation wavelength 955 nm (Rps. viridis) and 875 nm (Rb. sphaeroides), less than 10 % of the RC are excited per laser pulse. Probing pulses: 5-10 nm wide portion of a femtosecond white-light-continuum selected by a dispersion compensated spectrometer in front of the sample, parallel polarisations of exciting and probing pulses, probe intensities at least 30 times smaller than excitation intensities. The width of the instrumental response function was between 250 and 350 fs.

The time-resolved emission experiment was performed with the fluorescence up-conversion technique. (For details see Ref. 11). Excitation parameters were: λ_{exc} = 865 nm (Rb. sphaeroides), $t_p \simeq 200$ fs, less than 10 % of the RC absorb a photon. Up-conversion process: collinear type II phase matching in a 1mm BBO crystal; up-converted fluorescence emission 910-930 nm; width of the instrumental response function $\simeq 400$ fs.

In the standard nonadiabatic description of transient absorption spectroscopy one treats the molecular system as a set of electronic states where the vibrational levels of each state are in thermal equilibrium at some temperature T. As a consequence these states have well defined absorption properties. Transitions between the states are governed by reaction rates. The absorbance change is a sum of exponentials convolved with the instrumental response function. The number of exponentials is equal to the number of intermediate states populated during the reaction. The mathematical description of the absorbance change in the non-adiabatic theory is described in detail in Ref. 9. It has been shown in the literature [5] that non-adiabatic theory is justified at least at room temperature. As a consequence we analyse the

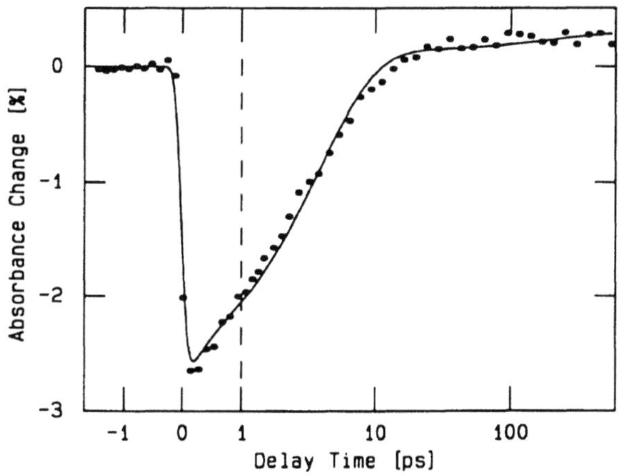

Figure 1. Transient absorption data from RC of Rps. viridis in the gain region (λ probe = 1050nm)

room temperature data by exponential functions. Due to the lack of a more reasonable theory we apply the same procedure - as a first order approximation - to the low-temperature data.

RESULTS AND DISCUSSIONS

Transient Absorption Spectroscopy of Reaction Centers of Rps. Viridis

The general features of transient absorption spectroscopy on Rps. viridis are similar to those found previously for Rb. sphaeroides [8, 9]. The decay of the excited electronic state P* of the special pair is seen at 1050 nm (Fig. 1). Here stimulated emission (gain) occurs which decays at later delay times with a time constant of approximately 3.5 ps. At early times the absorption changes more rapidly suggesting the existence of a faster kinetic component (See systematic deviations of data points from the model curve). This component is evident at measurements in the Q_x and Q_y absorption bands of the monomeric bacteriochlorophylls (Fig. 2a and b). Here a fast process with a time constant of 0.65 ± 0.3 ps occurs within a narrow spectral range. The other components observed have time constants of 3.5 ps and 200 ps.

The observation of three kinetic components with two time constants below 5 ps parallels the findings for Rb. sphaeroides. The experimental data strongly support the idea that the primary ET reactions of Rb. sphaeroides and Rps. viridis proceed via similar reaction models. While the qualitative agreement of the experimental results is striking it should be recalled that the subpicosecond kinetic component is somewhat faster in RC of Rps. viridis than in RC of Rb. sphaeroides.

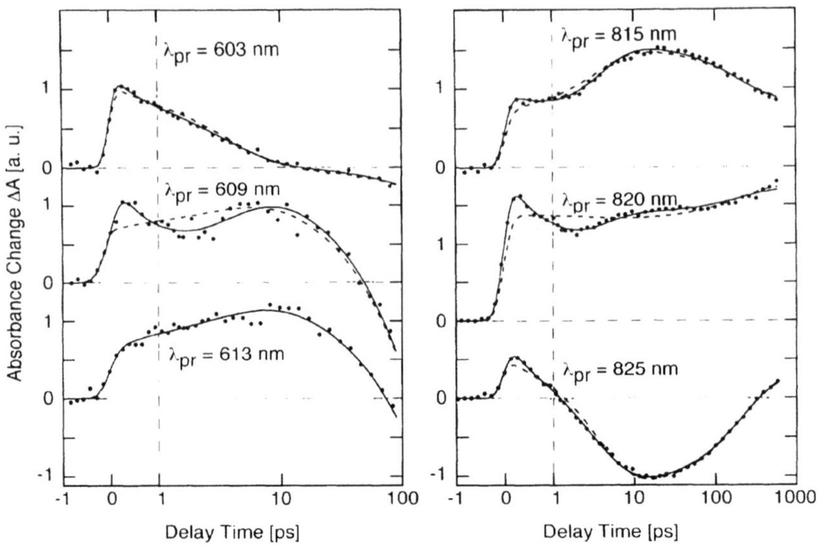

Figure 2. Transient absorption data (points) for RC from Rps. viridis recorded in the Q_x (left) and Q_y (right) absorption band of the accessory bacteriochlorophyll. The solid curves are calculated for a four component (0.65 ps, 3.5 ps, 200 ps, ∞) the broken curve for a three component model (3.5 ps, 200 ps, ∞).

Reaction Centers at Low Temperatures

A first set of experiments investigated the decay of the excited state P* via stimulated emission. At the low temperature of 25 K the signal points were close to a monoexponential model function with a time constant of $\tau_1 = 1.4 \pm 0.3$ ps. This transient was followed up to room temperature where the value of 3.5 ps was reached as discussed above. The data are in agreement with previous experimental studies [13-15]. Most interesting is the investigation of the temperature dependence of the fast kinetic component [12]. To this end we studied the transient absorption changes at probing wavelengths around 795 nm in the spectral range of the Q_y band of the bacteriochlorophyll at 25 K. The transient absorption data yielded the following results: One finds a complex time dependence of the absorbance change which excludes the possibility that there is only one, namely the 1.4 ps kinetic component. There exists an additional faster kinetic process with a time constant of 0.3 ± 0.15 ps. In addition there appear some weak oscillations similar to those reported recently by Voss et al. [15]. In a set of measurements we have recorded the temperature dependence of the absorbance change in a broader spectral range. We observe qualitatively similar transient absorption features at all temperatures.

Reaction Centers with Exchanged Bacteriochlorophyll a

In another set of experiments RC of Rb. sphaeroides were studied where the bacteriochlorophyll a molecule at the accessory position B_A and B_B were exchanged by [3-vinyl]-13^2-OH-bacteriochlorophyll a molecules [16]. The modification due to the 3-vinyl group is expected to change the redox potential of the BChl and as a consequence the energy of the radical pair state $P^+B_A^-$. This change should have pronounced consequences on the ET when the accessory BChl B_A is involved as an intermediate electron carrier. Indeed, one finds a strong change of the transient absorption data. The experimental data indicate that the RC's containing [3-vinyl]-13^2-OH-BChl a have a decay time of the excited electronic state P* of the special pair of 32 ps. On the other hand a long-lasting bleaching of the special pair absorption band shows that the exchange leads to RC's which are still photochemically active. In the [3-vinyl]-13^2-OH-BChl a containing RC's the 0.9 ps component is not visible. However, there are some indications that a related process exists which would have a longer time constant in the 5 ps domain.

REACTION MODELS

The structural arrangement of the RC supports the idea that the electron is transferred in several steps from the special pair P via B_A, H_A to Q_A (Model A of Fig. 3). The transient experimental data presented here do not give any contradiction against this reaction model. Far from it the analysis of the transient data using reaction model A yields exactly the spectra of the intermediates and would expect from in vitro measurements of the chromophores [17, 18]. This finding can be illustrated by Fig. 1. In this experiment the transient absorption at 1050 nm in the gain region was investigated. Surprisingly there was some faster initial decay of the signal (which was not seen in the short wavelength side of the gain). Data analysis using reaction model A indicates that the second intermediate I_2 must have an increased absorption in this spectral range. This observation fits well to the interpretation of I_2 being the radical pair state $P^+B_A^-$ as spectra of the bacteriochlorophyll b anion show a distinct absorption band around 1050 nm [17, 18].

However, most transient absorption data also fit to the two models B_1 and B_2 where the subpicosecond reaction is assumed to precede the 3.5 ps process: Here the intermediate I_2 is formed very fast. It decays with 3.5 ps in a second step. Calculating the absorption spectrum of I_2 for model B_1 and B_2 leads to the following characteristics: I_2 is similar to the electronically excited state P*. It also exhibits gain; thus it should be another excited electronic state of the special pair - we call it P**. Its further absorption properties differ only slightly from those of P*. The most straightforward interpretation of P** would be that P** is a vibrationally relaxed P* state (Model B_1). Here the electron will be transferred directly in a super-

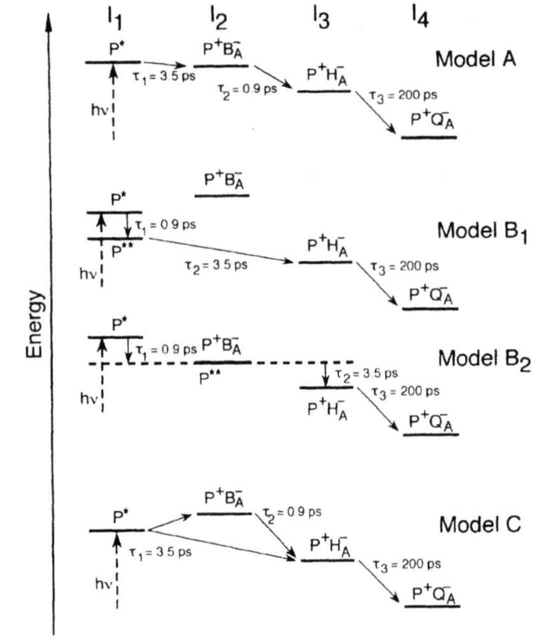

Figure 3. Schematic representation of possible reaction models for the primary photosynthetic ET. The time constants shown in the Figure represent the values for Rb. sphaeroides at room temperature.

exchange step from the special pair P to H_A. Somewhat different is the molecular interpretation for Model B_2, which is related to considerations given by H. Kuhn [19]. Model B_2 is based on the existence of an intermediate state $I_2 = P^{**}$ where the electron is delocalized over the special pair, the accessory BChl and the BPh. According to the experimental observations state $I_2 = P^{**}$ must be populated in the first 0.9 ps process. The slower 3.5 ps process is thought to be related to the trapping of the electron at the bacteriopheophytin H_A. Due to the delocalization of the electron in state P^{**} there is no need for a long- range superexchange ET in Model B_2.

In model C the energy of state $P^+B_A^-$ is the relevant parameter. Model C represents a combination of the stepwise ET (Model A) and the superexchange Model B_1 according to Bixon et al. [20, 5]. For an energy of state $P^+B_A^-$ close to the energy of P^* both reaction pathways may occur in parallel. As I_2 must have the spectral properties of a radical pair state $P^+B_A^-$ one obtains restrictions for the model parameters. We find a limit for the maximum yield of the direct superexchange transfer at room temperature of 10 %; the energy of state $P^+B_A^-$ is about 200 cm^{-1} (Rps. viridis) below that of P^*.

The experimental data obtained for RC at low temperatures and with exchanged bacteriochlorophylls allow to restrict furtheron the number of reaction models: The discussion of the two reaction Models B_1 and B_2 requires a subtile consideration of the experimental observations: In the pure superexchange picture of Model B_1 the fast kinetic component is related to vibrational relaxation in the excited state. From the theory of vibrational relaxation of polyatomic molecules and from a number of experiments (e. g. on amino acids [21]) it is well known that vibrational relaxation slows down at low temperatures. However, the fast reaction becomes considerably faster at low temperatures. This observation is incompatible with the interpretation of Model B_1. Additional arguments against vibrational relaxation come from experiments on modified RCs; e. g. on RCs where the monomeric BChl are exchanged by [3 vinyl]-13^2 OH-BChl and where the 3.5 ps time constant is increased to 32 ps. The molecular substitution leaves the special pair unaffected; as a consequence a P* vibrational process according to Model B_1 should be present and observable. However, the experiments do not exhibit the related 0.9 ps transient component.

The observed transient absorption data alone are not able to eliminate Model B_2. Additional information comes from hole-burning experiments (Johnson et al., [22]). In these experiments performed at very low temperatures narrow zero phonon holes were observed with a spectral width corresponding to a time constant of approximately 1 ps. From these data one can deduce that the first reaction process starting from the lowest vibrational level of P* is the slower, the 1.4 ps process. The faster 0.3 ps component must be (as it is not related with vibrational relaxation, see above) the second process in the reaction scheme. Since the important features of the reaction processes do not change strongly with temperature one may discard Model B_2 at room temperature as well.

The stepwise reaction Model A with the radical pair state $P^+B_A^-$ as a real intermediate with only a small contribution of a superexchange reaction is compatible with the extensive time resolved absorption data available today. At room temperature the stepwise ET is well described by theoretical studies giving reasonable values for the energetics in the RCs. However, the discussion of ET and absorption at low temperatures within the framework of adiabatic theory remains to be done.

NON-MONOEXPONENTIALITY OF THE FLUORESCENCE EMISSION

The experimental observations presented above fitted well into the reaction model A (or C) where only functionally necessary reaction steps are involved. We will show now that more detailed investigations exhibit an additonal kinetic component : Time resolved emission spectroscopy by fluorescence up conversion gives a valuable tool to study the properties of electronically excited emitting states without any interference of no emitting product states [11] allowing high sensitivity. An experimental result is given for RC from Rb. sphaeroides in Fig. 4 [11]. Here the decay of the emission is displayed in a semilogarithmic plot. The experimental data demonstrate that the decay of the fluorescence is not monoexponential: Apparently the 3.5 ps process deduced from the gain experiments must be split into a 2.2 ps and a 7 ps reaction. The amplitude ratio η = A(7 ps)/A(2.2 ps) is small, $\eta \simeq 0.25$. (Indications of such a biexponential decay of P* have been seen before. [9, 15]) Of special interest is the finding that the emission experiment does not give any indication for a subpico-

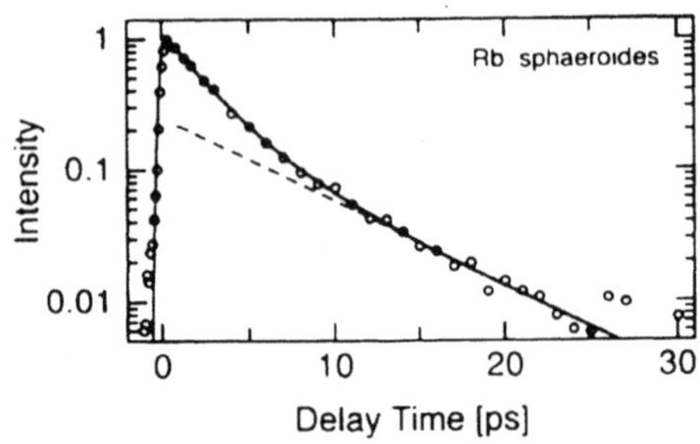

Figure 4. Time dependence of the fluorescence emission (910-930 nm) of RC from Rb. sphaeroides.

second kinetic component which one would expect, if the subpicosecond kinetic component is related to vibrational relaxation (Model B_1).

The observation of an additional time constant requires an extension of the reaction models: Trivial is the assumption of a functional heterogeneity of the sample. In this case one would deal with two components having a different speed of the primary ET reaction. For a homogeneous sample one has to assume that the longer emission decay time is related to a new intermediate state (we call it N). The experimental observation of N in emission indicates that it is coupled directly to P*. There are several possibilities to introduce the new state in a reaction model. We only want to discuss here the simplified situation where N is a not emitting state coupled only to P* while model A applies for the further reaction. In this case one can calculate the reaction rates to and from N (via the emission experiment) and the spectral properties of intermediate N from previously measured transient absorption data. This evaluation yields: The reaction from P* to N is slow with a rate of 1/13 ps while the reactive rate from P* to $P^+B_A^-$ is four times faster. The back reaction from N to P* is fast with a rate of 1/4.8 ps. The difference spectrum of state N shows spectral properties which are similar to those of $P^+B_A^-$. As a consequence one could speculate that N is the radical pair state $P^+B_B^-$ where the electron is transiently brought to the B branch. The further evaluation of the transient data shows that the spectra of the other intermediate states remain very similar to those obtained with the simplified reaction model A.

In conclusion, time resolved spectroscopy on reaction centers of the purple bacteria Rb. sphaeroides and Rps. viridis indicate that electron transfer of native RC occurs stepwise using the different chromophores of the A branch as real intermediate electron carriers. The observation of the biexponentiality could be taken as an indication that there is a transient population of the radical pair state $P^+B_B^-$ on the "inactive" B chromophore branch. Under this assumption the high asymmetry of the charge transfer would require a very slow electron transfer step from the monomeric bacteriochlorophyll B_B to the neighbouring bacteriopheophytin H_B.

Acknowledgements

The experiments were performed in collaboration with S. Buchanan, W. Kaiser, H. Michel, D. Oesterhelt and H. Scheer.

References

1. J. Deisenhofer, H. Michel, EMBO J. 8, 2149 (1989).

2. C.H. Chang, D. Tiede, J. Tang, U. Smith, J. Norris, M. Schiffer, FEBS Letters 205, 82 (1986).

3. M.E. Michel-Beyerle, M. Plato, J. Deisenhofer, H. Michel, M. Bixon, J. Jortner, Biochim. et Biophys. Acta 932, 52-70 (1988).

4. M.E. Michel-Beyerle, M. Bixon, J. Jortner, Chem. Phys. Lett. 151, 188-194 (1988).

5. M. Bixon, J. Jortner, M.E. Michel-Beyerle, Biochim. et Biophys. Acta 1056, 301 (1991).

6. J.L. Martin, J. Breton, A.J. Hoff, A. Migus, A. Antonetti, Proc. Natl. Acad. Sci. US 83, 957 (1986).

7. J. Breton, J.L. Martin, A. Migus, A. Antonetti, A. Orszag, Proc. Natl. Acad. Sci. US 83, 5121 (1986).

8. W. Holzapfel, U. Finkele, W. Kaiser, D. Oesterhelt, H.Scheer, H.U. Stilz, W. Zinth, Chem. Phys. Letters 160, 1 (1989).

9. W. Holzapfel, U. Finkele, W. Kaiser, D. Oesterhelt, H.Scheer, H.U. Stilz, W. Zinth, Proc. Natl. Acad. Sci. US 87, 5168 (1990).

10. K. Dressler, E. Umlauf, S. Schmidt, P. Hamm, W. Zinth, S. Buchanan, H. Michel, Chem. Phys. Letters 183, 270 (1991).

11. P. Hamm, D. Oesterhelt, R. Feick, H. Scheer, W. Zinth, to be published (1992).

12. C. Lauterwasser, U. Finkele, H. Scheer, W. Zinth; Chem. Phys. Letters 183, 471 (1991).

13. G.R. Fleming, J.L. Martin, J. Breton, Nature 333, 190 (1988).

14. N.W. Woodbury, M.Becker, D. Middendorf, W.W. Parson, Biochem. 24, 7516 (1985).

15. M.H. Voss, J.-C. Lambry, S.J. Robles, D.C. Youvan, J. Breton, J.-L. Martin, Proc. Natl. Acad. Sci. USA 88, 8885-8889 (1991).

16. U. Finkele, C. Lauterwasser, A. Struck, H. Scheer, W. Zinth, to be published in Proc. Natl. Acad. Sci. US (1992)

17. J. Fajer, D.C. Borg, A. Forman, D. Dolphin, R.H. Felton, J. Am Chem. Soc. 95, 2739-2741 (1973).

18. J. Fajer, M.S. Davis, D.C. Brune, L.D. Spaulding, D.C. Borg, A. Forman, Brookhaven Symp. Biol. 28, 74-104 (1976).

19. H. Kuhn, Phys. Rev. A 34, 3409-3425 (1986).

20. J. Bixon, J. Jortner, M.E. Michel-Beyerle, A. Ogrodnik, Biochim. et Biophys. Acta 977, 273-286 (1989).

21. T.J. Kosic, R.E.(Jr) Cline, D.D. Dlott, J. Chem. Phys. 81, 4932-4949 (1984).

22. S.G. Johnson, D. Tang, R. Jankowiak, J.M. Hayes, G. J. Small, J. Phys. Chem. 93, 5953 (1989) .

FEMTOSECOND OPTICAL CHARACTERIZATION OF THE EXCITED STATE OF *RHODOBACTER CAPSULATUS* D_{LL}

Marten H. Vos[1,2], Fabrice Rappaport[1], Jean-Christophe Lambry[1], Jacques Breton[2] and Jean-Louis Martin[1]

[1] Laboratoire d'Optique Appliquée, INSERM U275, Ecole Polytechnique-ENSTA 91120 Palaiseau, France

[2] Service de Biophysique, Département de Biologie, Centre d'Etudes Nucléaires de Saclay, 91191 Gif-sur-Yvette Cedex, France

INTRODUCTION

The precursor of photosynthetic electron transfer is the lowest singlet excited state of the pigment complex P^*. It is generally accepted that upon population of this state, P^* disappears and the radical pair $P^+H_L^-$ is formed in about 3 ps at room temperature (Martin et al.,1986, Woodbury et al.,1985, Holzapfel et al.,1990, Kirmaier and Holten, 1988) and somewhat faster, down to 700 fs, at low temperature (Fleming et al., 1988). However the precise nature of this reaction is subject to debate, especially the role of B_L, which is located between P and H_L (for a review see Martin and Vos, 1992). It has been proposed that $P^+B_L^-$ is a virtual intermediate in a super-exchange mechanism (Michel-Beyerle et al., 1988, Warshel et al., 1988). Based on the observation of transients faster than P^* decay, Zinth and coworkers proposed that $P^+B_L^-$ is a real intermediate, which decays with a faster rate than it is formed (Holzapfel et al., 1989, 1990).

Theoretically, primary electron transfer is usually described as a non-adiabatic process coupled to thermally equilibrated vibrational modes (Marcus and Sutin, 1985, Bixon and Jortner, 1986). It is thus assumed that thermalization of the vibrational modes relevant for primary electron transport occurs on a timescale faster than ~1 ps. Also any electronic relaxations in P^* are thought to occur much faster than primary electron transfer. One way to test these assumptions is to optically properly characterize the excited state P^*. To this aim we recently performed a set of femtosecond optical studies at low temperature (Vos et al., 1991, 1992). The kinetics were found to be modulated by oscillatory features with frequencies in the 15-70 cm^{-1} range. If these reflect vibrational motion this implies that low frequency modes are not all dephased (and hence not thermalized) on the timescale of electron transfer. Furthermore, the early time transient spectra at low temperature of fully functional reaction centers show many features in the Q_Y region of all pigments, including the bacteriopheophytins, indicating that the coupling between all pigments in the excited state is stronger

The Photosynthetic Bacterial Reaction Center II
Edited by J. Breton and A. Verméglio, Plenum Press, New York, 1992

than expected on the basis of calculations of the ground state spectra. Also a relaxation occurs on a timescale of ~100 fs. This relaxation comprises a redistribution of oscillator strength between the H band (bacteriopheophytin) and the B (bacteriochlorophyll) band.

Further insight in the inter-pigment interactions and the dynamics of the excited state may be obtained by studying reaction centers with a modified pigment composition. Here we present a spectral and kinetic characterization of the excited state of the D_{LL} mutant of *Rhodobacter capsulatus*, in which the bacteriopheophytin electron acceptor H_L has been genetically removed (Robles et al., 1990). This mutant has the additional advantage that the excited state decays (directly to the ground state) on a timescale of hundreds of picoseconds. This implies that the transient spectra of the excited state are directly obtained, avoiding the complications encountered in wild type reaction centers, where transient spectra of the excited state may considerably mix with dominant features of the charge separated state(s), even at short delay times.

The D_{LL} mutant has previously been used (Breton et al., 1990) to investigate the possible formation of radical pairs other than $P^+H_L^-$, which can normally not be accumulated to a significant extent. It was found, however, that P^* lives for hundreds of picoseconds before decaying to the ground state. It is noteworthy that a) H_M is not photoreduced within a few hundred picoseconds, indicating that electron transfer along the M branch is essentially unfavourable and b) $P^+B_L^-$ is not accumulated. If a scheme is adopted (cf. Holzapfel et al., 1990) in which $P^+B_L^-$ is formed prior to $P^+H_L^-$ one would expect this state to be trapped in the absence of H_L. So, the absence of a decay of P^* on the time scale of a few picoseconds pleads against such a model. It must be kept in mind however, that one cannot totally exclude the possibility that the redox potentials of the presumed states $P^+H_M^-$ and $P^+B_L^-$ are altered by the mutations, although no evidence for $P^+B_L^-$ formation was found in other mutants lacking H_L with a more native-like amino-acid composition near B_L (not shown).

In this paper, the properties of the D_{LL} mutant, a long-lived excited state without kinetic interference of charge separated states, are used to obtain a reliable optical characterization of the excited state. The results are discussed in comparison with functional reaction centers.

MATERIALS AND METHODS

Preparation of *R. capsulatus* D_{LL} chromatophores devoid of antenna pigments and of Q_A reduced reaction centers of *R. sphaeroides* strain R-26 was described before (Vos et al., 1992). The samples were diluted in glycerol (60% vol/vol) to an optical density of 0.5 at 870 nm in a cuvette with an optical path of 1 mm and were cooled in the dark to 10K in a convection cryostat.

Generation and amplification of the compressed broad band (25 nm width around 870 nm; the spectrum was cut off below 850 nm) pump pulses of 45 fs full width at half maximum, with a repetition rate of 30 Hz, is described elsewhere (17). The chirp in the probe beam, about 1 ps in the wavelength region of 750-850 nm, is compensated with two prisms to ±50 fs, such that the central wavelength part, around 800 nm, arrives first (t=0 fs is defined at the maximum of the cross-correlation function between the 870-nm pump and the probe at 800 nm). The remaining chirp still is comparable to the pulsewidth and hence spectra at early times are still moderately distorted. The absorption changes were detected with a pair of diodes for kinetic measurements and with an optical multichannel analyser, with a spectral resolution of ~1 nm, for spectral measurements. Unless otherwise indicated, the pump and probe beams were polarized in parallel. Typically 10-20% of the centers were excited upon each flash. The kinetics were independant of flash energy in this excitation range.

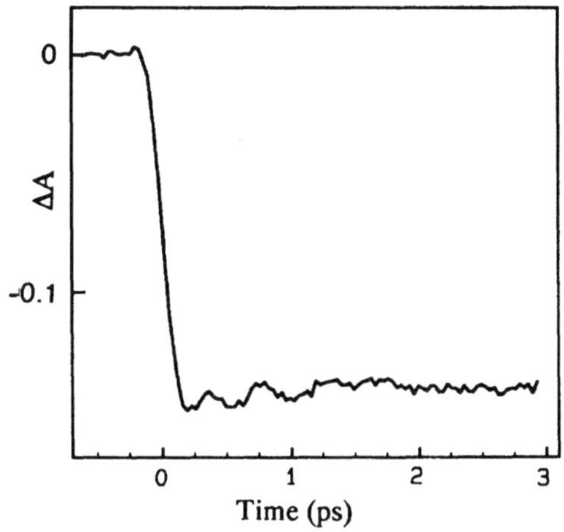

Fig. 1 Kinetics of stimulated emission (probe 930 nm) of *R. capsulatus* D_{LL} at 10K.

RESULTS AND DISCUSSION

The excited state P^* decays with a (1/e) time constant of ~200 ps at room temperature (Breton et al., 1990); at low temperatures similar decay times are observed. The overall kinetics in the Q_Y absorption region are the same as those in the stimulated emission (Breton et al., 1990), indicating that P^* decays back to the ground state in this time (see Introduction). At the timescale of a few picoseconds there is no overall kinetic evolution in the stimulated emission. However, at low temperature, the kinetics of the stimulated emission are modulated by oscillations (Fig. 1), which presumably reflect coherent vibrational motion of the protein, as discussed below.

To further study the nature of the excited state of the D_{LL} mutant we monitored transient spectra of this mutant at short delay times. Fig. 2 shows the spectra at 0 fs, 150 fs and 9 ps delay time at low temperature (10 K). The main features of the spectra are the bleaching of the main absorption band of P at 890 nm, an appearing band at 804 nm and the broad absorption increase at the blue side of the spectrum. The spectrum at 0 fs (time defined at 800 nm) is very similar to those at 150 fs and 9 ps. The lower or zero amplitude of the signal at the extremes of the spectrum are due to the chirp compensation, which is optimized for 800 nm in those spectra (see Materials and Methods). In reaction centers of *R. sphaeroides* R-26 (spectra shown in Fig. 3 for comparison) a separate band arises around 795 nm at t=0 fs (Vos et al., 1992). There is no clear sign of such a separate band in the transient spectra of the D_{LL} mutant of *R. capsulatus*. It should be noted that the relative absorption increase at t=0 fs in this region (compared to the 804 nm maximum) is somewhat higher than at later delay times. Kinetic measurements (not shown) indicate that there is indeed a very small, fast (~50 fs) decay of the absorption around 795 nm. However, the amplitude of this decay is much smaller (about 5 times) than that in *R. sphaeroides* R-26, measured under comparable conditions. The relaxation may well be part of the oscillations, which are also observed in this absorption region (Vos et al., 1991), or due to a pump/probe coherence effect. In conclusion it is seen that removal of H_L greatly reduces the initial absorption change in the 795 nm region and its subsequent relaxation. This indicates that this feature is related to a change in interaction in the entire pigment complex, in which H_L plays an essential role. In particular, one might envisage that upon formation of the

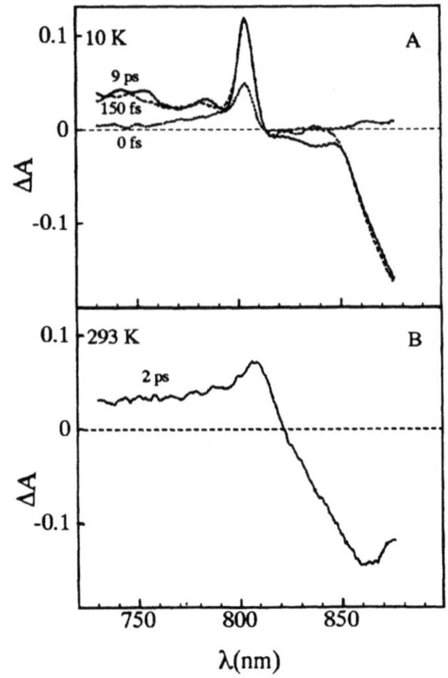

Fig. 2 Transient absorption spectra of *R. capsulatus* D_{LL} (A) at 293K, 2 ps and (B) at 10K, 0 fs (dotted), 150 fs (solid) and 9 ps (dashed).

excited state in H_L-containing reaction centers, oscillator strength is moved from the H_L band to the B band region. Early time transient spectra of wild type reaction centers of *R. capsulatus* will allow a more direct comparison with the D_{LL} mutant to further study this question.

The most marking feature of the transient spectra of the D_{LL} mutant is the appearing band at 804 nm. It should be noted that the presence of an absorption increase in this region is also visible at room temperature (Fig. 2), but much less pronounced, due to spectral congestion with broad P^* absorption at the blue side of the spectrum. More information regarding this band may be achieved

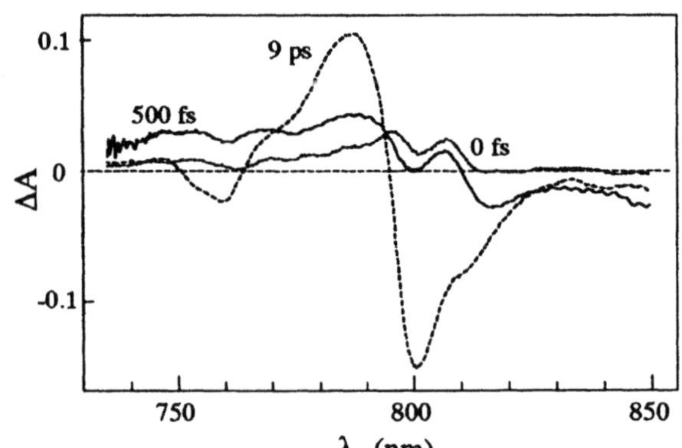

Fig. 3 Transient absorption spectra of *R. sphaeroides* R-26 at 10K at 0 fs (dotted), 500 fs (solid) and 9 ps (dashed).

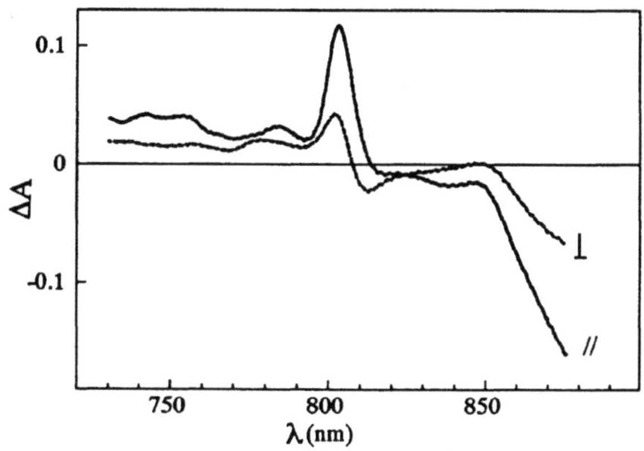

Fig. 4 Polarized transient absorption spectra of *R. capsulatus* D_{LL} at 10K. The probe pulse had a delay time of 9 ps and was polarized parallel (solid) and perpendicular (dashed) to the pump pulse.

from polarized spectra. Fig. 4 shows low temperature transient spectra for polarization parallel (as in all other figures) and perpendicular to the polarization of the excitation beam at a delay time of 9 ps. The bleaching near 815 nm is polarized perpendicularly, in agreement with the assignment to the higher exciton band of P of a negative feature at 813 nm in the steady state linear dichroism spectrum (Breton et al, 1990). More importantly, the polarization of the 804 nm band and the ground state P band are roughly parallel. This is similar to the appearing band at 805 nm in *R. sphaeroides* R-26. One way to interprete this band is that it is due to a monomer-like transition which arises from the breakdown of the dimeric interaction of P in the ground state. The appearance of this band in the B band region both in the excited state of the D_{LL} mutant and in all electronic states of P other than the ground state in *R. sphaeroides* R-26 (see Vos et al., 1992) supports this suggestion.

Oscillatory features

As mentioned above, details of the transient spectra at different early times differ due to oscillations in time. Especially, oscillatory features were observed near the (long time) isosbestic point of the spectra at the red side of the 804 nm band. Fig. 5 shows the kinetics at 812 nm. These kinetics are strongly modulated by oscillations with periods of about 450 fs (75 cm^{-1}) and of 2-3 ps (~15 cm^{-1}), similar to those observed in the stimulated emission region (Fig. 1). Among other possibilities, which will be discussed elsewhere, the appearance of such oscillations can be understood in terms of synchronized vibrational motion of the protein structure. In this picture, the absorption and emission spectra of the reaction center system are modulated by the relative orientation and distance of the pigments.

Protein vibrations in the observed frequency range (<100 cm^{-1}) are global protein modes (Go et al., 1983). Presumably many of such modes exist in the reaction center. It is remarkable that upon formation of the excited state oscillations at a few specific frequencies are observed in the kinetics. This may imply that a) only selected modes involve relative motion of pigments giving rise to spectral modulation and/or b) only a few modes couple strongly to the $S_0 \rightarrow S_1$ optical transition of P. Two recent observations support the latter possibility. A resonance Raman study with excitation directly in the P absorption region of *R. sphaeroides* R-26 shows enhancement of a few modes in the low frequency range (Shreve et al., 1991) (although non-enhanced modes may also be activated). Also, the spectrum of the energy correlation function obtained from molecular dynamics simulations of *Rps. viridis* exhibits low frequency modes (Tesch and Schulten, 1992) in the same frequency range as the oscillations observed in our experiments.

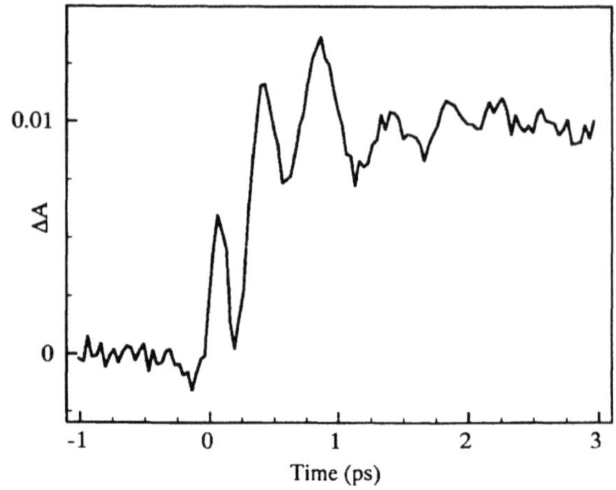

Fig. 5 Transient absorption kinetics (probe 812 nm) of *R. capsulatus* D_{LL} at 10K.

The oscillatory features decay in a few picoseconds, presumably due to anharmonicity (Vos et al., 1991) and elastic (pure dephasing) and inelastic (thermalization) collisions within the protein. It is interesting to note that the features reported here are very similar at 10K and 100K and still visible to some extent at 200K. It thus seems that the appearance of oscillations do not directly scale with $kT/\hbar\omega$. This may indicate that at room temperature the low frequency vibrational modes are not thermalized on the time scale of electron transfer. The above mentioned mechanisms (pure dephasing, anharmonicity and probably vibrational relaxation) may all be involved in the damping processes of the oscillations in the D_{LL} mutant.

The D_{LL} mutant does not perform charge separation. However, such oscillations have also been observed in *R. sphaeroides* R-26 (Vos et al., 1991) and in *R. capsulatus* wild type (to be published). The activation of very specific global modes and the conservation of their phase for a few ps suggests a functional importance of these modes for electron transfer. The frequency range of these modes (15-70 cm^{-1}) is very near to those of the modes (80-100 cm^{-1}) used to describe the thermodynamic properties of the primary electron transfer process in a non-adiabatic framework, where the modes are assumed to be thermalized (Bixon and Jortner, 1986, Fleming et al., 1988). Our results thus indicate that the latter assumption may not be justified. It is a challenging thought that charge separation may be a coherent and, if it is coupled to low frequency modes, near-adiabatic process. Parallel studies on the D_{LL} mutant and *R. capsulatus* wild type reaction centers are underway to get a deeper insight in the origin and significance of these oscillatory features.

Conclusions

In conclusion, our results indicate the presence of strong inter-pigment coupling in the excited state of reaction centers and an important role of the bacteriopheophytin electron acceptor H_L in this coupling. Excitation of P thus results in a delocalized electronic perturbation. This may explain the strong "electron-phonon" coupling of low frequency, global modes. The strong activation of specific modes in the neutral excited state suggests involvement of these modes in the primary electron transfer reaction.

ACKNOWLEDGMENTS

We thank S.G. Robles and D.C. Youvan for supplying the mutant sample. This work was supported by the Human Frontier Science Organization. M.H.V. was supported by a grant from the EEC Science Program during the course of this work.

REFERENCES

Bixon, M. and Jortner, J., 1986, *J. Phys. Chem.* 90:3795.

Breton, J., Martin, J.-L., Lambry, J.-C., Robles, S.J., and Youvan, D.C., 1990. *in* "Reaction Centers of Photosynthetic Bacteria", M.E. Michel-Beyerle, ed. pp. 293-302. Berlin, Springer

Fleming, G.R., Martin, J.-L. and Breton, J., 1988, *Nature* 333:190.

Holzapfel, W., Finkele, U., Kaiser, W., Oesterhelt, D., Scheer, H., Stilz, H.U., and Zinth, W., 1989, *Chem. Phys. Lett.* 160:1.

Holzapfel, W., Finkele, U., Kaiser, W., Oesterhelt, D., Scheer, H., Stilz, H.U., and Zinth, W., 1990, *Proc. Natl. Acad. Sci. USA* 87:5168.

Kirmaier, C. and Holten, D., 1988, *FEBS Lett.* 239:211.

Marcus, R.A. and Sutin, N., 1985, *Biochim. Biophys. Acta* 811:265.

Martin, J.-L., Breton, J. Hoff, A.J., Migus, A. and Antonetti, A., 1986, *Proc. Natl. Acad. Sci. USA* 83:957.

Martin, J.-L., and Vos, M.H., 1992, *Annu. Rev. Biophys. Biomol. Struct.* 21:199.

Michel-Beyerle, M.B., Plato, M., Deisenhofer, J., Michel, H., Bixon, M., and Jortner, J., 1988, *Biochim. Biophys. Acta* 932:52.

Robles, S.J., Breton, J. and Youvan, D.C., 1990, *Science* 248:1402.

Schulten, K. and Tesch, M., 1992, *J. Chem. Phys.*, in press.

Shreve, A.P., Cherepy, N.J., Franzen, S., Boxer, S.G., and Mathies, R.A., 1991, *Proc. Natl. Acad. Sci. USA* 88:11207

Vos, M.H., Lambry, J.-C., Robles, S.J., Youvan, D.G., Breton, J., and Martin, J.-L., 1991, *Proc. Natl. Acad. Sci. USA* 88:8885.

Vos, M.H., Lambry, J.-C., Robles, S.J., Youvan, D.G., Breton, J., and Martin, J.-L., 1991, *Proc. Natl. Acad. Sci. USA* 89:613.

Warshel, A., Creighton, S., and Parson, W.W., 1988, *J. Phys. Chem.* 92:2697.

Woodbury, N.W., Becker, M., Middendorf, D., and Parson, W.W., 1985., *Biochemistry* 24:7516.

ELECTRON TRANSFER IN *RHODOPSEUDOMONAS VIRIDIS* REACTION CENTERS WITH PREREDUCED BACTERIOPHEOPHYTIN BL

V.A. Shuvalov, A.Ya. Shkuropatov and A.V. Klevanik

Institute of Soil Science and Photosynthesis, Russian Academy of Sciences, Pushchino, Moscow region, 142292 Russia

INTRODUCTION

Bacterial reaction centers (RCs) contain three protein subunits (L, M and H), four bacteriochlorophylls (BChl), two bacteriopheophytins (H), two quinones (Q) and one atom of Fe [1]. Accordingly to spectral properties [2] and X-ray analysis [3] two BChls form a dimer, primary electron donor P. Two other BChls (B) and two H form two transmembrane cofactor chains located in L and M protein subunits and terminated by two Qs: P-BL-HL-QA and P-BM-HM-QB [3]. Only LA-chain is photochemically active. The excited state P* within 3-4 ps at 293K transfers an electron to (BLHL) acceptor complex [4-6]. An electron is further transfered to QA within 200 ps.

The formation of the state P+HL- was shown by ps measurements using nonselective [7,8] and selective [9] excitation of P in bacterial RCs as well as by the finding [10] the reaction CytP+HL-Q- $\rightarrow$ Cyt+PHL-Q- with very slow back reaction rate. Both measurements have revealed the similar optical spectra for the formation of HL-. The selective ps excitation of P has shown the possibility of the formation of P+BL- prior the formation of P+HL- at ~1 ps delay [9,11] since the spectra of ΔA at this delay included the bleachings of the P and BL bands without those of HL. Subpicosecond measurements of sequence of the electron transfer reactions in RCs have been interpreted as follows: the

The Photosynthetic Bacterial Reaction Center II
Edited by J. Breton and A. Verméglio, Plenum Press, New York, 1992

formation of the state P+BL- occurs with a time constant of 3.5 ps; this state is converted into P+HL- within ~0.6-0.9 ps [12,13].

The study of the photochemical reactions in the presence of reduced electron acceptors can give additional information about normal reactions in RCs with neutral acceptors if the electron transfer is not dramatically changed between P* and the earlier acceptors in the presence of prereduced latter ones. Using this approach the state P^F was discovered [14] which plays key role in the primary reactions in RCs with neutral acceptors. To find an acceptor functioning between P and HL the prereduction of HL may be also used. However the reduction of HL can induce additional effects, e.g. considerable increase of the relaxation rate of P* due to the excitation energy transfer to HL-.

It is accepted that there are two kinds of luminescence in ps-ns time domains related to the prompt fluorescence from primary excited state of P* and delayed fluorescence from P* which is secondary excited by the recombination of P+ and HL- [1,10].

The photoaccumulation of HL- in BChl a-containing RCs like *Chromatium minutissimum* leads to the decrease of the quantum yield of fluorescence of RCs [10]. This decrease has been found to be related to two processes: i) the decrease of the recombination luminescence due to the blocking the light-induced formation of P+HL-; ii) the quenching of excited state of P* by the excitation energy transfer to HL- having the absorption band at longer wavelength (~940 nm) than the absorption band of P (~870 nm).

In contrast to that in BChl b-containing *Rhodopseudomonas viridis* RCs the HL- band (~920 nm) is located at shorter wavelength than the P band (~960 nm). As a result the quantum yield of fluorescence of *R. viridis* RCs is increased when HL is reduced due to the increase of prompt fluorescence of P* which is larger than the decrease of recombination luminescence [15]. Therefore the study of the electon transfer reactions with prereduced HL in this case can reveal some intermediary steps between P* and HL. Earlier ns measurements [22] of *R.viridis* RCs in the presence of HL- at 90K have revealed 20-ns decaying component which was interpreted as the recombination of P+HM- although no considerable bleachings around 790 nm were registrated.

MATERIALS AND METHODS

Isolation procedure of *R. viridis* RCs, redox additions, low-temperature absorption spectra measurements were described earlier [16,17]. The ps ΔA were measured using YAG-Nd ps generator, pulse selector and amplifiers as described earlier [18]. The ps continuum as a measuring pulse was splitted into two beams. One of them was passed through the sample. Spectra of two beams were registered using OMA-2 (EG&G PARC, Princeton) to obtain the absorbance and difference absorbance spectra.

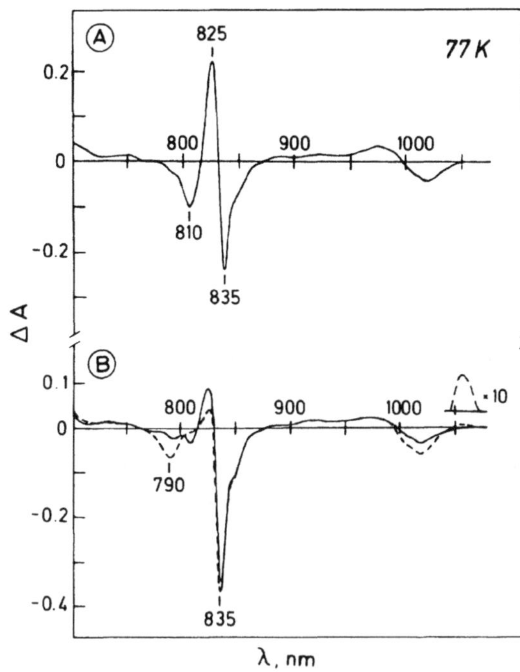

Fig.1.(A) The difference between absorption spectra of *R. viridis* RCs at -400 mV frozen down to 77K and then illuminated for 1 hour at 77K. Data from [19].

(B) The difference between absorption spectra of *R. viridis* RCs at -400 mV frozen in the dark down to 77K and: i)frozen down to 77K after 2 min of preillumination at 293K (solid) and ii) frozen down to 77K after 2 min of preillumination at 293K and then additionally illuminated at 77K for 1 hour (dashed). Data from [19].

RESULTS AND DISCUSSION

It has been found that at 293K the short illumination of *R.viridis* RCs by continuous light at low redox potential (prereduced quinones and cytochrome) causes the reduction of HL and partially HM [19] (Fig.1B). Further illumination of RCs at 77K is accompanied by the reduction of HM and partial reduction of BL (or BM) since the bleachings of HM and BL (BM) bands and the developing the characteristic band of radical anion of BChl b at 1060 nm [23] are observed. The illumination of *R.viridis* RCs frozen in the dark at low redox potential down to 77K induces the selective reduction of HL (Fig.1A) [19].

Taking into account these observations the study of ps events in the presence of HL- have been done at 77K in *R.viridis* RCs frozen in the dark with prereduced quinones and cytochrome and preilluminated at 77K to accumulate HL-. One can expect the conversion of the excited state of P* into charge separated state P+HM- or P+BL- (BM-) within ps time domain accordingly to the above mentioned observations. Previous ps measurements [20] were done at 293K under conditions which allow to reduce H and B molecules and to create additional quenching of P*.

Fig.2 shows the ps kinetics of ΔA at 1000 nm in *R. viridis* RCs with prereduced HL at 110K. One can see that the kinetics includes two phases with lifetimes of 150 ps and >1 ns. The initial amplitudes of two phases are almost equal.

Fig.3 shows the spectra of ΔA at different delays in two spectral regions: i) around the P band at 1000 nm (A); ii) around the bands of the H and B molecules in the 760-870 nm region (B).

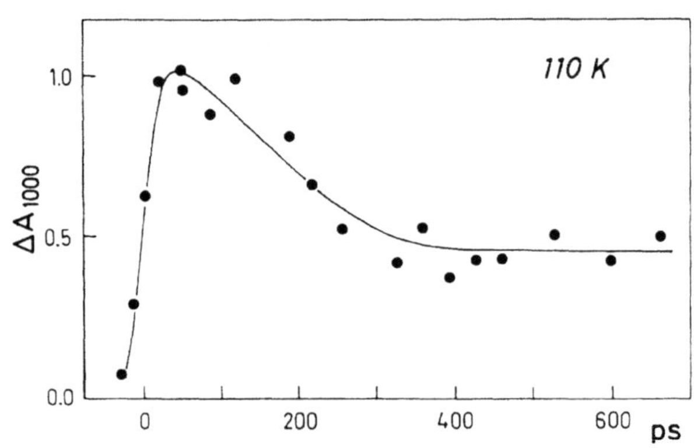

Fig.2. The ps kinetics of DA at 1000 nm of *R.viridis* RCs frozen in the dark at -400 mV down to 110K and then preilluminated for 10 min to reduce more than 95% of HL. Excitation by 25-ps pulses at 536 nm.

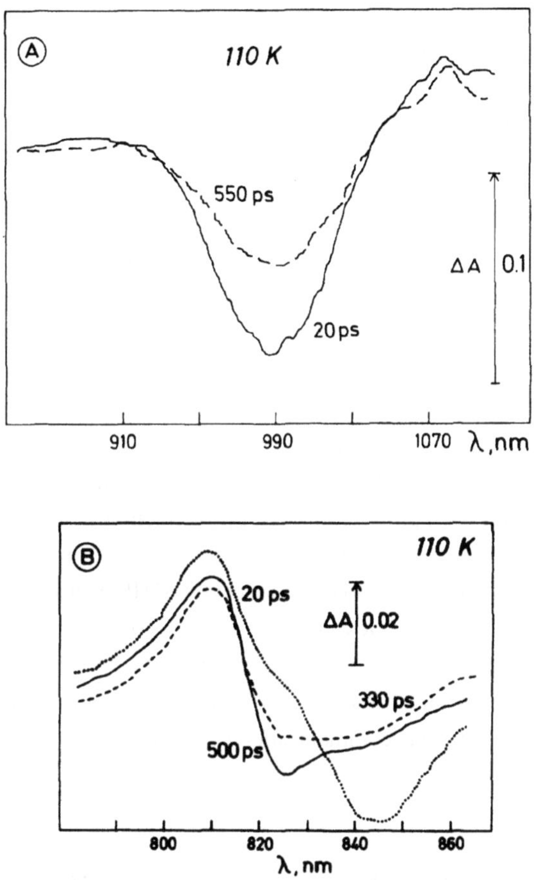

Fig.3. The spectra of ΔA at different ps delays (indicated by numbers near curves) of *R.viridis* RCs frozen in the dark at -400 mV down to 110K and then preilluminated for 10 min to reduce more than 95% of HL. Excitation by 25-ps pulses at 536nm.

One can see that at 20 ps the spectra mostly includes the bleachings of two excitonic bands of P at 990 and 845 nm. At latter delays the P bands are relaxed by ~50% and the bleaching of the B band at 825 nm (830-nm band shifted due to HL- formation) is developed. No notable bleachings of H bands at 790 and 810 nm are observed.

The bleachings of P bands at 20 ps can be assigned to the formation of excited state P*. The relaxation of this state occurs with time constant of 150 ps and is accompanied by the appearance of new state characterized by the bleachings of P and B bands. One can suggest that the ps formation of the state P+B- or P+HM- is observed since results obtained using accumulation method show the formation of HM- and of radical anion B- under similar conditions (Fig.1). At present it is impossible to exclude the ps formation of P+HM- although no notable bleaching of HM band is observed. On the other hand the formation of HM- observed by accumulation method might be due to slow (>1 ns) electron transfer from B-. We will call this ps state as P+B- assuming that P+HM- might be also included in that.

The quantum yield of the formation of P+B- is close to 0.5 since 50% of the bleaching at 1000 nm remains at 400 ps (Fig.2). It suggests that two processes compete to each other with approx. equal time constants (~300 ps): i) the decay of P* to the ground state; ii) the conversion of P* into P+B-.

The B band at 830 nm can be assigned to BL molecule since the BM absorbs light at 820 nm accordingly to the study using the treatment with NaBH4 [19]. This suggests that in the presence of HL- the formation of P+BL- with the time constant of ~300 ps is observed. The formation of P+BL- has been proposed [9,11-13] for electron transfer in RCs with neutral acceptors but the time constant for that (3-4 ps at 293K) is ~100 times faster. To explain the difference of rates one might assume that the reduction of HL relatively increases the energy level of P+BL- which is very close to P* in the presence of neutral HL [21]. The activationless formation of P+BL- with neutral HL may become activation one in the presence of HL- . At 110K this formation can be achieved via a tunnelling mechanism that takes much longer time (personal communication by W.W.Parson).

REFERENCES

[1] Parson, W.W. and Cogdell, R.J., *Biochim. Biophys. Acta* 416: 105-149 (1975)

[2] Norris, J.R., Scheer, H., Druyan, M.E. and Katz, J.J., *Proc. Natl. Acad. Sci. USA* 71: 4897-4900 (1974).

[3] Deisenhofer, J., Epp, O., Miki, K., Huber, R. and Michel, H., *Nature* 318: 618-624 (1985).

[4] Martin, J.-L., Breton, J., Hoff, A.J., Migus, A. and Antonetti, A., *Proc. Natl. Acad. Sci. USA* 83: 957-961 (1986).

[5] Woodbury, N.W., Becker, M., Middendorf, D. and Parson, W.W., *Biochem.* 24: 7516-7521 (1985).

[6] Shuvalov, V.A., "The Primary Conversion of the Light Energy at Photosynthesis (in Russian)", Nauka , Moscow (1990).

[7] Rockley, M.G., Windsor, M.W., Cogdell, R.G. and Parson, W.W., *Natl. Acad. Sci. USA* 72: 2251-2255 (1975).

[8] Kaufmann, K.J., Dutton, P.L., Netzel, T.L., Leigh, J.S.and Rentzepis, P.M., *Science* 188: 1301-1304 (1975).

[9] Shuvalov, V.A., Klevanik, A.V., Sharkov, A.V., Matveetz, Yu.A. and Krukov, P.G., *FEBS Lett.* 91: 135-139 (1978).

[10] Shuvalov, V.A. and Klimov, V.V., *Biochim. Biophys. Acta* 440: 587-599 (1976).

[11] Shuvalov, V.A. and Duysens, L.N.M., *Proc. Natl. Acad. Sci. USA* 83: 1690-1694 (1986).

[12] Chekalin, S.V., Matveetz, Yu.A., Shkuropatov, A.Ya., Shuvalov, V.A. and Yartsev, A.P., *FEBS Lett.* 216: 245-248 (1987).

[13] Holzapfel, W., Finkele, U., Kaiser, W., Oesterhelt, D., Scheer, H., Stilz, H.U. and Zinth, W., *Proc. Natl. Acad. Sci. USA* 87: 5168-5172 (1990).

[14] Parson, W.W., Clayton, R.K. and Cogdell, R.J., *Biochim. Biophys. Acta* 387: 265-278 (1975).

[15] Klimov, V.V., Shuvalov, V.A., Krakhmaleva, I.N., Klevanik, A.V. and Krasnovsky, A.A., *Biochem. (USSR)* 42: 519-530 (1978).

[16] Ganago, A.O., Shkuropatov, A.Ya. and Shuvalov V.A., *FEBS Lett.* 284: 199-202 (1991).

[17] Shkuropatov, A.Ya., Ganago, A.O. and Shuvalov, V.A., *FEBS Lett.* 287: 142-145 (1991).

[18] Shuvalov, V.A. and Klevanik, A.V., *FEBS Lett.* 160: 51-55 (1982).

[19] Shuvalov, V.A., Shkuropatov, A.Ya., Ismailov, M.A., in: "Proc. of 7th Intern. Congress on Photosynthesis", J. Biggins, ed., Providence. Vol.1, pp. 161-168 (1986).

[20] Holten, D., Windsor, M.W., Parson, W.W. and Thornber, J.P., *Biochim. Biophys. Acta* 501: 112-126 (1978).

[21] Parson, W.W., Creighton, S, Chu, Z.-T. and Warshel, A. in: "Proc. of 8th Intern. Congress on Photosynthesis", M. Baltschevsky, ed., vol. 1, pp 1.31-1.38, Stockholm, Kluwer Academic Publisher, Dordrecht (1989).

[22] Tiede, D.M., Kellogg, E.C., Kolaczkowski, S., Wasielewski, M.R., Ibid., vol.1, pp.1.129-132 (1989).

[23] Fajer, J., Davis, M.S., Brune, D.C., Spaulding, L.D., Borg, D.C. and Forman, A., in: "Brookhaven Symposia in Biology", J.M. Olson and G. Hind, eds., number 28, p.75-104 (1976).

FAST INTERNAL CONVERSION IN BACTERIOCHLOROPHYLL DIMERS

U. Eberl, M. Gilbert, W. Keupp, T. Langenbacher, J. Siegl, I. Sinning*, A. Ogrodnik, S.J. Robles†, J. Breton+, D.C. Youvan† and M.E. Michel-Beyerle

Institut für Physikalische und Theoretische Chemie
Technische Universität München
Lichtenbergstr. 4, D-8056 Garching (Germany)

*Max-Planck-Institut für Biophysik
Heinrich Hoffmann Str. 7, D-6000 Frankfurt 71 (Germany)

†Department of Chemistry
Massachusetts Institute of Technology
Cambridge, MA 02139 (USA)

+ SBE/DBCM, CEN Saclay
91191 Gif-sur-Yvette Cedex (France)

INTRODUCTION

The primary electron transfer in bacterial photosynthetic reaction centers (RCs) involves a fast (≃1-3ps) charge separation between the excited state of the primary electron donor (P) and an acceptor (either B_A or H_A) along the active pigment branch A. Although this reaction proceeds with a quantum yield close to unity, it must compete with loss channels which presently remain poorly understood. Under the conditions of isolated RCs two natural loss channels have to be envisaged: (i) internal conversion of $^1P^*$ and (ii) charge separation along the pigments B_B and H_B at the inactive B-branch. Most recently, the understanding of loss channels has become most relevant since a fast (<50ps), electric field induced loss of quantum yield[1] of $P^+H_A^-$ has been reported which so far is not included in any model for primary charge separation.

The investigation of loss channels in RCs becomes accessible whenever the usual pathway of primary charge separation via $P^+H_A^-$ and $P^+Q_A^-$ is impeded. Such blocking can be realized by (i) reducing H_A or (ii) removing H_A. Due to the bound cytochromes in RCs of *R. viridis* illumination under reducing conditions allows the photoaccumulation of H_A^- (i). Alternatively, in the mutant D_{LL} of *Rb. capsulatus* replacement of the D-helix of the M-polypeptide by the analogous helix of the L-subunit leads to a depletion of H_A[2](ii). It is interesting that in this case electron transfer from $^1P^*$ to the nearby B_A has not been observed[3,4] in femtosecond spectroscopy which revealed a long lifetime of $^1P^*$. On the other hand, earlier picosecond absorption measurements[5] on H_A reduced RCs showed recovery of the ground state P within 20ps reflecting the decay of either $^1P^*$ or P^+.

It is the goal of this paper to perform sensitive picosecond absorption and emission experiments on RCs from both, the symmetrized mutant D_{LL} of *Rb. capsulatus* and *R. viridis*, under the condition of blocked H_A. In the case of the D_{LL} mutant, the electronic states of the remaining cofactors will be further characterized in steady state Stark absorption and emission spectroscopy.

The Photosynthetic Bacterial Reaction Center II
Edited by J. Breton and A. Verméglio, Plenum Press, New York, 1992

MATERIALS AND METHODS

Reaction center preparation and characterization

(a) *R. viridis* with H_A reduced
Reaction centers of *R. viridis* were isolated as reported by Michel[6,7]. Reducing conditions were achieved by addition of 10mM Na-ascorbate and 1mg/ml Na-dithionite (final concentrations). Samples were mixed in a glass-box under a nitrogen atmosphere. All solutions were bubbled with nitrogen gas. Final concentrations were: 60% glycerol, 10mM Tris-HCl (pH8.0), 0.3% NGP (w/v), 5mM o-phenanthroline (stock solution in DMSO), 57μM reaction centers (for the absorption measurements) and 26μM reaction centers (for the fluorescence measurements). Samples were frozen in the dark and illuminated at 100K for 20min (λ>700nm, $150mW/cm^2$). The quantitative photoaccumulation of the state $PH_A^-Q_A^-$ was tested by steady state absorption measurements of the bleaching of the Q_y-band of H_A at 805nm.

(b) D_{LL} mutant of *Rb. capsulatus*
This mutant is genetically deprived of light harvesting pigments. Chromatophore membrane particles were prepared as described elsewhere[2,3,4].

Experimental Methods

(a) Stark spectroscopy in absorption
The Stark effect can be used as a sensitive spectroscopic tool to test whether the symmetrization of the protein environment in the D_{LL} mutant influences the electronic charge distribution of cofactors.

The Stark spectrum has been measured in a bath cryostat (liquid nitrogen, 77K) as described in ref.[8]. The samples (PVA film, thickness ≃ 100μm) were mounted between two glass slides coated with SnO_2 (surface resistance $100\Omega/cm^2$) with perpendicular orientation to the exciting light. A sinusoidal voltage (0-900V_{eff}) was applied to the conducting slides, yielding an electric field of ≃ $9\cdot 10^4$V/cm. The modulation of the transmitted light was detected with a photodiode and a lock-in amplifier (SR 530) using the second harmonic of the modulation frequency. The absorption spectra were measured in the Stark capacitor with a blank PVA film as reference.

The difference in dipole moments Δμ between ground and excited state can be calculated via the equation[9]:

$$f\Delta\mu\,[D] = \frac{1}{\sqrt{1+\sin^2\delta}}\,\frac{2.24}{E[10^5\ V/cm]}\left(\frac{\Delta A\,[OD]}{\nu \dfrac{d^2(A/\nu)}{d\nu^2}\,[cm^2]}\right)^{1/2}$$

ΔA, ν, E and f denoting the field induced absorption change, the wavenumber, the electric field and the correction factor for the internal field (commonly used f=1.2 [9]), respectively.

As in native RCs[8-13] the Stark effect in the Q_y bands of the D_{LL} mutant follows the second derivative of the absorption spectrum. Thus, the dominant contribution to the Stark effect originates from the interaction of the difference in dipole moments Δμ between ground and excited state and the externally applied electric field. Both, the Stark spectra of wildtype *Rb. capsulatus* and of the D_{LL} mutant at 77K are shown in Fig.1. Obviously, H_A - and therefore also its Stark band at 754nm - is missing in the D_{LL} mutant, as is known already from the absorption and LD spectra at 10K [3,4].

In the mutant the peak at 800nm is slightly red shifted by (3±1)nm. The values for Δμ of the bacteriochlorophylls (2.3D at ≃ 800nm) and the bacteriopheophytin H_B (3.3D to 3.6D at 743nm) are the same in the wildtype RC and mutant.

A more significant change is observed for the dimer band. In the mutant this band is blue shifted by about 16nm and shows a smaller Stark effect: by comparing with wild type RCs exhibiting Δμ≃6D a smaller value Δμ=(4.2±0.8)D is obtained for the dimer in the D_{LL} mutant. Obviously in the mutant the asymmetry in the electronic charge distribution of $^1P^*$ and therefore its charge transfer character are less pronounced. However, this conclusion rests on the assumption that the angle δ is the same in wild type and D_{LL} mutant of *Rb. capsulatus*.

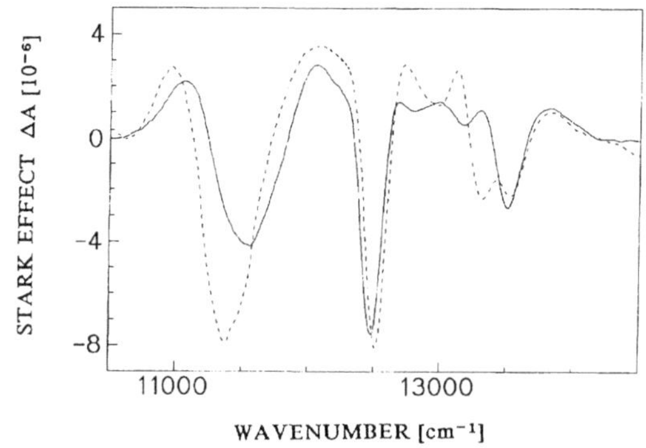

Fig.1. Stark spectrum of *Rb. capsulatus*, wild type (dashed line) and D_{LL} mutant (solid line), at 77K in the Q_y absorption bands. External electric field for both samples: $9 \cdot 10^4$V/cm.

The smaller value of $\Delta\mu$ in the mutant is not related to the environment of the RC since nearly the same $\Delta\mu$ is obtained for native RCs in the membrane and after isolation. The difference in dipole moments $\Delta\mu$ for the dimer in the D_{LL} mutant is still larger than that for the monomers, the electronic charge distribution of $^1P^*$ however is clearly less asymmetric than in the wild type dimers. This effect of the mutation is the result of the symmetrization of the protein environment around P.

(b) Transient difference absorption spectroscopy

Measurements with $\simeq$50ps resolution were performed in a spectrometer described in this volume[1].

(c) Steady state fluorescence

The electric field induced fluorescence change was measured in a single photon counting fluorimeter in orthogonal geometry as described in ref.[14]. The intensity of the exciting light at 840nm was about 250μW/cm^2, corresponding to $\simeq$ 0.7 turnovers/sec. The sample was embedded in a PVA film of $\simeq$ 90μm thickness sandwiched between two rectangular prisms and two mylar foils, yielding a rectangular cube. The mylar foils were coated with optically transparent electrodes supplying the electric voltage (5000V), and were oriented together with the film in the diagonal of the cube. Two orthogonal faces of the cube were positioned perpendicularly to the directions of excitation and emission. In order to determine fluorescence changes with a precision of better than ±0.04%, measuring times of more than 40h had to be envisaged. Such long term measurements required automatic control and on line numeric normalization procedures. The measurement has been carried out in several cycles at 90K.

The quantum yield of fluorescence Φ_f of the D_{LL} mutant at 277K has been determined following the procedure described in refs.[15,16]. The spectral distribution of the fluorescence of the D_{LL} mutant (in quanta per wavenumber) has been compared with the fluorescence standard Rhodamine 101 (Rh 101), assuming a quantum yield Φ_f(Rh 101) in ethanol of 0.91 [17]. The large error bars of the quantum yield $\Phi_f(D_{LL}) = (4.9 \pm 2.2) \cdot 10^{-2}$ result from the uncertainty about the amount of stray light in the sample due to the large membrane fragments (chromatophores) carrying the D_{LL} RCs.

(d) Time resolved fluorescence

The sample was excited at 864nm with 40ps pulses of a Hamamatsu PLP-01 laser diode at a repetition rate of 10MHz. The sample geometry was the same as in the steady state experiment. Fluorescence light was gathered with an aspheric condensor lens and focused on an iris diaphragm through two 920nm bandpass filters (bandwidth 20 nm) achieving an effective straylight rejection of $\simeq 10^{-7}$. Single photons were detected behind the diaphragm with a microchannel plate photomultiplier (Hamamatsu R2809U with S1-cathode) which was cooled to 190K in order to obtain a dark count rate of less than 2cps. The fluorescence decay was measured using the time correlated single photon counting technique. The total instrument response function had a FWHM less than 70ps.

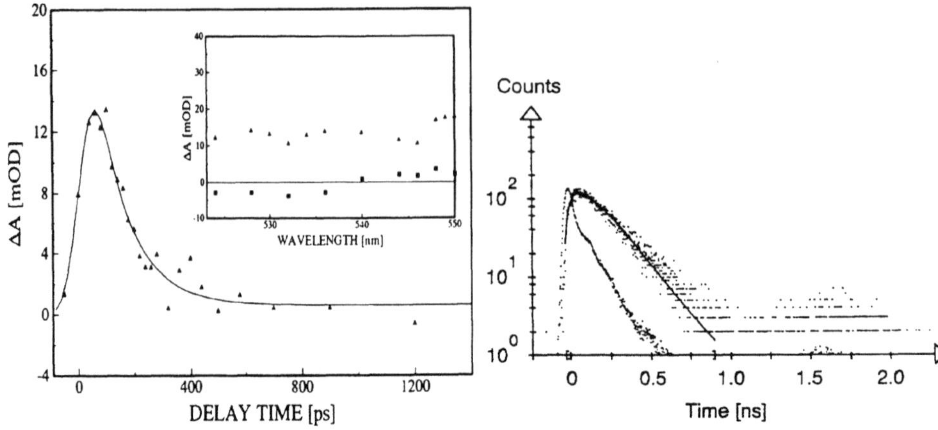

Fig.2. Time resolved, low temperature (80K) spectroscopy on RCs from *R.viridis* with H_A reduced.
(a) Decay of difference absorption monitored at 545nm. Insert: Difference absorption spectrum after 80ps (upper trace) and 1ns (lower trace). (b) Decay of fluorescence.

RESULTS

R. viridis

Fig.2a shows transient absorption measurements on doubly reduced RCs from*R. viridis* (state $PH_A^-Q_A^-$) at 80K in the Q_x bands of the two bacteriopheophytins between 520nm and 560nm. The difference spectrum at a delay time of 80ps (see Insert) shows a broad positive band, in contrast to unreduced RCs, where at ≃ 540nm a bleaching due to the reduction of H_A is observed. This behaviour reflects the facts that (i) charge separation to H_A is blocked due to a complete pre-reduction of H_A and (ii) H_B with its Q_x absorption peak at 533nm is not reduced within the time window of the measurement up to 1200ps. Throughout the broad band the positive feature decays with a time constant of (150±50)ps and (90±20)ps at 80K and 280K, respectively.

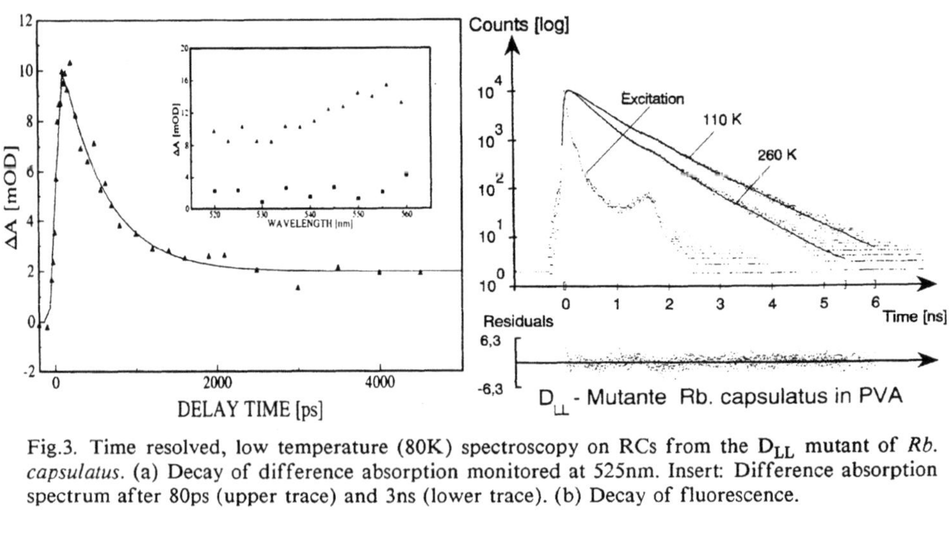

Fig.3. Time resolved, low temperature (80K) spectroscopy on RCs from the D_{LL} mutant of *Rb. capsulatus*. (a) Decay of difference absorption monitored at 525nm. Insert: Difference absorption spectrum after 80ps (upper trace) and 3ns (lower trace). (b) Decay of fluorescence.

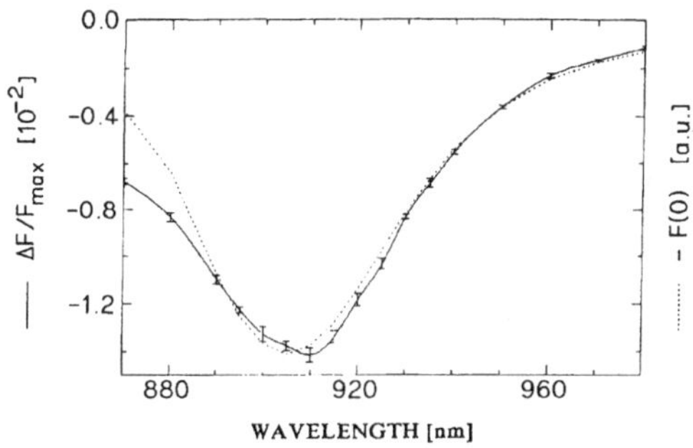

Fig.4. Low temperature (80K) fluorescence change $\Delta F = F(E)-F(0)$ of *Rb. capsulatus* in an external electric field ($5.4\cdot10^5$V/cm), normalized to the peak of the fluorescence signal (solid line). Negative lineshape of the fluorescence band (dashed line). Excitation at 840nm.

The same time constant of (130±20)ps at 05K is observed in ps-time resolved fluorescence measurements on H_A reduced RCs of *R. viridis* monitored at 1050nm (Fig.2b). This ensures that indeed the broad positive absorption band present after 80ps is due to the state $^1P^*$ and not to P^+, a discrimination which was not possible on the basis of previous experiments[5].

D_{LL} mutant from *Rb. capsulatus*

Transient absorption measurements on the D_{LL} mutant of *Rb. capsulatus* in the Q_x region of the bacteriopheophytins at (520-560)nm show a broad structureless feature similar to the one observed for *R. viridis* RCs in the state $PH_A^-Q_A^-$. Neither bleaching in the Q_x region of H_A nor in the one of H_B is observed. The positive difference absorption decays with time constants of (300±100)ps and (500±100)ps at 280K and 80K, respectively[18] (Fig.3a). These values are very well compatible with the results from the time-resolved fluorescence measurements. The temperature dependence of the decay rate is the same as in the doubly reduced RCs of *R. viridis*: in both cases the rate increases slowly with increasing temperature.

Time-resolved fluorescence traces of $^1P^*$ can be fit by two exponentials with time constants of 340ps and 720ps and a ratio of the amplitudes of 4:1 at 260K and with 440ps (70%) and 840ps (30%) at 80K (Fig.3b). The high temperature value is compatible with earlier published data of the decay of the stimulated emission with a time constant of ≃200ps at 295K[3]. Since these data refer to a shorter time window, they bear a larger uncertainty than the data presented here. Furthermore, the steady state quantum yield Φ_f of the fluorescence at 277K was determined as $\Phi_f \simeq (4.9\pm2.2)\cdot10^{-2}$ for the D_{LL} mutant, being 50-150 times larger than the one for wild type RCs. Assuming $\tau_0 \simeq 12$ns[19] for the intrinsic lifetime of $^1P^*$, this results in an average decay time of (580±260)ps for the excited dimer of the D_{LL} mutant.

The influence of an external electric field E on the decay characteristics of the excited dimer in the D_{LL} mutant is small[20]. Fig.4 shows the fluorescence change $(F(E)-F(0))/F_{max}$ between 870nm and 980nm after application of an external field of $5.4\cdot10^5$V/cm and excitation in the dimer band at 840nm (90K). The relative fluorescence change $\Delta F_{max}/F_{max}$ is -1.5% in the peak of the spectrum. It is by a factor of 3.8 larger than the Stark effect $\Delta A_{max}/A_{max}$ in the peak of the absorption spectrum and by a factor of 30 smaller and inverse in sign as compared to the one observed for native RCs. The fluorescence change does not reveal the shape of a typical Stark effect (second or first derivative), but follows quite well the negative lineshape of the fluorescence spectrum. This means firstly that the signal is dominated by a field induced change of the decay rate of $^1P^*$ and secondly that the value of $\Delta\mu_f$ between the fluorescing state and the corresponding Franck-Condon ground state cannot be larger than that for the absorption $\Delta\mu_a$ between the ground and the Franck-Condon excited state.

Excitation of other transition moments than the Q_y transition of P (like the Q_x or Q_y transitions of the monomeric pigments) leads to fluorescence changes which differ by less than 0.2%. Thus, the quantum yield of energy transfer in the RC can be assumed as independent from the external electric field.

DISCUSSION

The results on *R. viridis* with H_A reduced and on the D_{LL} mutant of *Rb. capsulatus* with H_A depleted show that blocking or removal of the electron acceptor H_A lead to a lifetime of the excited dimer $^1P^*$ of the order of (100-500)ps.

In the doubly reduced RCs of *R. viridis* and in the D_{LL} mutant, the decay rate of $^1P^*$ is slower than in the corresponding photoactive RCs by a factor of 30-100 and 100-350, respectively. This slow decay can be attributed to the internal conversion of $^1P^*$, since no spectral or kinetic evidence was found for a state between the primarily excited $^1P^*$ and the slowly repopulated ground state (see also ref.[3]). The room temperature value of (90±20)ps for *R. viridis* with H_A reduced is in contrast to previous measurements[5] yielding a dominant decay time of ≃20ps and a 100ps component with a smaller amplitude. Most likely, the fast component in ref.[5] is due to the high excitation intensity (>100 photons/RC, i.e. >300 times more than in the experiments presented here) giving rise to nonlinear processes like multi-photon excitation and/or annihilation.

With time constants of (90-500)ps the decay of $^1P^*$ in these blocked RCs is still significantly faster than the decay rate of isolated monomeric bacteriochlorophylls which is known to be ≃ (2-3.5)ns[17]. Presumably, the contribution of charge transfer states to $^1P^*$ in RC-dimers is responsible for such an increase of the internal conversion rate, as already discussed in refs.[21-24]. In dimers with pronounced charge transfer character like the heterodimer[22,23] and some chlorophyll-porphyrin dimers[24] an internal conversion rate as fast as $(30\text{-}40ps)^{-1}$ has been reported. Since in the D_{LL} mutant the charge transfer character of the excited state is less pronounced than in *R. viridis* - as reflected in the smaller value of $\Delta\mu$ derived from the Stark effect - this might be one possible explanation for the slowing of the decay of $^1P^*$ by a factor of ≃3 in the D_{LL} mutant as compared to *R. viridis* with H_A reduced.

Both RCs blocked at the site of H_A show a slightly activated temperature dependence of the decay rate which increases by a factor of 1.7 between 80K and 280K. This behaviour is expected from theoretical treatments of the internal conversion[25,26] : the energy gap law for the weak coupling regime of the electron-phonon-coupling predicts a small increase of the internal conversion rate with increasing temperature due to the larger number of excited vibrations at higher temperatures.

Another interesting feature of the decay of $^1P^*$ in the D_{LL} mutant is the slight field induced decrease of the steady state fluorescence. In principle, this may reflect an increase of the internal conversion rate by mixing-in of charge transfer states and/or a field induced electron transfer, e.g. B_B[1] at the inactive B-branch. The second possibility is presently tested in picosecond time resolved absorption experiments on the D_{LL} mutant in external electric fields.

The same decay rates as in fluorescence and stimulated emission of $^1P^*$ are observed in other regions of the absorption spectrum of the D_{LL} mutant like the Q_x region of the bacteriochlorophylls[27] and bacteriopheophytins or the Q_y region of the bacteriochlorophylls[3]. No evidence for a charge-separated state like $P^+H_B^-$ is found neither in the fs spectra nor in the measurements present here; the absence of bleaching in the region around 800nm[3] also excludes that the state $P^+B_A^-$ and/or $P^+B_B^-$ are created in a detectable amount prior to the decay of $^1P^*$ to the ground state. In conclusion, to be compatible with the data (e.g. the error bars of the ground state recovery), there can be no permanent charge separation, either to B_A on the active branch or to B_B or H_B on the inactive branch, faster than (1-2)ns in the D_{LL} mutant (and faster than (0.5-1)ns in *R. viridis* with H_A reduced).

This result is quite surprising in view of the fs time-resolved data on native RCs of *Rb. sphaeroides*[28] and *R. viridis*[29], which suggest B_A as a kinetic intermediate in the primary charge separation: according to Zinth et al. the primary electron transfer in *R. viridis* at room temperature occurs in 3.5ps from $^1P^*$ to $P^+B_A^-$ followed by a 0.65ps transfer from $P^+B_A^-$ to $P^+H_A^-$. In this picture, the absence of bleaching in the bacteriochlorophyll region and the long lifetime of the excited dimer in the "blocked" RCs can only be explained, if either the electronic coupling and/or the Franck-Condon factor between $^1P^*$ and $P^+B_A^-$ are drastically changed. In *R. viridis* with H_A reduced a considerable change in the free energy of $P^+B_A^-$ might be achieved by the two excess charges on H_A and Q_A. For the D_{LL} mutant a similar modification of either couplings and/or energetics with respect to B_A could result from symmetrization of the D-helices. e.g. from the replacement of the tyrosine residue at the position M210 by leucine. Nevertheless, it should be kept in mind that neither the absorption and the LD spectrum[3] nor the Stark spectrum reveal any significant changes for the bacteriochlorophyll monomers.

The long lifetime of the dimer in the "blocked" RCs sets also a severe restriction to the limit of unidirectionality of charge separation in the RC. For instance, the interpretation of previous measurements on doubly reduced RCs from *R. viridis*[30] would result in $k_A/k_B>1400:1$ when basing it to a lifetime of $^1P^*$ of $\simeq$150ps (this work) instead of 20ps[5].

CONCLUSIONS

The decay of the excited primary donor $^1P^*$ has been investigated in reaction center preparations where electron transfer was blocked at the position of the central electron acceptor H_A. Absorption and fluorescence experiments in the steady state and with a time resolution of 50ps yield the following results:

1. In the doubly reduced RCs from *R. viridis* (prepared in the state $PBH_A^-Q_A^-$) the primary donor $^1P^*$ decays with a time constant of (90±20)ps at 280K and (150±50)ps at 80K. This result borne out by a set of absorption/fluorescence measurements supersedes the recovery time 20ps of the 870nm absorption at room temperature reported earlier[5] on the basis of experiments performed under high excitation intensity.

2. A similarly fast, also slightly activated decay of $^1P^*$ is observed for the symmetrized mutant D_{LL} from *R. capsulatus* lacking H_A. For this mutat an analogous set of absorption/fluorescence measurements yields time constants of (300±100)ps at 280K and (500±100)ps at 80K.
   In agreement with these time-resolved measurements the steady state fluorescence yield of the D_{LL} mutant is determined to be $\simeq(4.9\pm2.2)\cdot10^{-2}$ at 277K, thus being 50-200 times larger than the one of RCs from wild type *Rb. capsulatus.*

3. So far there is no evidence for charge separation in one of the two reaction centers blocked at H_A.

4. In view of fluorescence quantum yield (see 2.) and also of the 12ns radiative decay rate of $^1P^*$, the short lifetime of $^1P^*$ in blocked RCs is attributed to fast internal conversion ($^1P^* \rightarrow P$). This interpretation is further supported by the slightly activated temperature dependence of the decay rate, a behaviour expected on theoretical grounds for internal conversion processes. Since it is not possible to explain the fast internal conversion within the frame of an energy gap argument, the contribution of charge transfer states is envoked. Indeed, the Stark absorption spectra of the dimers in both RCs, *R. viridis* and D_{LL} mutant of *Rb. capsulatus* point to a dipole moment change $\Delta\mu$ significantly exceeding the one of the monomer bacteriochlorophylls.

5. In contrast to RCs from *R. viridis* where charge separation is blocked by double reduction of H_A and Q_A, the membrane bound RC of the D_{LL} mutant could be cast into a PVA film, thus allowing the application of electric fields at low temperatures. Measurements of the steady state fluorescence yield show that the lifetime of $^1P^*$ increases by 1.5% in an external electric field of $\simeq7\cdot10^5$V/cm. This effect is by a factor of $\simeq$4 larger than the Stark effect measured in the peak of the Q_Y absorption band of the dimer P. In principle, this field induced decrease of the fluorescence may arise from two sources: (i) the increase of the internal conversion rate by mixing in charge transfer states and (ii) a field induced transfer to the inactive B-branch.

In purple bacteria fast internal conversion rates on the order of 100ps are inconsequential for the quantum yield of charge separation and therefore for the biological function of the reaction center. In the complete "machinery", reaction center and antenna, fast charge separation has to compete primarily with backtransfer of the excitation energy to the antenna usually proceeding on the time scale of 10ps. Only in antenna-reaction center systems where this backtransfer is strongly activated or in antenna free mutant charge separation should still be faster than internal conversion (i.e. faster than 100ps) in order to avoid a loss of quantum yield.

ACKNOWLEDGEMENTS

We are highly indebted to Professor Hartmut Michel at the Max-Planck-Institut fuer Biophysik in Frankfurt for stimulating and effective support of all the work based on reaction centers from *Rhodopseudomonas viridis.*

Financial support from the Deutsche Forschungsgemeinschaft (Sonderforschungsbereich 143), the Alfried Krupp von Bohlen und Halbach Stiftung (U.E. und M.G.) and the Human Frontiers Science Program is gratefully acknowledged.

REFERENCES

1. A. Ogrodnik, T. Langenbacher, U. Eberl, M. Volk and M.E. Michel-Beyerle , this volume.
2. S.J. Robles, J. Breton and D.C. Youvan (1990), *Science* 248:1402
3. J. Breton, J.L. Martin, J.C. Lambry, S.J. Robles and D.C. Youvan, in: Reaction Center of Photosynthetic Bacteria, (M.E. Michel-Beyerle, ed.) p.293, Springer Berlin (1990).
4. M.H. Vos, F. Rappaport, J.C. Lambry, J. Breton and J.L. Martin, this volume.
5. D. Holten, M. Windsor, W.W. Parson and J.P. Thornber, *Biochim. Biophys. Acta,* 501:112 (1978).
6. H. Michel, *J. Mol. Biol.* 158·567 (1982).
7. H. Michel, K.A. Weyer, H. Gruenberg, I. Dunger, D. Oesterhelt and F. Lottspeich, *EMBO J.* 5:1149 (1986).
8. H.P. Braun, M.E. Michel-Beyerle, J. Breton, S. Buchanan and H. Michel, *FEBS Lett.* 221:221 (1987).
9. D.J. Lockhart and S.G. Boxer, *Biochemistry* 26:664 (1987).
10. D. deLeeuw, M. Malley, G. Buttermann, M.Y. Okamura and G. Feher, *Biophys. J.* 37:111a (1982).
11. M. Lösche, G. Feher and M.Y. Okamura, *Proc. Nat. Acad. Sci. USA* 84:7537 (1987); M. Lösche, P.B. Madden, G. Feher and M.Y. Okamura, *Biophys. J.* 55:223a (1989).
12. D.S. Gottfried, M.A. Steffen and S.G. Boxer, *Biochim. Biophys. Acta* 1059:76 (1991).
13. T.J. DiMagno, E.Y. Bylina, A. Angerhofer, D.C. Youvan and J.R. Norris, *Biochemistry* 29:899 (1990).
14. A. Ogrodnik, U. Eberl, R. Heckmann, M. Kappl, R. Feick and M.E. Michel-Beyerle, *J. Phys. Chem.* 95:2036 (1991).
15. K.L. Zankel, D.W. Reed and R.C. Clayton, *Proc. Nat. Acad. Sci. USA* 61:1243 (1968).
16. C.A. Parker and W.T. Rees, *Analyst* 85:587 (1960).
17. J.C. Goedheer (1973), *Biochim. Biophys. Acta* 292:665, and later work until most recently: M. Becker, V. Nagarajan and W.W. Parson, *J. Am. Chem. Soc.*, in press.
18. Ogrodnik, A., Eberl, U., Haebele, T., Keupp, W., Langenbacher, T., Siegl, J., Volk, M. and Michel-Beyerle, M.E. (1992) *Biophys. J.* 61, 837a.
19. N.W. Woodbury and W.W. Parson (1984), *Biochim. Biophys. Acta* 767:345.
20. Eberl, U., Breton, J., Plato, M. and Michel-Beyerle, M.E. (1992) *Biophys. J.* 61:872a.
21. D.G. Johnson, W.A. Svec and M.R. Wasielewski (1988), *Isr. J. Chem.* 28:193.
22. L. McDowell, C. Kirmaier and D. Holten (1990), *Biochim. Biophys. Acta* 1020:239 (1990).
23. L. McDowell, C. Kirmaier and D. Holten (1991), *J. Phys. Chem.* 95:3379.
24. M.R. Wasielewski, D.G. Johnson, M.P.Nimczyk, G.L. Gaines III, M.P. O'Neil and W.A. Svec (1990), *J. Am. Chem. Soc.* 112:6482.
25. R. Englman and J. Jortner (1970), *Mol. Phys.* 18:145.
26. S.F. Fischer (1970), *J. Chem. Phys.* 53:3195.
27. Paper in preparation.
28. W. Holzapfel, U. Finkele, W. Kaiser, D. Oesterhelt, H. Scheer, H.U. Stilz and W. Zinth (1989), *Chem. Phys. Lett.* 160:1.
29. K. Dressler, E. Umlauf, S. Schmidt, P. Hamm, W. Zinth, S. Buchanan and H. Michel (1991), *Chem. Phys. Lett.* 183:270.
30. D.M. Tiede, E. Kellogg and J. Breton (1987), *Biochim. Biophys. Acta* 892:294.

PRIMARY CHARGE SEPARATION IN REACTION CENTERS: TIME-RESOLVED SPECTRAL FEATURES OF ELECTRIC FIELD INDUCED REDUCTION OF QUANTUM YIELD

Alexander Ogrodnik, Thomas Langenbacher,
Ulrich Eberl, Martin Volk and Maria E. Michel-Beyerle

Institut für Physikalische und Theoretische Chemie
Technische Universität München,
Lichtenbergstr.4, D-8046 Garching (Germany)

ABSTRACT

Electric field effects on reaction centers from *Rb. sphaeroides* have been detected in ps transient absorption at 90K. Upon application of a field of $7 \cdot 10^5$V/cm the bleaching of the bacteriopheophytin at the A-branch (H_A, 545nm) monitored after 30ps mirrors a reduction of the quantum yield of $P^+H_A^-$ formation by $\simeq$11% (P being the special pair). A similar field induced reduction of the quantum yield is observed when probing the bleaching of P at 870nm. However, this effect evolves with a time constant of $\simeq$900ps. We conclude that the loss channel employs an intermediate state with this lifetime before repopulating the ground state. Field induced formation of $P^+H_B^-$ (B-branch) was excluded from measurements in the Q_x transition of H_B at 533nm. Measurements in the blue and red wings of the 800nm Q_y band of the two bacteriochlorophylls (B_A and B_B) are compatible with the state $P^+B_B^-$ being the long-lived intermediate involved in the loss of quantum yield.

INTRODUCTION

The electrostatic energy of an electric dipole depends on the interaction between the dipole moment and the local electric field. Thus, an externally applied electric field can be used as an experimental tool to characterize (i) spectroscopic properties as well as (ii) the dynamics of electron transfer (ET) processes in donor/acceptor systems by shifting the free energies of the relevant dipole moments:

(i) The change of the extinction coefficient in an electric field (Stark effect) allows the determination of the difference in dipole moments and polarizabilities between the ground and excited state of an optical transition. In Stark effect measurements a large asymmetry ($\simeq$ 6D) was found for the excited state $^1P^*$ of the primary donor P in bacterial reaction centers (RCs)[1-5].

(ii) The large dipole moments which are involved in charge-separating systems can be used to manipulate electron transfer rates by significant amounts: for example, charge separation in the reaction center from the primary donor to the radical pairs $P^+B_A^-$, $P^+H_A^-$ and $P^+Q_A^-$ leads to electric dipole moments of 51D, 82D and 130D, respectively. The corresponding free energy shifts in an electric field of 10^6V/cm are 850cm^{-1}, 1380cm^{-1} and 2180cm^{-1}. Since ET processes comprising such states depend on their energy levels via the activation energy in the Franck-Condon factor and (especially in a superexchange ET) the energy variation of the electronic coupling, the ET rates can be changed in an external electric field[6-8].

The Photosynthetic Bacterial Reaction Center II
Edited by J. Breton and A. Verméglio, Plenum Press, New York, 1992

Electric field effects on ET rates have been investigated in detail for the recombination process $P^+Q_A^- \rightarrow PQ_A$ in oriented RCs embedded in stacked Langmuir Blodgett monolayers[9-10] or lipid bilayers[11-12] and in isotropically distributed RCs in polyvinyl alcohol (PVA) films[7]. The field dependence of the ET rate allows to infer on the corresponding free energy dependence of the rate in a "non-invasive" manner and without suffering from possible structural changes in the RC. These advantages cannot be matched in procedures which manipulate energy levels in the RC by site directed mutagenesis or pigment exchange[13].

The application of electric fields to elucidate the still controversial dynamics of primary charge separation can in principle focus on (i) absorption and (ii) fluorescence measurements:

(i) So far the electric field dependence of the primary ET process starting from the excited donor $^1P^*$ is only insufficiently known due to the low signal/noise ratio and the limited time window of the direct fs-measurement[14].

(ii) An alternative approach is based on the study of the electric field effect on the yield of the prompt fluorescence of $^1P^*$ which reflects the competition between the radiative decay rate and the field dependent primary ET[15]. The modulation of the fluorescence yield corresponds to the modulation of the primary ET rate in the electric field and the anisotropy of the fluorescence change contains the information about the direction of the dipole moment of the primary radical pair in the coordinate system of the RC[16-20]. Yet the attempts to infer on the mechanism of the primary ET from measurements of the stationary fluorescence in an external electric field are impeded by the fact that field dependent fluorescence signals with time constants two orders of magnitudes larger than the primary ET often dominate the stationary signal[20-21]. Therefore, also in this case the measurements have to be performed in real time of charge separation.

One of the most important questions concerning the function of the RC machinery is related to the unidirectionality of charge separation along the A-branch defining one of the two pigment branches. In this context the stimulating question arises, whether an external electric field shifting energies of large dipoles can be utilized to slow down charge separation along the A-branch and concomitantly speed up charge separation along the so far inactive B-branch. Such a "push-pull" mechanism is especially favoured by the mutual orientation of the primary cofactors since the orientation of the resulting dipole moments is almost antiparallel. As a result, the ET rates along the two branches may respond in an antagonistic way to the external field, thus violating the principle of unidirectionality.

On a slow time scale a field induced reduction of quantum yield has indeed been observed monitoring the state $P^+Q_A^-$[22]. Various attempts have been made to infer on field induced changes of the quantum yield of the earlier intermediate $P^+H_A^-$ but neither plausibility arguments[23] nor measurements on the microsecond time scale (monitoring the triplet $^3P^*$ [24]) are suited to prove field induced changes concerning processes prior to the formation of $P^+H_A^-$.

Thus, it is the goal of this paper to investigate the electric field effect on the quantum yield of $P^+H_A^-$ monitoring different spectral regions on the time scale of 30ps. Even though this temporal resolution does not allow to infer directly on the primary charge separation process, the high sensitivity of the measurements presented here might already be suited to detect small populations of early states favoured by the electric field.

EXPERIMENTAL DETAILS

Materials

Quinone depleted RCs of *Rb. sphaeroides R-26* obtained by modified standard procedures[25] were chosen in order to maximize the time window for the observation of electric field effects on the H_A^- anion. In the absence of quinone this state is trapped for ≃20ns at low temperature[25-26]. In quinone containing RCs any dynamics up to a few nanoseconds differing from that of $P^+Q_A^-$ formation will be buried under the 100ps signal reflecting the forward electron transfer from H_A^- to Q_A.

RCs were embedded in polyvinylalcohol (PVA) films of 80±10μm thickness which were dried for about 48h at 280K in a N_2 gas flow. Drying at such temperatures does neither cause unusual spectral shifts nor an increased fluorescence yield as reported for strongly dehydrated films[27]. Fs-time resolved measurements performed on identical samples did not reveal significant differences to samples in glycerol buffer solution at 80K[28]. The PVA-films were sandwiched between two glass slides coated with a conducting, optically transparent In-SnO_2 layer to which square pulses of up to 5.0kV with 30ms duration were applied.

Time resolved absorption spectroscopy

Transient absorption measurements were performed after exciting the samples at 600nm to less than 20% of saturation with 60ps pulses at an intensity of 0.25mJ/cm², so that effects from multiple excitation can be neglected. After an optical delay the absorption with and without excitation was probed in the regions 533nm-558nm, 850nm-880nm and 780nm-820nm. By detecting the relative energy of each excitation pulse, each single measurement of the absorbance change can be corrected for the fluctuating energy of the excitation.

To avoid major contributions from the Stark effect which in particular cannot be neglected in the long wavelength dimer band, the electric field was applied to the sample during the absorption measurements both with and without excitation. Each difference absorption measurement with the electric field applied was followed by one at zero field. 500 or 1000 of these measuring cycles were averaged to achieve a sufficiently small standard deviation for the absorption changes of $(3\text{-}5)\cdot 10^{-4}$OD. Detecting the absorption changes induced by switching the electric field in the absence of actinic light, the Stark effect on the absorption spectrum in the dimer band was determined ($\Delta A/A$=-1.5%) and utilized to calibrate the applied field ($7\cdot 10^5$V/cm), thus allowing the comparison with data from different laboratories[1-4].

RESULTS

The electric field dependence of the transient absorption of quinone depleted RCs has been monitored in different regions of the spectrum at 90K:

1. Transient absorption at 545nm

The bleaching at this wavelength relating to the maximum of the Q_x band of H_A, the bacteriopheophytin at the active branch, indicates the formation of H_A^-. Its change in an electric field is shown in Fig.1. Because of the 2ps charge separation, this band bleaches instantaneously on the time scale of this experiment and recovers slowly in accord with the known 20ns lifetime of

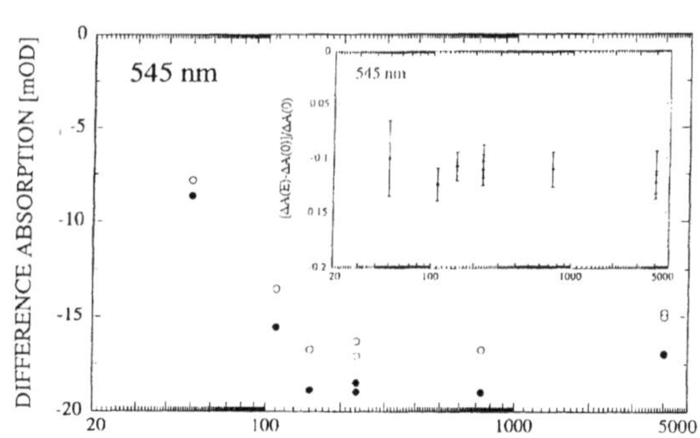

FIGURE 1. Kinetics of the transient absorption of *Rb. sphaeroides* R26 RCs (Q_A depleted) probed at 90K in the Q_x band of H_A at 545nm in the presence $\Delta A(E)$ (o) and the absence $\Delta A(0)$ (o) of an external electric field ($7\cdot 10^5$V/cm).
INSERT: Relative difference $(\Delta A(E)-\Delta A(0))/\Delta A(0)$.

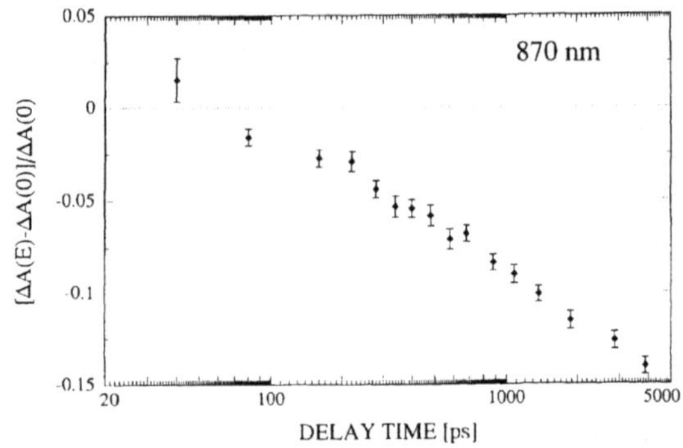

FIGURE 2: Kinetics of the electric field induced relative change of the groundstate bleaching of P $(\Delta A(E)-\Delta A(0))/\Delta A(0)$ probed at 870nm. Sample and conditions as in Fig.1.

$P^+H_A^-$[25-26] in Q_A^--depleted RCs. Application of an electric field of $7 \cdot 10^5$V/cm results in a reduction of the H_A bleaching amplitude by about 11±1.5 %. The effect is already fully developed in the rising edge of the signal (corresponding to an average delay of 30ps between excitation and probing) and remains constant within the first nanoseconds.

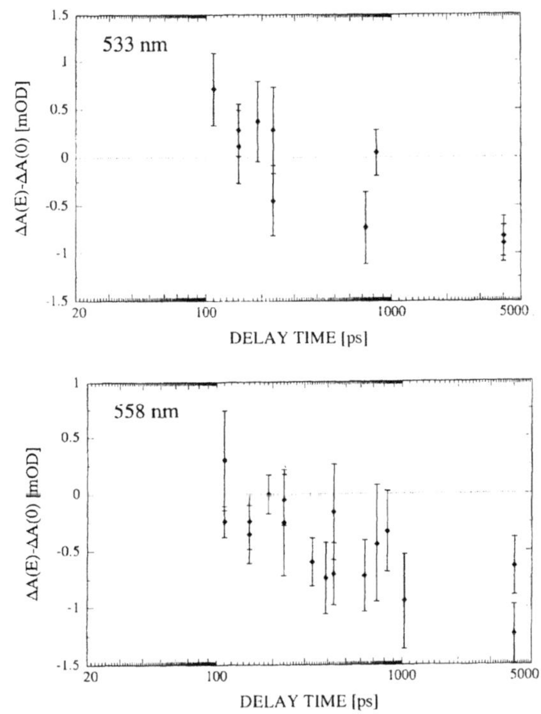

FIGURE 3: Kinetics of the electric field induced absolute change of the bleaching $\Delta A(E)-\Delta A(0)$ at (a) 533nm (Q_x band of H_B) and (b) 558nm (nonspecific absorption of $^1P^*$ and P^+). Sample and conditions as in Fig.1. The magnitude of the error bars is related to the small absorption signal.

2. *Transient absorption at 870nm*

The transient bleaching of the Q_y band of the primary donor at 870nm is depicted in Fig.2. In contrast to the effect on the H_A band (compare insert of Fig.1 with Fig.2), the electric field does not affect the 870nm signal at very early times but induces a 11% recovery of the ground state bleaching with a time constant of $\simeq$900ps. The further increase of the effect at later times, however, reflects already a field dependence of the triplet channel of the radical pair recombination[29].

The magnitude of the effect (e.g. probed at 870nm after 1ns) depends on the square of the field strength and is essentially the same at 90K and at 180K.
The difference in the time evolution of the electric field effects in the bleaching of H_A and P yields two important pieces of information:

(i) The electric field effects are not due to a field induced change of the efficiency of $^1P^*$ formation following excitation at 600nm, since in that case also the P signal should be affected instantaneously. Thus, our data prove that the field reduces the quantum efficiency of processes leading from $^1P^*$ to $P^+H_A^-$ by favouring a competing loss channel.

(ii) The time delay between charge separation and groundstate recovery shows that the loss channel involves an intermediate state which repopulates the ground state in a second much slower process.

Among the possible candidates for such an intermediate state are the radical pairs $P^+H_B^-$, $P^+B_B^-$, both located at the 'inactive' B-branch, and the state $P^+B_A^-$. While the bacteriochlorophyll monomers B_A and B_B defy spectroscopic distinction, such a distinction is well possible for the bacteriopheophytins based on the splitting of their Q_x bands peaking at 545nm (H_A) and 533nm (H_B).

3. *Transient absorption at 533nm and 558nm*

In contrast to the instantaneous electric field effect on H_A (Fig.1) no effect is detected at the Q_x absorption peak of H_B at 533nm (Fig.3a). In case the same amount of RCs that have failed to form $P^+H_A^-$ in the electric field would form the B-branch radical pair $P^+H_B^-$ instead, any field induced bleaching at 533nm should amount to the observed absorption increase of $\Delta A(E)-\Delta A(0)\simeq 2.5$mOD at 545nm, since the oscillator strength of both transitions is the same. Since this is not the case any significant field induced formation of $P^+H_B^-$ can be excluded.

Since a possible field effect on the Q_x absorption peak of H_B at 533nm (Fig.3a) cannot *apriori* be discriminated from effects on the superimposed $^1P^*$ or P^+ absorptions, possible field induced changes on these states were tested at 558nm (Fig.3b). While there is essentially no effect at early times, a slight reduction of the positive difference absorption signal at both 533nm and 558nm is observed after long delay times. At these wavelengths any time dependent change of the positive difference absorption is expected to reflect the decay of the absorption of both $^1P^*$ or P^+ and thus to parallel the dynamics of the groundstate recovery at 870nm in electric fields (Fig.2).

4. *Transient absorption between 770-820nm*

Fig.4a shows the relative change of the difference absorption spectrum induced by application of an electric field: 2ns after excitation the amplitude of the transient spectrum is reduced quite constantly throughout the range between 780nm to 820nm. In view of the difference spectrum (see insert) the constant negative amplitude of the electric field effect reflects a decrease of the electrochromic shift due to a decrease in quantum yield of $P^+H_A^-$. The constancy of the electric field effect is rooted in a field induced decrease of the absorption accompanied by a field induced increase of the absorption at the blue and red wings of the band, respectively.

The spectral characteristics of the electric field effect detected at earlier times, i.e. after 100ps, however, are different (Fig.4b). Similar to the effect observed at 545nm (Fig.1) the negative amplitude at $\lambda < 800$nm is related to the instantaneous electrochromic shift due to the field induced decrease in the quantum yield of $P^+H_A^-$. This effect is masked in the red wing of the band ($\lambda > 800$nm). This observation points to two antagonistic effects: (i) a field induced

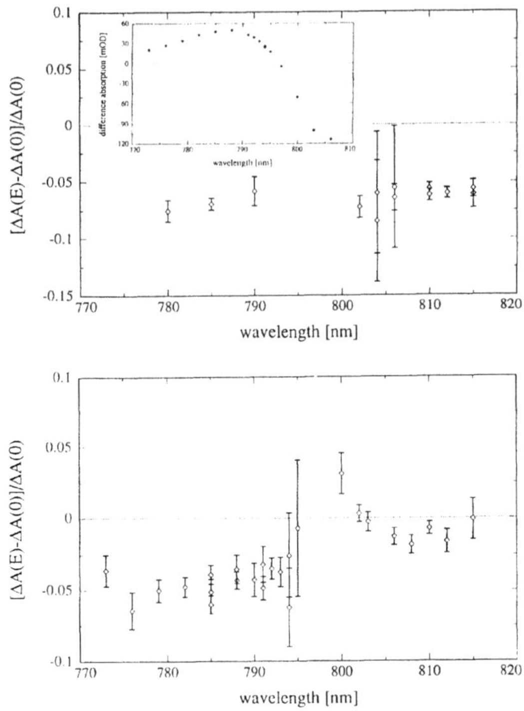

Fig.4a. Spectrum of the relative difference $(\Delta A(E)-\Delta A(0))/\Delta A(0)$ in the Q_y band of the monomeric bacteriochlorophylls with $E \simeq 6\ 10^5$V/cm probed after a time delay of 4ns. Insert: Spectrum of the difference absorbance probed after 100ps. Fig.4b: The relative difference as in Fig.4a probed at a delay time of 100ps.

decrease of the absorption of B due to bleaching and (ii) a concomitant increase of the absorption due to the smaller electrochromic blue shift after reduction in the H_A^- concentration.

The difference in the temporal evolution in the blue and red wings of the 800nm band is even more evident when directly tracing the electric field effect in the two regions as a function of time. At λ=780nm the electric field induced reduction of the positive transient absorption signal appears instantaneously, and remains constant up to 2ns, while at λ=820nm the effect evolves with a time constant of $\simeq$900ps. This long time constant is identical with the one observed for the groundstate recovery of P at 870nm (see Fig.2) which has been assigned to the decay of the intermediate state (e.g. P^+B^-) populated under the action of the external electric field.

The phenomenology of ps-time resolved absorption measurements on RCs in external electric fields can be summarized as follows:

(1) At 90K the quantum yield of $P^+H_A^-$ decreases by $\simeq$11% upon application of a field of $7 \cdot 10^5$V/cm when monitoring the bleaching of H_A 30ps after excitation. This finding reveals the action of a loss channel. In an isotropically excited sample this loss channel is found to be favoured on the average by the presence of an electric field.

The electric field induced reduction of the quantum yield of $P^+H_A^-$ detected here in absorption stands up for the behaviour of the *majority* of RCs in contrast to the electric field effect observed on the fluorescence signal[18,21,30] and cannot be confined to a rather small minority of RCs with particularly slow charge separation. This is a crucial advantage compared to the electric field effects observed on the fluorescence signal which by their very nature are not necessarily representative for the charge separation kinetics of the bulk of the sample.

(2) Since $P^+H_A^-$ formation occurs on a time scale of 2ps the present time resolution of 30ps is insufficient to distinguish whether the loss process is interfering before $P^+H_A^-$ formation or thereafter. In the latter case the process emanating to an intermediate state would have a time constant of less than 250 ps in order to produce a loss of 11% at our earliest detection time of 30ps. Since thereafter we observed no further loss of H_A^-, one would additionally have to assume that the loss process is stopped prior to our observation window. Qualitatively such a behaviour has earlier been attributed to the proposed nonrelaxed $P^+H_A^-$ states[31], which were considered to repopulate the groundstate very quickly prior to a relaxation process[32-33]. Such a loss mechanism, however, leading directly into the ground state, is incompatible with the slow time constant of the field induced recovery of the ground state absorption of P. Thus, the observed loss in quantum yield has to occur before formation of $P^+H_A^-$.

(3) The loss channel leads to an intermediate state which decays to the ground state with a time constant of $\simeq$ 900ps. It involves P (its groundstate absorption at 870nm remains bleached) but comprises neither H_A (545nm) nor H_B (530nm).

(4) Another relevant result on the possible nature of the intermediate is derived from differences in the electric field effect at the blue and red wings of the Q_y band of the two monomer bacteriochlorophylls. Subtracting the ns- from the 100ps-spectrum (i.e. Fig.4b - Fig.4a) provides the difference spectrum of the intermediate dissappearing with a time constant of $\simeq$900ps. This difference spectrum points to a specific decrease of the absorption at $\lambda > 800$nm in the electric field. This observation is compatible with the field induced population of a P^+B^- intermediate since (i) the negative radical ion B^- implies a symmetric decrease of the 800nm band and (ii) its positive counter ion P^+ introduces an electrochromic blue shift of the other B molecule. This blue shift compensates for the bleaching to be expected in the blue wing. In the red wing however, due to the reversed sign of the difference absorption (see insert in Fig.4a) the electrochromic shift and the bleaching of B are acting in the same direction.

DISCUSSION

QUANTIFICATION OF THE LOSS CHANNEL

Considering the extremely fast formation of $P^+H_A^-$ (2ps at 90K)[34], it is surprising to find a measurable reduction of its yield due to an electric field. Independent of the mechanism of this effect, our findings call for a fast loss channel which so far has escaped notice. In the following estimate of k_{loss} leading to the observed reduction of the quantum yield $\langle\Phi\rangle$ of charge separation, $\langle\Phi\rangle$ is an *orientational average*, since the rates determining $\Phi(E)$ depend on the orientation of each individual RCs with respect to the external field.

In the simplest case that only the charge separation rate $k_1(E)$ depends on the electric field the average quantum yield can be written as $\langle\Phi(E)\rangle = \langle k_1(E)/(k_1(E)+k_{loss})\rangle = 1-\langle\tau(E)\rangle k_{loss}$. The *averaged* lifetime of $^1P^*$, $\langle\tau(E)\rangle=\langle 1/(k_1(E)+k_{loss})\rangle$, can be obtained from the fs-ps experiments. In general, the $\langle\tau(E)\rangle$ will increase in randomly oriented samples, thus favouring k_{loss} in competing with the major quenching route of charge separation. According to transient absorption measurements $\langle\tau(E)\rangle$ does not increase by more than roughly 40±15% (value interpolated from data of ref.[14] at $1\cdot10^6$V/cm assuming a quadratic field dependence of k_1). This limit is also corroborated by time resolved fluorescence data[21,30,35]. With these data we can estimate $1/k_{loss}\simeq$ 6-14 ps. In case the fast loss rate persists in absence of a field, it should manifest itself in a quantum efficiency of $\simeq$0.67-0.85 deviating from the generally adopted value of 1.02±0.04[36].

A quantum yield of charge separation smaller than unity could, in principle, originate from some loss of Q_A in the RC. Apart from well-known difficulties in the determination of absolute quantum yields, the high value of unity reported in ref.[36] rests on the difference extinction coefficient of the 870nm band. This is derived from its incomplete bleaching under saturating illumination[37]. Considering that, in general, most RC preparations are lacking 15% Q_A[38] and therefore not contributing to the bleaching under steady state illumination, the difference extinction coefficient can be accordingly larger and consequently the absolute quantum yield of charge separation would have to be corrected to 0.85.

Alternatively, the above estimate of a quantum yield in zero field smaller than unity is irrelevant if both rates k_1 and k_{loss} are sensitive to the electric field, i.e. the field induced *decrease* of k_1 is accompanied by a field induced *increase* of k_{loss}.

POTENTIAL INTERMEDIATES INVOLVED IN THE LOSS OF QUANTUM YIELD

On the basis of the wavelength dependence of the temporal evolution of the electric field effect the possible intermediates $P^+H_A^-$ and $P^+H_B^-$ can be excluded in straightforward way. A different set of potential intermediates is addressed in the following question:

A conformationally relaxed, inactive state $^1P^*$ or a radical ion pair P^+B^- as long-lived intermediates?

Any conformational change of the protein surrounding P or a relaxation of the electronic state of $^1P^*$ could lead to changes of the electronic nature of the excited dimer or to changes within the radical pair state to be formed. By influencing the energies of the states involved and/or the geometry changing the electronic coupling between donor and acceptor, the charge separation rate may be changed. Thus, $^1P^*$ may exist in a photosynthetically *active* state with fast and an *inactive* state with slow charge separation including the possibility of dynamic transition between both. The ultimate lifetime of such an *inactive* state would be limited by the internal conversion rate repopulating the ground state, i.e. by the observed ≃900ps. In fact, in RCs lacking charge separation (D_{LL} mutant[21,39] and H_A-reduced *R. viridis* [21]) fast internal conversion rates of the excited bacteriochlorophyll dimer have been observed.

However, a serious argument against $^1P^*$ as long-lived intermediate is based on the comparison of the 800nm difference spectrum of the excited dimer in the D_{LL}[39] mutant and the analogous spectrum extracted from subtraction of the spectrum in Fig.4a from the one in Fig.4b. The latter one is qualitatively compatible with P^+B^- but not with $^1P^*$. A detailed analysis and simulations of these electric field induced difference spectra in the 800nm band will be published elsewhere[40].

In summary, the accumulated evidence of the electric field effects probed at 533nm-558nm, 780nm-820nm and 870nm is well compatible with the field induced population of P^+B^- but cannot discriminate between the two branches.

$P^+B_A^-$ or $P^+B_B^-$?

However, consistency with the data presented would be only given, if in ≃ 10% of the RCs $P^+B_A^-$ would adopt a lifetime of almost a nanosecond upon application of 7 10^5V/cm. In case of a two-step charge separation with electron transfer from $P^+B_A^-$ to $P^+H_A^-$ in less than 1ps[41], this would imply a field induced reduction of this rate by 3 orders of magnitude.

The most obvious loss channel profiting from an antagonistic energy shift of the two, almost antiparallel (140°) dipoles $P^+B_A^-$ and $P^+B_B^-$ could be charge separation to the B-branch of the RC. Such a loss channel is expected to be also field dependent, thus removing the necessity of low quantum yields of charge separation in zero field (see above). In view of our measurements in the 533nm-558nm region the formation of $P^+B_B^-$ would imply that subsequent charge separation to $P^+H_B^-$ cannot compete with recombination on the 900ps time scale. This requirement imposes severe restrictions to the relative energies of the radical pairs $P^+B_B^-$ and $P^+H_B^-$ at the low temperature (80K) of our measurements or to their electronic coupling matrix elements.

A field driven charge separation along the B-branch would be also consistent with the instantaneous reduction of quantum yield on $P^+H_A^-$ in other purple bacteria as e.g. *R. viridis* and its lack in RCs of *Chloroflexus aurantiacus*[29] where the sequence of pigments along the B-branch is P-H-H instead of P-B-H in RCs of purple bacteria. The value of such arguments derived from analogies among different species is certainly limited and might be doubted by the absence of field induced charge separation in the symmetrized D_{LL} mutant. This mutant lacks H_A and does not untergo charge separation in zero field[39]: in this case the long lifetime of $^1P^*$ of 500ps at 90K is not affected by a field induced loss channel of any kind[20,21].

A further identification of the mechanism underlying the electric field induced loss of quantum yield is expected from fs-ps spectroscopy in external fields, as from field induced changes on the near-IR P^+-band in RCs from *R. sphaeroides*, from field induced changes of the (well-resolved) Q_x absorption band in RCs from *R. sphaeroides* with of chemically exchanged bacteriochlorophyll monomers[42] and from DELFY-type absorption measurements[18] in the time domain of charge separation.

ACKNOWLEDGEMENTS

Financial support from the Deutsche Forschungsgemeinschaft (Sonderforschungsbereich 143) and the Alfried Krupp von Bohlen und Halbach Stiftung (U.E.) is gratefully acknowledged.

REFERENCES

1. M. Lösche, G. Feher and M.Y. Okamura, *Proc. Nat. Acad. Sci. USA* 84:7537 (1987).
2. D.J. Lockhart and S.G. Boxer, *Biochem.* 26:664 (1987).
3. H.P. Braun, M.E. Michel-Beyerle, J. Breton, S. Buchanan and H. Michel, *FEBS Lett.* 221:221 (1987).
4. D.J. Lockhart and S.G. Boxer, *Proc. Natl. Acad. Sci. USA* 85:107 (1988).
5. U. Eberl, J. Breton, M. Plato and M.E. Michel-Beyerle, *Biophys. J.* 61:872a (1992).
6. M. Bixon und J. Jortner, *J. Phys. Chem.* 92:7148 (1988).
7. S. Franzen, R.F. Goldstein und S.G. Boxer, *J. Phys. Chem.* 94:5135-5149 (1990).
8. S.H. Lin, C.Y. Yeh und G.Y.C. Wu, *Chem. Phys. Lett.* 166:195 (1990).
9. Z.D. Popovic, G.J. Kovacs, G.S. Vincett and P.L. Dutton, *Chem. Phys. Lett.* 116:405 (1985).
10. Z.D. Popovic, G.J. Kovacs, G.S. Vincett, G. Alegria and P.L.Dutton, *Chem. Phys.* 110:227 (1986).
11. A. Gopher, M. Schönfeld, M.Y. Okamura and G. Feher, *Biophys. J.* 48:311 (1985).
12. G. Feher, T.R. Arno and M.Y. Okamura, in: The Photosynthetic Bacterial Reaction Center. Structure and Dynamics. (Breton, J. and Vermeglio, A., eds.), pp.271-287, Plenum New York (1988).
13. M.E. Michel-Beyerle and H. Scheer, in: Reaction Centers of Photosynthetic Bacteria (Michel-Beyerle, M.E., ed.), pp.453, Springer 1990.
14. D.J. Lockhart, Ch. Kirmaier, D. Holten and S.G. Boxer, *J. Phys. Chem.* 94:6087 (1990).
15. D.J. Lockhart and S.G. Boxer, *Chem. Phys. Lett.* 144:243 (1988).
16. D.J. Lockhart, R.F. Goldstein and S.G. Boxer, *J. Chem. Phys.* 89:1408 (1988).
17. U. Eberl, A. Ogrodnik and M.E. Michel-Beyerle, *Z. Naturforsch.* 45a:763 (1990).
18. A. Ogrodnik, U. Eberl, R. Heckmann, M. Kappl, R. Feick and M.E. Michel-Beyerle, *J. Phys. Chem.* 95:2036 (1990).
19. A. Ogrodnik, U. Eberl, R. Heckmann, M. Kappl, R. Feick and M.E. Michel-Beyerle in: Reaction Centers of Photosynthetic Bacteria (Michel-Beyerle, M.E., ed.), pp. 157, Springer Berlin (1990).
20. U. Eberl, Thesis, Technical University Munich (1992).
21. A. Ogrodnik, U. Eberl, T. Häberle, W. Keupp, T. Langenbacher, J. Siegl, M. Volk and M.E. Michel-Beyerle, *Biophys. J.* 61:837a (1992).
22. Z.D. Popovic, G.J. Kovacs, P.S. Vincett, G. Alegria and P.L. Dutton, *Biochim. Biophys. Acta* 851:38 (1986).
23. C.C. Moser, G. Alegria, M.R. Gunner and P.L. Dutton, *Isr. J. Chem.* 28:133 (1988).
24. S. Franzen, K.-Q. Lao, B. Stanley and S.G. Boxer, *Biophys. J.* 61:874a (1992).
25. A. Ogrodnik, M. Volk, R. Letterer, R. Feick and M.E. Michel-Beyerle, *Biochim. Biophys. Acta* 936:361 (1988).
26. D.E. Budil, V. Kolaczkowski and J.R. Norris, in: Progress in Photosynthesis Research, (Biggins,J. ed.) Vol.I pp.245, Martinus Nijhoff Dordrecht (1987).
27. R.K. Clayton, *Biochim. Biophys. Acta* 504:255 (1978).
28. W. Zinth, private communication.
29. M. Volk, Thesis, Technical University Munich (1991).

30. A. Ogrodnik, U. Eberl, W. Keupp, M. Kappl and M.E. Michel-Beyerle, *Molecular Crystals and Liquid Crystals* (1992), in press.
31. N.W. Woodbury and W.W. Parson, *Biochim. Biophys. Acta* 767:345 (1984).
32. S.L. Logunov and V.Z. Pashchenko, *Sov. J. Quantum Electron.* 19:88 (1989).
33. R.A. Goldstein and S.G. Boxer, *Biochim. Biophys. Acta* 977:78 (1989).
34. J. Breton, J.-L. Martin, G.R. Fleming and J.-C. Lambry, *Biochemistry* 27:8276 (1988).
35. A. Ogrodnik and M.E. Michel-Beyerle, in: Photoprocesses in Transition Metal Complexes, Biosystems and other Molecules; Experiment and Theory, ed. E. Kochanski (Plenum press, New York, 1992), in press.
36. C. Wraight and R. Clayton, *Biochim. Biophys. Acta* 333:246 (1973).
37. S.C. Straley, W.W. Parson, D.C. Mauzerall and R.K. Clayton, *Biochim. Biophys. Acta* 305:597 (1973).
38. M. Volk, G. Aumeier, T. Häberle, A. Ogrodnik, R. Feick and M.E. Michel-Beyerle, *Biochim. Biophys. Acta* (1992), in press.
39. J. Breton, J.-L. Martin, J.-C. Lambry, S.J. Robles and D.C. Youvan in: Reaction Centers of Photosynthetic Bacteria, (M. E. Michel-Beyerle, ed.) pp. 293 Springer Berlin (1990).
40. G. Bieser, Diploma Thesis Technical University Muenchen (1992).
41. W. Holzapfel, U. Finkele, W. Kaiser, D. Oesterhelt, H. Scheer, H.U. Stilz and W. Zinth, *Chem. Phys. Lett.* 160:1 (1989).
42. A. Struck, E. Cmiel, M. Fischer, A. Müller, W. Schäfer and H. Scheer, *FEBS Lett.* 268:180 (1990).

ELECTRIC FIELD EFFECTS ON THE QUANTUM YIELDS AND KINETICS OF FLUORESCENCE AND TRANSIENT INTERMEDIATES IN BACTERIAL REACTION CENTERS

Steven G. Boxer, Stefan Franzen, Kaiqin Lao, David J. Lockhart, Robert Stanley, Martin Steffen, and Jonathan W. Stocker

Department of Chemistry
Stanford University
Stanford, California 94305-5080

INTRODUCTION

The application of external electric fields alters the rates of electron transfer reactions principally by changing the free energy of the reaction [1]. Effects can be measured directly on the kinetics and indirectly by measuring the quantum yields of intermediates. Information can also be obtained using competing processes such as emission for a photoinduced electron transfer reaction. In addition to providing information on the basic mechanism of an electron transfer reaction, the observed electric field effects can be related to the regulation of charge separation processes by the transmembrane potential.

Ideally these field effects would be measured for a polar, oriented sample giving only a single projection of the change in dipole moment responsible for the field effect on the applied field direction. This would give a simple shift in the rate of electron transfer and a direct mapping between the free energy and rate. This situation has only been realized for one case, that of $P^+Q_A^-$ charge recombination at room temperature in a single lipid bilayer [2,3]. Several studies have been presented for non-oriented (isotropic) samples, which are the simplest to prepare, but this requires more sophisticated methods of data analysis [4-8]. A new approach involving covalent attachment of site-specificially modified RCs is currently being developed in our lab [9].

The Photosynthetic Bacterial Reaction Center II
Edited by J. Breton and A. Verméglio, Plenum Press, New York, 1992

The reaction scheme for the initial charge separation and recombination processes in isolated Rb. sphaeroides (R26) reaction centers (RCs) is shown in Figure 1. In the absence of Q_A, charge recombination occurs from the P^+H^- state to form ^3P or ground state PH. There has been a great deal of experimental and theoretical work on the initial charge separation step, $^*P$ to P^+H^-, focussing on whether this reaction occurs with the intermediate formation of P^+B^-, or whether P^+B^- serves instead to mediate electron transfer between $^*P$ and P^+H^- by superexchange. The conclusion from fs/ps transient absorption is unclear, and strong positions favoring one or two-step mechanisms have been offered [10-12]. From the perspective of a group not doing these kinds of measurements, it is unclear that the issue can be resolved by transient absorption because of the difficulty of assigning the absorption spectra of intermediate states and because non-exponential, even oscillatory, kinetics have been observed under some circumstances [10]. For this reason alternative, complementary approaches, including the application of electric fields and site-directed mutagenesis (in a sense, the modification of internal electric fields), may be useful.

Some time ago, we demonstrated that the steady-state fluorescence from $^*P$ is enhanced upon application of an electric field [4] and that the fluorescence from an isotropic sample becomes polarized in an applied field [5]. These data were interpreted in terms of the orientations of the dipole moment whose formation competes with fluorescence, for example P^+P^-, P^+B^- or P^+H^-. The directions of these dipole moments, within clearly stated limits, were obtained from the x-ray structure of RCs [13-15]. This led to the conclusion that the observed angle between the fluorescence transition moment and the electric dipole moment whose formation competes with fluorescence is incompatible with direct formation of P^+B^- from $^*P$. Assuming that only P^+B^- and P^+H^- need to be considered, the data are more consistent with direct P^+H^- formation from $^*P$. These data have been quantitatively confirmed by Ogrodnik and co-workers [16], who also extended the original experiment by exciting the sample with linearly polarized light. The purpose of this extension is to further distinguish the projections of the P^+B^- and P^+H^- dipoles on the fluorescence transition moment. The results were again inconsistent with P^+B^- formation and are more consistent with P^+H^- formation.

At this meeting Ogrodnik and co-workers suggested that the change in fluorescence upon application of a field measured by steady-state fluorescence contains contributions from fluorescence components other than the prompt fluorescence which competes directly with primary charge separation [see paper by these authors in this volume]. It is difficult to obtain definitive information from single-photon counting experiments because of the limited time resolution, so the best approach to measuring the electric field effect is to measure it directly on the prompt fluorescence, detected by fluorescence upconversion as demonstrated by the Flemming and Zinth groups at this meeting [see papers by these authors and their co-workers in this volume]. In the next

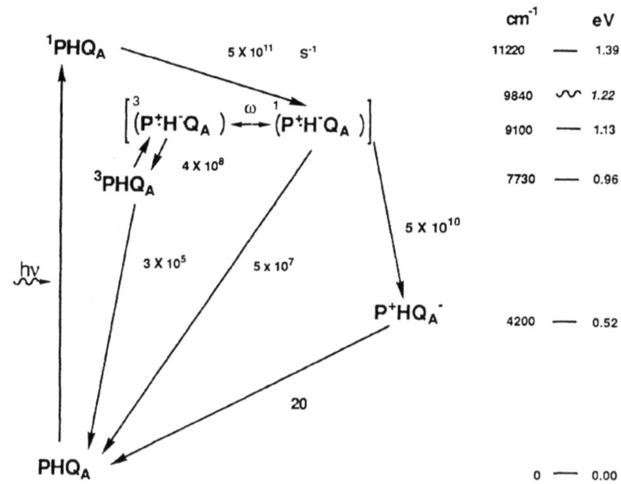

Figure 1. Reaction scheme for the initial charge separation and recombination reactions in isolated reaction centers. When Q_A is present, $P^+Q_A^-$ formation occurs; when Q_A is absent or pre-reduced, ^3P is formed in competition with recombination from $^1(P^+H^-)$. The rates are approximate. The energetics of each state estimated from various approaches are shown on the right.

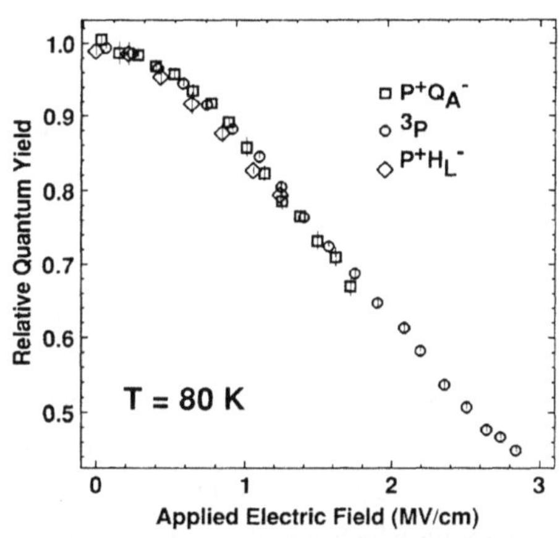

Figure 2. Effect of an applied electric field on the relative quantum yield of formation of the $P^+Q_A^-$, P^+H^-, and ^3P states. The relative quantum yields of P^+H^- and ^3P is measured in Q_A-depleted RCs. All experiments are performed in dry PVA films at 80 K. Relative quantum yields are obtained by measuring the change in absorption in the absence of the field and then with the field on during formation of the state of interest. Given the sample capacitance and resitance, it possible to turn the field on during formation of $P^+Q_A^-$ and ^3P and off during their decay, but this is not possible for P^+H^-. Methods for dealing with transient Stark effects are described in detail elsewhere [28].

section, we critically review some of the issues that arise in the steady-state experiments and describe qualitatively several new results and cautions. This analysis is followed by a brief report of measurements of the effects of applied fields on the quantum yields of formation of the P^+H^-, $P^+Q_A^-$ and ^3P states by transient absorption.

ANALYSIS OF ELECTRIC FIELD EFFECTS ON FLUORESCENCE

In the original paper on the analysis of electric field effects on the anisotropy induced in the steady-state fluorescence [5] we were careful to distinguish the approximations involved in the experimental observations from those needed to interpret the data. Although discussed in detail, we review here some of the issues. The principal assumption is that the electric-field-dependent process which leads to the anisotropy in the fluorescence is a process which competes with the prompt fluorescence. Two possible candidates are the primary electron transfer step and internal conversion. The steady-state fluorescence and the change in fluorescence may be contaminated with other contributions, but it was assumed that the anisotropy in the change in fluorescence is due to one of these competing processes. Thus, even if fluorescence components which are long-lived, such as those found by Ogrodnik et al. in their samples, contribute significantly to both the fluorescence and the change in fluorescence, the anisotropy of the change in fluorescence is the important observable, and this has not yet been reported in a time-resolved measurement. In our original analysis, we suggested that the field dependence of the electron transfer rate could be substantial since the P^+X^- (X = B or H) dipole is quite large. We could not rule out a contribution from internal conversion, involving, e.g., internal charge-transfer states of P. The work of Kirmaier et al. on the heterodimer mutants [17,18] strongly suggests that if the charge transfer character of $^*P$ is large, then non-radiative channels to the ground state become quite efficient. Bixon and Jortner [19] predicted electric field effects on the primary charge separation which are much larger than what was observed. Although there are quantitative problems with their analysis [1,8], one expects substantial electric field effects, especially if two steps are involved [6].

A second key assumption which was discussed in the original paper is that the steady-state fluorescence is not dominated by delayed fluorescence. The precise origins of delayed fluorescence, especially components with lifetimes shorter than tens of ns and are not dependent on magnetic fields, is not yet clear [20,21]. It is generally believed that all components involve recombination from ion-pair states, P^+X^-. Because the delayed, recombination fluorescence involves an activated process, it is reasonable to expect that the field dependence should be very strong, in contrast to the approximately quadratic dependence which is observed [5]. Furthermore, the contribution in Q-containing (P^+H^- short-lived) vs. Q-depleted (P^+H^- much longer-lived) should be very different, but little difference was observed except at the highest applied field strengths [5]. We now know (see next section) that the situation is more complicated, because a substantial population of RCs, with a specific orientation relative to the applied field, fails to go forward in an electric field and returns to the ground state. Thus, the population which is capable of producing recombination fluorescence is not isotropic, and RC orientational subpopulations which would produce a large increase in delayed fluorescence in an applied field have already returned to the ground state. That is, both quantum yield failure and delayed fluorescence may be most sensitive to the same orientations of dipoles in the field. Thus, it remains a viable possibility that the steady-state change in fluorescence and its anisotropy have significant contributions from delayed, recombination fluorescence. This is being tested directly by measuring the prompt component in a field.

One of the difficult aspects of electric field effect measurements is that it is necessary to go to quite high electric fields to observe effects, and this is most easily done in PVA films. However, RCs in PVA, especially when very dry, have somewhat altered properties. Although, the electric-field-dependent properties may be the same in dry PVA and other media, this is a matter for concern. For this reason, we have perfected methods for applying quite large electric fields (as high as 1 MV/cm) to frozen glycerol/buffer glass samples [22]. This has many advantages including: ease of sample preparation, small sample requirements, conditions which are comparable to those used most commonly in low temperature transient absorption experiments, and assurance that the RCs are not partially oriented. Similar results, both for the change in fluorescence and its anisotropy were obtained in glycerol/buffer and PVA under conditions where both can be measured. There are, however, some interesting differences. We have observed that for WT RCs in PVA, prolonged irradiation leads to radical changes in some of the properties of the RCs. These effects, which we collectively term photoconversion, include: (i) An increase in fluorescence. (ii) A change in the sign of the electric field enchancement of fluorescence. (iii) A loss in $P^+Q_A^-$ quantum yield (following prolonged irradiation, the quantum yield of $P^+Q_A^-$ formation is reduced by more than 80%). (iv) The rate of changes (i)-(iii) with illumination are approximately identical. The fractional changes are initially quite rapid, then the remaining process occurs over a longer period. (v) These changes are completely reversible upon warming to room temperature and re-cooling, and they occur whether or not a field is applied during illumination. (vi) Careful measurements of the absorption and Stark effect spectra of samples during and following photoconversion indicate that no intermediates such as H^- are accumulated. (vii) Photoconversion occurs very slowly if Q_A is removed; however, Q_A^- accumulation is not the origin of the effect, because ^3P is not observed upon photoconversion. (viii) Photoconversion occurs quite slowly for WT in PVA; the rate is much slower in glycerol/buffer glasses; therefore glycerol/buffer is a better medium for studying electric field effects. Surprisingly, the rate of photoconversion is extremely rapid in some mutants such as the Rb. sphaeroides TyrM210 to Phe and the beta mutants in PVA [23]. "Rapid" means on the order of seconds with tens of mW/cm^2 illumination, so that an experimenter unaware of this process, might think that the electric field effect on the steady-state fluorescence for these mutants is very different from WT when it is not. Because the rate of photoconversion is much slower in glycerol/buffer, electric field effects for these mutants can only be measured and compared meaningfully in this medium. These effects will even impact such simple measurements in PVA as the dependence of the change in fluorescence on field strength. If one always scans from low to high field in time, the apparent change in fluorescence with field may actually be a change due to the length of irradiation (e.g. it may appear that the effect is saturating at high field).

We have recently investigated the D_{LL} mutant which lacks H_L [24]. It has been shown that $^*P$ lives for several hundred ps in this mutant with no evidence for P^+B^- formation or electron transfer down the inactive M-branch [25,26]. Assuming that the only change is the loss of H_L, this experiment is the strongest evidence favoring the one-step mechanism: in the two-step mechanism, removal of H_L should not affect the initial step, but this step was in fact observed to be profoundly affected. No evidence has been presented suggesting that this mutant is in any other significant way different from WT. Its absorption [26] and electroabsorption [24] spectra are very similar to WT, except for the clear loss of H_L and other minor changes which can be ascribed to individual changes in the vicinity of P and B_L (each of which we have independently examined [24]). We have examined the fluorescence and change in fluorescence in an applied electric field from this mutant RC in whole cells and chromatophore membranes which contain no antenna (the strain was generously provide by D. Youvan). The fluorescence is sometimes badly contaminated with impurity fluorescence, presumably from free or aggregated bacteriochlorophyll in the membranes, nonetheless, it is evident that the fluorescence quantum yield from $^*P$ is much greater for D_{LL} than WT, as expected. Upon application of a field there is a small decrease in fluorescence. This is the opposite of what is observed for WT. In combination with the observation that electron transfer is not occuring from $^*P$ in this mutant, these results are consistent with the interpretation that the enhancement of fluorescence for WT is due primarily to an effect on the initial electron transfer reaction. It does not rule out the possibility that delayed fluorescence contributes to the effect for WT, because no delayed component is expected for the D_{LL} mutant. The results also suggest that the field effect on the intrinsic fluorescence is quite small.

The analysis of the electric field induced fluorescence anisotropy depends on our estimates for the directions of the dipole moments. What matters is the change in dipole moment between the final state (P^+X^-) and the initial state ($^*P$). If we assume that that the dipole moment of $^*P$ is small relative to the P^+X^-, then the difference dipole moment is just the product state dipole moment. At the present time we know very little about the absolute direction of the dipole moment of $^*P$, although some indirect information can be obtained from the spectral lineshape of the change in fluorescence [27]. There are also uncertainties in obtaining estimates for the directions of the product state dipole moments. The simplest approach is to use the vectors connecting the geometric centers of P, B and H. Detailed information on the asymmetry of the charge distribution in P^+ was presented at this meeting [see papers by Lubitz, Feher and their co-workers in this volume]. Although a substantial asymmetry is observed, it is important to note that the epr and endor data are for fully relaxed P^+ observed long after its formation. What matters for the anisotropy experiment (assuming the fluorescence is prompt fluorescence) is the charge distribution on the ps timescale. The asymmetry could be even greater or the hole could be symmetrically delocalized. At this time we have no further information on the charge distribution, and therefore we prefer to simply use the geometric center of P. Obviously there is no experimental information on B^-, and only limited data are available for H^-, again only on a timescale which is long compared to the relevant timescale.

Ogrodnik et al. have presented time-resolved fluorescence data in an applied electric field at this meeting. The basic conclusion appears to be that the change in fluorescence upon application of a field has contributions from several components including the prompt component. Given the time resolution and signal-to-noise of these experiments, it is difficult to obtain quantitative information on the amplitude of the prompt component, and the origins of the longer-time signals are not known. Furthermore, mechanistic information is not obtained from the amplitude of the change in fluorescence, but rather from its anisotropy in an applied field. Finally, it is important to stress that any electron transfer mechanism, especially one involving two-steps for the formation of $P^+H_L^-$, should depend upon an applied electric field. Unless two effects of opposite sign are fortuitously cancelling each other, the prompt fluorescence should be sensitive to an applied field. It is incumbent upon those who are trying to understand this mechanism to explain or predict the results of electric field effect experiments.

EFFECTS OF APPLIED ELECTRIC FIELDS ON THE QUANTUM YIELDS OF INTERMEDIATES

As discussed in the introduction section, it is also possible to measure the effects of applied electric fields directly on the kinetics of primary charge separation. This is a difficult experiment with an isotropic sample [6], and although improved experiments are currently in progress, no new results are available yet. In lieu of such direct measurements, indirect information can be obtained from measurements of field effects on the quantum yield of intermediates. The results are discussed in detail elsewhere

[28] and have been presented earlier [29]. Ideally, it would be desirable to turn on the field during the formation of an intermediate and then turn it off during the measurement of the amplitude. In this way a relative quantum yield (field on normalized to field off) can be obtained without the complication of the Stark effect on the absorption feature used to probe the quantum yield. The latter is especially troublesome if measurements are made using the special pair Q_y band which exhibits a large Stark effect [30,31]. With our current equipment it is possible to gate the field on and off in approximately 30 us. Referring to the reaction scheme in Figure 1, it is clear that by synchronizing the excitation flash with this field pulse, it is possible to have the field on during the formation of $P^+Q_A^-$ and off during its decay (exactly the inverse of the strategy used to study the effect of a field on the $P^+Q_A^-$ decay [7]). However, the formation of $P^+Q_A^-$ involves at least two steps (k_1 and k_Q). At low temperature in Q_A-depleted RCs, ^3P decays in about 100 microseconds; consequently ^3P can be formed in a field and its quantum yield determined in the absence of a field. ^3P formation likewise involves at least two electron transfer steps, plus the radical pair singlet-triplet mixing process; however, the formation of ^3P from the triplet radical pair (rate constant k_T) is different from that for $P^+Q_A^-$ formation (k_Q). Finally it is possible to measure the quantum yield of the P^+H^- state itself. Given the decay rate of P^+H^- in Q_A-depleted RCs (about 20ns), the field is on both during the formation and decay of this state. Furthermore, given the limited time resolution of our ns transient absorption instrumentation (determined by the pulse-width of the excitation laser, about 8 ns), we can only determine the relative quantum yield of P^+H^- quite long after its formation. Therefore, there is an imperfect match between these experiments and the earlier ps kinetic measurements [6].

The results of relative quantum yield measurements for all three states in PVA at 77K are shown in Figure 2. Noting the caveats in the previous paragraph, several observations emerge. (i) The applied field causes a substantial net loss of quantum yield, denoted quantum yield failure in the following (note that the absolute zero-field quantum yield is not measured, so these are relative quantum yields). The quantum yield failure is quite large, exceeding 50% at the highest fields measured. This is striking because the samples are not oriented, which implies that both populations whose free energies are shifted up and down are likely contributing to the quantum yield failure. Furthermore, it is quite possible that as the field strength increases, different internal dipoles (see below) contribute to the quantum yield failure. Note that both the magnitude of the dipole and its effectiveness in enhancing a decay process determine its contribution to the observed effect. If field-dependent branching pathways to the ground state exist, at infinite field one expects 100% quantum yield failure, even for an isotropic sample. (ii) Similar effects are obtained in glycerol/buffer glasses for field strengths where the effects can be measured (less than 1 MV/cm). Although the experimental limitation of applying very large fields to glasses precludes comparisons at the highest fields, it appears that the effects are not affected appreciably by the medium. (iii) The effects on the P^+H^-, $P^+Q_A^-$ and ^3P states follow the same variation with applied field to within the experimental error. Examination of the reaction scheme (Figure 1) suggests that some common loss channel must exist en route to P^+H^- as we detect it at 10 ns. Even though our measurement of P^+H^- is at 10 ns, because k_Q is about $(100\ ps)^{-1}$ and the $P^+Q_A^-$ quantum yield failure parallels the other states, it appears that the loss channel is earlier than 100 ps.

The most obvious interpretation of this data is that the rate of the initial charge separation step and/or some non-radiative process within $^*P$ are affected by the applied field. Assuming an intrinsic excited state lifetime of several hundred ps in the absence of electron transfer (obtained from stimulated emission and absorption measurements on the D_{LL} mutant [26]), a loss of quantum yield of about 15% at 1 MV/cm would lead to a very large increase in the fluorescence quantum yield, much larger than the observed 40% increase [4,5]. The measurement of ps transient absorption is less definitive because the data to date are not of such high signal-to-noise [6]. It was very difficult to measure any change at all on the ps timescale, and although this is consistent with the observed effect on the fluorescence quantum yield, it is not possible to rule out a change in integrated quantum yield which might be compatible with the quantum yield failure data.

In order to obtain further information, an experiment analogous to the electric field induced fluorescence quantum yield anisotropy experiment [5] can be performed on the population which fails to produce intermediates in a field. The idea is that some change in dipole(s) is responsible for the sensitivity of the quantum yield to the applied electric field. Because this dipole has an orientation in the molecular axis system (as does the absorption transition moment used to probe the quantum yield failure), the population which returns to the ground state by the failure route will not be isotropic with respect to the applied field. Experimental details and methods of analysis are presented in detail elsewhere [28]. The angle dependent information can be obtained either by measuring the angle dependence of the transient absorption or by analyzing the lineshape of the absorption of the recovered P in an applied field. The latter is known as the dynamic Stark effect, and its analysis is similar to that published earlier for fluorescence lineshapes [27]. The Stark effect has a well known dependence on orientation in an applied field. The Stark spectrum for a completely isotropic sample is obtained before photoexcitation, followed by the Stark effect spectrum of the transient, quantum yield failed population of P. Irrespective of how the experiment is performed, the data are consistent with a dominant process involving an angle greater than the magic angle (54.7^o) between the P870 absorption transition dipole moment and the dipole moment which is effective in causing quantum yield failure.

We can consider the three most obvious choices for dipoles which are responsible for quantum yield failure. (i) The P^+P^- internal charge-transfer dipole within $^*P$, whose angle can be estimated from the absorption Stark effect [30,31] to be about 38° relative to the P870 transition dipole moment (note: this is an experimentally observed value, not a value deduced from the x-ray structure and assumptions about the charge distributions; however, it is possible that other internal dipoles are more effective in causing non-radiative decay). (ii) The P^+B^- dipole whose angle can be estimated to be less than the magic angle relative to the P870 transition moment for most reasonable guesses of the charge distributions in P^+ and B^-. (iii) The P^+H^- dipole whose angle is estimated to be greater than the magic angle relative to the P870 transition moment for most reasonable guesses of the charge distributions in P^+ and H^-. The results to date are not consistent with (i), and likely not with (ii) as the dominant process. We stress that as the field is increased, more than one process may contribute. The experimental situation can be somewhat improved by photoselective excitation of the RC, and this is currently in progress.

We note finally that this experiment has several advantages over the analogous fluorescence experiment. (i) The absorption transition dipole moment direction is known quite accurately [32]. Although the fluorescence transition moment has been shown to be nearly parallel to the absorption moment, this involves an additional experiment with additional and compounding sources of error. (ii) A transient absorption measurement is sensitive to the bulk of the sample, therefore contributions from small contaminating populations which are always a threat with fluorescence measurements are avoided. (iii) The interpretation of the field effect on the fluorescence requires information on $\mu(^*P)$, whereas the quantum yield failure experiment depends on the ground state dipole moment $\mu(P)$. It is likely that $\mu(^*P) > \mu(P)$, therefore neglect of this dipole in calculating the needed difference for the process is probably a better approximation. It is important to note that with the time resolution currently available, we are only able to evaluate the direction of the dipole moment responsible for the quantum yield failure on the ns timescale, and so cannot make direct comparisons with the data on the prompt fluorescence. It is quite possible that different states or pathways affect the quantum yields of fluorescence and various intermediates.

Acknowledgments This work is supported in part by the NSF Biophysics Program. R. Stanley is the recipient of an NIH National Research Service Award; M. Steffen is an MSTP trainee supported by NIGMS grant GM07365; J. Stocker is the recipient of an NSF Predoctoral Fellowship.

REFERENCES

1. Boxer, S. G., Goldstein, R. A., Lockhart, D. J., Middendorf, T. R., and Takiff, L. (1989) J. Phys. Chem. 93, 8280.

2. Gopher, A., Schonfeld, M., Okamura, M.Y., Feher, G. (1985) Biophys. J. 48, 311.
3. Feher, G.; Arno, T. R.; and Okamura, M. Y. in The Photosynthetic Bacterial Reaction Center: Structure and Dynamics Breton, J.; Vermeglio, A., Eds.; Plenum Press, New York and London, 1988, pp. 271-287
4. Lockhart, D.J. and Boxer, S.G. (1988) Chem. Phys. Lett. 144, 243.
5. Lockhart, D.J., Goldstein, R.F., Boxer, S.G. (1988) J. Chem. Phys. 88, 1408.
6. Lockhart, D. J., Kirmaier, C., Holten, D., Boxer, S.G. (1990) J. Phys. Chem., 94, 6987.
7. Franzen, S.; Goldstein, R. F.; Boxer, S. G. J. Phys. Chem. 1990, 94, 5135.
8. Franzen, S.; K.-Q. Lao; Boxer, S. G. Chem. Phys. Lett., in press.
9. Boxer, S.G., Stocker, J., Franzen, S., Salafsky, J. (1992) in Molecular Electronics Science and Technology, Ari Aviram, ed., Am. Inst. of Physics Conf. Proc., American Institute of Physics, N.Y., 262, p.236.
10. Vos, M.H., Lambry, J-C., Robles, S.J., Youvan, D.C., Breton, J. and Martin J-L. (1991) Proc. Natl. Acad. Sci. 88, 8885.
11. Kirmaier, C.; Holten D. Proc. Natl. Acad. Sci. U.S.A. 1990, 87, 3552.
12. Holzapfel, W., Finkele, U., Kaiser, W., Oesterhelt, D., Scheer, H., Stilz, H. U., and Zinth, W. (1990) Proc. Natl. Acad. Sci. USA, 87, 5168.
13. Deisenhofer, J.; Epp, O.; Miki, K.; Huber, R.; Michel, H. (1984) J. Mol. Biol. 180,385.
14. Allen, J. P.; Feher, G.; Yeates, T. O.; Komiya, H.; Rees, D. C. (1987) Proc. Natl. Acad. Sci. 84, 6162.
15. Chang, C.-H.; Tiede, D.; Tang, J.; Smith, U.; Norris, J.; Schiffer, M. (1986) FEBS Lett. 205, 82.
16. Ogrodnik, A., Eberl, U., Heckmann, R., Kappl, M., Feick, R. and Michel-Beyerle (1991) J. Phys. Chem. 95, 2036.
17. Kirmaier, C., Holten, D., Bylina, E.J., and Youvan, D. C. (1988) Proc. Natl. Acad. Sci. U.S.A. 85, 7562.
18. Schenk, C. C., Gaul, D. F., Steffen, M. A., Boxer, S. G., McDowell, L., Kirmaier, C, and Holten, D. (1990) in Reaction Centers of Photosynthetic Bacteria (Michel-Beyerle, M. -E., ed.) p.229, Springer-Verlag, Berlin
19. Bixon, M.; Jortner, J. J. Phys. Chem. 1986, 90, 3795.

20. Woodbury, N.W. and Parson, W.W. (1984) Biochim. Biophys. Acta, 767, 345; Woodbury, N.W. and Parson, W.W. (1986) Biochim. Biophys. Acta, 850, 197.

21. Goldstein, R.A., Boxer, S.G. (1989) Biochim. Biophys. Acta, 977, 78.

22. Hammes, S., Mazzola, L., Boxer, S.G., Gaul, D., Schenck, C. (1990) Proc. Natl. Acad. Sci., 87, 5682.

23. Steffen, M. and Boxer, S.G., to be published.

24. Stocker, J.; Steffen, M.; Boxer, S.G., submitted for publication.

25. Robles, S.J., Breton, J., and Youvan, D.C. (1990) Science 248, 1402

26. Breton, J., Martin, J.-L., Lambry, J.-C., Robles, S.J., and Youvan, D.C. (1990) in Reaction Centers of Photosynthetic Bacteria (Michel-Beyerle, M. -E., ed.) p. 293, Springer-Verlag, Berlin.

27. Lockhart, D.J., Hammes, S.L., Franzen, S., Boxer, S.G. (1991) J. Phys. Chem., 95, 2217.

28. Lao K.-Q.; Franzen, S.; Lambright, D. G.; Boxer S. G. J. Phys. Chem., submitted

29. Franzen, S., Lao, K-Q., Stanley, R., Boxer, S.G. (1992) Biophys. J. 61, A153.

30. Lockhart, D. J.; Boxer, S. G. Biochemistry 1987, 26, 644.

31. Lösche, M.; Feher, G.; Okamura, M. Y.(1987) Proc. Natl. Acad. Sci., 84, 7537.

32. Zinth, W.; Sander, M.; Dobler, J.; Kaiser, W. in Antennas and Reaction Centers of Photosynthetic Bacteria 1985, Springer-Verlag Series in Chemical Physics Vol. 42, p.97-102

RADICAL PAIR DYNAMICS IN THE BACTERIAL PHOTOSYNTHETIC REACTION CENTER

M. Bixon,[1] Joshua Jortner,[1] and M.E. Michel-Beyerle[2]

[1]School of Chemistry, Tel Aviv University
Ramat Aviv, 69978 Tel-Aviv, Israel

[2]Institut für Physikalische Chemie, Technische Universität München
Lichtenbergstrasse 4, D-8046 Garching b. München, Germany

I. INTRODUCTION

The electron-hole recombination processes in the ion radical pairs P^+B^-H and P^+BH^- across the A branch of the bacterial photosynthetic reaction center (RC) are taking place in the same molecular framework in which the primary charge separation occurs. Consequently, the rate constants for the recombinations are correlated to the rate constants of the primary processes, and their investigation can help in the elucidation of the primary charge separation mechanism.

The primary charge separation processes result in the formation of the P^+BH^- radical pair in the singlet state. Interconversion into the triplet state is induced by hyperfine interactions. Both singlet and triplet radical pair states can undergo recombination, the triplet recombines to the triplet state $^3P^*BH$ and the singlet recombines to the ground state P, with the rate constants k_T and k_s respectively.[1-5] Provided that local configurational relaxations are negligible, the relative geometrical structure of the molecules and ions of the prosthetic groups involved remains basically the same for the primary and for the recombination processes. Therefore, direct intermolecular electron couplings, which depend basically on intermolecular distances and relative orientations, are expected to be similar in all these processes. The observed differences in the rate constants for the forward and back electron transfer are due mainly to the different energetics[7,8] (manifested in the Franck-Condon factors). As the interpretation of the radical pair dynamics should be consistent with the description of the primary dynamics,[6] recombination dynamics provides circumstantial evidence concerning the primary process.

II. THEORETICAL BACKGROUND

The conventional expression for the nonadiabatic ET rate is given by:

$$k = \frac{2\pi}{\hbar}V^2F \qquad \text{(II.1)}$$

where V is the electronic coupling matrix element between the initial and final state and F is the thermal averaged Franck-Condon factor.

As is well known[9-11] the classical Marcus expression for the high temperature Franck-Condon factor

The Photosynthetic Bacterial Reaction Center II
Edited by J. Breton and A. Verméglio, Plenum Press, New York, 1992

$$F = (4\pi\lambda k_B T)^{-1/2} \exp\left\{-\frac{(\Delta G + \lambda)^2}{4\lambda k_B T}\right\} \tag{II.2}$$

should be modified in the inverted region, where $-\Delta G >> \lambda$. This modification comes through because the available excess energy can be transformed into intramolecular vibrational excitations of the products. The excitations involve the high frequency intramolecular vibrational modes which are modified by the ET process, and therefore have nonvanishing Franck-Condon factors. In principle, one has to take into account all the relevant internal modes with their specific frequencies and overlap integrals. In practice it is impossible to obtain such detailed information, and therefore one employs a one mode model with an effective high frequency ω_c and (reduced) shift S_c between the initial and final states. In this approximation the averaged FC factor takes the following form:[9-11]

$$F = (4\pi\lambda k_B T)^{-1/2} e^{-S_c} \sum_{n=0}^{\infty} \frac{S_c^n}{n!} \exp\left\{-\frac{(\Delta G + \lambda - n\hbar\omega_c)^2}{4\lambda k_B T}\right\} \quad . \tag{II.3}$$

The medium modes which are responsible for the reorganization energy are assumed to have low frequencies ω_m, and are treated in the high temperature approximation, that is $\hbar\omega_m << k_B T$. At sufficiently low temperatures, when that condition breaks down, one has to take into account the quantum mechanical nature of the medium modes, and modify the expressions accordingly.

The formal expressions for the rate constants involve several parameters for which there is no precise information, so that their values are obtained by a self-consistent fitting to the experimental data. The treatment rests on the following approximations.

(1) The electronic coupling depends only on the relative geometry of the molecules involved, not on the specific electronic states. The electronic coupling V_{PB} for $^1P^*BH/^1(P^+B^-H)$ is the same as the coupling PBH/P^+B^-H while the coupling V_{BH} for $^1(P^+B^-H)/^1(P^+BH^-)$ is the same as $^3(P^+B^-H)/^3(P^+BH^-)$.

(2) No protein medium and prosthetic groups relaxation effects on the couplings and the energies. $\Delta G(^3P^* - P^+BH^-)$ measured from radical pair dynamics on the ns time scale is taken to have the same value in the primary process, which occurs on the ps time scale. The same assumption is also invoked for the couplings. The value for the electronic matrix element V_{PB} which is used in the superexchange expression for the radical recombination rates (on a time scale of ~10ns) is the same as the one used for the charge separation step (on a time scale of ~3 ps).

(3) High-frequency intramolecular modes and shifts are represented by a single effective frequency.

III. RECOMBINATION OF THE P^+B^-H RADICAL PAIR

To a first approximation the electronic couplings and reorganization energies are taken to depend mainly on the relative geometry of the donor-acceptor system and not on the specific electronic states. Under this assumption the electronic coupling V_{PB} for the radical pair $P^+B^-H \rightarrow PBH$ recombination rate k_R to the ground state is equal to the coupling in the forward reaction $^1P^*BH \rightarrow P^+B^-H$ with the rate k_1. The forward and the recombination rates are, of course, different because of the very large difference between the respective free energies of reaction. The forward free energy of reaction is in the range[6] $\Delta G_1 = -100$ - -600 cm^{-1}, whereas the free energy for k_R is about -10500 cm^{-1}, placing the recombination reaction deep in the inverted region. Accordingly, the ratio k_R/k_1 of the recombination rate to the forward rate is equal to the ratio of the corresponding Franck-Condon factors. This ratio depends on the free energies of reaction, as well as on several parameters which can be assumed to have the same values for the two reactions. (a) The reorganization energy λ_{PB} for the charge transfer between the close neighbors P and B. From the analysis of the primary rates[6] one deduces a value of $\lambda_{PB} \simeq 800 - 1000$ cm^{-1}. (b) The frequency and shift of an effective internal high frequency mode. On the basis of the analysis of the recombination rates in the radical pair P^+BH^- (section IV) we take the values $\hbar\omega_c \simeq 1500$ cm^{-1} and $S_c \simeq 1$.

The calculated ratio between the Franck-Condon factors and between the rates is $k_R/k_1 = 300 - 500$. With a forward time constant of k_1 = 3ps we estimate a recombination rate $k_R \simeq$ 1ns.

The recombination of the radical pair P^+B^-H is too slow to compete effectively with the direct ET to the bacteriopheophytin, and therefore this process has no influence on the primary charge separation kinetics and its quantum yield. It may become an important decay channel in some mutants. This analysis is consistent with the sequential electron transfer mechanism in the mutant (M)L214H.[6]

IV. RECOMBINATION OF THE P^+BH^- RADICAL PAIR IN THE SINGLET AND TRIPLET CHANNELS

The two recombination reactions of P^+BH^- (to the triplet state and the ground singlet state) are vastly different in their energetics. The triplet recombination has a free energy gap $\Delta G \simeq -1300 cm^{-1}$, being nearly activationless, whereas the recombination to the ground state has $\Delta G \simeq -9000 cm^{-1}$,[7-8] which places it deep in the inverted region. On the other hand, the two reactions have nearly the same initial and final charge distributions, and therefore should have the same value for the medium reorganization energy λ. The weak temperature dependence of k_T, together with information about the primary processes, brackets its value to the range $1500cm^{-1} < \lambda < 2000cm^{-1}$.

The detailed experimental investigations[2-5,12,13] of the dynamics of the P^+BH^- radical pair provide us with information regarding the recombination rates in the triplet and singlet channels, and the exchange interaction J, i.e., the energy difference between the triplet and singlet states of P^+BH^-. In the triplet channel the free energy of recombination is $\Delta G_T \simeq -1300 cm^{-1}$ and the rate k_T is temperature independent ($k_T \simeq 5 \cdot 10^8 s^{-1}$).[12,13] The singlet recombination into the ground state has a free energy of $-9500 cm^{-1}$ (being deep in the inverted region), and is slightly activated ($k_s \simeq 2 - 4 \cdot 10^7 s^{-1}$ at T = 300K).[12,13]

The coupling responsible for the recombination has both direct V^{direct} and superexchange V^{super} contributions, $V^{eff} = V^{direct} + V^{super}$. The superexchange contribution for the recombination is given by the expression:

$$V^{super} \simeq \frac{V_{PB}V_{BH}}{\delta E} \simeq \frac{\alpha V_{PB}{}^2}{\Delta G_{BH} + \lambda_{BH}} \tag{IV.1}$$

where $\alpha = V_{BH}/V_{PB}$, ΔG_{BH} is the energy gap between P^+BH^- and P^+B^-H, and λ_{BH} is the P^+BH^-/P^+B^-H reorganization energy. Employing the parameters used for the sequential description of the primary process[6] V_{PB}=20-40 cm^{-1}, $\alpha \simeq 2$, $\Delta G_{BH} \simeq 1500 cm^{-1}$ and $\lambda_{PB} \simeq 1000 cm^{-1}$, we estimate

$$V^{super} = 0.4cm^{-1} - 1.6cm^{-1} \quad . \tag{IV.2}$$

On the other hand, if one assumes the unistep superexchange model for the primary process, the required basic electronic couplings are considerably larger, i.e., $V_{PB} \simeq 75 cm^{-1}$, $\alpha \simeq 3$, while $\Delta G_{PB} = 2000 cm^{-1}$, we estimate

$$V^{super} \simeq 5 cm^{-1} \quad . \tag{IV.3}$$

V^{direct} for P^+BH^-/PBH and for $^3P^+BH^-/^3P^*BH$ is not known. On the basis of the assumption that the coupling depends mainly on the relative geometry and not on the specific electronic states we have the same coupling for $k_T \propto V^2_T$ and $k_s \propto V^2_s$, with $V_T \approx V_s \approx V^{eff}$.

The parameters required for the description of the recombination reactions can be extracted by directly analyzing the corresponding experimental data.[12,13] The recombination in the triplet channel appears to be weakly temperature dependent with a rate of $k_T \simeq (2ns)^{-1}$. Assuming the process to be activationless one obtains from Eqs. (II.1) and (II.2) an electronic coupling $V_T \simeq 1cm^{-1}$. From this analysis it appears that V_T extracted from the experimental recombination data is consistent with the sequential description of the primary process (Eq. IV.2), rather than with the unistep model (Eq. IV.3). Assuming that the coupling in the singlet and the triplet channels are similar, one has to blame the Franck-Condon factors for the difference between the rates k_s and k_T ($k_T/k_s \simeq 10-20$).

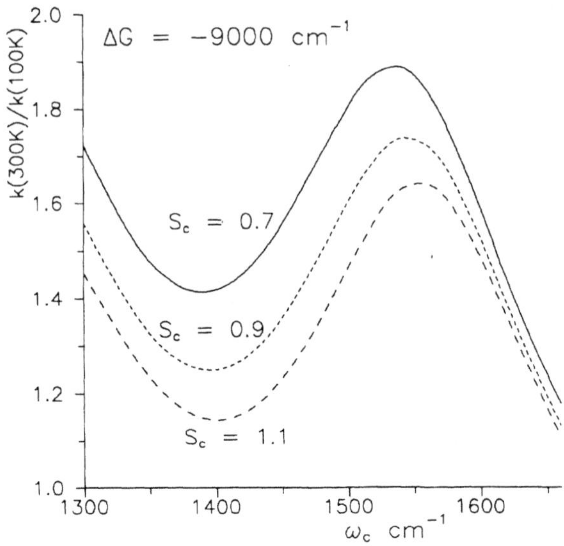

Figure 1. The dependence of the ratio of the ET rate at T = 300K and at T = 100K (for $\Delta G = -9000\ cm^{-1}$) on the frequency ω_c and the shift S_c for the high-frequency mode. Note the weak temperature dependence for a strongly exothermic ET in the inverted region.

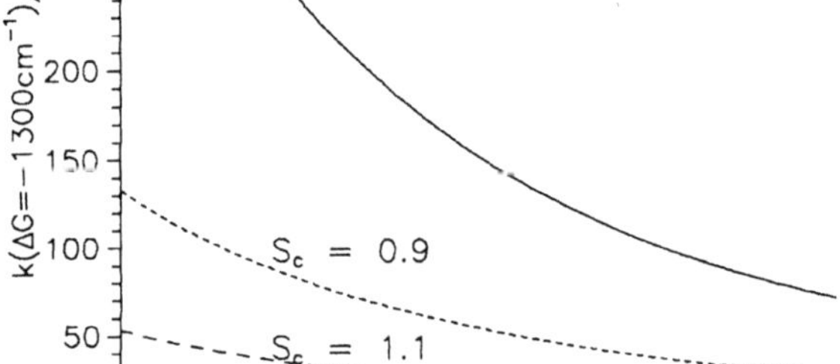

Figure 2. The dependence of the ratio of the ET rates for $\Delta G = -1300\ cm^{-1}$ (pseudoactivationless) and $\Delta G = -9000\ cm^{-1}$ (inverted region) on the frequency and shift for the high-frequency mode.

The singlet recombination is slightly activated with a rate constant $k_s \simeq (25ns)^{-1}$ at 285K and $k_s \simeq (50ns)^{-1}$ at 120K. The free energy of this reaction is about $-9000\ cm^{-1}$ with a reorganization energy of $1600\text{-}1800 cm^{-1}$. Numerical calculations using Eq. (II.3) (Figs. 1 and 2) show that it is possible to reproduce the observed slight activation and the ratio of ~20 between the triplet and singlet rates, by using an averaged intramolecular frequency of $\hbar\omega_c = 1500\ cm^{-1}$ and a shift of $S_c \simeq 1$.

V. THE EXCHANGE INTEGRAL OF THE RADICAL PAIR P^+BH^-

The exchange integral J is another observable of the P^+BH^- radical pair, which should be accounted for by the model in a self-consistent way. J is defined as the energy gap between the singlet and triplet states. It is experimentally determined[2-5,12,13] to be $J \simeq 1\cdot10^{-3} cm^{-1}$, being nearly temperature independent. The deviation from degeneracy (J=0) results from the interactions with other electronic states, which means that one has to take into account corrections to the Born-Oppenheimer approximation. J can be represented by $J = \delta E_s - \delta E_T$ where the singlet (δE_s) and the triplet (δE_T) shifts are due mainly to interactions with the close-lying excited electronic states $^1P^*BH$ and $^3P^*BH$. In the radical pair P^+B^-H the singlet and triplet states are nearly degenerate, therefore its interaction with P^+BH^- will not result in a contribution to J. On the other hand, the state P^+B^-H may play the role of a superexchange mediator for the interaction with $^1P^*BH$.

In general, one has to treat the interactions between the vibronic states of P^+BH^-, and the manifold of multimode vibrational states corresponding to each of the electronic singlet ($^1P^*$) and triplet ($^3P^*$) states. At low temperatures only the low vibrational states of P^+BH^- are populated and the radical pair dynamics correspond only to these states. Accordingly, we shall focus on the exchange integral for the ground vibrational state of P^+BH^-.

The calculation of δE_s can be performed by employing the simple first-order perturbation theory, because the energies of the vibronic manifold of $^1P^*BH$ and of P^+B^-H are much above those of P^+BH^- (i.e., typical energies are $\langle\Delta E\rangle \simeq 2000 + \lambda \simeq 3500cm^{-1}$). In general, a first order perturbation theory, combined with the partitioning method, results in the following expression for the energy shift:

$$\delta E = \sum_i \delta E(i) \tag{V.1}$$

$$\delta E(i) = -\frac{1}{E_v(i)-E(P^+BH^-)}\left[V_{i,P^+H^-} + \frac{V_{i,P^+B^-}V_{BH}}{E_v(P^+B^-H) - E(P^+BH^-)}\right]^2 , \tag{V.2}$$

where the couplings and energies (E_v denotes a vertical energy) are calculated in the equilibrium configuration of the P^+BH^- state with energy $E(P^+BH^-)$. The main contribution for the singlet shift is from the state $^1P^*$,

$$\delta E_s(^1P^*) = -\frac{1}{E_v(^1P^*BH)-E(P^+BH^-)}\left[V_{PH} + \frac{V_{PB}V_{BH}}{E_v(P^+B^-H) - E(P^+BH^-)}\right]^2 \tag{V.3}$$

The combination

$$\bar{V}_{PH} = \left[V_{PH} + \frac{V_{PB}V_{BH}}{E_v(P^+BH^-) - E(PB^+H^-)}\right] \tag{V.4}$$

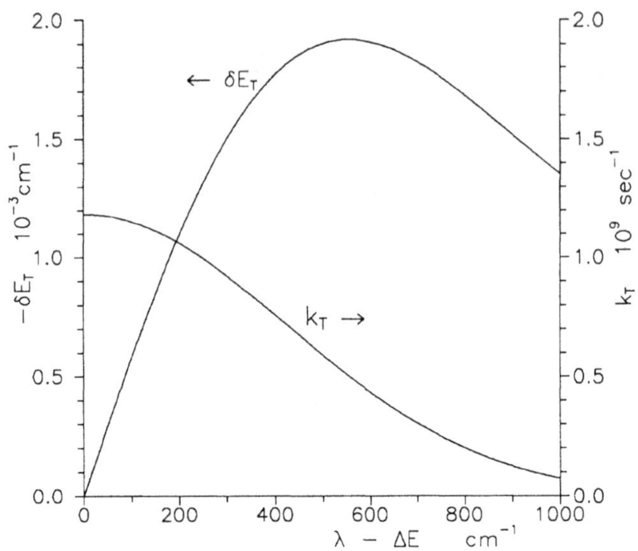

Figure 3. The dependence of the triplet shift δE_T and the triplet recombination rate k_T of the radical pair on λ-ΔE_T, where ΔE_T is the energy gap between $^3P^*BH$ and $^3(P^+BH^-)$, while λ is the reorganization energy between these potential surfaces.

gives the effective coupling (combination of direct and superexchange) between $^1P^*BH$ and P^+BH^-, in the equilibrium configuration of P^+BH^-. It is apparent from Eqs. (IV.1) and (IV.2) and the analysis in section IV that the effective coupling can be identified with V_s and its value is $\bar{V}_{PH} \simeq 1 cm^{-1}$. Thus the singlet shift is approximated as:

$$\delta E_s \simeq \frac{\bar{V}^2_{PH}}{\langle \Delta E \rangle} \simeq \frac{1}{3500} \simeq 0.3 \cdot 10^{-3}\ cm^{-1} \quad . \tag{V.5}$$

It is very dangerous to use the first order perturbation theory for the triplet shift. The ground vibrational state of $^3P^*BH$ is lower by $\simeq$1300cm^{-1} than that of P^+BH^-. Therefore, the vibrational ground state of $^3(P^+BH^-)$ interacts with a quasi-continuous manifold of the $^3P^*BH$ vibronic states. A model calculation of the relevant vibronic interactions is performed by considering a multimode harmonic system characterized by a set of frequencies $\{\hbar\omega_i\}$ with dimensionless shifts $\{s_i\}$. Invoking the Condon approximation we evaluate the coupling of the P^+BH^- vibrational ground state with a vibronic state of $^3P^*BH$ to be the product of a constant electronic matrix element V_T and a vibrational overlap integral. The energy dependent vibronic interaction is characterized by the sum of the couplings squared per unit energy interval of the vibronic manifold. As the electronic coupling matrix element is assumed to be the same for all vibronic states, one has to evaluate the energy dependence of SFC[E], the sum of Franck-Condon factors per unit energy interval. A good approximation for this quantity is given by a Gaussian distribution centered on the reorganization energy, $E_{max} = \lambda$, and a second moment $\mu_2 = (\lambda \cdot \hbar\omega_{eff})^{1/2}$, where ω_{eff} is an effective average frequency of the low frequency medium modes which contribute to the reorganization energy.

$$SFC[E] = (2\pi\lambda\hbar\omega_{eff})^{-1/2} \exp\left\{ -\frac{(E-\lambda)^2}{2\lambda\hbar\omega_{eff}} \right\} \quad . \tag{V.6}$$

According to the standard theory[15,16] a state interacting with a continuum (or quasi-continuum) acquires a shift δE and a width Γ, which is related to the rate of transition into the quasi-continuum by $k=\Gamma/\hbar$. The width acquired by $^3(PB^+H^-)$ is proportional to the recombination rate in the triplet channel at low temperature (k_T) and is given by:

$$\Gamma = \hbar k_T = 2\pi\, V^2{}_T \cdot SFC[\Delta E] \quad , \tag{V.7}$$

where SFC[ΔE] is the sum of FC factors per unit energy interval of the $^3P^*BH$ vibronic manifold, at the ground state energy of $^3P^+BH^-$ (ΔE). The shift δE_T is given by the principal value integral:

$$\delta E_T = P \int_0^\infty \frac{V^2{}_T \cdot SFC[E]}{E - \Delta E}\, dE \quad . \tag{V.8}$$

Both k_T and the triplet shift δE_T emerge from the same formalism and calculations and therefore they are correlated. Fig. 3 presents k_T and δE_T as function of the difference between the reorganization energy and the energy gap. This calculation used the following parameters: $V_T=1cm^{-1}$, $\lambda=1600cm^{-1}$ and $\hbar\omega_{eff}=100cm^{-1}$. Except close to the origin, the contribution of δE_T to the exchange integral is much larger than that of its singlet counterpart. With the parameters that are used in Fig. 3, both k_T and J are compatible with the experimental data. From the analysis we also obtain that $|\delta E_T| > |\delta E_s|$, while $\delta E_T < 0$. Accordingly, the triplet state is located below the singlet, so that $J > 0$.

VI. CONCLUSIONS

The electronic coupling matrix element, which emerges from the analysis of the three observables k_s, k_T, J is $V_T \simeq V_s \simeq 1\ cm^{-1}$, being considerably smaller than the coupling estimated from superexchange interaction in the unistep primary model ($V \simeq 5cm^{-1}$ according to Eq. IV.3). This result sets limits $V_{PB} \leq 40\ cm^{-1}$ and $V_{BH} \leq 80\ cm^{-1}$ for the electronic interactions between the prosthetic groups. These moderate electronic coupling terms favor the sequential mechanism (at $T \simeq 300K$) for the primary charge separation in the photosynthetic RC. Of course, this conclusion rests on the assumption of the absence of marked configurational relaxation of the protein and of the prosthetic groups upon electron transfer. It is interesting to note that while the singlet and the triplet recombination channels of P^+BH^- include a major contribution of superexchange interaction via the physically mediating P^+B^-H state, the primary charge separation (at least at room temperature) is dominated by sequential electron transfer via the chemically mediated P^+B^-H intermediate.

REFERENCES

1. W.W. Parson and R.J. Codgell, *Biochim. Biophys. Acta* 416:105-149 (1975).
2. A. Ogrodnik, H.W. Krüger, H. Orthuber, R. Haberkorn, M.E. Michel-Beyerle, and H. Scheer, *Biophys. J.* 39:91-99 (1982).
3. J.R. Norris, M.K. Bowman, D.E. Budil, J. Tang, C.A. Wraight, and G.L. Closs *Proc. Natl. Acad. Sci. USA* 79:5532-5536 (1982).
4. C.E.D. Chidsey, C. Kirmaier, D. Holten, and S.G. Boxer, *Biochim. Biophys. Acta* 766:424-437 (1984).
5. A.J. Hoff, *Photochem. Photobiol.* 43:727-745 (1986).
6. M. Bixon, J. Jortner, and M.E. Michel-Beyerle (this volume).
7. A. Ogrodnik, M. Volk, R. Letterer, R. Feick, and M.E. Michel-Beyerle, *Biochim. Biophys. Acta* 936:361-371 (1988).
8. R.A. Goldstein, L. Takiff, and S.G. Boxer, *Biochim. Biophys. Acta* 934:253-263 (1988).
9. S. Efrima and M. Bixon, *Chem. Phys.* 13:447-460 (1976).
10. J. Jortner, *J. Chem. Phys.* 64:4860 (1976).
11. M. Bixon and J.Jortner, *J. Phys. Chem.* 95:1941-1944 (1991).

12. A. Ogrodnik, N. Remy-Richter, M.E. Michel-Beyerle, and R. Feick, *Chem. Phys. Lett.* 135:576-581 (1987).
13. V. Volk, Ph.D. thesis, TU München (1991).
14. M. Bixon, M.E. Michel-Beyerle, and J. Jortner, *Israel J. Chem* 28:155-168 (1988).
15. M. Bixon and J. Jortner, *J. Chem. Phys.* 48:715 (1968).
16. W.M. Gelbart and J. Jortner, *J. Chem. Phys.* 54:2070 (1971).

THE PRIMARY CHARGE SEPARATION IN BACTERIAL PHOTOSYNTHESIS. WHAT IS NEW?

M. Bixon,[1] Joshua Jortner,[1] and M.E. Michel-Beyerle[2]

[1]School of Chemistry, Tel Aviv University
Ramat Aviv, Tel-Aviv 69978, Israel

[2]Institut für Physikalische Chemie, Technische Universität München
Lichtenbergstrasse 4
D-8046 Garching, b. München, Germany

I. PROLOGUE

The basic structure-function relation in photobiology, which pertains to the mechanism of the primary charge separation in photosynthetic reaction centers (RC), is not yet elucidated, although the first structure of the bacterial photosynthetic RC was determined eight years ago. The major difficulty involved in the understanding of this central energy conversion process in biology is that it requires information on energetics, electronic interactions and nuclear dynamics in electronically excited states, which cannot be readily inferred from the structural data in the ground electronic state. This state of affairs reflects on the intrinsic limitations of structure to infer an excited state dynamics.

The primary charge separation from $^1P^*$ to H in photosynthetic bacteria and some of their mutants is characterized by:[1,2] (i) An ultrafast rate in the psec time domain. (ii) Non-Arrhenius weak temperature dependence from room temperature down to 4K. (iii) Unidirectionality across the A branch of the RC. The ultrafast primary charge separation from $^1P^*$ to H over a center-to-center distance of 17Å is too fast to be accounted for by direct exchange, and the role of physical and/or chemical mediation of electron transfer (ET) has to be invoked. An obvious candidate for physical or/and chemical mediation is the accessory bacteriochlorophyl (B) which is in close contact with P and H.

The energy of the ion pair state P^+B^- is expected to be close to $^1P^*$, however, our ignorance of the detailed energetics (and other dynamic parameters) precludes an a-priori theoretical prediction of the primary ET mechanism and one has to rely on a phenomenological a-posteriori analysis of the experimental data in terms of ET theory. All mechanisms proposed for the primary ET attribute a special role to B, which involve: (I) A one-step superexchange mechanism.[3-13] (II) Two-step ET with P^+B^-H as a real chemical intermediate.[14-19] (III) The parallel sequential-superexchange mechanism,[20] which provides a unified scheme that includes both the unistep and sequential mechanism as limiting cases. Are these mechanisms relevant for the primary ET events in the RC, and what are the biophysical implications of the mechanism of the primary ET?

II. EXPERIMENTAL INFORMATION

The wealth of experimental information obtained using advanced techniques naturally separates into room temperature data, which are of direct biological relevance, and low (cryogenic)-temperature information, which is of central diagnostic value. The room temperature (T = 295-300K) data reveal that:

The Photosynthetic Bacterial Reaction Center II
Edited by J. Breton and A. Verméglio, Plenum Press, New York, 1992

(1) The lifetime of $^1P^*$, as interrogated by stimulated emission, involves a major contribution of τ_1 = 2.6-3.5 psec[21-28] for both *Rb.sphaeroides*[21-27] and *R.viridis*.[28]

(2) The kinetics of the ion pair P^+BH^- (and presumably also of P^+B^-H), which was monitored by time-resolved absorption, is characterized by two lifetimes which for *Rb.sphaeroides* are τ_1 = 3.5±0.4 psec and τ_2 = 1.1±0.4 psec according to Zinth et al.[24-26] This observation was confirmed by Fleming, Norris and their collaborators[27] who obtained τ_1 = 2.6±0.3 psec and τ_2 = 1.25±0.25 psec. For *R.viridis* τ_1 = 3.5±0.4 psec and τ_2 = 0.65±0.2 psec.[28]

(3) The kinetic data (1) and (2) are consistent with the sequential mechanism II with the branching ratio F_{seq} between the sequential and superexchange channels being $F_{seq} > 0.8$.[24-27]

(4) Sparse information indicates that τ_1 is independent on the initial excitation energy of $^1P^*$ (in the range 850-880 nm), presumably precluding vibronic dependence of τ_1.[21-27]

(5) Long-time tails were reported in the time evolution of $^1P^*$, as interrogated by stimulated emission and by fluorescence upconversion.[29-31] The decay of P^* in wild type (WT) RCs can be fit by a biexponential[29-31] with[29,30] τ_1 = 2.6 psec and $\tau_3 \simeq$ 12 psec, and with relative amplitudes of $A_3/A_1 \sim 0.15$. This biexponential pattern prevails also for some mutants with A_3/A_1 being larger.[29,30] The most straightforward (but not exclusive) interpretation involves structural heterogenuity resulting in a continuous or a bimodal distribution of electronic couplings and/or energy gap.

(6) Local mutagenesis of the Tyrosine M210, and of the phenylalanine L183, which are located in close proximity to P, B_A and H_A and to P, B_B and H_B, respectively, results in marked changes in τ_1.[29,32-34] Of considerable interest is the TyrM210 → phenylalanine mutant, which retards τ, with $\tau_1 \simeq$ 20 psec (single exponential fit)[29,32,33] or $\tau_1 \simeq$ 6 psec, $\tau_3 \simeq$ 30 psec and $A_3/A_1 \simeq 0.5$ (two-exponential fit),[29] apparently revealing the manifestation of energetic destabilization of the P^+B^-H state. Rather baffling is the local mutant phenylalanine L183 → tyrosine, where τ_1 is shortened relative to the WT RC.[33,34] It is still questionable whether this effect can be explained in terms of the energetics of the P^+B^-H state, or whether static configurational changes due to mutations have to be invoked. The global mutation of the protein strands which leads to the D_{LL} mutant of *Rb.capsulatus* with the absence of the H_A prosthetic group, does not reveal ET from $^1P^*$ to P^+B^- [35] (in apparent contradiction to (3)), which may be blamed on structural and/or energetic changes of P^+B^-.

The low temperature kinetic data are still controversial regarding both facts and interpretation.

(7) The lifetime of $^1P^*$, as interrogated by stimulated emission[21-23] and by hole burning,[36] is τ_1 = 1.2-1.4 psec at 10K.

(8) τ_1 constitutes the appearance time of P^+BH^-, which was monitored by time-resolved absorption.

(9) The width of the zero-phononon line of $^1P^*$, as determined by hole burning[36] at 10K corresponds to the lifetime τ_1. Accordingly, the primary ET occurs from the spectroscopic state.

(10) The time-resolved dynamics of the RC contains supershort (subpsec) and decay oscillating components.[26,37,38]

(11) Upon cooling of the RC of *Rb.sphaeroides* the lifetime τ_2 gradually shortens,[26] reaching the value of τ_2 = 300±150 fsec at 25K.[26] On the other hand, the value of τ_2 = 90 fsec was reported at 10K.[38]

(12) An oscillatory contribution[37,38] (with a maximal relative amplitude of ~ 0.02) is exhibited following excitation by a 60 fsec pulse at 10K and at 100K, with a period of 0.6 psec and a damping time of 0.8 psec (at 10K), being presumably due to vibrational coherence.

The interpretation of observations (10)-(12) is still a matter of controversy. Of considerable interest is the effect of dephasing of vibrational coherence on ET dynamics. Obviously, excitation of a nuclear wavepacket, which constitutes a coherent superposition of vibrational levels of $^1P^*$, will exhibit an oscillatory time evolution, reflecting the excitation conditions and the excited state level structure. However, the microscopic depletion rates of various vibronic levels of $^1P^*$ are not affected by coherence effects.

The nature of the electronically excited state involved in the primary ET can correspond to: (a) A spectroscopic vibronic state $^1P^*$. (b) A superposition of coherently excited vibronic $^1P^*$ states. (c) A lower vibronic state(s) of $^1P^*$ reached by vibrational excitation of (a) or (b). (d) A charge transfer $P^{\pm}$ state of the dimer reached by internal conversion from the spectroscopic state. Observation (12) supports options (a), (b) or (c). Accordingly, primary ET occurs from the spectroscopic state(s). The distinction between (a) and (b) pertains only to the optical preparation, while no distinction can be made between (a) and (c) due to observation (4). Finally, observation (9) excludes possibility (d). Thus, ET occurs from the spectroscopic $^1P^*$ state(s).

On the basis of the foregoing analysis we propose that the ET dynamics from a single vibronic state or a thermally equilibrated manifold of $^1P^*$ can be described in terms of a sequential-superexchange parallel kinetic scheme.

III. ULTRAFAST ACTIVATIONLESS ET RATES

The primary ET lifetimes τ_1 and τ_2 correspond to some of the fastest ET observed at room temperature and constitute the fastest ET times recorded at cryogenic temperatures. It is imperative to enquire whether such ultrafast lifetimes are amenable to theoretical description in terms of the conventional nonadiabatic multiphonon theory, which expresses ET rates in the familiar form

$$k = \left(\frac{2\pi}{\hbar}\right) V^2 F \tag{1}$$

where V is the electronic coupling and F the Franck-Condon factor. The discussion of the applicability of Eq. (1) for the description of the primary ET lifetimes should also rest on the empirical (experimental) observation that the relevant ET processes are activationless or pseudoactivationless. This assertion rests on the small (free) energy gaps for the relevant ET processes (i.e., the gap between $^1P^*BH$ and P^+BH^- being -2000 cm^{-1}), and on "intelligent guesses" of the medium reorganization energy, which are consistent with the weak temperature dependence of the primary ET rates. The conditions for the applicability of Eq. (1) are:

(1) The applicability of the nonadiabatic limit. The Landau-Zenner parameter $\gamma_{LZ} = 2\pi V^2/\hbar\omega_m(\lambda_m k_B T)^{1/2}$ is $\gamma_{LZ} < 1$, where $\omega_m \simeq 100$ cm^{-1} is the (mean) protein frequency and $\lambda_m \simeq 1000\text{-}2000$ cm^{-1} is the protein medium reorganization energy. For these characteristic values of the RC medium nonadiabatic ET prevails for $V < 70$ cm^{-1} at $T = 300K$. We shall subsequently demonstrate the validity of this relation, providing a-posteriori identification for the applicability of the nonadiabatic limit for the primary (and all other) ET processes in the RC.

(2) Insensitivity of the ET rates to medium dynamics. This state of affairs can be realized under two circumstances. (i) Fast medium vibrational dynamics. The times $\langle\tau\rangle$ of medium relaxation and excitation are fast on the time scale of the microscopic ET rates, with the ET process being rate determining. Molecular dynamics[39] simulations of the response of the RC to charge redistribution gives $\langle\tau\rangle \simeq 100$ fsec, which is temperature independent in the range 300K-10K, and implies that $k >> (100 \text{ fsec})^{-1}$. This strong condition is on the verge of its applicability for $\tau_2 = 300$ fsec at low temperatures. (ii) Activationless ET.[40] In this case the microscopic rates are insensitive to the details of the distribution of the initial vibrational states. This condition applies for the primary ET processes, which are not controlled by medium dynamics.

From this analysis we conclude that: (1) Eq. (1) is applicable for primary ET. (2) The upper limit for the magnitude of the activationless ET rate is given by the adiabatic rate $k \simeq \omega_m/2\pi$, which for $\omega_m \simeq 100$ cm^{-1} is $k^{-1} \simeq 300$ fsec. Thus the low-temperature $\tau_2 = 300\pm150$ fsec value reported by Zinth et al[26] is on the verge of the upper limit of the applicability of the conventional ET theory.

IV. THE PARALLEL PRIMARY ET

The phenomenological unified kinetic scheme for the primary ET is[20]

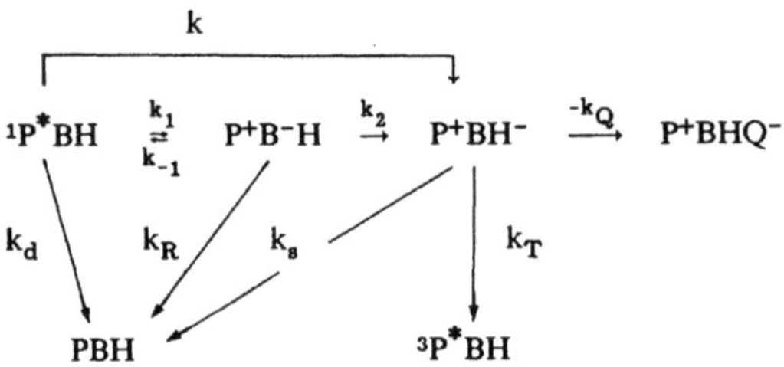

(2)

The four microscopic primary ET rate constants are:

The superexchange rate $k = \left(\frac{2\pi}{\hbar}\right) (V_{PB}V_{BH}/\delta E)^2 F(\lambda,\omega_m,S_c,\hbar\omega_c,\Delta G,T)$ (3)

The sequential rate $k_1 = \left(\frac{2\pi}{\hbar}\right)^2 V_{PB}^2\ F(\lambda_1,\hbar\omega_m,S_c,\hbar\omega_c,\Delta G_1,T)$ (4)

The reverse rate $k_{-1} = k_1 \exp(-\Delta G_1/k_B T)$ (5)

The ultrafast rate $k_2 = \left(\frac{2\pi}{\hbar}\right) V_{PB}^2\ F(\lambda_2,\hbar\omega_m,S_c,\hbar\omega_c,\Delta G_2,T)$ (6)

where V_{PB} and V_{BH} are the electronic coupling terms between $^1P^*BH/P^+B^-H$ and P^+B^-H/P^+BH^-, respectively and $\delta E = \Delta G_1+\lambda_1$ is the vertical energy difference between the potential surfaces of $^1P^*BH$ and P^+B^-H at the minimum of the former (Fig. 1). ΔG, ΔG_1 and ΔG_2 are the (free) energy gaps. The medium vibrational modes are characterized by the average frequency ω_m and reorganization energies λ, λ_1 and λ_2. The high-frequency intramolecular vibrations are represented by a single effective vibrational mode ω_c and an effective reduced shift S_c. The Franck-Condon factors are

$$F(\lambda,\hbar\omega_m,S_c,\hbar\omega_c,\Delta G,T) = \sum_{n=0}^{\infty} F_m(\Delta G-n\hbar\omega_c)F_c(n) \quad (7)$$

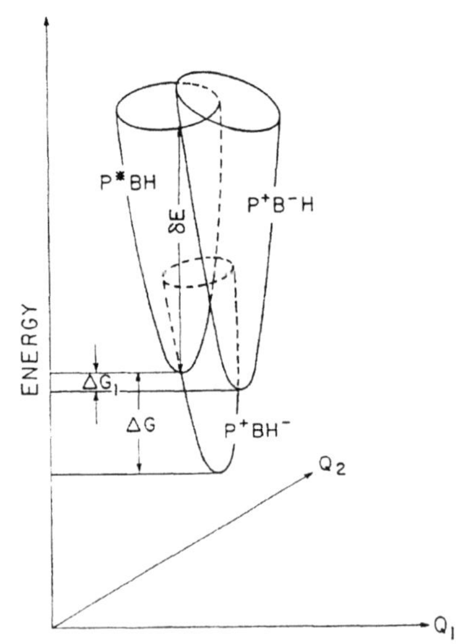

Figure 1. Schematic potential energy surfaces for primary ET in the RC.

with the medium contributions

$$F_m(\Delta E) = (\hbar\omega_m)^{-1} \left(\frac{\bar{v}+1}{\bar{v}}\right)^{p/2} \exp\left[-S_m(2\bar{v}+1)\right] I_p\left(2S_m[\bar{v}(\bar{v}+1)]\right)^{1/2} \tag{8}$$

where $\bar{v} = [\exp(\hbar\omega_m/k_BT)-1]^{-1}$ is the thermal population of the mode, $p = \Delta E/\hbar\omega_m$, $S_m = \lambda/\hbar\omega_m$ and $I_p(\cdot)$ is the modified Bessel function of order p. The intramolecular high-frequency contribution is

$$F_c(n) = \exp(-S_c)\, S_c^n/n! \tag{9}$$

The energy damping rates are: k_d = (300 psec)$^{-1}$ for the internal conversion of $^1P^*$, k_R = (600 psec)$^{-1}$ for the recombination of P^+B^-H, k_T = (2 nsec)$^{-1}$ and k_S = (20 nsec)$^{-1}$ for the triplet and singlet recombination of P^+BH^-.

The unified scheme, Eq. (2), should not be considered as an empirical scheme for the fit of experimental kinetic data. Rather, the lifetimes emerging from the kinetic analysis should be consistent with the electronic, energetic and nuclear parameters. In our analysis we have mapped the kinetic observables over a range of the physically 'acceptable' parameters. The following "reasonable" parameters were chosen for wild type RCs:

(1) Energetics. ΔG = 2000 cm^{-1}, with $\Delta G = \Delta G_1+\Delta G_2$. The energy ΔG_1, which essentially determines the mechanism of primary ET was varied in the range ΔG_1 = -700 cm^{-1} to 0.
(2) Vibrational frequencies. $\hbar\omega_m$ = 95 cm^{-1}, $\hbar\omega_c$ = 1500 cm^{-1}.
(3) Nuclear reorganization energies. λ = 1800 cm^{-1}, λ_1 = 800 cm^{-1} and λ_2 = 1200 cm^{-1} were taken assuming that k, k_1 and k_2 are nearly activationless. The intramolecular coupling in S_c = 1.
(4) Electronic couplings. The parameters $\bar{V}_{PB} = V_{PB}\exp(-S_c/2)$ and $\alpha = V_{BH}/V_{PB}$ were determined by requiring that the two lifetimes should be equal to the experimental values, i.e., τ_1 = 3.3 psec and τ_2 = 0.9 psec at 300K (Fig. 2).
(5) Temperature dependence. Lowering of the temperature causes an anisotropic contraction of the protein matrix, resulting in small configurational changes between the prosthetic groups, which go a far way in modifying the electronic couplings. To account for thermal contraction $\bar{V}_{PB}$ and α were chosen at 25K to fit the experimental data τ_1 = 1.4 psec and τ_2 = 0.3 psec (Fig. 2).

V. ANALYSIS OF THE PRIMARY ET IN WT RCc

We shall now consider the relevant observables emerging from the modelling, which are displayed vs the free energy gap ΔG_1 at several temperatures. The available kinetic information allows for a preliminary determination of energetic and electronic parameters.

(1) Two relaxation times for the primary ET. The modelling is consistent with the experimental observations from 300K[24-28] down to 4K.[26]
(2) The amplitude ratio A_1/A_2 in the decay of $[P^*] = A_1\exp(-t/\tau_1) + A_2\exp(-t/\tau_2)$ (Fig. 3). Experimentally the decay of P^* does not reveal a short time component,[24-28] i.e., $A_1/A_2 > 5$. Thus, according to Fig. 3, $\Delta G_1 \leq -100$ cm^{-1}.
(3) Analysis of the temperature independence of the magnetic data for the P^+BH^- radical pair[41] indicates that $\Delta G_1 \geq -600$ cm^{-1}.
(4) The electronic couplings. At T = 300K the fitted $\bar{V}_{PB}$ and α values vary in the range $\bar{V}$ = 17-32 cm^{-1} and α = 2.2-1.0, while at 25K $\bar{V}$ = 19-90 cm^{-1} and α = 2.4-1.2 (Fig. 2). The fitted α values are lower than the value α = 4 inferred from the theoretical intermolecular overlap approximation.[10] The magnetic data for the P^+BH^- radical pair[42] indicate that $\bar{V}_{PB} \leq 40$ cm^{-1}, which corresponds to $-700 \leq \Delta G_1 \leq -100$ cm^{-1} at 300K and -700 cm^{-1} $\leq \Delta G_1 \leq -300$ cm^{-1} at 25K (Fig. 2).
(5) Energy domain. From points (2) and (3) we conclude that in WT *Rb.sphaeroides* -600 cm^{-1} $\leq \Delta G_1 \leq -100$ cm^{-1}, spanning an uncertainty range of ~ 2 kcal mole^{-1}. Inhomogeneous broadening (e.g. δ ~ 300 cm^{-1} for the energy of $^1P^*$)[36] will smear the kinetic data over a finite ΔG_1 domain.

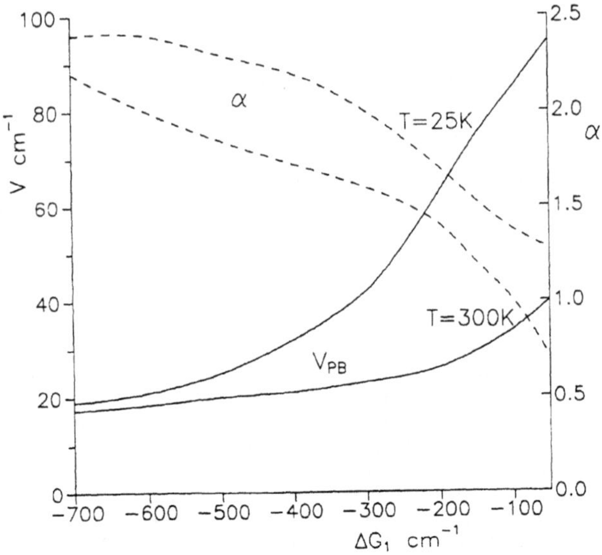

Figure 2. The fitted dependence of the electronic coupling $\bar{V}_{PB} = V_{PB}\exp(-S_c/2)$ and of the ratio $\alpha = V_{BH}/V_{PB}$ on the free energy gap at 300K and at 25K.

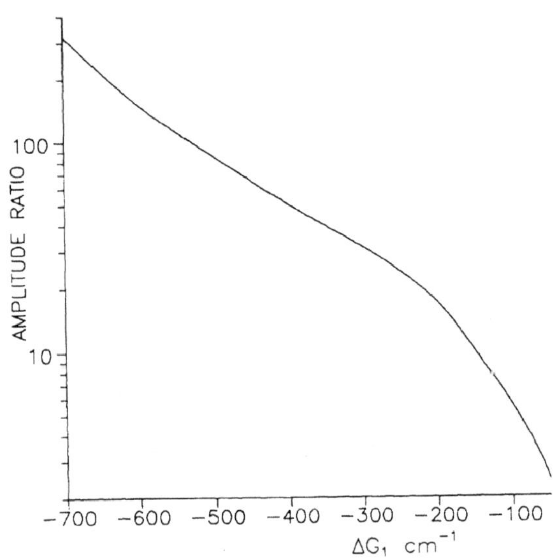

Figure 3. The ΔG_1 dependence of the amplitude ratio A_1/A_2 for the decay of $^1P^*$ at 300K.

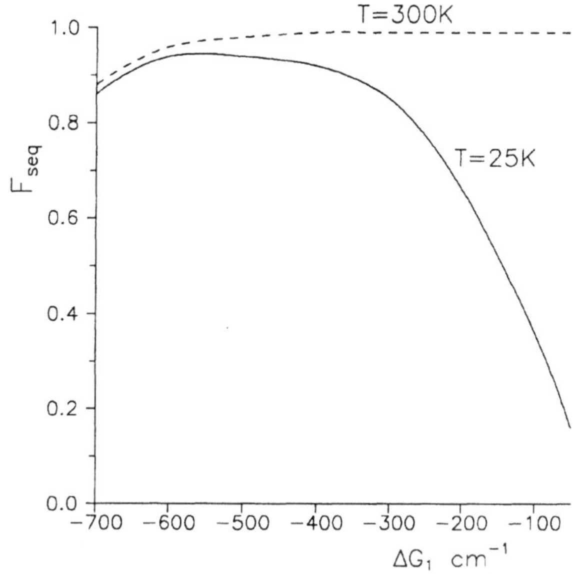

Figure 4. The ΔG_1 dependence of the branching ratio F_{seq} at 300K and at 25K.

(6) Mechanisms. The branching ratio F_{seq} for the sequential channel (Fig. 4) at T = 300K and T = 25K reveals the following mechanisms: (I) The sequential mechanism prevails at all temperatures for $\Delta G_1 \leq -300$ cm^{-1}. (II) A sequential mechanism operates at high temperatures and a superposition of sequential and superexchange mechanisms appears at low temperatures (-300 cm$^{-1} \leq \Delta G_1 \leq -100$ cm^{-1}). The permissible energy domain (point (5)) corresponds to ranges (I) and/or (II). Of course, inhomogeneous broadening effects may result in the simultaneous prevalence of ranges (I) and (II).

From the foregoing analysis we infer that the sequential mechanism dominates at room temperature, with a minor involvement of superexchange (F_{seq} = 0.96-0.99), while at cryogenic temperatures either exclusive sequential or parallel sequential-superexchange mechanisms may be operative.

VI. SOME IMPLICATIONS OF MUTAGENESIS

Functional and structural control of ET by mutagenesis provides, in principle, a powerful tool for the modification of the electronic couplings and nuclear Franck-Condon factors, allowing to infer on the role of single AARs and/or of prosthetic groups on ET dynamics. However, this information is still limited by our ignorance regarding structural modifications induced by mutagenesis.

A test for our parallel model for primary ET involves the (M)L214H mutant,[43,44] where the H group along the A branch is replaced by B, with the order of the prosthetic groups along this branch being PB_1B_2. The experimental information[44] reveals that the decay time of $^1P^*B_1B_2$ is τ_1 = 6.4±0.8 psec at T = 295K and $\tau_2 \simeq$ 2.5 psec at T = 5K, the decay time of the ion pair (presumably $P^+B_1B_2^-$) is τ_I = 325±25 psec at T = 295K and the quantum yield of $P^+B_1B_2Q^-$ is Y = 0.60 at T = 295K and Y = 0.25 at T = 5K. It is possible to adopt the same kinetic scheme adopted for the description of primary ET in the WT RC to this mutant

$$
{}^{1}P^{*}B_{1}B_{2} \xrightarrow{k_1} P^{+}B_{1}^{-}B_{2} \underset{k_{-2}}{\overset{k_2}{\rightleftharpoons}} P^{+}B_{1}B_{2}^{-} \xrightarrow{k_Q} P^{+}B_{1}B_{2}Q^{-}
$$

$$
300\ \text{psec} \downarrow \qquad \downarrow 600\text{psec} \qquad \downarrow 20\text{nsec} \tag{10}
$$

The only change in the parameters is the energetics of the second step (for k_2 and k_{-2}). The radical pair $P^+B_1B_2^-$ has a considerably higher energy than P^+BH^-. The experimental results can be reconciled by taking the $P^+B_1^-B_2$ and $P^+B_1B_2^-$ states as quasidegenerate with $\Delta G \simeq \Delta G_1 \simeq -400\ cm^{-1}$. This analysis results in $k_1 = (6.4\ psec)^{-1}$, $k_2 = (3\ psec)^{-1}$ and $k_{-2} \simeq k_2$ at T = 300K, while at 25K $k_1 = (2.5\ psec)^{-1}$, $k_2 = (1000\ psec)^{-1}$, with the second step becoming a bottleneck at low temperatures. At room temperature the back transfer from $P^+B_1B_2^-$ to $P^+B_1^-B_2$ is the cause for the long relaxation time τ_1 = 325 psec which is considerably longer than the Q reduction time k_Q^{-1} = 200 psec. This model eliminates some inconsistencies, which arise if the experimental data for this mutant are interpreted in terms of unistep ET ${}^1P^*B_1B_2 \rightarrow P^+B_1B_2^-$. Such a scheme requires a recombination rate of $P^+B_1B_2^-$ to be $(1\ nsec)^{-1}$, being ten times faster than in the WT RC, and an activated ET to Q with $k_Q = (600\ psec)^{-1}$ at 295K and $k_Q = (3200\ psec)^{-1}$ at low temperatures.[44] Our scheme eliminates these difficulties.

VII. EPILOGUE

The kinetic optimization principle for the primary ET in the photosynthetic RC requires that the primary process is fast, and being characterized by a high ($Y \simeq 1$) overall quantum yield, effectively competing with energy waste processes, i.e., (i) backtransfer of electronic energy from ${}^1P^*$ to the antenna pigments, (ii) internal conversion of ${}^1P^*$ to the electronic ground state and (iii) ion pair recombinations of P^+B^-H and of P^+BH^-. Thus the operative optimization conditions are $\tau_{ET} \lesssim 10$ psec and $Y \geq 0.95$ for the primary charge separation. This state of affairs is realized over the entire ΔG_1 range of interest and for a wide range of V_{PB} electronic coupling (Fig. 5), where the domain above the photosynthetic

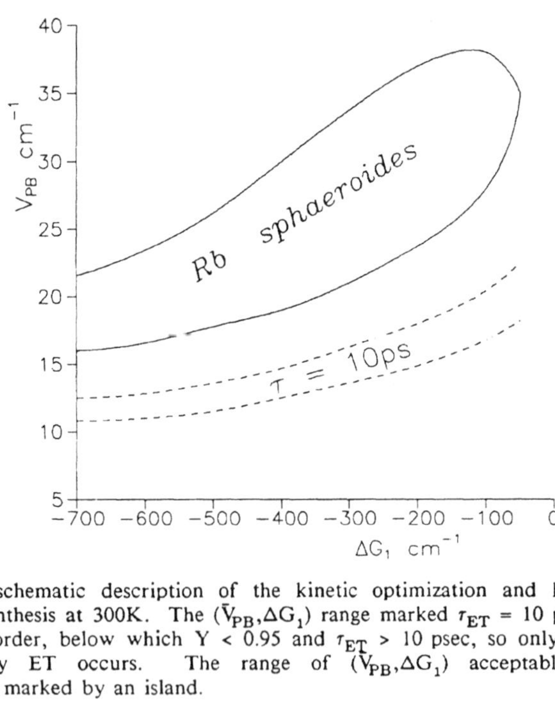

Figure 5. A schematic description of the kinetic optimization and kinetic stability in bacterial photosynthesis at 300K. The $(\bar{V}_{PB}, \Delta G_1)$ range marked τ_{ET} = 10 psec constitutes the photosynthetic border, below which Y < 0.95 and τ_{ET} > 10 psec, so only above it fast and efficient primary ET occurs. The range of $(V_{PB}, \Delta G_1)$ acceptable parameters for *Rb.sphaeroides* is marked by an island.

border ($\tau_{ET} \approx 10$ psec) marks the photosynthetic domain. The range of acceptable parameters for *Rb.sphaeroides* (at T = 300K) is marked by the island in the photosynthetic domain in Fig. 5.

Fast and efficient primary Et requires:

(a) The electronic coupling V_{PB} to be sufficiently large. Translated into structural terms it requires sufficiently close P-B edge-to-edge distance. Of course, intermolecular electronic overlap limits V_{PB}.

(b) Optimization of the nuclear Franck-Condon factor(s), corresponding to nearly activationless ET.

(c) Principle (b) requires a sufficiently negative value of ΔG_1 (for a surprisingly small value of λ_1). This will optimize F for the forward reaction and minimize the backreaction.

These conditions result in a delicate interplay between sequential (e.g., moderate V_{PB}, small λ_1 and sufficiently negative ΔG_1) and superexchange mechanism (e.g., larger V_{PB} and V_{BH}, larger λ and small negative or positive ΔG_1). The delicate balance of the energetics determines the mechanism of the primary ET. From our analysis it appears that at room temperature in WT and some mutants of the RC, the sequential ET channel dominates and the direct unistep channel is negligible. At low temperatures some interesting issues remain open.

The parallel-sequential-superexchange mechanism proposed herein is of intrinsic interest for the general nature of the primary charge separation in RC, being stable for variations of ΔG_1 over a range of 500 cm^{-1} (2 Kcal $mole^{-1}$). This stability of the primary mechanism provides a safety valve to insure the occurrence of ET in an inhomogeneously broadened system and, more important, to insure the prevalence of ultrafast and efficient ET in different bacterial and plant photosynthetic RCs.

REFERENCES

1. "The Photosynthetic Bacterial Reaction Center. Structure and Dynamics," J. Breton and A. Vermeglio, eds., Plenum NATO ASI Series, New York (1988).
2. "Reaction Centers of Photosynthetic Bacteria", M.E. Michel-Beyerle, ed., Springer Verlag, Berlin (1990).
3. N.W. Woodbury, M. Becker, D. Middendorf, and W.W. Parson, *Biochemistry* 24:7516-7521 (1985).
4. J. Jortner and M.E. Michel-Beyerle. *In*: "Antennas and Reaction Centers of Photosynthetic Bacteria," M.E. Michel-Beyerle, ed., Springer, Berlin (1985) pp. 345-354.
5. S.F. Fischer, I. Nussbaum, and P.O.J. Scherer. *In*: "Antennas and Reaction Centers of Photosynthetic Bacteria," M.E. Michel-Beyerle, ed., Springer, Berlin (1985) pp. 256-263.
6. A. Ogrodnik, N. Remy-Richter, M.E. Michel-Beyerle, and R. Feick, *Chem. Phys. Lett.* 135:576-581 (1987).
7. J.R. Norris, D.E. Budil, D.M. Tiede, J. Tang, S.V. Kolaczkowski, C.H. Chang, and M. Schiffer. *In*: "Progress in Photosynthetic Research," J. Biggins, ed., Martinus Nijhoff, Dordrecht (1987) Vol. I, pp. 1.4.363-1.4.369.
8. M.E. Michel-Beyerle, M. Plato, J. Deisenhofer, H. Michel, M. Bixon, and J. Jortner, *Biochim. Biophys. Acta* 932:52-70 (1988).
9. M. Bixon, J. Jortner, M. Plato, and M.E. Michel-Beyerle. *In*: "The Photosynthetic Bacterial Reaction Center. Structure and Dynamics," J. Breton and A. Vermeglio, eds., Plenum NATO ASI Series, New York (1988), pp. 399-420.
10. M. Plato, K. Möbius, M.E. Michel-Beyerle, M. Bixon, and J. Jortner, *J. Am. Chem. Soc.* 110:7279-7285 (1988).
11. M. Bixon, M.E. Michel-Beyerle, and J. Jortner, *Isr. J. Chem.* 28:155-168 (1988).
12. M. Bixon, J. Jortner, M.E. Michel-Beyerle, and A. Ogrodnik, *Biochim. Biophys. Acta* 977:273-286 (1989).
13. R.A. Friesner and Y. Won, *Biochim. Biophys. Acta* 977:99-122 (1989).
14. R.A. Marcus, *Isr. J. Chem.* 28:205-213 (1988).
15. R. Haberkorn, M.E. Michel-Beyerle, and R.A. Marcus, *Proc. Natl. Acad. Sci. USA* 70:4185-4188 (1979).
16. R.A. Marcus, *Chem. Phys. Lett.* 133:471-477 (1987).

17. S.V. Chekalin, Ya.A. Matveetz, A.Ya. Shkuropatov, V.A. Shuvalov, and A.P. Yartzev, *FEBS Lett.* 216:245-248 (1987).
18. R.A. Marcus, *Chem. Phys. Lett.* 146:13-22 (1988).
19. S. Creighton, J.-K. Hwang, A. Warshel, W.W. Parson, and J. Norris, *Biochem.* 27:774-781 (1988).
20. M. Bixon, J. Jortner, and M.E. Michel-Beyerle, *Biochim. Biophys. Acta* 1056:301-315 (1991).
21. J.-L. Martin, J. Breton, A.J. Hoff, A. Migus, and A. Antonetti, *Proc. Natl. Acad. Sci. USA* 83:957-961 (1986).
22. J. Breton, J.-L. Martin, A. Migus, A. Antonetti, and A. Orszag, *Proc. Natl. Acad. Sci. USA* 83:5121-5125 (1986).
23. J. Breton, J.-L. Martin, G.R. Fleming, and J.-C. Lambry, *Biochem.* 27:8276-8284 (1988).
24. W. Holzapfel, U. Finkele, W. Kaiser, D. Oesterhelt, H. Scheer, H.U. Stilz, and W. Zinth, *Chem. Phys. Lett.* 160:1-7 (1989).
25. W. Holzapfel, U. Finkele, W. Kaiser, D. Oesterhelt, H. Scheer, H.U. Stilz, and W. Zinth, *Proc. Natl. Acad. Sci. USA* 87:5168-5172 (1990).
26. C. Lauterwasser, U. Finkele, H. Scheer, and W. Zinth, *Chem. Phys. Lett.* 183:471 (1991).
27. C.K. Chan, T.J. DiMango, L.X.Q. Chen, J. Norris, and G.R. Fleming, *Proc. Natl. Acad. Sci. USA*i 88:11202 (1991).
28. K. Dressler, E. Umlauf, S. Schmidt, P. Hamm, W. Zinth, S. Buchanan, and H. Michel, *Chem. Phys. Lett.* 183:270 (1991).
29. G.R. Fleming and J. Norris (this volume).
30. X. Xie, M. Du, S.J. Rosenthal, T.J. DiMagno, M.E. Schmidt, J.R. Norris, and G.R. Fleming, Ultrafast Phenomena VIII, Antibas, France (June 1992).
31. P. Hamm and W. Zinth, Ultrafast Phenomena VIII, Antibas, France (June 1992).
32. U. Finkele, C. Lauterwasser, W. Zinth, K.A. Gray, and D. Oesterhalt, *Biochem.* 29:8517 (1990).
33. M. Schiffer, C.K. Chan, C.H. Chang, T.J. DiMagno, G.R. Fleming, S.L. Nance, J.R. Norris, S.W. Snyder, M.C. Thurnauer, D. Tiede, and D.K. Hanson (this volume).
34. J. Norris and G.R. Fleming (private communication and to be published).
35. S.J. Robles, J. Breton, and D.C. Youvan, *Science* 248:1402 (1990).
36. S.G. Johnson, D. Tang, R. Jankowiak, J.M. Hayes, G.J. Small, and D.M. Tiede, *J. Phys. Chem.* 94:5849 (1990).
37. M.H. Vos, J.C. Lambry, S.J. Robles, D.C. Youvan, J. Breton, and J.L. Martin, *Proc. Natl. Acad. Sci. USA* 88:8885 (1991).
38. M.H. Vos, J.C. Lambry, S.J. Robles, D.C. Youvan, J. Breton, and J.L. Martin, *Proc. Natl. Acad. Sci. USA* (in press, 1992).
39. K. Schulten and M. Tesch, *Chem. Phys.* 158:421-446 (1991).
40. M. Bixon and J. Jortner, to be published.
41. A. Ogrodnik, M. Volk, and M.E. Michel-Beyerle, reference 1, p. 177 (1990), and to be published.
42. M. Bixon, J. Jortner, and M.E. Michel-Beyerle (this volume).
43. V. Nagarajan, W.W. Parson, D. Gaul, and C. Schenk, *Proc. Natl. Acad. Sci. USA* 87:7888 (1991).
44. C. Kirmaier, D. Gaul, R. DeBey, D. Holten, and C.C. Schenk, *Science* 251:922 (1991).

Multi–Mode Coupling of Protein Motion to Electron Transfer in the Photosynthetic Reaction Center: Spin – Boson Theory Based on a Classical Molecular Dynamics Simulation[1]

Dong Xu and Klaus Schulten*

Beckman Institute and Department of Physics
University of Illinois at Urbana-Champaign
405 North Mathews Ave., Urbana, IL 61801, USA

Abstract

We present a quantum mechanical description (spin–boson model) for electron transfer rates in the photosynthetic reaction center of *Rh. viridis* which assumes for the protein nuclear motion a broad distribution of vibrational modes rather than a few discrete modes. We demonstrate that linear coupling of electron transfer to such modes is in agreement with molecular dynamics simulations. We establish that the multi-mode coupling between electron transfer and protein motion can be described in surprisingly simple terms and requires as input only that one monitors the fluctuations of the so–called energy gap function $\Delta E(t)$, i.e., the energy required for a virtual transfer of the electron. The amplitude and the relaxation time of the associated correlation function are the essential parameters which determine electron transfer rates in the framework of the spin–boson model. We present the temperature and redox energy dependence of these rates which are found in agreement with observations and also with Marcus theory at high temperature, even though the latter assumes coupling to a single mode.

1 Introduction

The electron transfer in the photosynthetic reaction center of *Rh. viridis* involves three initial steps: $P_S H_L Q_A Q_B \rightarrow P_S^+ H_L^- Q_A Q_B \rightarrow P_S^+ H_L Q_A^- Q_B \rightarrow P_S^+ H_L Q_A Q_B^-$. Different from Arrhenius behavior of most chemical reactions, one observes that electron transfer rates vary little with temperature, in some instances even increase when temperature is lowered. In order to explain the redox and temperature dependence of electron transfer rates, previous interpretations have assumed that quantum mechanical behavior of electron transfer arises through a small number of nuclear degrees of freedom, such as one or two, particularly strongly coupled to the electron transfer re-

[1]Part of this paper is excerpt from [1]

The Photosynthetic Bacterial Reaction Center II
Edited by J. Breton and A. Verméglio, Plenum Press, New York, 1992

action. In most cases, the respective theories can fit experiments very well, however, they rest on artificial parameters [2, 3, 4, 5, 6].

The coupling between electron transfer and a protein as a medium is due to the Coulomb interaction, which is long range and encompasses a very large volume. Correspondingly, the coupling involves small, additive contributions of many motions of a protein rather than only a few dominant contributions. In fact, MD simulations reported in [7, 8, 9] revealed that the coupling involves essentially all nuclear degrees of freedom of the protein investigated, the photosynthetic reaction center. All components of the protein contribute rather evenly to the coupling between electron transfer and medium as simulations with and without cut-off of Coulomb interactions revealed.

One of the reasons why few-mode description of electron transfer in proteins are widely accepted is that it can be treated quantum mechanically, and that it fits experimental data over a wide temperature domain. In case of multi-mode coupling, one might be very discouraged by the fact that all degrees of freedom need to be described quantum mechanically, at least all those degrees of freedom for which holds $k_BT \leq \hbar\omega_\alpha$ where ω_α is the frequency connected with the respective nuclear motion. However, in case one describes electron transfer as a 2-state process, assumes linear coupling as well as harmonic motion of the protein atoms, the resulting stochastic quantum system can be described in a rather straight forward way. Following earlier work by Onuchic et al. [10, 11, 12], we applied the spin–boson model to biological electron transfer[1]. We will demonstrate below that the resulting description is actually rather simple. The theory outlined employs classical simulations to obtain the relevant parameters. We will calculate the electron transfer rates as a function of redox energy and temperature [7].

2 Spin – Boson Model of Electron Transfer

A detailed review of the spin–boson model can be found in [13]. In case of electron transfer in proteins, the spin–boson model can be related to a simple microscopic picture, namely, the well-known Marcus energy diagram[14, 15]. In this diagram, the free energy of both reactant and product states is described by a one–dimensional harmonic potential with identical force constants f. We assume the reactant and product free energy curves have the functional form,

$$E_R = \frac{1}{2} f q^2 , \quad E_P = \frac{1}{2} f (q - q_P)^2 - \epsilon_o . \tag{1}$$

In the above equations, q represents schematically the nuclear configuration of the protein and q_P, ϵ_o represent the shift of the equilibrium position after the electron transfer. As pointed out in [3] and [7], the potential functions originate from a dependence on thousands of nuclear coordinates, which define a many-dimensional potential-energy surface. The spin–boson model goes beyond the Marcus model in that it allows one to represent the multitude of degrees of freedom coupled to the electron transfer through an ensemble of harmonic oscillators of various frequencies.

In analogy to Marcus theory, we also postulate that the nuclear degrees of freedom of the medium are too inert to change during the electron transfer step, i.e., the Born–Oppenheimer approximation can be applied for the nuclear degrees of freedom. Let us assume that the protein matrix coupled to electron transfer can be represented through N different oscillators where N is of the order of magnitude of the number of atoms in the protein, i.e., about 10^4 in the case of the photosynthetic reaction center. We denote the frequencies of these oscillators by ω_α, $\alpha = 1, 2, \ldots, N$ and the associated vibrational coordinates by q_α, $\alpha = 1, 2, \ldots, N$. Let us assume further that each mode is coupled to the electron transfer such that the harmonic potentials in the reactant state ($E_{R,\alpha}$) and in the product state ($E_{P,\alpha}$) differ as follows ($\alpha = 1, 2, \ldots, N$)

$$E_{R,\alpha} = \frac{1}{2} m_\alpha \omega_\alpha^2 q_\alpha^2 \,, \quad E_{P,\alpha} = \frac{1}{2} m_\alpha \omega_\alpha^2 (q_\alpha - q_{o,\alpha})^2 - \epsilon_{o,\alpha} \,. \tag{2}$$

In this notation m_α are effective constants which do not need to be individually identified as we will see below.

The total energy in the reactant and product states is then

$$E_R = \sum_{\alpha=1}^{N} \left(\frac{p_\alpha^2}{2 m_\alpha} + E_{R,\alpha} \right) \,, \quad E_P = \sum_{\alpha=1}^{N} \left(\frac{p_\alpha^2}{2 m_\alpha} + E_{P,\alpha} \right) \,, \tag{3}$$

where p_α is the momentum operator of the oscillator. The spin–boson Hamiltonian combines these energies with a quantum mechanical 2-state Hamiltonian as follows

$$H_{sb} = \begin{pmatrix} E_R & V \\ V & E_P \end{pmatrix} \,, \tag{4}$$

where V accounts for the electron coupling between reactant and product states, the coupling originating from tunneling of the electron between electron donor and acceptor.

After some algebra [1], one can separate the above Hamiltonian into three parts and a constant $C\,\mathbb{1}$

$$\hat{H}_{sb} = \hat{H}_{el} + \hat{H}_{osc} + \hat{H}_{coupl} + C\,\mathbb{1} \,. \tag{5}$$

The first term is a simple two–state Hamiltonian, resembling a spin operator

$$\hat{H}_{el} = V \sigma_x + \frac{1}{2} \epsilon \sigma_z \,, \tag{6}$$

where σ_x, σ_z are Pauli matrices. ϵ accounts for the energy difference of reactant and product states and results from the sum of the redox energy of all modes

$$\epsilon = \sum_{\alpha=1}^{N} \epsilon_{o,\alpha} \,. \tag{7}$$

The second term in (5) represents the medium thermal motion described through an ensemble of independent linear oscillators (bosons)

$$\hat{H}_{osc} = \sum_\alpha \left(\frac{\hat{p}_\alpha^2}{2 m_\alpha} + \frac{1}{2} m_\alpha \omega_\alpha^2 x_\alpha^2 \right) \mathbb{1} \tag{8}$$

where x_α denotes the spatial coordinate

$$x_\alpha = q_\alpha - \frac{q_{o,\alpha}}{2} \,. \tag{9}$$

The third term in (5) represents the coupling between the vibrational degrees of freedom and the two–state system. The coupling is linear in x_α and diagonal in the two-state system

$$\hat{H}_{coupl} = \frac{1}{2} \sigma_z \sum_\alpha c_\alpha x_\alpha \,. \tag{10}$$

Here c_α describes the strength of the coupling of the electron transfer to the α-th oscillator, in this case,

$$c_\alpha = m_\alpha \omega_\alpha^2 q_{o,\alpha} \,. \tag{11}$$

The last term in (5)

$$C\,\mathbb{1} \;=\; \frac{1}{2}\sum_{\alpha=1}^{N}\left(\epsilon_{o,\alpha} \;+\; \frac{1}{4}m_\alpha\omega_\alpha^2 q_{o,\alpha}^2\right)\,\mathbb{1}\;, \tag{12}$$

is a constant which does not affect the electron transfer rates and, thus, can be omitted. The above derivation is not only a formal mathematical transformation, but also a shift from the multi-mode Marcus diagram to an equivalent physical picture, i.e., the spin–boson model.

Even though there appear many variables and parameters in the equations above, ultimately the spin–boson model, as advertised in [13], is characterized completely by a well defined average property of the system, the spectral function $J(\omega)$

$$J(\omega) \;=\; \frac{\pi}{2}\sum_{\alpha=1}^{N}\frac{c_\alpha^2}{m_\alpha\omega_\alpha}\,\delta(\omega \;-\; \omega_\alpha)\,. \tag{13}$$

$J(\omega)$ can be assumed to be a smooth function determined by few parameters. These parameters can be determined from a classical molecular dynamics simulation. Once one knows $J(\omega)$ and V, one can calculate all the properties for the spin–boson system. As a matter of fact, the parameter V is not very important; it appears only in a prefactor V^2 which multiplies the electron transfer rate. The simple dependence results from an application of Fermi's golden rule, an approximation which appears to be valid in case of electron transfer [7].

3 The Relation between the Spin – Boson Model and Classical Molecular Dynamics Simulations

Previous investigations of the coupling of electron transfer and protein thermal motions have been based on classical descriptions (see [7] and references therein). The nuclear motions were complemented by a quantum mechanical description for the electron transfer, described by a 2-state model. The coupling to the classical protein motion yields a fluctuating diagonal contribution for the 2-state Hamiltonian as given in (6). This contribution can be determined as the energy difference $\Delta E(t)$ between reactant and product states at each instant in time. Figure 1 provides a good illustration how $\Delta E(t)$ relates to the spin–boson model in the photosynthetic reaction center of *Rh. viridis*: the figure shows on the left hand side the protein atoms, rendered in grey, in which are embedded, rendered in black, the prosthetic groups involved in electron transfer: hemes, chlorophylls, pheophytines and quinones. The electron transfer is coupled to a wide range of protein atoms through long range Coulomb forces. The right hand side indicate one of the electron transfer reactions; $\Delta E(t)$ represents the energy gap between reactant and product states which strongly fluctuates due to the coupling to the thermal motion of the protein. The two states of electron transfer are described as the spin operator [see Eq. (6)], while the thermal vibrations of the protein correspond to the bosons in the spin–boson model.

As long as one can assume that the Hamiltonian is temperature independent, e.g., that the protein structure and, hence, the coupling terms as well as the spectral function $J(\omega)$ do not change with temperature, one can expect that the classical simulations allow one to determine a suitable quantum mechanical model. For this purpose, one carries out a classical simulation at high temperature, characterizes $J(\omega)$ corresponding to the simulated $\Delta E(t)$ and employs the the resulting $J(\omega)$ at all temperatures. Since at physiological temperatures ($T\;=\;300\,\mathrm{K}$), the majority of frequencies of modes satisfy the property $\hbar\omega_\alpha/k_BT << 1$ one can assume the classical limit to be realized at $T\;=$

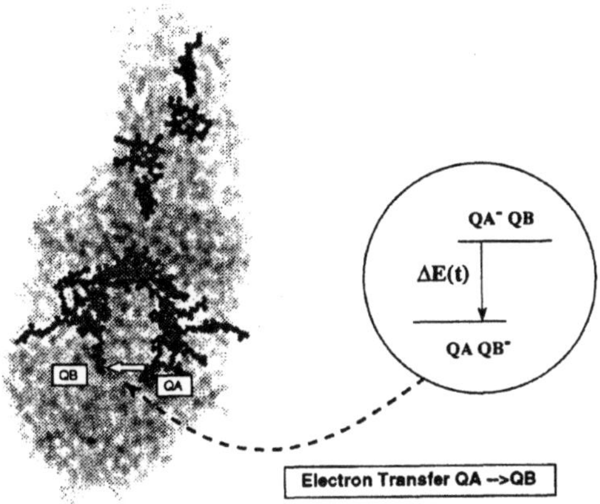

Figure 1. This figure shows on the left hand side the protein atoms of the photosynthetic reaction center of *Rhodopseudomonas viridis* (in grey) and the prosthetic groups involved in electron transport (in black). One can recognize in the upper part of the protein complex four heme groups. The center contains a sandwich complex of two chlorophylls from which stretch to both sides each a chlorophyll, a pheophytine and, towards the bottom, a quinone. These prosthetic groups conduct electrons, the electron movement being accompanied by a response of the thermal motion of the (grey) protein atoms. Indicated is also on the right hand side one of the electron transfer reactions, $Q_A \rightarrow Q_B$; $\Delta E(t)$ represents the energy gap between reactant and product states which depends strongly on the thermal motion of the protein. In the spin–boson model, the two states of electron transfer are described as the spin operator, while thermal vibrations of the protein are accounted for by the boson operators.

300 K. At this temperatures, quantum descriptions and classical descriptions should then coincide and, therefore, the classical simulations should allow one to determine a suitable characteristic function $J(\omega)$.

$J(\omega)$ in the spin–boson system can be characterized in molecular dynamics simulations through the energy–energy correlation function, as discussed in [7, 8]:

$$C(t) = \langle\, (\,\Delta E(t) - \langle \Delta E \rangle\,)\,(\,\Delta E(0) - \langle \Delta E \rangle\,)\,\rangle\,, \tag{14}$$

$J(\omega)$ and $C(t)$ are related to each other through the Fourier cosine transform. It holds [1, 16, 17]

$$\frac{J(\omega)}{\omega} = \frac{1}{k_B T} \int_0^\infty dt\, C(t) \cos \omega t\,, \tag{15}$$

$$C(t) = \frac{2 k_B T}{\pi} \int_0^\infty d\omega\, \frac{J(\omega)}{\omega} \cos \omega t\,. \tag{16}$$

If one monitors in a classical MD simulation, the normalized correlation function, i.e., $C_1(t) = C(t)/C(0)$, and the rms-deviation from the mean of $\Delta E(t)$ i.e.,

$$\sigma = \sqrt{\langle \Delta E^2 \rangle - \langle \Delta E \rangle^2}\,, \tag{17}$$

then one can use the following expression [1] to determine $J(\omega)$

$$\frac{J(\omega)}{\omega} = \frac{\sigma^2}{k_B T} \int_0^\infty dt\, C_1(t) \cos \omega t\,. \tag{18}$$

In the simulations reported in [7] for the electron transfer $P_S \rightarrow H_L$, $C_1(t)$ exhibits an approximate exponential decay with a relaxation time $\tau = 94\ fs$. The simulation in [7] also provided $\sigma = 3.9$ kcal/mol at a temperature $T = 300$ K. For the sake of simplicity we will assume that the energy–energy correlation function is well represented by a mono-exponential function $e^{-\frac{t}{\tau}}$. The relationship (18) then yields

$$J(\omega) = \frac{\sigma^2 \omega}{k_B T} \int_0^\infty dt\, e^{-t/\tau} \cos \omega t = \frac{\eta \omega}{1 + \omega^2 \tau^2} ; \tag{19}$$

$$\eta = \frac{\sigma^2 \tau}{k_B T} = 25.15\, h . \tag{20}$$

where h is Planck's constant.

4 Calculation of Electron Transfer Rates

In order to determine the electron transfer rate $k(\epsilon, T)$, we start from the expression provided in [13]

$$k(\epsilon, T) = \left(\frac{2V}{\hbar}\right)^2 \int_0^\infty dt \cos\left(\frac{\epsilon t}{\hbar}\right) \cos\left[\frac{Q_1(t)}{\pi \hbar}\right] \exp\left[-\frac{Q_2(t)}{\pi \hbar}\right] . \tag{21}$$

Evaluation of this expression requires first an evaluation of the time-dependent functions $Q_1(t)$ and $Q_2(t)$ which are defined in terms of integrals over $J(\omega)$ as follows

$$\begin{aligned} Q_1(t) &= \int_0^\infty d\omega\, \omega^{-2} J(\omega) \sin \omega t \\ Q_2(t) &= 2 \int_0^\infty d\omega\ \omega^{-2} \sin^2\left(\frac{\omega t}{2}\right) \coth\left(\frac{\beta \hbar \omega}{2}\right) J(\omega) \end{aligned} \tag{22}$$

where $\beta = 1/k_B T$. Using the expression (19) for $J(\omega)$ one obtains $Q_1(t)$ analytically

$$Q_1(t) = \int_0^\infty d\omega \frac{\eta \sin \omega t}{\omega (1 + \omega^2 \tau^2)} = \frac{\eta \pi}{2} \left[1 - \exp\left(-\frac{t}{\tau}\right)\right] \tag{23}$$

The electron transfer rate is then

$$\begin{aligned} k(\epsilon, T) = & \left(\frac{2V}{\hbar}\right)^2 \int_0^\infty dt \cos\left(\frac{\epsilon t}{\hbar}\right) \cos\left[\frac{\eta}{2\hbar}\left(1 - e^{-t/\tau}\right)\right] \times \\ & \times \exp\left[-\frac{2\eta}{\pi \hbar} \int_0^\infty d\omega \frac{\sin^2\left(\frac{\omega t}{2}\right)}{\omega(1 + \omega^2 \tau^2)} \coth\left(\frac{\beta \hbar \omega}{2}\right)\right] . \end{aligned} \tag{24}$$

To simplify this expression we define $x = t/\tau$, $y = \omega \tau$, and $\gamma = \eta/h$. This yields the final expression

$$\begin{aligned} k(\epsilon, T) = & \left(\frac{2V}{\hbar}\right)^2 \tau \int_0^\infty dx \cos\left(\frac{\epsilon \tau}{\hbar} x\right) \cos\left[\gamma \pi \left(1 - e^{-x}\right)\right] \times \\ & \times \exp\left[-4\gamma \int_0^\infty dy \frac{\sin^2\left(\frac{xy}{2}\right)}{y(1 + y^2)} \coth\left(\frac{\hbar}{2 k_B \tau} \cdot \frac{y}{T}\right)\right] . \end{aligned} \tag{25}$$

The stated numerical values of $V/\hbar$ in [7] is 5 ps^{-1}. Equations (25) allow one, in principle, to evaluate the electron transfer rate $k(\epsilon, T)$. However, straightforward numerical quadrature of (25) is very time consuming since it involves a double integral. One can use some faster, albeit approximate expressions for the exp[$\cdots$] factor in the integrand of (25). We define

$$q_2(x) \;=\; \int_0^\infty dy\, \frac{\sin^2\left(\frac{xy}{2}\right)}{y\,(1\,+\,y^2)} \coth\left(\alpha y\right) \;,\quad \alpha \;=\; \frac{\hbar}{2k_B T T}\,. \tag{26}$$

As demonstrated in [1], $q_2(x)$ is a monotonously increasing function of x. Hence the main contribution to (25) stems from the region of small x. When x is small, there are at least two ways to approximate $q_2(x)$ in a very simple form. One is to calculate some sample points at small x and at a certain temperature for $q_2(x)$, and then fit all the points into the form $A\,x^\delta$ with $1 < \delta < 2$. The other way, suggested by A. Szabo (private communication), is based on the analytical expansion of $q_2(x)$ which holds for small x

$$q_2(x) \;\approx\; \frac{x^2}{4}\,\left[\,f(\alpha) \;-\; \ln x\,\right] \tag{27}$$

where

$$f(\alpha) \;=\; \int_0^\infty \frac{y\,dy}{1\,+\,y^2}(\coth(\alpha y) \;-\; 1)\,. \tag{28}$$

The second approach is very useful to derive further approximate properties for $k(\epsilon, T)$.

Since $q_2(x)$ is ϵ–independent one can use the same numerical approximation for all ϵ values considered. Hence, for a given temperature obtaining $k(\epsilon, T)$ at all different ϵ values requires one to evaluate $q_2(x)$ only once. Then (25) becomes

$$k_{appr}(\epsilon, T) \;=\; \left(\frac{2V}{\hbar}\right)^2 \tau \int_0^\infty dx \cos\left(\frac{\epsilon\tau}{\hbar}\,x\right)\, \cos\left[\gamma\pi\,\left(1\,-\,e^{-x}\right)\right]\, e^{-4\gamma q_2(x)}\,. \tag{29}$$

Obviously, the numerical procedure chosen is much less time consuming than evaluating (25) by double quadrature.

5 High and Low Temperature Limit

5.1 High Temperature Limit

The expression (21) of the electron transfer rate together with the functional behavior of $Q_2(t)$ suggests that one may employ the method of steepest descent, at least in the high temperature limit, for an approximate evaluation. This approximation is based on a quadratic expansion of $Q_2(t)$ around its minimum at $t \;=\; 0$. The procedure requires one to determine the quantity

$$\mu \;=\; \left.\frac{d^2}{dt^2}\, Q_2(t)\right|_{t=0} \tag{30}$$

The expression for $Q_2(t)$ in (22) yields

$$\mu \;=\; \int_0^\infty d\omega J(\omega) \coth\left(\frac{\beta\hbar\omega}{2}\right)\,. \tag{31}$$

Unfortunately, for many choices of $J(\omega)$ this expression diverges and the steepest descent method cannot be applied. However, we note that the divergence of (31) is due to $\omega \;\to\; \infty$ contributions to the integral over $J(\omega)$. Since the number of modes in a protein are finite, the divergence in (31) is due to an artificial analytical form of $J(\omega)$.

If one would assume a cut-off frequency ω_c, i.e., replace $J(\omega)$ by $J(\omega)\theta(\omega - \omega_c)$, a divergence would not arise in (31). One may, hence, assume that the second derivative (30) actually exists, approximate

$$Q_2(t) \approx \frac{1}{2}\mu t^2 , \tag{32}$$

and employ this in a steepest descent method.

At a sufficiently high temperature, contributions to the integral in (21) arise only in a vicinity of $t = 0$ in which (32) is small. In this case, one can approximate $Q_1(t)$ in (22) linearly around $t = 0$

$$Q_1(t) \approx \nu t ; \quad \nu = \left. \frac{d}{dt} Q_1(t) \right|_{t=0} \tag{33}$$

where

$$\nu = \int_0^\infty d\omega \frac{J(\omega)}{\omega} . \tag{34}$$

By using the approximations (32) and (33) in (21), if ϵ is not close to 0, one obtains [10, 11, 1]

$$k(\epsilon, T) \approx \frac{2\pi V^2}{\hbar} \frac{1}{\sqrt{2\pi\delta^2}} \exp\left[-\frac{(\epsilon - \epsilon_m)^2}{2\delta^2} \right] . \tag{35}$$

where

$$\delta^2 = \frac{\hbar\mu}{\pi} = \frac{\hbar}{\pi} \int_0^\infty d\omega \, J(\omega) \coth\left(\frac{\beta\hbar\omega}{2}\right) \tag{36}$$

$$\epsilon_m = \frac{\nu}{\pi} = \frac{1}{\pi} \int_0^\infty d\omega \frac{J(\omega)}{\omega} . \tag{37}$$

At a high enough temperature, i.e., $T > 100$ K, according to our numerical calculations, one can show further [1]

$$\delta = \sigma ; \quad \epsilon_m = \frac{\sigma^2}{2k_B T} . \tag{38}$$

According to (37), ϵ_m is actually temperature independent. Hence, one can rewrite (35) into a form which agrees with the rate predicted by the classical Marcus theory

$$k_M(\epsilon, T) = \frac{2\pi V^2}{\hbar} \frac{1}{\sqrt{2\pi f k_B T q_o^2}} \exp\left[-\frac{(\epsilon - \frac{1}{2} f q_o^2)^2}{2k_B T f q_o^2} \right] \tag{39}$$

where

$$f q_o^2 = 2\epsilon_m = \left. \frac{\sigma^2}{k_B T} \right|_{T=300K} . \tag{40}$$

5.2 Low Temperature Limit

At low temperatures, one can employ (28) for $\alpha = \frac{\hbar}{2k_B \tau T} \to \infty$, to approximate $q_2(x)$ further. It can be verified

$$\lim_{\alpha\to\infty} f(\alpha) = \frac{\pi^2}{12\alpha^2} . \tag{41}$$

The value of the integral in (29) results mainly from contribution of small x, Accordingly at low temperatures, we can assume the overall integrand to be dominated by the

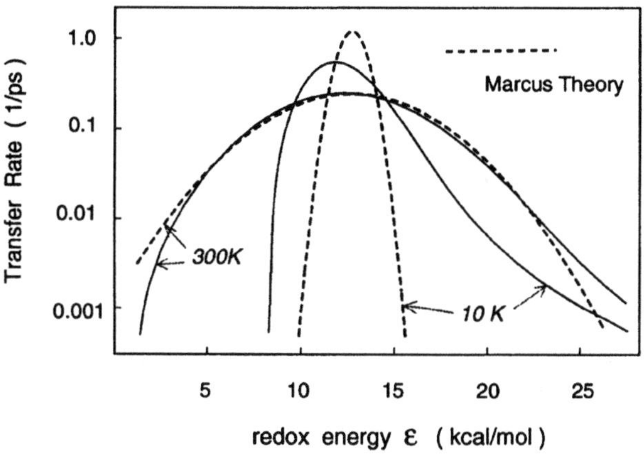

Figure 2. Comparison of electron transfer rates $k(\epsilon, T)$ shown as a function of ϵ evaluated in the framework of the spin–boson model (solid lines) and by Marcus theory (dashed lines) at temperatures 10 K and 300 K. The functions are centered approximately around ϵ_m.

interval in which $\gamma\pi^2 x^2/12\alpha^2$ is small . Therefore, one can apply (27) to expand the exponential part of (29),

$$e^{-4\gamma q_2(x)} = \exp\left(\gamma x^2 \ln x - \frac{\gamma\pi^2 x^2}{12\alpha^2}\right) \tag{42}$$

$$= \exp(\gamma x^2 \ln x)\left[1 - \left(\frac{\gamma\pi^2 x^2}{12}\right)\left(\frac{2k_B\tau T}{\hbar}\right)^2\right].$$

Then the electron transfer rate at $T \to 0$ can be expressed

$$k(\epsilon, T) \approx k(\epsilon, 0) - k_1(\epsilon)\left(\frac{2k_B\tau T}{\hbar}\right)^2, \tag{43}$$

where

$$k_1(\epsilon) = \left(\frac{2V}{\hbar}\right)^2 \tau \int_0^\infty dx \cos\left(\frac{\epsilon\tau}{\hbar}x\right)\cos\left[\gamma\pi\left(1 - e^{-x}\right)\right]\left(-\frac{\gamma\pi^2 x^2}{12}\right)\exp(\gamma x^2 \ln x). \tag{44}$$

From (43), one concludes that at low temperatures, the electron transfer rate is actually changing very slowly. This behavior has been found in many observations [5, 18].

6 Results

In Figure 2 we present the calculated electron transfer rates $k(\epsilon)$ as a function of the redox energy difference ϵ for temperatures $T = 10$ K and $T = 300$ K, and compare the results to transfer rates predicted by the Marcus theory. One can observe that at high temperatures, the rate evaluated from the Marcus theory in a wide range of ϵ agrees well with those evaluated from the spin–boson model at $T = 300$ K, a behavior which is expected from the high temperature limit derived above. However the Marcus theory and the spin–boson model differ significantly at $T = 10$ K. The rate as a function of ϵ at low temperatures for the spin–boson model is asymmetrical. This result agrees with observations reported in [6] which show a distinct asymmetry with respect to ϵ_m at low temperatures. Such asymmetry is not predicted by the models of Marcus and Hopfield [3, 4, 2].

If one makes the assumption that biological electron transfer systems evolved their ϵ-values such that rates are optimized, one should expect that electron transfer rates in the photosynthetic reaction center are formed through a choice of $\epsilon \rightarrow \epsilon_{max}$, such that $k(\epsilon_{max})$ is a maximum. In Fig. 3 we present corresponding maximum transfer rates, $k(\epsilon_{max})$ as well as non-optimal values for $\epsilon = \epsilon_{max} \pm \delta$, where $\delta = 2.5$ kcal/mol. Experimental data of electron transfer processes in the photosynthetic reaction center show increases similarly to those presented in Fig. 3 [19, 20, 21, 18]. However, Figure 3 demonstrates also that electron transfer at ϵ-values slightly off the maximum position can yield a different temperature dependence than that of $k(\epsilon_{max}, T)$, namely temperature independence or a slight decrease of the rate with decreasing temperature. Such temperature dependence has also been observed for biological electron transfer [18]. As Nagarajan et al. reported in [18] the temperature dependence of the transfer rate resembles that of $k(\epsilon_{max}, T)$ in photosynthetic reaction centers of native bacteria and in (M)Y210F mutants with tyrosine at the 210 position of the M–unit replaced by phenylalanine. However, a replacement of this tyrosine by isoleucine ((M)Y210I-mutant) yields a transfer rate which decreases like $k(\epsilon_{max} - \delta, T)$ shown in Fig. 3. This altered temperature dependence should be attributed to a shift of the redox potentials, i.e., $\epsilon_{max} \rightarrow \epsilon_{max} - \delta$.

7 Summary

The key new aspect of our investigation is two-fold: first, we base all model parameters on molecular dynamics simulations; second, the spin–boson model allows one to account for a very large number of vibrations quantum mechanically. We have demonstrated that the spin–boson model is well suited to describe the coupling between protein motion and electron transfer in biological redox systems. The model, through the spectral function $J(\omega)$, can be matched to correlation functions of the redox energy

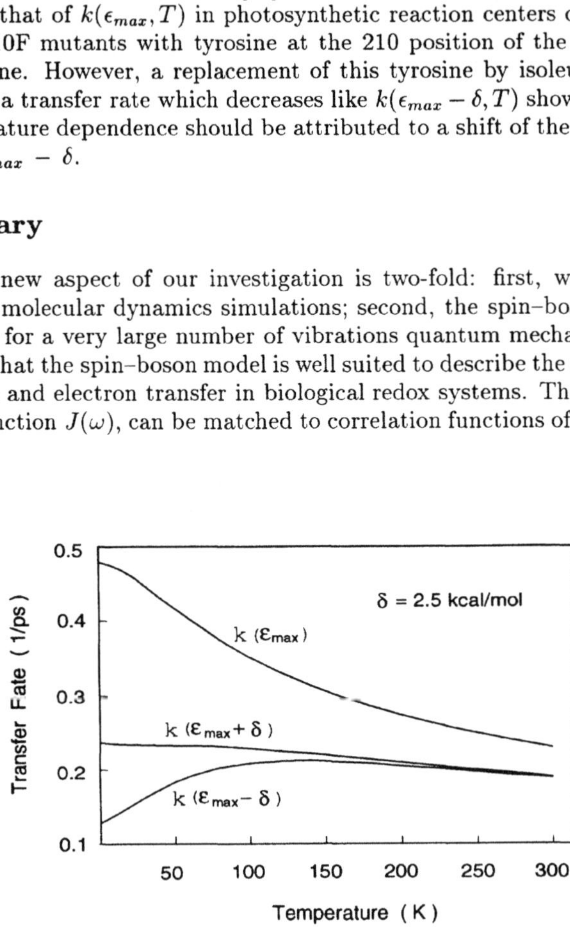

Figure 3. Comparison of the temperature dependence of the maximum transfer rate of $k(\epsilon_{max})$ and off-maximum value $k(\epsilon_{max} \pm \delta)$, where $\delta = 2.5$ kcal/mol. $k(\epsilon_{max}, T)$ represents the fastest transfer rate of the system, the rates $k(\epsilon_{max} \pm \delta, T)$ are slower since their ϵ–values deviate from the optimal value ϵ_{max}.

differences $\Delta E(t)$ through the relationships (15, 18) where $\Delta E(t)$ can be determined through a classical molecular dynamics simulation. We have demonstrated that the expressions for the electron transfer rates resulting from the spin–boson model can be evaluated numerically for a wide range of redox energy differences ϵ and temperatures T. The input parameters involved in the calculations are from molecular simulations rather than from an artificial fit. Hence even though the spin–boson model may not yield qualitatively different predictions from models involving a small number of vibrational modes coupled to the electron transfer, it certainly makes the role of the medium surrounding an electron transfer reaction appear in a new light: essentially all medium motions are coupled significantly to the reaction.

The main result regarding the electron transfer rates evaluated is that for a spectral function consistent with molecular dynamics simulations the spin–boson model at physiological temperatures predicts transfer rates in close agreement with those predicted by the Marcus theory. However, at low temperatures deviations from the Marcus theory arise. The resulting low temperature rates are in qualitative agreement with observations. The spin–boson model explains, in particular, in a very simple and natural way the slow rise of transfer rates with decreasing temperature, as well as the asymmetric dependence of the redox energy.

The combination of simulation methods and analytical theory has proven to be a promising approach to investigate biological redox processes. Neither approach by itself can be successful since, on the one hand, proteins are too heterogeneous and ill understood to be molded into simple models, on the other hand, simulation methods are blind, leaving one with too much information and as a result, with none. The present example, connecting a single simulated observable, the medium redox energy contribution $\Delta E(t)$, with a model, the spin–boson model, which does not contain superfluous or undetermined parameters, most likely can be extended to other important protein reactions.

Acknowledgements

We thank A. Leggett for directing us towards the spin–boson model and for helpful advice. We are also grateful to A. Szabo for pointing out the analytical approximation for $q_2(x)$. This work has been supported by the National Institutes of Health (grant P41-RR05969). Computer time for this project has been made available by the National Center for Supercomputing Applications funded by the National Science Foundation.

References

[1] Dong Xu and Klaus Schulten. Coupling of protein motion to electron transfer in a photosynthetic reaction center: Investigating the low temperature behaviour in the framework of the spin–boson model. *Chem. Phys.*, 1991. submitted.

[2] J. J. Hopfield. Electron transfer between biological molecules by thermally activated tunneling. *Proc. Natl. Acad. Sci. USA*, 71:3640–3644, 1974.

[3] R. A. Marcus and N. Sutin. Electron transfers in chemistry and biology. *Biochem. Biophys. Acta*, 811:265–322, 1985.

[4] H. Sumi and R. A. Marcus. Dynamical effects in electron transfer reactions. *J. Chem. Phys.*, 84:4894–4914, 1986.

[5] G. R. Fleming, J. L. Martin, and J. Breton. Rate of primary electron transfer in photosynthetic reaction centers and their mechanistic implications. *Nature.*, 333:190–192, 1988.

[6] M. R. Gunner and P. Leslie Dutton. Temperature and ΔG_o dependence of the electron transfer from BPh^- to Q_a in a reaction center protein from *rhodobacter sphaeroides* with different quinones as Q_a. *J. Am. Chem. Soc.*, 111:3400–3412, 1989.

[7] K. Schulten and M. Tesch. Coupling of bulk atomic motion to electron transfer: Molecular dynamics and stochastic quantum mechanics study of photosynthetic reaction centers. *Chemical Physics*, 158:421–446, 1991.

[8] M. Nonella and K. Schulten. Molecular dynamics simulation of electron transfer in proteins — theory and application to $Q_A \rightarrow Q_B$ transfer in the photosynthetic reaction center. *J. Phys. Chem.*, 95:2059–2067, 1990.

[9] Herbert Treutlein, Klaus Schulten, J.Deisenhofer, H.Michel, Axel Brünger, and Martin Karplus. Chromophore-protein interactions and the function of the photosynthetic reaction center: A molecular dynamics study. *Proc. Natl. Acad. Sci. USA*, 89:75–79, 1991.

[10] A. Garg, J. N. Onuchic, and V. Ambegaokar. Effect of friction on electron transfer in biomolecules. *J. Chem. Phys.*, 83:4491–4503, 1985.

[11] J. N. Onuchic, D. N. Beratan, and J. J. Hopfield. Some aspects of electron-transfer reaction dynamics. *J. Phys. Chem.*, 90:3707–3721, 1986.

[12] J. N. Onuchic. Effect of friction on electron transfer: The two reaction coordinate case. *J. Chem. Phys.*, 86:3925–3943, 1987.

[13] A. J. Leggett, S. Chakravarty, A. T. Dorsey, M. P. A. Fisher, A. Garg, and W. Zwerger. Dynamics of the dissipative two-state system. *Rev. Mod. Phys.*, 59:1–85, 1985.

[14] R. A. Marcus. On the energy of oxidation-reduction reactions involving electron transfer. I. *J. Chem. Phys.*, 24:966–978, 1956.

[15] R. A. Marcus. Electrostatic free energy and other properties of states having nonequilibrium polarization. II. *J. Chem. Phys.*, 24:979–989, 1956.

[16] Ilya Rips and Joshua Jortner. Dynamic solvent effects on outer-sphere electron transfer. *J. Chem. Phys.*, 87:2090–2104, 1987.

[17] J. S. Bader and D. Chandler. Computer simulation of photochemically induced electron transfer. *Chem. Phys. Lett.*, 157:501–504, 1989.

[18] V. Nagarajan, W. W. Parson, D. Gaul, and C. Schenck. Effect of specific mutations of tyrosine-(m)210 on the primary photosynthetic electron-transfer process in rhodobacter sphaeroides. *Proc. Natl. Acad. Sci. USA*, 87:7888–7892, 1990.

[19] M. Bixon and J. Jortner. Coupling of protein modes to electron transfer in bacterial photosynthesis. *J. Phys. Chem.*, 90:3795–3800, 1986.

[20] J. L. Martin, J. Breton, J. C. Lambry, and G. Fleming. The primary electron transfer in photosynthetic purple bacteria: Long range electron transfer in the femtosecond domain at low temperature. In J. Breton and A. Vermeglio, editors, *The Photosynthetic Bacterial Reaction Center: Structure and Dynamics*, pages 195–203, New York and London, 1988. Plenum Press.

[21] C. Kirmaier and D. Holten. Temperature effects on the ground state absorption spectra and electron transfer kinetics of bacterial reaction centers. In J. Breton and A. Vermeglio, editors, *The Photosynthetic Bacterial Reaction Center: Structure and Dynamics*, pages 219–228, New York and London, 1988. Plenum Press.

PULSED ELECTRIC FIELD INDUCED REVERSE ELECTRON TRANSFER FROM GROUND STATE $BChl_2$ TO THE CYTOCHROME c HEMES IN *Rps. viridis*

Guillermo Alegria, Christopher C. Moser
and P. Leslie Dutton

Johnson Research Foundation
Department of Biochemistry and Biophysics
University of Pennsylvania
Philadelphia, Pa. 19104

INTRODUCTION

Energy storage and utilization by living organisms is intimately linked to the transfer of reducing equivalents from a reductant to an oxidant through a series of membrane-bound redox proteins. The chemiosmotic mechanism of energy transduction [1] relies upon two essential geometric constraints: the separation of aqueous compartments by electrically tight and proton impermeant membranes, and the asymmetric arrangement of redox components across these membranes. The bioenergetic membranes of mitochondria, chloroplast and microorganisms convert redox potential free-energy into a transmembrane electric potential ($\Delta\Psi$) when electrons are transferred alone, into a concentration gradient (ΔpH) when electron transfers are coupled neutrally to proton movement across the membrane, and into both $\Delta\Psi$ and ΔpH when electron transfers are coupled to H^+ ion movement. This complementary action of the electron transfer chains builds up an electrochemical free energy gradient of protons ($\Delta\mu_{H^+} = \Delta\Psi + RT\log\Delta pH$) which is used to drive ATP synthesis and transport. The reversible nature of these coupled reactions has long been recognized *in vivo* by phenomenon of "reversed electron flow" [2] concomitant with ATP hydrolysis or $\Delta\Psi$ or ΔpH dissipation. This has been observed as the net reduction of oxidized low potential redox cofactors or substrates with concomitant oxidation of reduced high potential cofactors or substrates over nearly all the respiratory chain and much of the photosynthetic systems. However, observation of these events at a molecular level and on a single turnover basis is not as straightforward as, for instance, following the dominant exotermic reactions initiated by light or oxygen. Reverse electron flow is usually experimentally perceived as an increase in the steady state levels of reduction of low redox potential components upon addition of ATP [3] or generation of $\Delta\Psi$ or ΔpH [4].

The Photosynthetic Bacterial Reaction Center II
Edited by J. Breton and A. Verméglio, Plenum Press, New York, 1992

In this paper we address the attendant problems by using as a test system the electron transfer between the bacteriochlorophyll dimer ($BChl_2$) and the four hemes of the cytochrome c polypeptide subunit of the reaction center-cytochrome c complex of *Rps. viridis* . In the native state only the reaction between the $BChl_2$ and the first heme of the array (C_{559}) contributes significantly to transmembrane charge separation since the outer three hemes extend into the water phase [5]. Thus, only the first of the four electron transfer steps is significantly electrogenic and this alone therefore expected to be affected by the imposition of an external $\Delta\Psi$. Strategies adopted in the present work circumvent this limitation by construction of electro-deposited films of reaction center cytochrome c complex in phospholipids, followed by partial dehydration and incorporation into a capacitor with optically transparent electrodes. Under these conditions, it was predicted that the application of sufficiently large voltages to the capacitor would induce electron transfer to occur from the $BChl_2$ to the hemes by changing the normally endothermic character of these reactions ($+\Delta G$) into exothermic processes ($-\Delta G$).

Our study includes the effect of the application of external fields on films in which most of the cytochrome c complement is oxidized and the $BChl_2$ reduced. We have observed that modest fields induce the reduction of the closest heme (C_{559}) while higher electric fields resulted in transport to the furthest and lowest potential heme (C_{553}). In both cases the electron transfer process took place on a sub millisecond time scale. Removal of the voltage resulted in rapid return to the starting state. This is the first time that reverse electron transfer has been activated by voltages applied directly from an external electric source. This demonstration shows the feasibility of electric field induced reverse electron transfer, and holds promise as an unusually flexible alternative for pulse activation of ground state electron transfer.

MATERIALS AND METHODS

Preparation of electro-deposited films

Rps. viridis reaction centers were isolated from chromatophores according to standard chromatographic procedures. Reaction center-phospholipid vesicles were prepared as described before [6].

Electro-Deposited (ED) films were prepared on Indium Tin oxide (ITO) coated slides closely following the procedure described in ref. [7]. ED films were allowed to dry in a constant 12% relative humidity (R.H.) chamber for 24 hrs prior to the completion of the capacitive cell which consisted of vacuum evaporation of 5000 Å of polyethalene (PE) and 50 to 90 Å of Pt. Typical capacitor areas were from 50 to 80 mm^2. Capacitances ranged from 600pf to 1.2 nf , as measured with a B&K Precision digital capacitance meter. Film thicknesses were measured with a Talysurf apparatus, which probes the film surface with a fine diamond tip coupled to a mechanic-electric transducer. Typical values ranged from 3 to 6 mμ (± 40 Å). A schematic illustration of a finished capacitor is presented in figure 1.

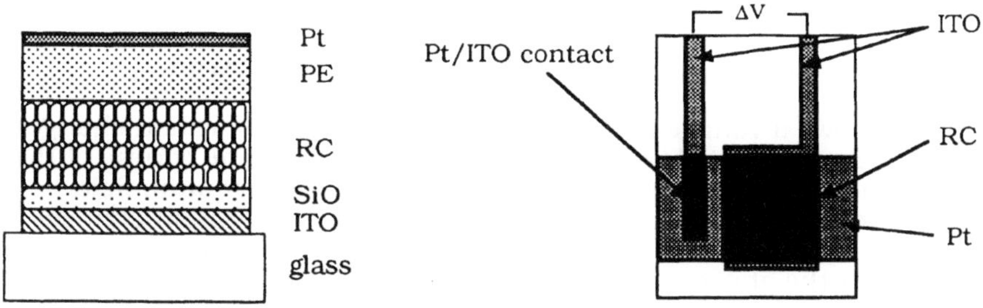

Figure 1. Schematic representation of a semi-transparent capacitive cell containing an electro-deposited film of *Rps. viridis* reaction centers.

Spectroscopic analysis

Standard and polarized absorption spectra of the ED films were obtained with the instrumentation described previously [6]. Time-resolved voltage-induced absorption changes were monitored with photomultiplier tubes in a single beam configuration in the 400-1000 nm range while measurements in the 1250-1400 nm region were performed with a liquid nitrogen cooled germanium diode. A more detailed description of the instrumentation and data collection procedures will be presented in a future publication.

All spectroscopic measurements were done at approximately 23% R.H. by keeping the slide in a sealed cuvette under a constant flow of argon humidified in a saturated NaCl solution.The film capacitance values were measured periodically during data collection .

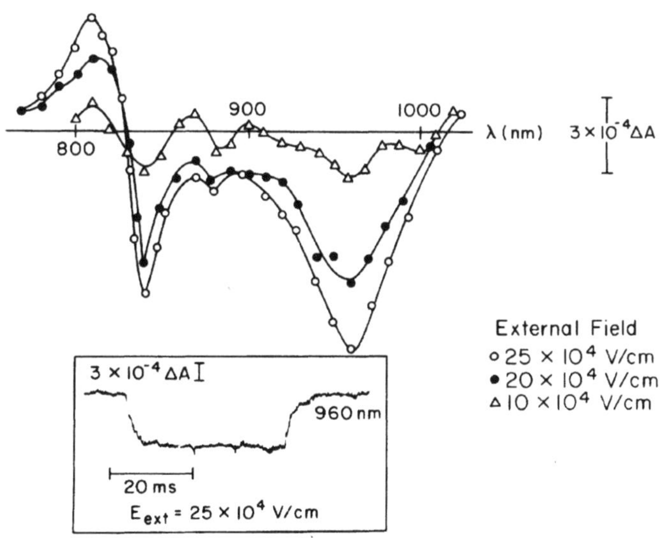

Figure 2. Electric field induced absorption changes in the near IR region of the spectrum in an ED film of *Rps. viridis*. Each point was averaged approximately 200 times. The inset shows the time course of the absorption change at 960nm upon application of a 40ms high voltage pulse (external field of 25 x 10^4 V/cm).

RESULTS

Field effects in the Reaction Center absorption spectrum

Figure 2 shows a voltage profile of the field-induced absorption changes in an ED film of *Rps. viridis* RCs. The inset shows the time course of the absorption changes at 960 nm corresponding to the maximum field applied. Notice that while at low electric fields the absorption changes involve increases and decreases averaging to approximately zero, at higher fields the absorption changes are identical to those produced by photo-bleaching of $BChl_2^+$. This strongly suggests that the field is inducing the oxidation of $BChl_2^+$.

A critical test for the notion that the external electric fields are inducing the formation of $BChl_2^+$ and not simply an electrochromic Stark effect, is the measurement of the field-induced changes in the 1300 nm band, since it is known to arise solely from $BChl_2^+$ and is absent in the starting $BChl_2$ state. Figure 3 shows the field profile of the absorption changes from 1230 to approximately 1400 nm in the same film as in figure 2. It is clear that the changes at low fields match those reported for the photo-bleaching of $BChl_2^+$, however, at the highest field applied the band appears to have a structure absent in the light-induced spectrum whose origin is still unclear.

Field effects in the cytochrome *c* α and Soret bands

Figure 4 shows the field induced absorption changes in the α and Soret bands of the cytochrome c hemes for the same film as in the previous figures. It can be observed that the field has produced absorption changes consistent with the reduction of a cytochrome. In addition, the absorption decrease in the 600 nm region confirms the formation of $BChl_2^+$. The inset shows the time course of the absorption changes at 553nm.

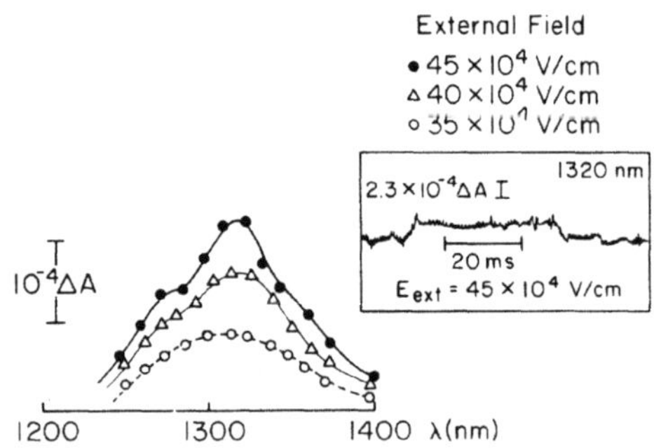

Figure 3. Electric field-induced absorption changes in the 1300 nm band in an ED film of *Rps. viridis* RCs. Inset shows the time course of the change at 1320nm upon application of a 45 x 10^4 V/cm external field.

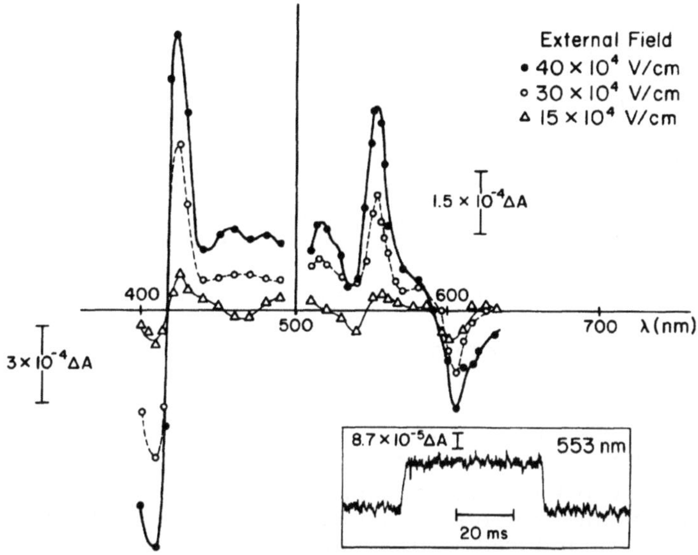

Figure 4. Electric field-induced absorption changes in the α and Soret bands of the cyt-c in an ED film of *Rps. viridis.* The inset shows the time course of the change at 553nm.

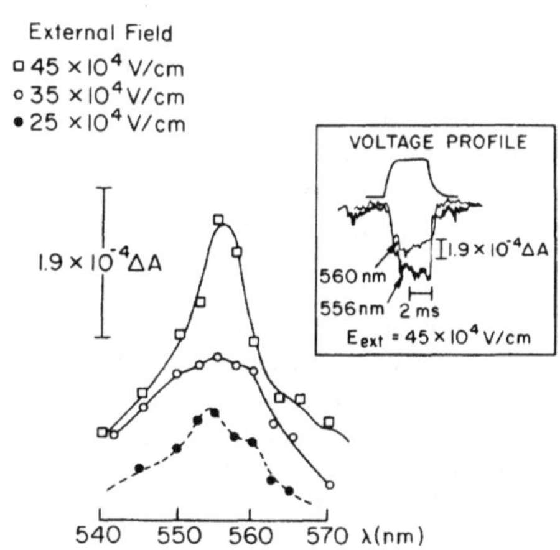

Figure 5. Field profile of the absorption changes in the α– band in the same film as in figure 4. The inset shows the time course of the change upon application of a 4ms pulse at 560 and 555nm.

Figure 5 shows the detailed field profile for the absorption changes in the α band of the cytochrome c. Notice that at low fields there is a broad absorbance increase suggesting reduction of more than one species of cytochrome c. At higher fields the position of the maximum shifts to the blue. Finally, at the highest field, the maximum seems to shift to approximately 553nm and the shoulder at 560nm has disappeared. The inset shows the time course of the applied voltage and of the absorption changes upon application of an external field of 45 x10^4 V/cm at two different wavelengths, 560nm and 555nm. Notice that the absorption change at 560nm decreases during the time when the voltage pulse remains constant while the change at 555nm increases during the same time interval. This result is reminiscent of the light-induced changes associated with the re-reduction of the photo-oxidized heme 559 by heme 556 [8,9] in *Rps. viridis* reaction centers in solution.

DISCUSSION AND CONCLUSIONS

Figures 2-5 provide a very clear demonstration that in our ED films the high voltage has induced electron donation from the reduced $BChl_2$ to the oxidized low potential cytochrome c hemes of the cyt c subunit. Additional evidence supports the notion of a field-induced redox reaction; the field induced absorption changes do not follow a quadratic dependence on the field. Indeed, the field dependence of the 960nm change seemed to follow a Nernstian behavior (data not shown). These field induced absorption changes are at least an order of magnitude larger than Stark changes observed in *Rb. sphaeroides* multilayer capacitors. In these preparations the reaction center is lacking cytochrome c and field induced electron transfer cannot take place.

We calculated the concentrations of $BChl_2^+$ and ferrocytochrome generated at a given field using the absorption changes at 960nm and 551nm (extinction coefficients for $BChl_2$ / $BChl_2^+$ and C_{red}/ C_{ox} of $123mM^{-1}cm^{-1}$ and $23mM^{-1}cm^{-1}$, respectively). For a field of 25×10^4 V/cm these numbers are 4.27×10^{12} $BChl_2^+/cm^2$ and 4.1×10^{12} C_{red}/cm^2 which are in excellent agreement and provide further support to the notion that the field triggers the reverse electron transfer reaction from $BChl_2$ to ferricytochrome(s) c.

An intriguing aspect of the results presented in figures 4 and 5 is the identification of the hemes involved in the redox reaction. At low fields the α-band spectrum suggests that reduction includes the highest potential heme with its maximum at 559nm; however, there is also a clear contribution centered at approximately 555nm. The inset of figure 5 shows that at high fields, the signal at 560nm is initially dominant but then, even though the field is constant, decreases concomittantly with an increase of the 555nm signal during the same time interval. This points out to the possibility that the field is inducing the electron transfer from heme 559 to either heme 556 or 553 and that the spectrum at low fields is begining to reveal that shift. As the external field is increased, the maximum shifts to approximately 553nm suggesting the reduction of the lowest potential heme.

Preliminary calculations of the free energy shifts between the hemes and $BChl_2$ introduced by the external field through a field-dipole interaction suggest that, given the free energy differences in solution and the range of field strengths accessible in our experiments, reduction of the lowest potential heme can only be accounted for if the free energy gap between the dimer and the heme in the dry state at zero field is considerably smaller than its value in solution. The cytochrome c subunit *in vivo* is known to be immersed in water, therefore it is not surprising that placing the RC in an environment with a significantly different dielectric constant can have an effect on the electrostatic interactions that determine the redox properties of the hemes. A detailed analysis of these effects will be presented in a future publication.

In summary, the present work demonstrates for the first time that some of the light-induced electron transfer reactions in photosynthetic reaction centers can be induced to occur in the reverse direction in the dark by application of external electric fields.
The observed reversible electron transfer reaction from $BChl_2$ to both, high potential and low potential cytochrome c hemes, can be rationalized in terms of an external field modulation of the free energy for the reaction through a field-dipole interaction. Preliminary analysis of the observed electron transfer from $BChl_2$ to the low potential heme 553 suggests that the midpoint potential of the heme may have suffered a considerable shift to more positive values under low humidity conditions.

Acknowledgment

This work was supported by Grant GM 41048-02 from the United States Public Health Service.

REFERENCES

1. Mitchell, P. (1961) Nature, **191**;144-148.
2. Chance, B. and Hollunger, G. (1961) **236**;1577-1584.
3. West, I.C., Mitchell, P. and Rich, P.R. (1988) Biochim. Biophys. Acta **933**;35-41.
4. Robertson, D.E. and Dutton, P.L. (1988) Biochim. Biophys. Acta **935**;273-291.
5. Dracheva, S.M., Drachev, L.A., Konstantinov, A.M., Semenov, A.Yu., Skulachev, V.P., Arutjunjan, A.M., Shuvalov, V.A., and Zaberezhnaya, S.M. (1988) Eur. J. Biochem. **171**;253-264.
6. Alegria, G. and Dutton, P.L. (1991) Biochim. Biophys. Acta **1057**;239-257.
7. Kononenko, A.A., Lukashev,E.P., Maximychev, A.V., Chamorovsky, S.K., Rubin, A.B., Timashev, S.F. and Chekulaeva, L.N. (1986) Biochim. Biophys. Acta **850**;162-169.
8. Shopes, R.J., Levine, L.M.A. , Holten, D. and Wraight, C.A. (1987) Photosynth. Res. **12**;165-180.
9. Kaminskaya, O., Konstantinov, A.A. and Shuvalov, V.A. (1990) Biochim. Biophys. Acta **1016**;153-164.

STRUCTURAL CHANGES FOLLOWING THE FORMATION OF $D^+Q_A^-$ IN BACTERIAL REACTION CENTERS: MEASUREMENT OF LIGHT-INDUCED ELECTROGENIC EVENTS IN RCs INCORPORATED IN A PHOSPHOLIPID MONOLAYER

Peter Brzezinski,[1] Melvin Y. Okamura,[2] and George Feher[2]

[1]Present address: Department of Biochemistry and Biophysics
Chalmers University of Technology, S-412 96 Göteborg, Sweden

[2]Department of Physics, University of California, San Diego
La Jolla, CA 92093-0319, USA

1. INTRODUCTION

The primary process of photosynthesis that takes place in the bacterial reaction center (RC) entails a light-induced excitation of a bacteriochlorophyll dimer (D) followed by a sequential transfer of an electron from D to the primary and secondary acceptors Q_A and Q_B [reviewed in (1)]. To obtain a large quantum yield the ratio of the forward electron transfer rate to the the recombination rate must be large. In RCs from *Rhodobacter sphaeroides* this ratio is $10^2 - 10^3$, making the charge-separation a very efficient process. It has been postulated that a conformational change accompanying charge separation may aid in producing the high ratio of forward to back reaction (2-4).

Conformational changes associated with light induced charge separation have been discussed by several authors (5-11). A particularly drastic effect was reported by Kleinfeld et al. (3) who compared the rate of electron transfer in RCs that were frozen under illumination with those frozen in the dark. They observed an increase in the rate of the electron transfer $Q_A^-Q_B \rightarrow Q_AQ_B^-$ by several orders of magnitude when RCs were frozen under illumination. In addition, the charge recombination $D^+Q_A^- \rightarrow DQ_A$ was slower (and non-exponential) in RCs frozen under illumination than in RCs frozen in the dark. The results were interpreted in terms of a structural rearrangement following the charge separation $DQ_A \rightarrow D^+Q_A^-$.

In this work we have studied light-induced voltage changes following charge separation in RCs oriented in a lipid monolayer. The measurements are sensitive to charge displacement in the RC arising from electron and proton transfer as well as from conformational changes in the RC. The technique we used was first described by Trissl et al. (12).

A preliminary account of this work has been presented (13).

2. MATERIALS AND METHODS

2.1 The Experimental Setup

The experimental setup is similar to the one described previously (14,15); a

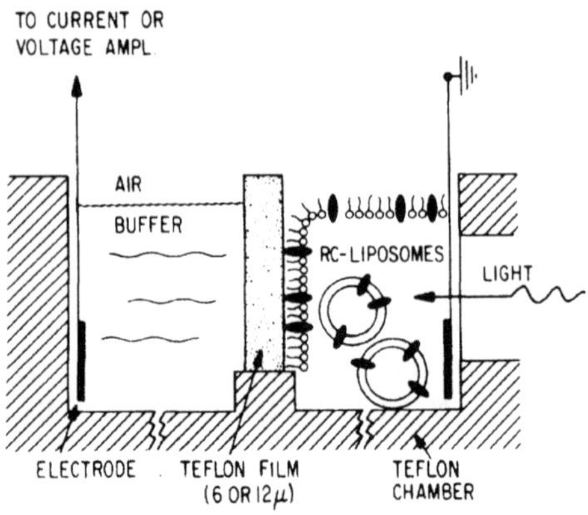

Figure 1. Schematic representation of the experimental set-up used to measure the electrogenicity of electron transfer steps (After refs. 14,15).

schematic representation of it is shown in Fig. 1. A black-Teflon chamber consisting of two aqueous compartments is separated by a 6 μm Teflon film (area ~ 0.3 cm^2, Sunders, CA). An RC-lipid monolayer was formed at the water surface in one of the cell compartments. The liquid level in the compartment was slowly raised, allowing the monolayer to adhere to the Teflon. The opposing cell compartment was filled with a 50 mM KCl solution. Light-induced voltage changes were measured across electrodes immersed in the cell compartments.

2.2 Reactants

RCs from *Rhodobacter sphaeroides* R-26 were purified with the solubilizing detergent LDAO as described (16). RCs with one quinone, i.e. Q_A, were prepared by the method of Okamura et al. (17). Cytochrome c (horse heart type VI, Sigma, MO) was more than 95% reduced with hydrogen gas, using platinum black (Aldrich, WI) as a catalyst (18). Soybean lecithin (Sigma, MO) was purified as described in (19).

2.3 Preparation of Liposomes Containing RCs

Lipid vesicles containing RCs were prepared as described (20). Briefly, the phospholipids were dispersed to a final concentration of 10 mg lipid/ml in a buffer composed of 10 mM KCl, 2.5 mM sodium citrate, 2.5 mM PIPES (1,4-piperazinediethanesulfonic acid), 2.5 mM HEPPS [4-(2-hydroxyethyl)-1-piperazinepropanesulfonic acid], 2.5 mM CHES (2-(cyclohexylaminoethanesulfonic acid)) and 2.5 mM CAPS [3-(cyclohexylamino) propanesulfonic acid][1] (all buffers except sodium citrate were from Calbiochem-Behring, CA). To prepare RCs with two quinones, phospholipid and ubiquinone (UQ_{10}, Sigma, MO) were dried together from a hexane solution prior to dispersion in the buffer at pH 8. An external quinone-to-RC ratio of ≥50:1 was required to obtain 85-95% of the RCs with an active Q_B. RCs in 0.025% LDAO were added to a final concentration of 5 μM. The solution was sonicated for ~ 10 minutes at 22 ± 2°C until it was optically clear.

2.4 Preparation of the RC-Lipid Monolayer

The vesicle solution was diluted to a final concentration of 1 mg lipid/ml in the buffer described above and supplemented with 10 mM $CaCl_2$. In the presence of Ca^{2+} the vesicles break and a monolayer is formed at the surface of the liquid (21,22).

[1]The effective pH range of the composite buffer was 4 < pH < 11.

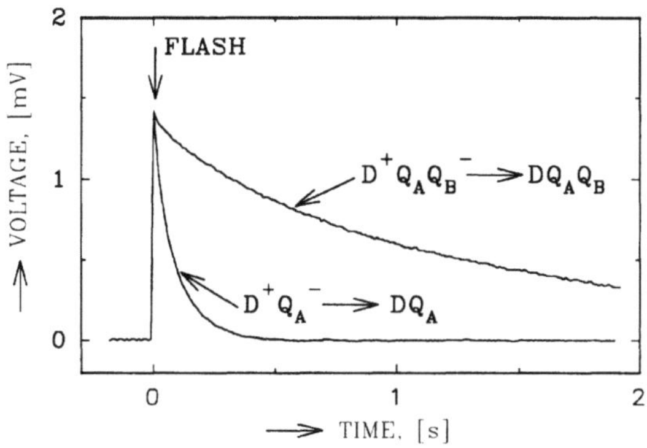

Figure 2. Voltage changes following pulsed illumination of RCs incorporated in a lipid monolayer. The initial increase in voltage is due to the charge separation $DQ_AQ_B \rightarrow D^+Q_A^-Q_B$. The electron transfer $D^+Q_A^-Q_B \rightarrow D^+Q_AQ_B^-$ proceeds on a time scale much shorter than shown. The decrease in voltage, with a rate constant of ~ 1 s^{-1}, is due to the charge recombination $D^+Q_AQ_B^- \rightarrow DQ_AQ_B$. Upon addition of 100 μM Terbutryne or in one-quinone RCs the charge-recombination rate constant was ~ 10 s^{-1}, characteristic of the recombination $D^+Q_A^-Q_B \rightarrow DQ_AQ_B$. The recombination rates are the same as observed in RCs in solution (see Fig. 3).

Conditions: pH 8.0, T = 20 °C, buffer as described in Materials and Methods.

2.5 Electrical Measurements

The electrical signals were detected using Calomel (Ingold Electrodes, MA) or silver/silver chloride electrodes that were shielded from actinic light. Measurements were carried out in a way similar to those described (23,24). The light-induced voltage changes were measured using an operational amplifier (LF356, National Semiconductors in the time range 100 ns – 10 μs or OP128, Burr-Brown in the time range 10 μs – 10 s). To improve the signal-to-noise ratio the signals were passed through an RC filter with a variable time constant and averaged using a Nicolet model 1180 transient recorder.

2.6 Light Sources

Pulsed illumination of the RC-lipid layer was provided by a pumped dye laser (Phase R, Model DL 1200V, NH) using the dye Rhodamine (Phase R, NH), which had a maximal output energy at 590 nm. The pulse had an energy of ~ 0.1 J and a width of ~ 0.4 μs.

3. EXPERIMENTAL RESULTS

3.1 The Basic Recombination Times of Q_A^- and Q_B^- with D^+

Following pulsed illumination of the RC-lipid interfacial layer we observed a rapid (>10^7 s^{-1}) increase in the voltage due to the initial charge separation $DQ_A \rightarrow D^+Q_A^-$, followed by a decrease in voltage due to the charge recombination (see Fig. 2). When the Q_B site was occupied the recombination rate constant was ~ 1 s^{-1}, consistent with the charge recombination of $D^+Q_AQ_B^-$ (see Fig. 3). Upon addition of 100 μM Terbutryne, which blocks electron transfer from Q_A^- to Q_B, or when RCs containing ~ 0.8 UQ_{10}/RC were used the rate constant increased to ~ 10 s^{-1}, consistent with the charge recombination of $D^+Q_A^-$ (Fig. 3).

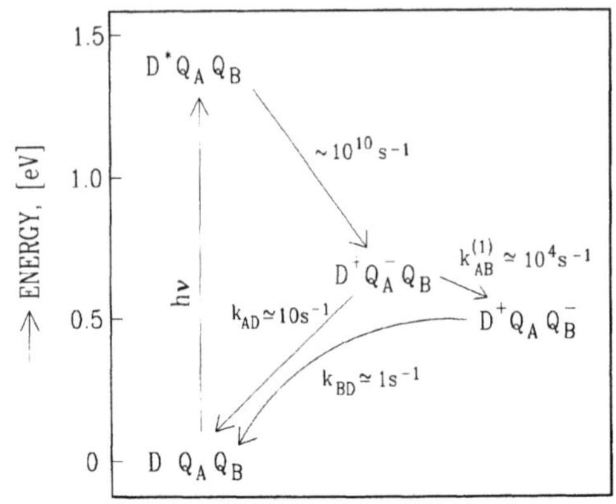

Figure 3. Electron-transfer reactions in the RCs of *Rb. sphaeroides*. Following the absorption of a photon, an electron is transferred from the excited donor, via an intermediate pheophytin acceptor (omitted for simplicity), to Q_A and then to Q_B. Transfer rates are given for room temperature and rounded to the nearest power of 10. The recombination rate k_{AD} is ~ 10^3 times smaller than the forward electron-transfer rate $k_{AB}^{(1)}$. The recombination k_{BD} proceeds via the intermediate state $D^+Q_A^-Q_B$.

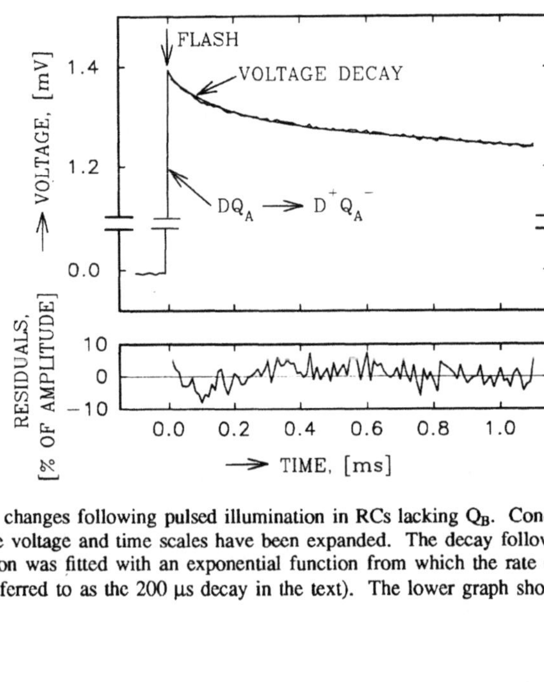

Figure 4. Voltage changes following pulsed illumination in RCs lacking Q_B. Conditions are the same as in Fig. 2 except the voltage and time scales have been expanded. The decay following immediately after the charge separation was fitted with an exponential function from which the rate constant of ~ 5500 s^{-1} was determined (referred to as the 200 μs decay in the text). The lower graph shows the residuals of the fit.

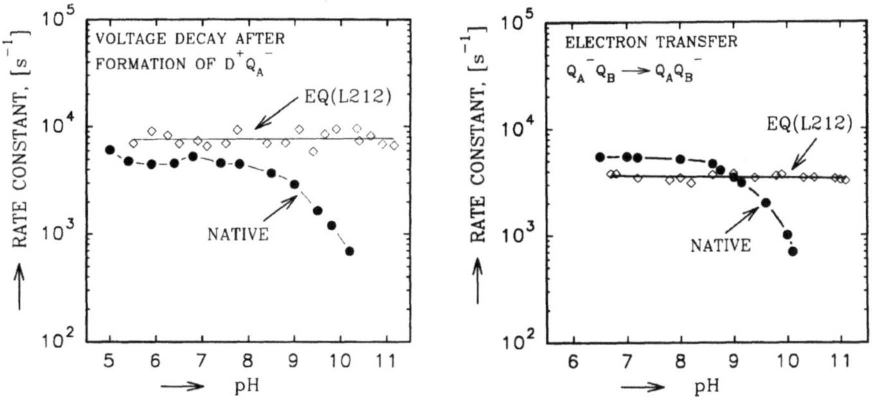

Figure 5. (A) pH dependence of the rate constant of the 200 μs voltage decay measured in native (●) and EQ(L212) mutant (◇) RCs lacking Q_B.

(B) pH dependence of the electron-transfer rate $Q_A^-Q_B \rightarrow Q_AQ_B^-$ in native (●) and EQ(L212) mutant (◇) RCs [data from ref. 25]. Conditions the same as in Fig. 2.

3.2 The 200 μs Decay

In Fig. 4 the voltage changes following pulsed illumination of the RC-lipid layer are shown on an expanded time and voltage scale. After the voltage change associated with the initial charge separation we observed a small decrease (decay) of the voltage with a rate constant of 5000 ± 1000 s^{-1}, i.e. a 200μs decay time (at pH 8.0). The decay rate was the same as that for the electron transfer $Q_A^-Q_B \xrightarrow{k_{AB}^{(1)}} Q_AQ_B^-$ (see Fig. 3). However, the same decay was observed in both RCs containing Q_B and in RCs lacking Q_B (in the presence of 100 μM Terbutryne or 0.8 UQ_{10} per RC). Therefore, the voltage decay *cannot be associated with the electron transfer* $Q_A^-Q_B \rightarrow Q_AQ_B^-$.

The same decay was also observed after addition of excess cytochrome c^{2+} (~ 1 mM), which reduced D^+ at a rate larger than 5000 s^{-1}, i.e. before the onset of the 200 μs decay. Thus, the 200 μs decay following the charge separation $DQ_A \rightarrow D^+Q_A^-$ was not associated with the $D \rightarrow D^+$ transition and must, therefore, be associated with the $Q_A \rightarrow Q_A^-$ transition.

The sign of the voltage decay was consistent with a negative charge moving in the direction from the cytoplasmic towards the periplasmic side of the RC (or a positive charge moving in the opposite direction). The magnitude of the voltage decay was 9±2% of the voltage change due to the $D^+Q_A^-$ charge separation. Assuming a homogeneous dielectric constant in the RCs, this corresponds to a displacement of one electronic charge by ~ 2 Å (i.e., 9% of the distance between D^+ and Q_A^-) in a direction perpendicular to the plane of the membrane. If several charges participate, a correspondingly smaller displacement would be required. The origin of the charge displacements could, for instance, be a structural change (see Summary and Discussion Section).

3.3 pH Dependence of the 200 μs Decay[1]

The rate of the voltage decay was essentially pH independent in the range 5<pH<8 and decreased above pH ~ 8 (Fig. 5A). The decay rate followed the pH dependence of

---

[1]The characteristic time of 200 μs is observed at pH8 and T = 20°C. However, we continue to refer to the phenomenon as the "200 μs decay" even when the characteristic time changes under different conditions.

$k_{AB}^{(1)}$ in the entire measured pH range (Fig. 5B). The decrease of $k_{AB}^{(1)}$ above pH ~ 8 has been attributed to the deprotonation of Glu-L212 (25) because in mutant RCs in which Glu-L212 was also replaced with its non-protonable analog Gln [(EQ(L212) mutant)], $k_{AB}^{(1)}$ was also pH independent above pH 8 (Fig. 5B). To test if the deprotonation of Glu-L212 above pH ~ 8 had an effect on the rate of the voltage decay we measured the voltage changes in the EQ(L212) mutant RCs. We found the 200 μs decay in the mutant RCs to be pH independent in the entire measured pH range (Fig. 5A). This shows that the electrogenic event in the RCs which generates the 200 μs voltage decay was influenced by the protonation state of Glu-L212.

3.4 Temperature Dependence of the 200 μs Decay

The temperature dependence of the decay rate was measured between 0 and 30°C (Fig. 6). The enthalpy of activation ($\Delta H^{\#}$) was calculated from the slope of the solid line to be 0.4 ± 0.1 eV. This value is similar to $\Delta H^{\#}$ of $k_{AB}^{(1)}$, which was determined to be ~ 0.56 eV (26).

4. SUMMARY AND DISCUSSION

We have measured voltage changes following the charge separation $DQ_A \rightarrow D^+Q_A^-$ in bacterial RCs containing one quinone. Following an increase in voltage associated with the charge separation $D^+Q_A^-$ we observed an ~ 9% decrease (decay) of the voltage with a characteristic decay time of ~ 200 μs. The decay rate, as well as its pH and temperature dependencies were very similar to those of the electron transfer $Q_A^-Q_B \xrightarrow{k_{AB}^{(1)}} Q_AQ_B^-$. In addition, we found that the pH dependence of the decay rate in the Glu L212 → Gln mutant was similar to the pH dependence of $k_{AB}^{(1)}$ in the mutant. Experiments with RCs having one quinone and in the presence of cytochrome c led us to conclude that the 200 μs decay is associated with the $Q_A \rightarrow Q_A^-$ transition and is *not* a consequence of electron transfer from Q_A^- to Q_B. Moreover, the conformational change associated with the decay may be a requirement for the electron transfer to occur. The question that needs to be answered: What is the origin of the 200 μs decay? We do not have a definitive answer at present. Several possible mechanisms are discussed below.

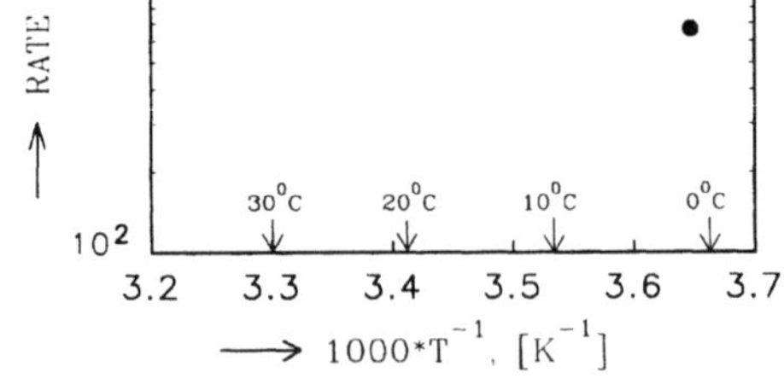

Figure 6. Temperature dependence of the rate constant of the 200 μs voltage decay shown in Fig. 4. The solid line represents a least-square fit to the data points. From the slope of the line the enthalpy of activation, $\Delta H^{\#} \simeq 0.4$eV, was determined. Conditions the same as in Fig. 2.

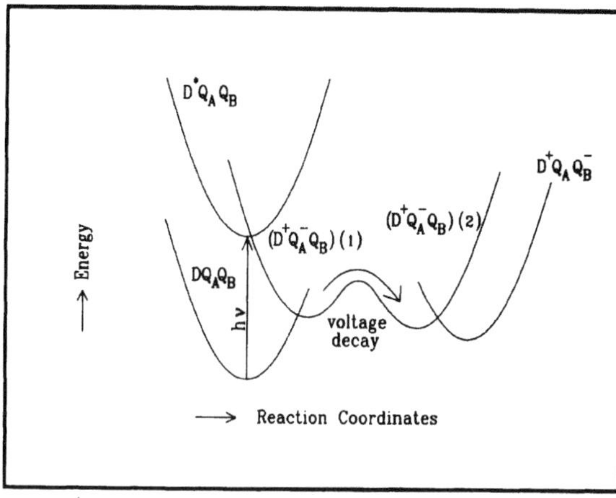

Figure 7. Schematic representation of a phenomenological model that explains the experimental observations. Following the absorption of a photon by the RC an electron is transferred from the excited donor (D^*) to Q_A forming the charge-separated species $D^+Q_A^-Q_B$. This species can exist in two different states (1) and (2). The transition from state (1) to state (2) is postulated to be electrogenic, giving rise to the observed 200 μs voltage decay. The electron transfer $Q_A^-Q_B \rightarrow Q_AQ_B^-$ can only take place from state (2).

4.1 Movement of Q_A^-

A voltage change can arise if Q_A^- moves in the quinone pocket towards D^+ after the charge separation. Kleinfeld et al. (3) interpreted the results of their recombination kinetics in terms of Q_A^- moving away from D^+ upon charge separation, which is not consistent with the sign of the observed voltage decay. They base their conclusions on the decreased rate of charge recombination, k_{AD}, when RCs were frozen under illumination (i.e., in the charge-separated state). Although a movement of Q_A^- away from D^+ is the simplest interpretation of their results, in the absence of a detailed quantitative theory of the k_{AD} mechanism, other explanations of their results cannot be excluded.

An inspection of the RC structure shows that Q_A can move up to ~ 2 Å without static hindrance (27); the allowed displacement is essentially parallel with the membrane surface and is therefore not expected to be electrogenic.

4.2 Proton Uptake or Release

Since protons carry a charge, proton transfer within the RCs or between the solution and the RCs can produce a voltage change. Following the charge separation $DQ_A \rightarrow D^+Q_A^-$, there is a proton uptake associated with the $Q_A \rightarrow Q_A^-$ transition (28,29). The proton uptake upon forming Q_A^- takes place on the same time scale as the voltage decay (30), which suggests that these events may be related. The sign of the observed voltage change requires that proton transfer proceeds from the periplasmic side of the RC towards Q_A^-. Evidence against this process is provided by the observation that the decay was independent of the charge state of D which is nearer to the periplasmic side than Q_A. An alternative possible mechanism involves a conformational change, upon formation of Q_A^-, that would bring negative charges closer to protonatable amino acids, allowing them to become protonated.

4.3 A General, Thermodynamic, Model that Explains the Changes in Kinetics Associated with the Charge Separation

It is clear from the preceding sections that we have been unable so far to definitively identify the molecular process that gives rise to the 200 μs decay. We resort, therefore, to a phenomenological model to explain the observed experimental results.

The model, shown in Fig. 7, is based on the assumption that the charge-separated state can exist in two conformations designated by $(D^+Q_A^-Q_B)(1)$ and $(D^+Q_A^-Q_B)(2)$, to be abbreviated henceforth as states (1) and (2). We postulate that the observed 200 μs decay is associated with the transition from state (1) to (2). The electron transfer from Q_A^- to Q_B, ($k_{AB}^{(1)}$), can occur only from state (2). This model is in accord with the following observations:

i) The time constant of the 200 μs decay is the same as that of $k_{AB}^{(1)}$. This is expected if the transition from (1) to (2) is the rate-limiting step in the transfer of an electron from Q_A^- to Q_B.

ii) The 200 μs decay is observed in the absence of Q_B. In the model Q_B plays no role in the transition between (1) and (2).

iii) The effect of Glu L212 on the 200 μs decay. The protonation state of Glu L212 affects the relative stability of states (1) or (2), thereby affecting the transfer kinetics between the two.

iv) The freezing out of $k_{AB}^{(1)}$ in RCs cooled in the dark (3). The transition from (1) to (2) is thermally activated. RCs cooled in the dark are trapped in state (1), unable to overcome the barrier at low temperatures.

v) RCs cooled under illumination are able to transfer an electron from Q_A^- to Q_B at low temperatures. RCs cooled under illumination are trapped in state (2) from which an electron can be transferred to Q_B in an essentially activationless process.

In the model described, we have assumed for simplicity only two discrete states. However, it is likely that there will be a distribution of states. Evidence for a distribution comes from the observed non-exponential charge recombination kinetics (3,8,9).

The states in the above model were postulated on an ad hoc basis. The challenge now is to identify the structural differences between state (1) and (2). One approach would be to obtain the X-ray structure of the RCs in the charge-separated state. Another would be to start with the known static structure and to explore possible arrangements following charge separation by molecular dynamics simulations.

ACKNOWLEDGEMENTS

We thank E. Abresch for the preparation of reaction centers, A. Messinger and R.A. Isaacson for their technical assistance, and P. McPherson, M. Paddock, and J. Onuchic for helpful discussions. The original contributions of Y. Blatt who initiated the monolayer work in our laboratory is gratefully acknowledged. The work was supported by a grant from the National Science Foundation (DMB89-15631) and a fellowship (to P.B.) from the Swedish Natural Science Research Council.

REFERENCES

1. G. Feher, J.P. Allen, M.Y. Okamura, and D.C. Rees, Structure and function of bacterial photosynthetic reaction centres, *Nature* 339:111 (1989).
2. A. Warshel, Role of the chlorophyll dimer in bacterial photosynthesis, *Proc. Natl. Acad. Sci. USA* 77:3105 (1980).
3. D. Kleinfeld, M.Y. Okamura, and G. Feher, Electron-transfer kinetics in photosynthetic reaction centers cooled to cryogenic temperatures in the charge-separated state: evidence for light-induced structural changes, *Biochemistry* 23:5780 (1984).
4. G. Feher, M.Y. Okamura, and D. Kleinfeld, Electron transfer reactions in bacterial photosynthesis; charge recombination kinetics as a structure probe, in: "Protein Structure: Molecular and Electronic Reactivity," R. Austin, E. Buhks, B. Chance, D. Devault, P.L. Dutton, H. Frauenfelder, and V.I. Goldanskii, eds., Springer Verlag, New York (1987).
5. P.P. Noks, E.P. Lukashev, A.A. Kononekko, P.S. Venediktov and A.B. Rubin, On possible role of macromolecular components in the functioning of the photosynthetic reaction centres of purple bacteria, *Mol. Biol. (Moscow)* 11:1090 (1977).

6. H. Arata, and W.W. Parson, Enthalpy and volume changes accompanying electron transfer from P-870 to quinones in *Rhodopseudomonas sphaeroides* reaction centers, *Biochim. Biophys. Acta* 636:70 (1981).

7. H. Arata, and W.W. Parson, Delayed fluorescence from *Rhodopseudomonas sphaeroides* reaction centers; enthalpy and free energy changes accompanying electron transfer from P-870 to quinones, *Biochim. Biophys. Acta* 638:201 (1981).

8. N.W. Woodbury, and W.W. Parson, Nanosecond fluorescence from isolated photosynthetic reaction centers of *Rhodopseudomonas sphaeroides*, *Biochim. Biophys. Acta* 767:345 (1984).

9. C. Kirmaier, D. Holten, and W.W. Parson, Temperature and detection-wavelength dependence of the picosecond electron-transfer kinetics measured in *Rhodopseudomonas sphaeroides* reaction centers. Resolution of new spectral and kinetic components in the primary charge-separation process, *Biochim. Biophys. Acta* 810:33 (1985).

10. P. Parot, J. Thiery, and A. Vermeglio, Charge recombination at low temperature in photosynthetic bacteria reaction centers, in: "The Photosynthetic Bacterial Reaction Centre", J. Breton, and A. Vermeglio, eds., Plenum Press, New York (1988).

11. E. Nabedryk, K.A. Bagley, D.L. Thibodeau, M. Bauscher, W. Mäntele, and J. Breton, A protein conformational change associated with the photoreduction of the primary and secondary quinones in the bacterial reaction centers, *FEBS Lett.* 266:59 (1990).

12. H.-W. Trissl, A. Darszon, and M. Montal, Rhodopsin in model membranes: charge displacements in interfacial layers, *Proc. Natl. Acad. Sci. USA* 74:207 (1977).

13. P. Brzezinski, M.L. Paddock, A. Messinger, M.Y. Okamura, and G. Feher, Electrogenic change after the formation of $D^+Q_A^-$ in bacterial reaction centers, *Biophys. J. (abstr.)* 61:A101 (1992).

14. Y. Blatt, A. Gopher, M. Montal, and G. Feher, Photovoltages from reaction centers incorporated in interfacial lipid layers, *Biophys. J.* 41:A121 (1983); A. Gopher, Y. Blatt, M. Schönfeld, M.Y. Okamura, G. Feher, and M. Montal, The effect of an applied electric field on the rate of charge recombination in reaction centers incorporated in planar lipid bilayers, *Biophys. J.* 48:311 (1985).

15. G. Feher, and M.Y. Okamura, Structure and function of the reaction centers from *Rhodopseudomonas sphaeroides*, in: "Advances in Photosynthesis Research," C. Sybesma, ed., M. Nijhoff, W. Junk, The Netherlands (1984).

16. G. Feher, and M.Y. Okamura, Chemical composition and properties of reaction centers, in: "The photosynthetic Bacteria," R.K. Clayton and W.R. Sistrom, eds., Plenum Press, New York (1978).

17. M.Y. Okamura, R.A. Isaacson, and G. Feher, The primary acceptor in bacterial photosynthesis: the obligatory role of ubiquinone in photoactive reaction centers of *Rhodopseudomonas sphaeroides*, *Proc. Natl. Acad. Sci. USA* 72:3491 (1975).

18. P. Rosen, and I. Pecht, Conformational equilibria accompanying the electron transfer between cytochrome c (P551) and azurin from *Pseudomonas aeruginosa*, *Biochem.* 15:775 (1976).

19. Y. Kagawa, and E. Racker, Partial resolution of the enzymes catalyzing oxidative phosphorylation, *J. Biol. Chem.* 246:5477 (1971).

20. A. Gopher, Y. Blatt, M. Schönfeld, M.Y. Okamura, G. Feher, and M. Montal, The effect of an applied electric field on the rate of charge recombination in reaction centers incorporated in planar lipid bilayers, *Biophys. J.* 48:311 (1985).

21. H.G. Schindler, Exchange and interactions between lipid layers at the surface of a liposome solution, *Biochim. Biophys. Acta* 555:316 (1979).

22. H.G. Schindler, Formation of planar bilayers from artificial or native membrane vesicles, *FEBS Lett.* 122:77 (1980).

23. M. Schönfeld, M. Montal and G. Feher, Functional reconstitution of photosynthetic reaction centers in planar lipid bilayers, *Proc. Natl. Acad. Sci. USA* 76:6351 (1979).

24. G. Feher T.R. Arno, and M.Y. Okamura, The effect of an electric field on the charge recombination rate of $D^+Q_A^- \rightarrow DQ_A$ in reaction centers from *Rhodobacter sphaeroides* R-26, in: "The photosynthetic Bacterial Reaction Center," J. Breton, and A. Vermeglio, eds., Plenum Press, New York (1988).

25. M.L. Paddock, S.H. Rongey, G. Feher, and M.Y. Okamura, Pathway of proton transfer in bacterial reaction centers: replacement of Glu^{212} in the L subunit by Gln inhibits quinone (Q_B) turnover, *Proc. Natl. Acad. Sci. USA* 86:6602 (1989).

26. D. Kleinfeld, M.Y. Okamura, and G. Feher, Electron transfer in reaction centers of *Rhodopseudomonas sphaeroides*. I. determination of the charge recombination pathway of $D^+Q_AQ_B^-$ and free energy and kinetic relations between $Q_A^-Q_B$ and $Q_AQ_B^-$, *Biochim. Biophys. Acta* 766:126 (1984).

27. J.P. Allen, G. Feher, T.O. Yeates, H. Komiya, and D.C. Rees, Structure of the reaction center from *Rhodobacter sphaeroides* R-26: protein-cofactor (quinones and Fe^{2+}*) interactions, Proc. Natl. Acad. Sci. USA* 85:8487 (1988).

28. P.H. McPherson, M.Y. Okamura, and G. Feher, Light-induced proton uptake by photosynthetic reaction centers from *Rhodobacter sphaeroides* R-26, I. protonation of the one-electron states $D^+Q_A^-$, DQ_A^-, $D^+Q_AQ_B^-$, and $DQ_AQ_B^-$, *Biochim. Biophys. Acta* 934:348 (1988).

29. P. Maróti, and C.A. Wraight, Flash-induced H^+ binding by bacterial photosynthetic reaction centers: influences of the redox states of the acceptor quinones and primary donor, *Biochim. Biophys. Acta* 934:329 (1988).

30. P. Maróti, and C.A. Wraight, Anamalous kinetics of flash-induced H^+-ion binding in reaction centers from *Rb. sphaeroides, Biophys. J. (abstr.)* 55:428a (1989).

CHARGES RECOMBINATION KINETICS IN BACTERIAL PHOTOSYNTHETIC REACTION CENTERS: CONFORMATIONAL STATES IN EQUILIBRIUM PRE-EXIST IN THE DARK

Barbara Schoepp, Pierre Parot, Jean Lavorel
and André Verméglio

DPVE/SBC C.E. CADARACHE 13108 Saint-Paul-lez-Durance
FRANCE

INTRODUCTION

In whole bacterial cells (or in chromatophores) treated with orthophenantrolin, or in purified reaction centers, RCs, prepared with only one quinone acceptor, Q_A, the $P^+Q_A^-$ state resulting from the photochemical charge separation only decays by recombination ($P^+Q_A^- \longrightarrow PQ_A$). Several years ago, we have studied the complex wavelength dependence of the $P^+Q_A^-$ decay at cryogenic temperatures and revealed its biphasicity for RCs and chromatophores of *Rhodobacter sphaeroides* and *Rhodospirillum rubrum*. In order to explain this biphasicity we proposed that the RCs were present in two conformational states[1,2,3]. On the other hand, for these same species, at room temperature, the decay is monophasic, suggesting that under this condition the two conformational states can equilibrate rapidly[4]. Numerous instances of polyphasic kinetics in bacterial systems have been documented. Biphasicity of charge recombination from $P^+Q_A^-$ or $P^+Q_B^-$ at room temperature has been reported in *Rhodopseudomonas viridis* RCs[5,6,7] and *Rhodobacter sphaeroides* RCs where the native ubiquinone has been replaced by anthraquinone[8]. The above results have been interpreted as originating from two distinct conformational states pre-existing the excitation flash. In the nanosecond time range, Woodbury and Parson[9] have observed polyphasic fluorescence decay. Careful analysis of the formation and decay of the P^+I^- state shows that these events appear with a range of time constants depending on the detecting wavelength[10]. In *Chloroflexus aurantiacus* at low temperature, the decay of P^* is clearly non-exponential[11]. A proposed possible explanation of this heterogeneity could be a rapid interconversion among a distribution of conformational states in the excited state P^*. In the present work, we have studied the temperature (between 10K and 300K), pre-illumination and wavelength dependence of the recombination reaction in order to : 1) find out whether the RC conformational states are formed by the flash or pre-exist in the dark and 2) inquire at what temperature these states can equilibrate rapidly.

MATERIALS AND METHODS

Several bacterial species were used for preparing RCs and/or chromatophores : *Rhodobacter sphaeroides* strains R26 and 2-4-1, *Rhodospirillum rubrum* strain G9. Light-induced absorption changes were measured with a home-built apparatus as described previously[1]. Due to critical temperature phenomena of the water-glycerol mixture, the recordings were disturbed near 250K when the temperature was raised too slowly, the samples becoming most of the time milk-white; this effect can explain some uncertainty in our results in the above temperature range. Data were visualised and recorded using a digital oscilloscope Tektronix 7D20 under control of a Compaq 386/33 microcomputer. When needed, the sample was pre-illuminated at 900 nm, with a Sylvania 800 W lamp filtered through a 900 nm interference filter (Pomfret Research Optics Inc., 50 nm bandwidth). The kinetic data were processed with a program (AT) implementing a modified Simplex algorithm for analyzing a sum of exponentials (see Appendix 1).

RESULTS AND DISCUSSION

Biphasicity

The significance of the biphasic decay at low temperature as indicating the occurence of two distinct kinetic components, i.e. two conformational states, is most clearly demonstrated when observing the kinetics in the vicinity of the isobestic point of the light-induced difference spectrum (800 or 762 nm, see Fig. 1) : depending on the precise observation wavelength, a variety of complex kinetics are obtained with components of diverse signs (*e.g*. Fig. 2) and magnitudes.

Remarkably, the fit is always optimal with two components and their time constants are independent of the observation wavelength. Actually, at some specific wavelength, one may even almost single out the fast or the slow component and thus observe it in isolation (*e.g*. Fig. 4b below). One can get such complex kinetics around the isobestic points up to about 200K ; above that temperature range, the spectroscopic distinction between the components gets blurred and the biphasicity may only be studied in the 810 nm band (or in the dimer's long wavelength absorption band).

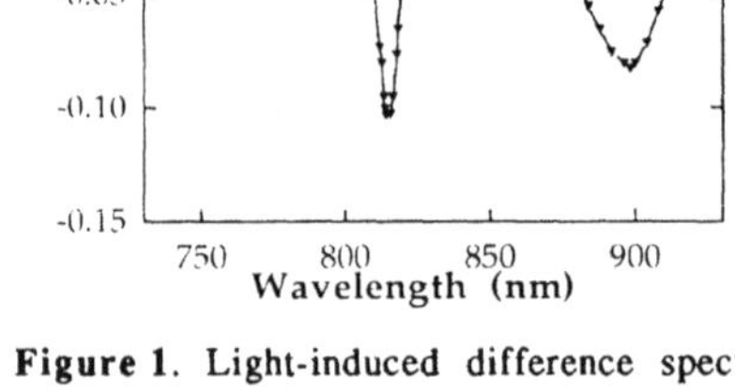

Figure 1. Light-induced difference spectrum measured at 10K for a suspension of reaction centers prepared from *Rhodobacter sphaeroides* 2-4-1. Arrows indicate wavelength of complex kinetics signals. The absorbance changes were induced by an actinic laser flash at 590 nm.

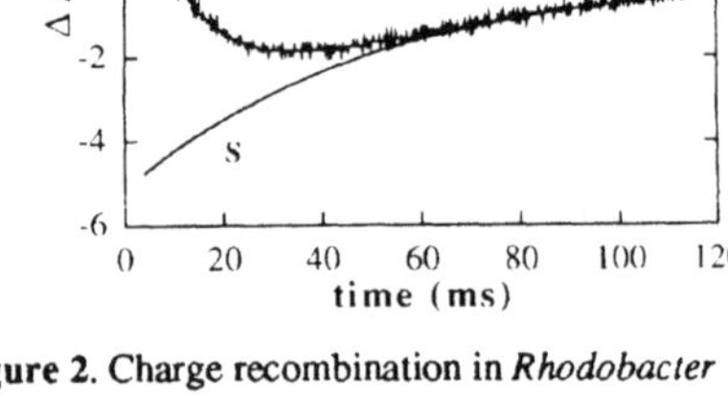

Figure 2. Charge recombination in *Rhodobacter sphaeroides* R26 reaction centers at 70K. The analysis resolved two exponential components (F : fast, S : slow) in the complex ΔA signal at 801 nm. F : time constant = 10 ms, amplitude = 60 % ; S : time constant = 30 ms, amplitude = 40 %.

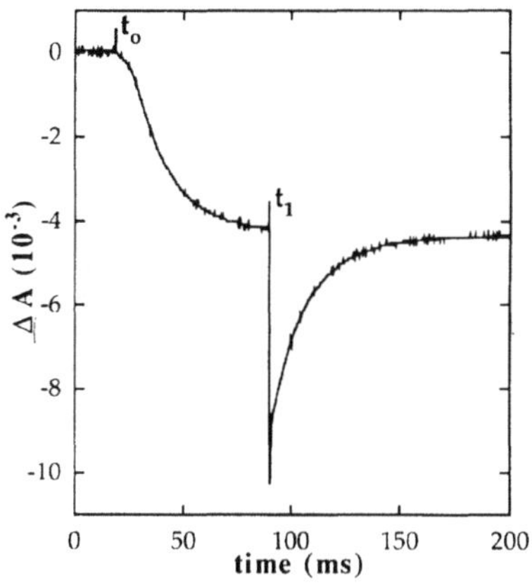

Figure 3a. Flash-induced absorption changes in the presence of a continuous background illumination for a suspension of reaction centers prepared from *Rhodobacter sphaeroides* R26. Background was turned on at time t_0; laser flash was triggered at time t_1.

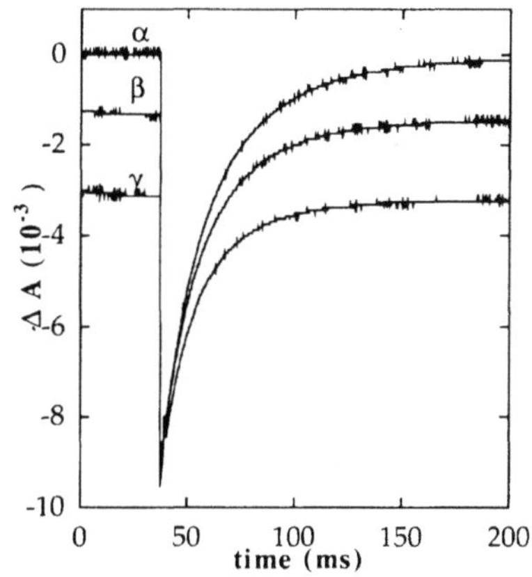

Figure 3b. Charge recombination kinetics without (α) and with background illumination (relative intensities β : 40 %, γ : 100 %) for *Rhodobacter sphaeroides* R26 reaction centers.The relative amplitudes and time constants of the slow and fast phases are given in Table 1.

Are the two conformational states formed by the flash or do they pre-exist in the dark? Effect of background continuous illumination

A simple kinetic model for explaining the biphasicity of the low temperature charge recombination (model A), is to assume that the reaction center at rest takes on a single form (say R), but that the photochemical charge separation induces two states (say F^* and S^*) with different decay rates, both recombining to R. In other words, model A assumes that the states distribution is created by light activation. Alternatively, one might suppose (model B) that the two states on the contrary pre-exist before flash excitation, R being replaced by two distinct forms (say F and S). We thought that double flashes or flash over background illumination experiments could help in selecting the right model. For technical reasons, only the second type of experiment was performed (see Fig.3a).

TABLE 1. Recombination kinetics following a saturating flash in the presence of background illumination of variable intensity. The data are relative to the decays in Fig. 3b. I : Background relative intensity ; τ : time constant (ms) ; % : amplitude per cent of total.

	Slow		Fast	
I	τ	%	τ	%
0 %	76	21	21	79
40 %	78	16	20	84
100 %	81	7	18	93

Clearly, assuming that S* is indeed very slow, firing a flash over a non-saturating steady-state background will induce almost a fast decay in model B, or the same biphasic decay relatively as without background in model A (see Appendix 2 for a more precise qualification). Figure 3b illustrates the effect of a variable background intensity on a subsequent flash-induced recombination.

The decays are analyzed in Table 1 showing that the relative amplitude of the slow component decreases when the background intensity increases. The effect is best seen in the vicinity of the isobestic point (Figure 4 (a and b)) where the disappearance of the slow component is directly evidenced. The above results are rather in favor of model B (pre-existing states), since the amplitude of the slow component following a flash induced absorption changes in the presence of background illumination does not depend on the background light intensity in model A (see Appendix 2).

Effect of pH and illumination during cooling

We had noted previously[1] that the recombination was not affected by pH when RCs were cooled in darkness and that, as also observed by Kleinfeld[4] , the recombination time was increased and became pH dependent when RCs were cooled under strong continuous illumination. These observations were repeated more carefully with *Rhodospirillum rubrum* chromatophores at 10K (Table 2). Illumination during cooling induces a threefold increase of both time constants at pH 7.8 compared to darkness and still more at pH 11 ; also it enhances somewhat the relative S amplitude.

The pH dependence of the overall time constants for RCs cooled under illumination seems to obey reasonably well the relation $\tau = (H^+)^{0.045}$ proposed by Kleinfeld. The half time of

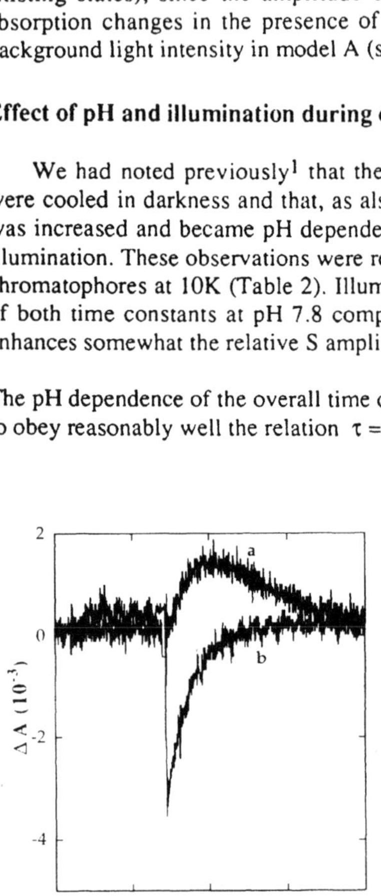

Figure 4a. Charge recombination kinetics observed in *Rhodobacter sphaeroides* R26 reaction centers close to the isobestic point (801 nm) with (a) and without (b) background illumination. Notice that the slow component (positive) in curve a is almost absent in curve b.

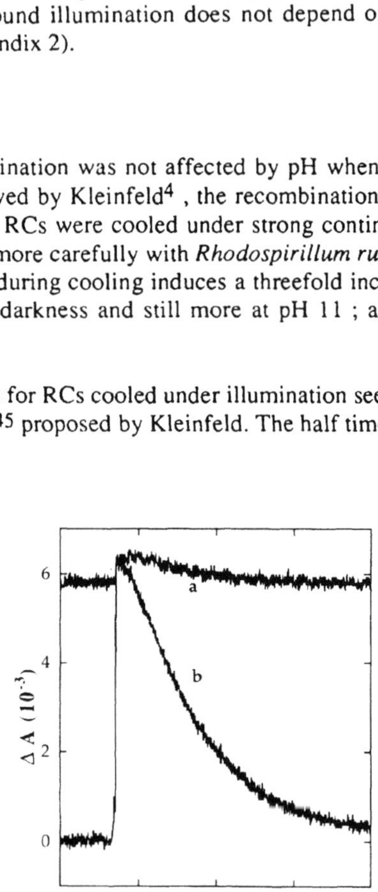

Figure 4b. Kinetics of the slow component of the same reaction centers as in Fig 4a, observed at 800.5 nm (see Fig. 5a) with (a) and without (b) background illumination. The slow component amplitude is strongly depressed in curve a ; a slight inflexion at the start of the decay is due to the fast component.

TABLE 2. Effect of pH and/or light conditions during cooling on the recombination kinetics in *Rhodospirillum rubrum* G9 chromatophores at 10 K. Time constants (ms) in columns under the labels "Fast" and "Slow" ; S : amplitude of slow phase per cent of total.

cooled	pH	Fast	Slow	S
in darkness	7.8 or 11	7.7	21.1	37 %
illuminated	7.8	16.6	71.7	47 %
illuminated	11	22.3	125.4	46 %

the fast component is less sensitive to pH than the slow one. This result is in agreement with the proposal of Franzen *et al*[12] who suggested from an electric field effect on the charge recombination at low temperature that the slow recombination involves a proton which is not present for the fast recombination process.

Perhaps not unrelated to the above is the observation that continuous illumination at low temperature tends to accumulate states with longer recombination time constants, particularly for S* (not shown).

From the above series of experiments, we conclude that two conformational states pre-exist at low temperature. The relative contribution of these states and the decay kinetics depend on the pH and pre-illumination during cooling.

Temperature dependent equilibrium of two conformational states

About the simplest kinetic model (model C, see Appendix 3) for explaining that the decay, biphasic at cryogenic temperatures, becomes monophasic at room temperature is to assume that the two conformational states are actually always in equilibrium. In the first case, the situation is practically frozen (at least in the time range of recombination), whereas, in the second case, the partial reactions of the equilibrium do take part in the kinetics of the decay. Such a model may be pictured as follows,

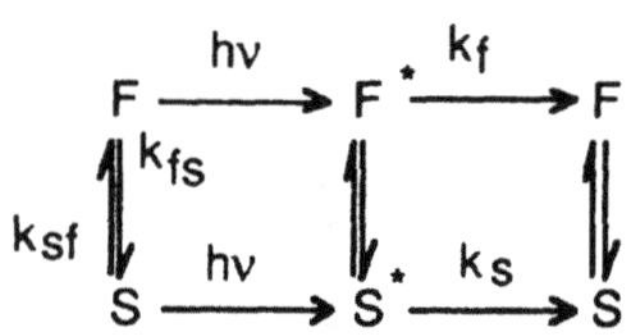

(1)

where F and S stand for the fast- and slow-recombining state of the reaction center respectively, and the starred symbols for their corresponding photoactivated versions; k_f and k_s are the recombination rate constants with $k_f > k_s$; it is assumed that the equilibrium rate constants k_{sf} and k_{fs} are not modified by the light-activation. The mathematics of the model is given in Appendix 3. Several conclusions of the theory are worth emphasizing. When the rate constants of the partial reactions in the equilibrium become commensurate with k_f and k_s, each state decays biphasically, the phases time constants being function of all 4 rate constants (k_f, k_s and the equilibrium rate constants k_{sf} and k_{fs}). Further, when the equilibrium becomes fast compared to recombination, the two states tend to follow a common monophasic recombination decay (with the fast phase vanishingly small). Notice that, when the equilibrium is frozen, model C is equivalent to model B and also that the experiment does not provide enough information to allow a complete description of the system in general: the numerical analysis yields 3 parameters (2 time constants and the relative amplitude distribution), while the model depends on 4 unknowns (k_f, k_s and the equilibrium rate constants k_{sf} and k_{fs}).

Figure 5 (a and b) illustrates the behaviour of the 2 components in *Rhodobacter sphaeroides* R26 reaction centers when the temperature is varied between 10K and 300K. While the time constants rise smoothly in the above temperature range (Fig. 5a), the amplitudes are barely modified up to about 150K, when the fast component start decreasing to disappear altogether at room temperature (Fig. 5b). The abrupt transition at 200K has to be compared to the disappearance of the complex kinetics observed near the isobestic point (see fig.1) as already mentioned. This transition might suggest the effect of a phase transition on the conformational equilibrium. However, in a semi-quantitative simulation (not shown) of the temperature dependence of the model, a similar transition of the fast component amplitude is easily revealed, suggesting on the contrary that this effect is of pure kinetic origin. At any rate, the fact that the temperature behaviour of time constants and amplitudes are largely unrelated (no abrupt transition for the former) also agrees with the model in which the distribution of states and the mechanism of recombination are controlled by distinct parameters.

CONCLUSIONS

All in all, our results seem more definitely in favor of the model of a pre-existing distribution of states in equilibrium (model C) rather than of the model of a flash-induced states distribution (model A). Also, the shift from biphasic to monophasic when temperature is changed from cryogenic to room is a natural consequence of the inner structure in model C, while on the contrary it may only result from an *ad hoc* assumption in model A.

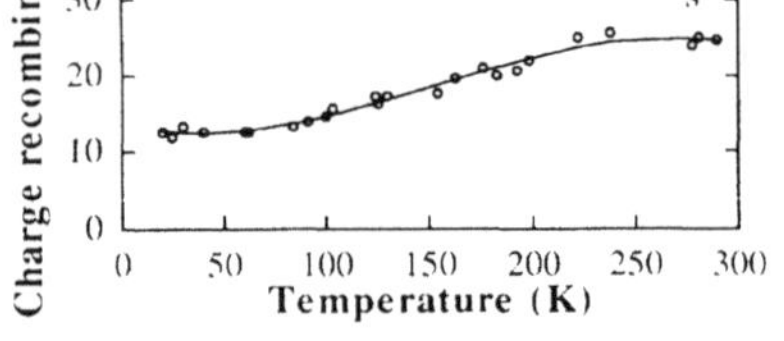

Figure 5a. Temperature dependence of the two components (S : slow, F : fast) half-times of the charge recombination for *Rhodobacter sphaeroides* R26 reaction centers ; half-time= (0.693)*(time constant)

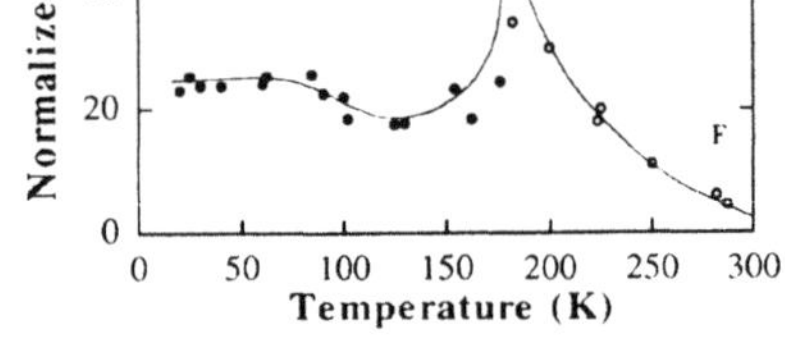

Figure 5b. Temperature dependence of the two components (S : slow, F : fast) relative amplitudes of the charge recombination for *Rhodobacter sphaeroides* R26 reaction centers.

The model of equilibrium of states as proposed here is but a formal framework aimed at accounting for the kinetics of recombination in our system. It is very likely that the hypothesis of conformational states - even if too general - will be the key notion to endow this abstract model with concrete characters. In this respect, the temperature dependence of the average recombination time constant bears probably witness of a contraction of the RC protein structure during cooling and of the strong (exponential) dependency of the recombination rate upon the distance between reactants P^+ and Q_A^- [13,14]. No doubt that further work on this system, in particular as regards pH effects, light conditions during cooling and effects of continuous illumination, should help in revealing a more realistic picture of the recombining bacterial reaction center.

APPENDIX

1. The AT program. The fitting procedure uses the elementary Simplex algorithm[15] to least-squares determine τ_i in the model function,

$$\Sigma(\alpha_i \cdot \exp(-\tau / \tau_i)) \qquad \text{(A1)}$$

the index i running over the assumed number of components, while α_i are solved by a standard linear technique[16]. As the procedure is iterative, a substantial saving in computation time is achieved by grouping the experimental data additively in a minimum number of packs (n components require at least 2·n packs). In order to minimize the effect of experimental noise, several independent runs of the program are performed each with a set of τ_i seeds randomly chosen ; the solution is defined as the average of the best-fitting sets (τ_i, α_i). Incidentally, looking at the standard deviation of (τ_i, α_i) and at the fitting error, it is not too difficult to determine n, the optimal number of components.

2. Decay after flash over background illumination

In model A (light-induced distribution), the differential equations are :

$$d(F^*)/dt = k^*_f \cdot I \cdot R - k_f \cdot F^* \qquad \text{(A2)}$$

$$d(S^*)/dt = k^*_s \cdot I \cdot R - k_s \cdot S^* \qquad \text{(A3)}$$

where the photochemical rate constants are distinct for each component ($k^*_f \cdot I$ and $k^*_s \cdot I$, **I** = light **I**ntensity) and R is the dark stable form of the RC, common to F^* and S^*. The solution of this system is not simple ; suffice it to say that each component has a biphasic decay and that their steady-state ratio is :

$$(F^*/S^*)_{ss} = (k_s \cdot k^*_f)/(k_f \cdot k^*_s) \qquad \text{(A4)}$$

There is pile-up of the slow component but it is independent of light intensity.

Models B and C are equivalent for low temperature experiments and the corresponding differential equations are :

$$d(F^*)/dt = k^* \cdot I \cdot F - k_f \cdot F^* \qquad \text{(A5)}$$

$$d(S^*)/dt = k^* \cdot I \cdot S - k_s \cdot S^* \qquad \text{(A6)}$$

where $k^* \cdot I$ is the photochemical rate constant common to both components, and

$F + F^* = F_0$, $S + S^* = S_0$, where F_0 and S_0 stand for the two components concentrations at rest. At steady-state, under background illumination, the ratio of components is :

$$(F^*/S^*)_{ss} = [(k^* \cdot I + k_s)/(k^* \cdot I + k_f)] \cdot (F_0/S_0) \quad (A7)$$

showing that the slow component has piled up at steady-state and conversely has been depleted in the phases of the flash-induced decay. Besides, the latter is somewhat shortened, as such :

$$\tau_f = 1/(k_f + k^* \cdot I) \qquad (A8)$$

$$\tau_s = 1/(k_s + k^* \cdot I) \qquad (A9)$$

3. Model C: Equilibrium of two states. See Equ.1.

The signal in the recombination kinetics after a flash is a linear combination (spectroscopic coefficients) of the two states F^* and S^* obeying the differential equations :

$$d(F^*)/dt = -(k_f + k_{fs}) \cdot F^* + k_{sf} \cdot S^* \qquad (A10)$$

$$d(S^*)/dt = -(k_s + k_{sf}) \cdot S^* + k_{fs} \cdot F^* \qquad (A11)$$

where k_{fs}, k_{sf} are the rate constants in the equilibrium partial reactions: F^* ---> S^*, S^*---> F^* respectively. In general, when k_{fs}, k_{sf} are commensurate with k_f, k_s, each component decays biphasically with time constants :

$$\tau_1 = 1/[\ (k_f + k_s + k_{fs} + k_{sf}) - ((k_f + k_{fs} - k_s - k_{sf})^2 + 4 \cdot k_{fs} \cdot k_{sf}))^{1/2}\ ] \qquad (A12)$$

$$\tau_2 = 1/[\ (k_f + k_s + k_{fs} + k_{sf}) + ((k_f + k_{fs} - k_s - k_{sf})^2 + 4 \cdot k_{fs} \cdot k_{sf}))^{1/2}\ ] \qquad (A13)$$

with therefore $\tau_1 < \tau_2$; the amplitudes of the fast and slow phases being in general different for each component. When the equilibrium is "frozen" (cryogenic temperatures), k_{fs} and k_{sf} may be neglected in Equs A12, A13, then simply F^* decays with time constant :

$$\tau_f = 1/k_f \qquad (A14)$$

and S^* with time constant :

$$\tau_s = 1/k_s \qquad (A15)$$

On the contrary, assuming an infinitely fast equilibrium (room temperature), one still gets two time constants common to both components but one may show that the fastest phase is negligibly small. Therefore, both components decay monophasically with time constant :

$$\tau_{rt} = (k_{fs} + k_{sf})/(k_f \cdot k_{sf} + k_s \cdot k_{fs}) \qquad (A16)$$

REFERENCES

1. P. Parot, J. Thiery and A. Verméglio, Charge recombination at low temperature in photosynthetic bacterial reaction centers : evidence for two conformational states, *Biochim. Biophys. Acta* 893 : 534 (1987).

2. P. Parot, J. Thiery and A. Verméglio, Charge recombination at low temperature in photosynthetic bacteria reaction centers, *in* : "The Photosynthetic Bacterial Reaction Center Structure and dynamics" : 251, J. Breton and A. Verméglio eds., Plenum Publishing Corporation, New-York-London (1988).
3. P. Sebban, P. Parot, L. Baciou, P. Mathis and A.Verméglio, Effects of low temperature and lipid rigidity on the charge recombination process in *Rps. viridis* and *Rb. sphaeroides* reaction centers, *Biochim. Biophys. Acta* 1057 : 109 (1991).
4. D. Kleinfeld, M.Y. Okamura and G. Feher, Electron-transfert in photosynthetic reaction centers cooled to cryogenic temperatures in charge-separated state : evidence for light-induced structural changes, *Biochemistry* 23 : 5780 (1984).
5. L. Baciou, E. Rivas and P. Sebban, $P^+Q_A^-$ and $P^+Q_B^-$ charge recombinations in *Rps. viridis* chromatophores and in reaction centers reconstituted in phosphatidylcholine liposomes. Existence of two conformational states of reaction centers and effects of pH and *o*-phenanthroline, *Biochemistry* 29 : 2966 (1990).
6. P. Sebban and C. Wraight, Heterogeneity of the $P^+Q_A^-$ recombination kinetics in reaction centers from *Rps.viridis* : the effects of pH and temperature, *Biochim. Biophys. Acta* 974 : 54 (1989).
7. J-L. Gao, R.J. Shopes and C. Wraight, Heterogeneity of kinetics and electron transfert equilibria in the bacteriopheophytin and quinone electron acceptors of reaction centers from *Rps. viridis*, *Biochim. Biophys. Acta* 1056 : 259 (1991).
8. P. Sebban, PH effect on the biphasicity of the $P^+Q_A^-$ charge recombination kinetics in the reaction centers from *Rb. sphaeroides*, reconstituted with anthraquinones, *Biochim. Biophys. Acta* 936 : 124 (1988).
9. N.W. Woodbury and W.W. Parson, Nanosecond fluorescence from isolated photosynthetic reaction centers of *Rps. sphaeroides*, *Biochim. Biophys. Acta* 767 : 345 (1984).
10. C. Kirmaier and D. Holten , Evidence that a distribution of bacterial reaction centers underlies the temperature and detection-wavelength dependence of the rates of the primary electron-tranfert reactions, *Proc. Natl. Acad. Sci. USA* 87 : 3552 (1990).
11. M. Becker, V. Nagarajan, D. Middendorf, W.W. Parson, J.E. Martin and R.E. Blankenship, Temperature dependence of the initial electron-transfert kinetics in photosynthetic reaction centers of *C. aurantiacus*, *Biochim. Biophys. Acta* 1057 : 299 (1991).
12. S. Franzen, R.F. Goldstein and S.G. Boxer, Electric field modulation of electron transfert reaction rates in isotropic systems : long-distance charge recombination in photosynthetic reaction centers, *J. Phys. Chem.* 94 : 5135 (1990).
13. J.J. Hopfield, Electron transfert between biological molecules by thermally activated tunneling, *Proc. Nat. Acad. Sci. USA* 71 : 3640 (1974).
14. J. Jortner, Dynamics of electron transfert in bacterial photosynthesis, *Biochim. Biophys. Acta* 594 : 193 (1980).
15. S.N. Deming and S.L. Morgan , Simplex optimisation of variables in analytical chemistry, *Anal. Chem.* 45 : 278A (1973).
16. J. Thiery, Personal communication.

PROTEIN RELAXATION FOLLOWING QUINONE REDUCTION IN *RHODOBACTER CAPSULATUS:* DETECTION OF LIKELY PROTONATION-LINKED OPTICAL ABSORBANCE CHANGES OF THE CHROMOPHORES

David M. Tiede[1] and Deborah K. Hanson[2]

[1]Chemistry Division
[2]Biological and Medical Research Division
Argonne National Laboratory
Argonne, Illinois, 60439

INTRODUCTION

The formation of quinone anions in bacterial reaction centers produces distinctive optical absorption shifts of the bacteriopheophytin and bacteriochlorophyll chromophores[1,2]. The absence of a direct molecular contact between the quinones and these chromophores[3-5] suggests that the electrochromic shifts may be due to an electrostatic interaction between the optical transition dipoles of the chromophores and the electric field associated with the quinone anions. In this case, the electrochromism induced by the quinone anions potentially provides an opportunity to examine the propagation of electric fields, and hence the local dielectric, within the reaction center protein.

One intriguing feature is that Q_A^- and Q_B^- induce characteristic, but different absorbance changes*. Absorbance changes associated with Q_A and Q_B anions have been characterized in *Rhodobacter sphaeroides*[1] and *Rhodopseudomonas viridis*[2] by trapping reaction centers in the PQ_A^- or PQ_B^- states following photo-excitation in the presence of exogenous electron donors. The most prominent change produced by formation of Q_A^- is a red-shift of an optical transition on the low-energy side of the bacteriopheophytin Q_y absorption band, that is presumably due to H_L. In addition, Q_A^- formation is also associated with blue-shifts of the bacteriochlorophyll dimer Q_y band and of a transition on the low-energy edge of the accessory bacteriochlorophyll band. In contrast, Q_B^- produces a red-shift of a transition on the high-energy side of the bacteriopheophytin band that is

*Abbreviations: Q_A, the first quinone acceptor, which is a ubiquinone in *Rb. sphaeroides* and *Rb. capsulatus*, and menaquinone in *Rps. viridis* (see ref. 3 for review); Q_B, the second quinone acceptor, which is a ubiquinone in all three species; H_L, H_M, the bacteriopheophytins associated with the L and M protein subunits; P, the bacteriochlorophyll dimer.

The Photosynthetic Bacterial Reaction Center II
Edited by J. Breton and A. Verméglio, Plenum Press, New York, 1992

likely to be H_M, but causes much smaller changes in the accessory bacteriochlorophyll band, and has no effect on the dimer absorption.

The stronger perturbation of H_L by Q_A^- and H_M by Q_B^- makes intuitive sense because the crystal structures[3-5] show that each quinone is closely associated with a single bacteriopheophytin. However there is no obvious structural feature to explain the difference in perturbation of the bacteriochlorophyll absorptions by Q_A^- and Q_B^-.

INDO calculations with the *Rps. viridis* coordinates show that a through-space electrostatic interaction is a plausible mechanism for explaining the absorption shifts of the bacteriopheophytins by the quinone anions[6, 7]. However, the smaller electrochromic shifts predicted for the accessory bacteriochlorophylls due to Q_A^- formation[6, 7] are of opposite sign from that observed, and electrochromic shifts for the bacteriochlorophyll dimer were not calculated. Electrochromic shifts of the reaction center have been characterized by the application of externally applied electric fields (Stark spectroscopy). The bacteriochlorophyll dimer shows an unusually large Stark effect compared to the other pigments in the reaction center, which has been attributed to a large charge-transfer character of the excited state, $^*P$[8-10]. This raises the interesting possibility that the Q_A^- induced electrochromic shifts in the accessory bacteriochlorophyll region may be dominated by underlying transitions with high charge-transfer character.

In this paper we present time-resolved spectra associated with Q_A and Q_B anion formation in chromatophores of *Rhodobacter capsulatus*. Significantly, these results show that when initially formed, Q_A^- and Q_B^- produce equivalent electrochromic shifts on the reaction center bacteriochlorophylls. Subsequent protein relaxation events, in part involving proton transfer, attenuate this effect. The attenuation is nearly complete for Q_B^-. Electrochromism induced by quinone anion formation provides a vehicle for calibrating the propagation of electrostatic potentials in the reaction center protein, and may provide explanations for anomalies in low temperature electron transfer reactions with the quinones.

MATERIALS AND METHODS

Rb. capsulatus and *Rb. sphaeroides* R-26 chromatophores were prepared by breakage of cells in 50 mM Tris, pH 7.8 in a French Press followed by differential centrifugation. Cells and chromatophores were stored frozen. The "wild-type" *Rb. capsulatus* strain lacking the light-harvesting II complex was constructed by complementing the U43 deletion strain with the pU2922 plasmid[11]. Chromatophores were resuspended in 10 mM Tris, pH 7.8.

Single wavelength absorbance transients at room temperature were recorded on an instrument of local design, using a 1-2 mJ, 20 ns wide 590 nm laser pulse from a YAG-pumped, rhodamine dye laser. Chromatophores were diluted to have an absorbance of about 0.4 cm^{-1} at 802 nm. The kinetic traces shown here were the average of 8 to 16 transients, with at least 45 s between laser flashes.

RESULTS AND DISCUSSION

Spectra of Trapped PQ_A^- and PQ_B^- States in *Rb. capsulatus*

Difference spectra (PQ^- - PQ) associated with formation of the trapped PQ_A^- and PQ_B^- states in *Rb. capsulatus* chromatophores are shown in figure 1. The spectra associated with the PQ_A^- and PQ_B^- states were recorded 60 ms and 400 ms following laser excitation as described in the figure legend. The patterns of absorbance changes are completely analogous to those associated with the formation of the quinone anions in reaction centers of *Rb. sphaeroides*[1] and *Rps. viridis*[2]. This is further evidence that the characteristic difference spectra associated with the PQ_A^- and PQ_B^- states reflect common features of bacterial reaction centers.

Kinetics of the $Q_A^-Q_B$ to $Q_AQ_B^-$ Electron Transfer

The trapped PQ^- spectra suggest wavelengths for monitoring the $Q_A^-Q_B$ to $Q_AQ_B^-$ electron transfer. The difference between the two spectra in figure 1 gives the absorbance changes expected to accompany electron transfer. An isosbestic point for Q_A anion formation is seen at 760 nm. This provides a useful wavelength for monitoring the $Q_A^-Q_B$ to $Q_AQ_B^-$ electron transfer. Figure 2 shows absorbance transients measured at this wavelength with chromatophores of *Rb. capsulatus*. The unresolved, instantaneous rise is associated with the formation of P^+. The slower absorbance increase can be fit with a monoexponential rise time of about 20 μs. This kinetic rise can be assigned to the reaction time for the $Q_A^-Q_B$ to $Q_AQ_B^-$ electron transfer.

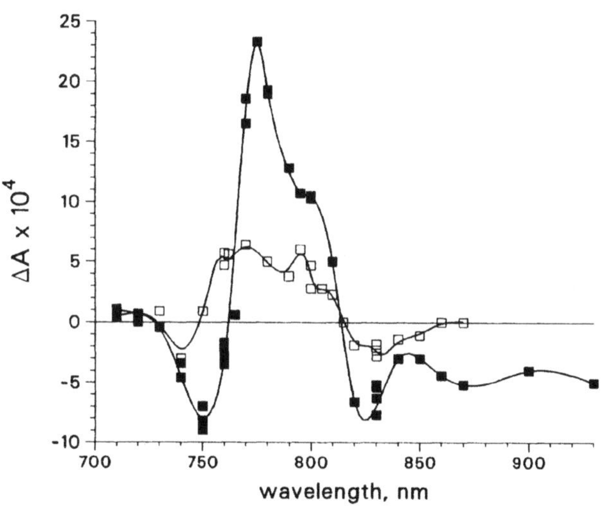

Figure 1. PQ^- - PQ difference spectra measured in chromatophores of *Rb. capsulatus*. The filled squares mark the absorbance changes associated with the formation of PQ_A^-, which was generated following a laser flash using 200 μM ferrocene as an electron donor to P^+, and with 3 mM orthophenanthroline to block electron transfer to Q_B. The plot shows the absorbance changes measured 60 ms following excitation. The empty squares mark the absorbance changes associated with the formation of PQ_B^-, which was measured with the addition of 200 uM ferrocene, and as the absorbance difference measured 400 ms following the first laser flash and that measured 400 ms following a second flash spaced. The two flashes were spaced 400 ms apart. The first flash generates PQ_B^- and the second flash removes this by formation of the double reduced quinone.

Transient Spectra Associated with the $P^+Q_B^-$ State

The transient spectra associated with quinone reduction are mostly obscured by the generally much larger optical changes produced by the formation of P^+.

$$PQ_AQ_B \underset{50 - 100\ \text{ms}}{\overset{200\ \text{ps}}{\Leftrightarrow}} P^+Q_A^-Q_B \underset{1 - 2\ \text{s}}{\overset{20\ \mu\text{s}}{\Leftrightarrow}} P^+Q_AQ_B^-$$

However, the contribution of P^+ can eliminated by subtracting the transient spectra associated with the initial $P^+Q_A^-Q_B$ state from the final $P^+Q_AQ_B^-$ state to yield a $Q_AQ_B^-$ - $Q_A^-Q_B$ difference spectrum.

Figure 3 shows two $Q_AQ_B^-$ - $Q_A^-Q_B$ difference spectra. The spectrum marked by the filled squares was obtained using absorbance of the $P^+Q_AQ_B^-$ state measured 100 μs following the flash. Even though the kinetic trace in figure 2 shows that the $Q_A^-Q_B$ to $Q_AQ_B^-$ electron transfer is essentially complete by this time, the lineshape of the 100 μs $Q_AQ_B^-$ - $Q_A^-Q_B$ spectrum differs markedly from the one predicted by the trapped PQ^- states shown in figure 1. Instead of absorbance changes centered near 860 nm, 810 nm and 760 nm, only the single bandshift near 760 nm is seen. This suggests that only the bacteriopheophytin absorption bands respond immediately to the electron transfer from Q_A to Q_B.

Following $Q_A^-Q_B$ to $Q_AQ_B^-$ electron transfer and the prompt response of the bacteriopheophytin absorption bands, other slower relaxation processes cause an attenuation of the electrochromic shifts. For example, this can be seen at longer times in the absorbance transient measured at 760 nm, shown in figure 4. Following electron transfer, a decay is seen with a lifetime of about 450 μs. This component is not seen in the decay

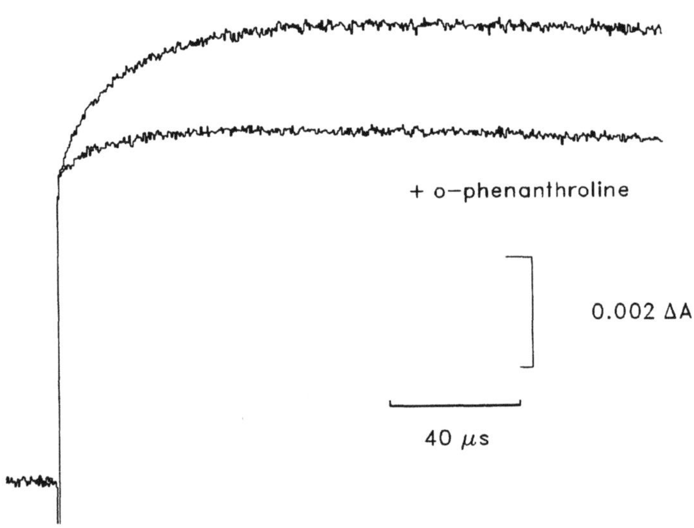

Figure 2. Absorbance transients measured at 760 nm with chromatophores of *Rb. capsulatus*. Chromatophores, A(802 nm) = 0.94 cm^{-1}, were suspended in 10 mM Tris, pH 7.8. $Q_A^-Q_B$ to $Q_AQ_B^-$ electron transfer was blocked in the second trace by the addition of 3 mM orthophenanthroline.

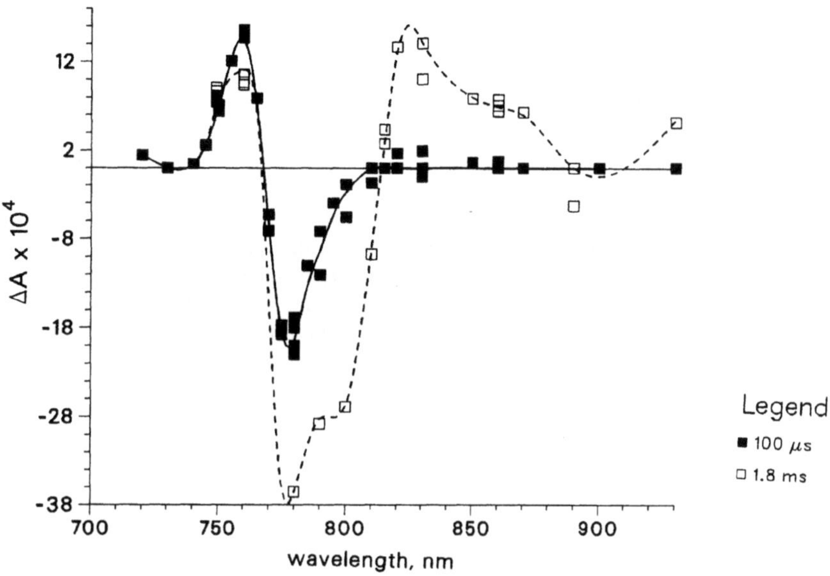

Figure 3. Transient $Q_AQ_B^-$ - $Q_A^-Q_B$ difference spectra in chromatophores of *Rb. capsulatus*. The spectrum marked by the filled squares is the difference between the absorbance of the $P^+Q_AQ_B^-$ state measured at 100 μs following laser excitation and the initial $P^+Q_A^-Q_B$ state, measured at 2 μs. The unfilled squares mark the absorption difference between the $P^+Q_AQ_B^-$ state measured 1.8 ms after laser excitation, minus that of the initial $P^+Q_A^-Q_B$ state. This difference included a correction for the decay of P^+ during as described in the text and illustrated in figure 4.

of transients detected at 860 nm and 450 nm, which directly monitor the concentration of the P^+ and P^+Q^- (sum of both Q_A^- and Q_B^-) states. In figure 4 the 860 nm transient was normalized to the amplitude of the instantaneous change due to P^+ at 760 nm, as shown on an expanded time-scale in figure 1. The difference between the 760 nm and normalized 860 nm transients is the bottom trace in figure 4. This difference identifies the additional kinetic components present in the 760 nm transient that are not associated with the lifetime of P^+.

A similar subtraction of the 860 nm transients can be done at other wavelengths. By normalizing the 860 nm transient to the amplitude of the instantaneous absorbance change, subsequent absorbance changes not associated with the lifetime of P^+ are measured with respect to the absorbance of the initial $P^+Q_A^-Q_B$ state. This subtraction procedure reveals kinetic components that are not seen in the P^+ decay. Residual transients obtained in this manner are shown in figure 5 at selected wavelengths . For example, the residual transient at 830 nm exhibits no prompt response to the $Q_A^-Q_B$ to $Q_AQ_B^-$ electron transfer, but the slow rise suggests that a decay of this bandshift occurs with a lifetime about 400 μs. Significantly, residual transients are not seen at 815 nm. This is consistent with this wavelength being an isosbestic point for both Q_A/Q_A^- and Q_B/Q_B^-, but not P/P^+. The absence of residual kinetic components in the electrochromism attributable uniquely to P^+ indicates that the slow kinetic components reflect relaxations associated with quinone anion formation and not P^+. A plot of the absorbance changes remaining 1.8 ms after excitation and corrected for P^+ decay are shown by the unfilled squares in figure 3. This spectrum reflects the $P^+Q_AQ_B^-$ following the protein relaxation events. This difference spectrum is compatible with the one predicted from the trapped PQ^- states in figure 1.

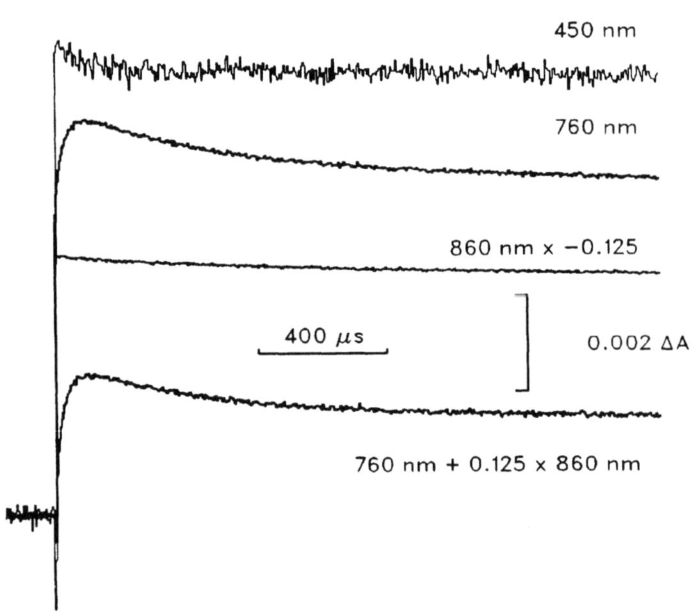

Figure 4. Absorbance transients measured with chromatophores of *Rb. capsulatus*. The sample and measuring conditions were identical to those in figure 2, except for the slower time scale.

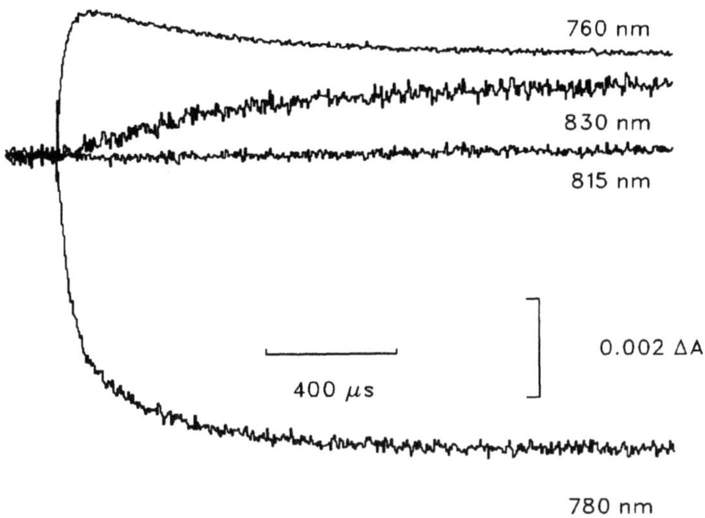

Figure 5. Absorbance transients remaining after subtraction of the decay kinetics for P^+, measured at 860 nm. These traces show the kinetic components which are present at selected wavelengths, but absent in the 860 nm transients.

The relaxation kinetics measured at 760 nm show a strong pH dependence. These relaxation components are not discernable in the 760 nm transients below pH 4 or above pH 10, although relaxation components are still seen in the bacteriochlorophyll absorbance bands. The pH sensitivity suggests that part of the relaxation involves proton transfer. Further information on the mechanism of this relaxation is suggested by a series of photosynthetically competent mutants that contain an alanine residue instead of a glutamic acid residue at L212 (see Schiffer, M. et al., this volume). In these cases a relaxation component with a 400 - 500 μs lifetime is missing, suggesting the involvement of glutamic-L212 in this relaxation.

Transient Spectra Associated with the $P^+Q_A^-$ State

Evidence for protein relaxation following the formation of Q_A anion is also found when electron transfer to Q_B is blocked. Figure 2 shows a small absorbance increase at 760 nm following the formation of the initial of $P^+Q_A^-$ state in the presence of 3 mM orthophenanthroline. This could represent residual $Q_A^-Q_B$ to $Q_AQ_B^-$ electron transfer, but a plot of the amplitude of this component (unfilled circles, figure 6) does not match the difference spectrum seen for the initial $Q_AQ_B^-$ state, and its risetime is slightly faster. Slower relaxation components are also seen in the presence of orthophenanthroline. In this case it is clear that decay kinetics of P^+ and Q_A^- ought to match. However, when the 860 nm transients are normalized and subtracted from the kinetic traces at other wavelengths, residual kinetic components are seen. The amplitudes of the residual absorbance changes measured 1.8 ms after excitation are plotted as the filled circles in figure 6. This spectrum differs from the $P^+Q_AQ_B^-$ measured at 1.8 ms (re-plotted in figure 6, unfilled squares). These experiments suggest that the residual kinetics do not arise from an incomplete block of the $Q_A^-Q_B$ to $Q_AQ_B^-$ electron transfer, but arise from protein relaxation that alters the electrochromic shifts induced by Q_A anion formation.

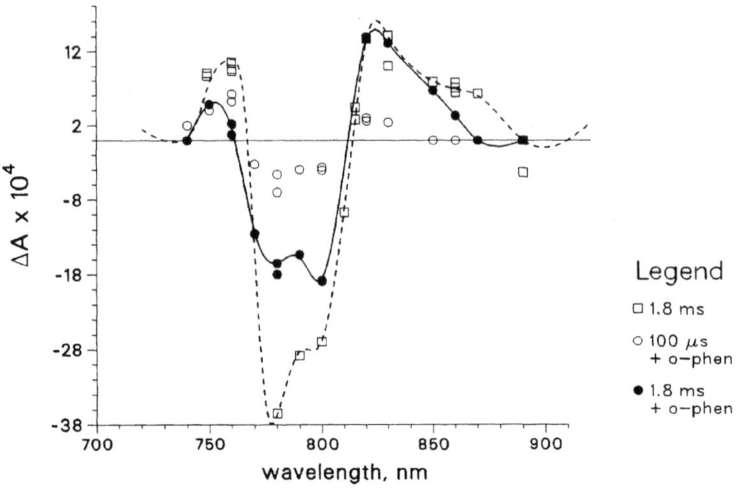

Figure 6. Difference spectra associated with the time-evolution of the $P^+Q_A^-$ state. $Q_A^-Q_B$ to $Q_AQ_B^-$ electron transfer was blocked by the addition of 3 mM orthophenanthroline. The unfilled circles mark the difference between the absorbance measured 100 μs following excitation minus that measured at 2 μs. The filled circles mark the absorption difference remaining at 1.8 ms after subtracting the 860 nm transient, normalized to the amplitude to the initial change as described in the text.

DISCUSSION

These experiments have revealed a time-evolution in the optical absorption spectra associated with the Q_A and Q_B anion states. This time-evolution is not due to electron transfer, but can be interpreted as due to a charge reorganization in the vicinity of the quinone anions that leads to a stabilization of charge and attenuation of the electrostatic field propagation through the protein.

Only the bandshifts in the bacteriopheophytin absorption region are found to respond promptly to the electron transfer from Q_A to Q_B. These absorbance transients suggest a reaction time of 20 μs for the electron transfer from $Q_A^-Q_B$ to $Q_AQ_B^-$ in *Rb. capsulatus* chromatophores. This is discernably faster than the 40 μs to 50 μs reaction time measured by ourselves and others[12,13] in *Rb. sphaeroides* chromatophores, but equivalent to that measured in *Rps. viridis* cells and chromatophores[14]. We find that the Q_A to Q_B electron transfer rate slows down upon extraction of reaction centers from the membrane with detergent. This is analogous to kinetic differences observed for this reaction with *Rb. sphaeroides* reaction centers in chromatophores and in the isolated state[12,13].

In contrast to the prompt responses of the bacteriopheophytin absorption bands to the $Q_A^-Q_B$ to $Q_AQ_B^-$ electron transfer, electrochromic shifts in the bacteriochlorophyll absorption bands are initially unaltered by the electron transfer. A through-space electrostatic interaction would suggest that initially both quinone anions produce equivalent electric field increases at the bacteriochlorophylls. Following the formation of the $P^+Q_AQ_B^-$ state, relaxation processes with a response time of about 400 μs are found to attenuate of the electrochromic shifts in the bacteriochlorophyll band region. A similar relaxation is seen for the $P^+Q_A^-$ state when its lifetime is extended by blocking electron transfer to Q_B. However, the quenching of the electrochromic shift is found to be greater for Q_B^- than Q_A^-.

Several mechanisms are possible for the quenching of the electrochromic shifts induced by the quinone anions, including movement of the quinone anion away from the original binding site or a reorganization of protein charge in the vicinity of the quinone anions. Either mechanism will affect electric field propagation in the reaction center. These mechanisms have also been considered to explain decays of photovoltages generated with reaction centers in planar membranes (Brzezinski, P. et al., this volume).

The pH dependence of the relaxation kinetics in the bacteriopheophytin bands implies that at least part of the quenching mechanism involves proton transfer. The pH dependence of the relaxation events parallel the pH dependence for proton uptake on the first flash[15,16]. This suggests that the relaxation may be related to the redistribution of protons in the reaction center. The absence of the 400 μs relaxation component in reaction center revertants that contain a L212Glu → Ala mutation suggests that this residue is involved in the relaxation mechanism.

The protein reorganization can be expected to stabilize the quinone anions states. The degree of stabilization will depend upon the time scale and temperature of the experiment. The eventual freezing-out of the protein relaxation at low temperature can be expected to leave the quinone anion in a high energy state. This may provide an explanation for the anomalies in the temperature dependence of the electron transfer reactions with the quinones.

ACKNOWLEDGEMENTS

This work was supported by the U.S. Department of Energy, Office of Basic Energy Sciences, Division of Chemical Sciences (DMT) and Office of Health and Environmental Research (DKH), under Contract W-31-109-Eng-38.

REFERENCES

1. A. Vermeglio and R.K. Clayton, Kinetics of electron transfer between the primary and the secondary electron acceptor in reaction centers from *Rhodopseudomonas sphaeroides*, *Biochim. et Biophys. Acta* 461:159 (1977).
2. R.J. Shopes and C.A. Wraight, The acceptor quinone complex of *Rhodopseudomonas viridis* reaction centers, *Biochim. et Biophys. Acta* 806:348 (1985).
3. J.P. Allen, G. Feher, T.O. Yeates, H. Komiya and D.C. Rees, Structure of the reaction center from *Rhodobacter sphaeroides* R-26: The protein subunits, *Proc. Natl. Acad. Sci. USA* 84:6162 (1987).
4. C.-H. Chang, O. El-Kabbani, D.M. Tiede, J. Norris and M. Schiffer, Structure of the Membrane-bound protein photosynthetic reaction center from *Rhodobacter sphaeroides*, *Biochem.* 30:5352 (1991).
5. J. Deisenhofer, O. Epp, K. Miki, R. Huber and H. Michel, Structure of the protein subunits in the photosynthetic reaction centre of *Rhodopseudomonas viridis* at 3 Å resolution, *Nature* 318:618 (1985).
6. L.K. Hanson, M.A. Thompson, M.C. Zerner and J. Fajer, Theoretical models of electrochromic and environmental effects on bacterio-chlorophylls and -pheophytins in reaction centers, *in*: "NATO Advanced Research Workshop on the Structure of the Photosynthetic Bacterial Reaction Center," J. Breton and A. Vermeglio, ed.,Plenum Press,New York(1987).
7. L.K. Hanson, J. Fajer, M.A. Thompson and M.C. Zerner, Electrochromic effects of charge separation in bacterial photosynthesis: theoretical models, *J. Amer. Chem. Soc.* 109:4728 (1987).
8. T.J. DiMagno, E.J. Bylina, A. Angerhofer, D.C. Youvan and J.R. Norris, Stark effect in wild-type and Heterodimer-containing reaction centers from *Rhodobacter capsulatus*, *Biochem.* 29:899 (1990).
9. D.J. Lockhart and S.G. Boxer, Magnitude and direction of the change in dipole moment associated with excitation of the primary electron donor in *Rhodopseudomonas sphaeroides* reaction centers, *Biochem.* 26:664 (1987).
10. M. Losche, G. Feher and M.Y. Okamura, The Stark effect in reaction centers from *Rhodobacter sphaeroides* R-26 and *Rhodopseudomonas viridis*, *Proc. Natl. Acad. Sci. USA* 84:7537 (1987).

11. E.J. Bylina, R.V.M. Jovine and D.C. Youvan, A genetic system for rapidly assessing herbicides that compete for the quinone binding site of photosynthetic reaction centers, *Bio/Technology* 7:69 (1989).
12. A. Vermeglio and J. Breton, Electron transfer between primary and secondary electron acceptors in chromatophores and reaction centers of photosynthetic bacteria, *in*: "Function of Quinones in Energy Conserving Systems," B.L. Trumpower, ed.,Academic Press,New York(1982).
13. A.R. Crofts and C.A. Wraight, The electrochemical domain of photosynthesis, *Biochim. Biophys. Acta* 726:149 (1983).
14. W. Leibl and J. Breton, Kinetic properties of the acceptor quinone complex in *Rhodopseudomonas viridis*, *Biochem.* 30:9634 (1991).
15. P. Maroti and C.A. Wraight, Flash-induced H^+ binding by bacterial photosynthetic reaction centers: influences of the redox states of the acceptor quinones and primary donor, *Biochim. Biophys. Acta* 934:329 (1988).
16. P.H. McPherson, M.Y. Okamura and G. Feher, Light-induced proton uptake by photosynthetic reaction centers from *Rhodobacter sphaeroides* R-26. I. Protonation of the one-electron states $D^+Q_A^-$, DQ_A^-, $D^+Q_AQ_B^-$, and $DQ_AQ_B^-$, *Biochim. Biophys. Acta* 934:348 (1988).

STUDY OF REACTION CENTER FUNCTION BY ANALYSIS OF THE EFFECTS OF SITE-SPECIFIC AND COMPENSATORY MUTATIONS

Marianne Schiffer,[1] Chi-Kin Chan,[2] Chong-Hwan Chang,[1] Theodore J. DiMagno,[2] Graham R. Fleming,[2] Sharron Nance,[1] James Norris,[2,3] Seth Snyder,[3] Marion Thurnauer,[3] David M. Tiede,[3] and Deborah K. Hanson[1]

[1]Biological and Medical Research Division
Argonne National Laboratory, Argonne, IL 60439

[2]Department of Chemistry
The University of Chicago, Chicago, IL 60637

[3]Chemistry Division
Argonne National Laboratory, Argonne, IL 60439

INTRODUCTION

The protein portion of the reaction center (RC) complex is composed of three subunits: the intermembrane L and M chains, and the cytoplasmic H polypeptide. The cofactors are within the transmembrane region; they are related by approximate twofold symmetry as are the homologous L and M chains [1-4]. The cofactors consist of a bacteriochlorophyll dimer that is the primary electron donor, two bacteriochlorophyll monomers, two bacteriopheophytins, a non-heme iron atom, and two quinones which serve as the final electron acceptors.

Other than its obvious architectural role, the detailed contribution of the protein to the electrochemistry of the RC remains obscure. Clearly individual amino acids can influence the spectral properties of the cofactors, electron and proton transfer rates, and ultimately, the directionality of electron and proton transfer through the complex. For this study, we are focusing on residues that break the twofold symmetry of the structure at positions that may critically affect electron and/or proton transfer. Through site-specific mutagenesis, we aim to understand how the protein modifies the chemical properties of the primary and secondary quinones and how it contributes to the observed unidirectionality of electron transfer through the otherwise symmetrical RC complex.

AMINO ACIDS THAT AFFECT THE PROPERTIES OF THE QUINONES

Although Q_A and Q_B in the RCs of *Rhodobacter capsulatus* and *Rhodobacter sphaeroides* are identical ubiquinone (UQ_{10}) molecules, their *in situ* chemical properties are quite different, presumably as a result of the protein environment in which each is located. As the primary quinone acceptor, Q_A receives a single electron, is not protonated, and cannot

The Photosynthetic Bacterial Reaction Center II
Edited by J. Breton and A. Verméglio, Plenum Press, New York, 1992

be reduced further; the electron is transferred from Q_A to Q_B within 100 μs. As the terminal acceptor, Q_B accepts electrons from two successive light-induced turnovers of Q_A, followed by the uptake of two protons from the cytoplasm [5]. Q_B has no direct contact with the aqueous environment, thus protons have to travel to it through the protein matrix. The quinol form diffuses out of the RC and is replaced by a quinone from the membrane pool [6]. In this section, we describe the characterization of site-specific mutants of *R. capsulatus* that symmetrize or reverse the residues found at four positions in the quinone binding pockets, and the characterization of revertant strains which carry mutations that compensate for the site-specific alterations, restoring electron and proton transfer.

Structural analysis of the quinone binding sites. Table 1 lists the amino acid residues that form the surface of the cavity in which each quinone is located in the purple bacteria *Rhodopseudomonas viridis* [7], *R. sphaeroides* [8,9], *R. capsulatus* [10], *Rhodospirillum rubrum* [11], and the green bacterium *Chloroflexus aurantiacus* [12,13].

Table 1. Amino acids that surround the quinones *

	Q_B						Q_A				
	vir	sph	cap	rub	aur		vir	sph	cap	rub	aur
L186	A	A	A	A	L	M213	L	L	L	L	L
L189	L	L	M	L	Cys	M216	A	M	M	M	M
L190	His	His	His	His	His	M217	His	His	His	His	His
L193	L	L	L	L	L	M220	Thr	Thr	Thr	Thr	Thr
L194	I	V	V	I	I	M221	I	I	I	I	I
L212	Glu	Glu	Glu	Glu	Glu	M246	A	A	A	A	A
L213	Asn	Asp	Asp	Asn	Asn	M247	A	A	A	A	Gln
L215	Tyr	F	Tyr	Tyr	F	M249	F	F	F	F	F
L216	F	F	F	F	F	M250	W	W	W	W	W
L220	V	V	M	I	Gln	M254	I	M	M	M	M
L222	Tyr	Tyr	Tyr	Tyr	Tyr	M256	F	F	F	F	F
L223	Ser	Ser	Ser	Ser	Ser	M257	Asn	Asn	Asn	Asn	Asn
L224	I	I	V	V	V	M258	A	A	A	A	A
L225	G	G	G	G	G	M259	Thr	Thr	Thr	Thr	Asn
L226	A	Thr	Thr	Thr	Glu	M260	I	M	M	M	A
L229	I	I	I	I	V	M263	V	I	I	I	I
L230	His	His	His	His	His	M264	His	His	His	His	His
L232	L	L	L	V	L	M266	W	W	W	W	W
M232	Glu	Glu	Glu	Glu	Glu						

*Polar residues are shown using the three-letter code.

Twelve of the 18 M-chain residues that define the Q_A site are conserved in all five organisms. Only five of the residues forming the Q_A site are polar, and all are conserved in the four species of purple bacteria. The Q_A sites of *R. sphaeroides, R. capsulatus,* and *Rsp. rubrum*, which bind UQ_{10}, are identical. In *Rps. viridis* and *C. aurantiacus*, the quinone in the Q_A site is menaquinone. The largest number of amino acid substitutions is found in *Rps. viridis*, where four large hydrophobic residues are replaced with smaller ones, thereby increasing the volume of the binding pocket by 110 $Å^3$.

Of the L-chain residues that form the Q_B site, only nine out of 19 are conserved in all five species. Five of these residues are polar, while four are hydrophobic. At five positions in the Q_B site, conservative substitution of one aliphatic hydrophobic residue for another occurs. At three of these positions (L194, L224, and L229), the β-forked residues isoleucine and valine are found and thus might be required for function. At position L213, both asparagine (polar but not charged) and aspartic acid (charged) occur. Glutamic acid M232 forms part of the Q_B binding site; this residue has no equivalent in the L chain. Q_B is UQ_{10} in the four purple bacteria and is menaquinone in *C. aurantiacus*. H175Glu, not listed in Table 1, forms the bottom of the Q_B cavity on the cytoplasmic side. H175Glu is part of a circular patch of charged residues that forms the interface between the L, M, and H subunits. The center of this region is located below the Q_B site; the patch extends to the Q_A site. None of the acidic residues H175Glu, L212Glu, nor L213Asp appear to participate in salt bridge formation [1].

Although the quinone binding sites share several identical or functionally similar amino acid residues at symmetry-related positions, the Q_B site is significantly more polar than the Q_A site. Residues M246 and M247 in the Q_A site are conserved hydrophobic alanines in all but *C. aurantiacus*, while the equivalent residues in the Q_B site at L212 and L213 are always either acidic or polar (L212Glu, L213Asp or Asn).

Double mutant L212Glu-L213Asp → Ala-Ala

In order to probe the function of L212Glu and L213Asp in both electron and proton transfer, we constructed a double mutant of *R. capsulatus* in which these residues were replaced by alanines found in the Q_A site. We used the deletion strain and system of plasmids that has been described previously [14-16]. This double mutant, in which the quinone binding sites have been partially symmetrized, does not grow under photosynthetic conditions (PS^-), but optical spectra of dark-grown cells showed absorptions that are characteristic of normal amounts of properly assembled light harvesting and RC complexes. EPR spectra of this mutant displays wild-type characteristics of the bacteriochlorophyll dimer. Also, the time constant of the stimulated emission decay of the singlet excited state of the dimer is similar to that observed for the wild-type [17]. These data show that RCs of the double mutant assemble properly and function in primary charge separation events.

Isolation and Genetic Characterization of Photocompetent Revertants. By incubating plates under photosynthetic conditions, photocompetent (PS^+) revertants of the plasmid-borne L212Ala-L213Ala mutant were selected. To date, genetic characterization has defined five classes of revertants and suppressors (Table 2). The mutations that restore the PS^+ phenotype are plasmid-borne in Classes 1-4 and are chromosomal in Class 5. None of the reversion events restored a polar residue at position L212. In fact, our results show that there is no absolute requirement for polar residues at either L212 or L213 position.

Classes 1 and 2. Only two of the 24 PS^+ strains characterized so far carry a reversion at one of the original sites and in both, L213Ala has reverted through a GCC → GAC transversion to the wild-type aspartic acid residue. No other sequence changes were found.

The second class of PS^+ strains retains the original double mutation at L212-213 and carries a third mutation at L225 that acts as an intragenic suppressor of the double mutation. This amino acid, normally a glycine residue that is conserved in the above five species, has been changed by a transition to aspartic acid (GGC → GAC). It is located on the opposite side of the Q_B binding pocket from L213. Molecular modeling of the L225Gly → Asp substitution using the *R. sphaeroides* RC structure [1] showed that L225Asp could point towards Q_B, its oxygen atoms occupying essentially the same space as those of L213Asp. No other changes were found.

In order to test whether the mutation observed in each class of revertant was sufficient to restore photocompetence when present in concert with the remaining site-specific mutations, we engineered combinations of revertant and wild-type genes of the *puf* operon for complementation analysis [15]. In Class 1 & 2 strains, the L gene carrying the mutations described above was sufficient to confer photocompetence when coupled with a wild-type M gene. Since photocompetence is a plasmid-borne trait in both classes (Table 2), the compensatory mutation cannot be in the H gene, and the mutations described can be directly correlated with the PS$^+$ phenotypes.

Table 2. Revertants and suppressors of the double mutant in the Q_B site *

	Strains	L212	L213	L225	M43	M231	Cotransfer of PS$^+$ with Plasmid
Wild-Type		Glu	Asp	Gly	Asn	Arg	
Double Mutant		Ala	Ala	-	-	-	
Class 1 (Rev.)	LL4,6	Ala	-	-	-	-	Yes
Class 2 (Sup.)	LL1-3; UV7,8 HL1,2,4-7	Ala	Ala	Asp	-	-	Yes
Class 3 (Sup.)	LL7,8	Ala	Ala	-	-	Leu	Yes
Class 4 (Sup.)	HL8	Ala	Ala	-	Asp	-	Yes
Class 5 (Sup.)	UV1-6; LL5; HL3	Ala	Ala	-	-	-	No

*(-) indicates identity with the wild-type sequence. Spontaneous mutants selected under low light (LL) or high light (HL) growth conditions; UV-induced mutants received 30 s of ultraviolet exposure before HL selection.

Class 3. Sequence analysis, coupled with a complementation test, has shown that the representatives of Class 3 carry a mutation in the M chain that acts as an intergenic suppressor of the L212Ala-L213Ala double mutant phenotype. These strains retain the alanine substitutions at L212-213, and amino acid M231, which is normally an arginine residue that is involved in conserved ion pair interactions with H125Glu and H232Glu (Figure 1, [1]), has been changed by a transversion to leucine (CGC → CTC). Plasmid constructions that coupled the L gene from these strains with the wild-type M gene did not confer the PS$^+$ phenotype upon transfer of the chimeric plasmid to the U43 deletion strain. When both the L and M genes isolated from these strains were combined with the remainder of the wild-type plasmid, cotransfer of the PS$^+$ phenotype with the chimeric plasmid was observed. No other sequence changes were found in the L or M genes. These results confirm that the sequence alteration at M231 compensates for the L212Ala-L213Ala double mutation that is still present in the L gene.

Class 4. Preliminary sequence analysis determined that the L212Ala-L213Ala double mutation was still present in this strain. Chimeric plasmids were constructed in order to map the second-site suppressor mutation [18]. These constructs determined that the M gene isolated from the revertant plasmid was necessary to confer photocompetence in strains carrying the double mutant L gene. Sequencing of the M gene derived from the revertant plasmid determined that a transition (GAA → GGA) had occurred, resulting in the M43Asn

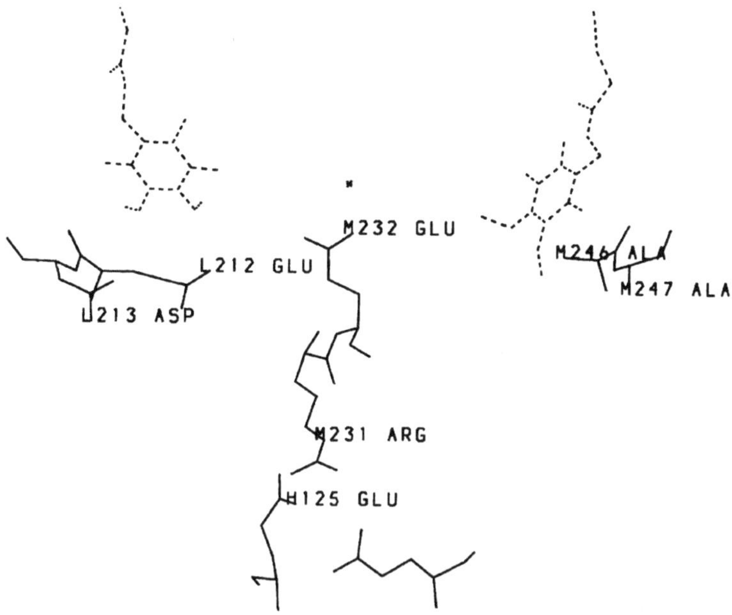

Figure 1. Molecular model, using the *R. sphaeroides* structure [1], showing the interactions between amino acids described in the text. Q_A (right) and Q_B (left) are represented by dashed lines; numbers refer to the *R. capsulatus* sequence [10]. In the site-specific mutants described, the amino acids at L212, L213, M246, and M247 were symmetrized or reversed, yielding PS^- mutants. The substitution of Leu for Arg at M231 acts as an intergenic suppresssor of the L212Ala-L213Ala site-specific double mutant, restoring photocompetence.

→ Asp substitution. No other sequence changes were found. Thus, the replacement of M43Asn by Asp serves to suppress the PS^- phenotype caused by the loss of acidic residues at L212 and L213.

Kinetics of the $Q_A^-Q_B \rightarrow Q_AQ_B^-$ reaction. The kinetics of electron transfer from Q_A to Q_B following a single laser flash were determined in chromatophores by following the photoinduced electrochromic shift of the L-side bacteriopheophytin, observed at 760 nm [19]. Figure 2 shows these transients measured in chromatophores of the wild-type and revertant strains, corrected for the instantaneous rise and slow decay of P^+ (Tiede and Hanson, this volume). The transients detected following formation of the initial $P^+Q_A^-$ state in the double mutant differed dramatically from those shown in Figure 2. The relatively slow rise and decay detected in the double mutant suggested forward and back rate constants of about 110 s^{-1} and 50 s^{-1}, respectively, for the $Q_A^-Q_B \leftrightarrow Q_AQ_B^-$ equilibrium. It is difficult to be certain that the transients measured in the double mutant are due to electron transfer between the quinones because the time scale of these transients is comparable to that for protein relaxation events associated with the $P^+Q_A^-$ state (Tiede and Hanson, this volume), and the kinetics were not sensitive to inhibitors such as *o*-phenanthroline or atrazine. However, the loss of sensitivity to inhibitors may be a consequence of the amino acid replacements in the Q_B site. A lower sensitivity to inhibitors was also seen for some of the revertant strains. In addition, the transient spectra associated with these kinetics were consistent with Q_A to Q_B electron transfer.

If we assume that the transients detected in the double mutant are due to $Q_A \rightarrow Q_B$ electron transfer, the derived rate constants would suggest a free energy drop of about −20 meV between Q_A and Q_B, compared to about −80 meV in the wild-type [5]. The second electron transfer, $Q_A^-Q_B^- \rightarrow Q_AQ_B^{2-}$, does not occur in the double mutant, presumably because of the interruption of the proton transfer pathway. Similar results have been obtained with other mutants constructed at these sites [20-22]. Figure 2 shows that the $Q_A^-Q_B \rightarrow Q_AQ_B^-$ kinetics in the Class 1, 2, and 3 strains are restored to levels similar to that of the wild-type. Furthermore, quinone oscillations and multiple cytochrome turnovers were all found to be restored in these strains (not shown). These measurements provide assays for rate-limiting proton transfer to the Q_B site. These results show that both electron and proton transfer functions are restored in the Class 1, 2, and 3 strains.

Summary and Conclusions. Our results show that neither L212Glu nor L213Asp is obligatory for efficient light-induced electron or proton transfer in *R. capsulatus*. Second-site mutations, located within the Q_B binding pocket or at a more distant site, can compensate for mutations at L212 and L213. We have shown through genetic characterization, in each case, that the single additional mutation is solely responsible for restoring the photosynthetic phenotype.

The mutations found in Class 1 and 2 strains demonstrate that acquisition of a single negatively charged residue within the Q_B binding site (at position L213, or on the other side of the binding pocket at position L225) is sufficient to restore normal redox, electron and proton transfer functions to the complex.

The substitution of Asp for Asn at residue M43, which is located 9 Å from Q_B, acts as an intergenic suppressor of the L212Ala-L213Ala mutations (Class 4). The Asp,Asn combination of residues at the L213 and M43 positions is conserved in five species of photosynthetic bacteria whose RC sequences are known. In *R. capsulatus* and *R. sphaeroides*, the pair is L213Asp-M43Asn [8-10]. But, the RCs of *Rps. viridis*, *Rsp. rubrum*, and *C. aurantiacus* reverse the combination to L213Asn-M43Asp[7, 11-13]. In this respect, the Q_B site of the above suppressor strain (L212Ala-L213Ala-M43Asp) resembles that of the latter three species in that it couples an uncharged residue at L213 with an acidic residue at M43. These RCs, in which L213 is an amide, must employ an alternative proton transfer pathway. The observation that the M43Asn → Asp mutation in *R. capsulatus* compensates for the loss of both acidic residues at L212 and L213 suggests that M43Asp is involved in a new proton transfer route in this strain that resembles the one normally used in RCs of *Rps. viridis, Rsp. rubrum*, and *C. aurantiacus*.

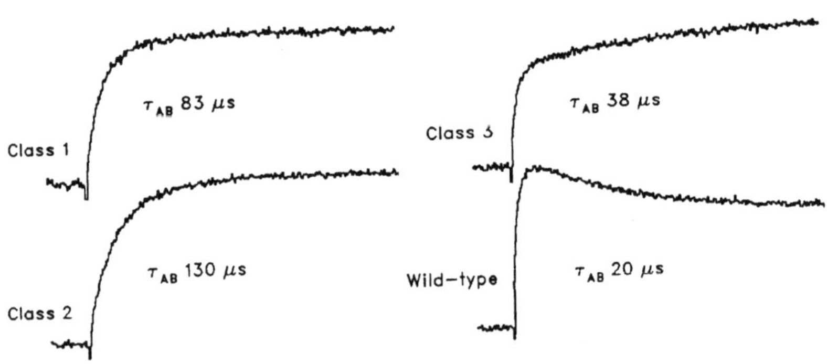

Figure 2. Kinetics of transfer of the first electron from Q_A to Q_B in chromatophores following a single laser flash, observed at 760 nm, corrected for the instantaneous rise and slow decay of P^+. Lifetimes (τ) of the $Q_A^-Q_B \rightarrow Q_AQ_B^-$ reaction in PS^+ revertants (Class 1, 2, and 3) of the L212Ala-L213Ala double mutant strain resemble that of the wild-type.

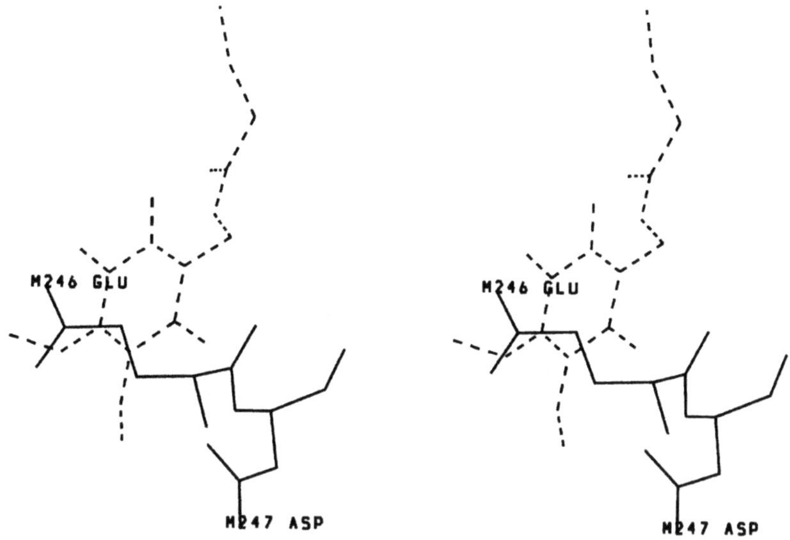

Figure 3. Stereo model, based on the *R. sphaeroides* structure [1], that represents the M246Glu-M247Asp mutant in the Q_A site. The mutant residues were oriented to minimize interference with neighboring residues.

More distant changes in charged residues of the reaction center can also compensate for the mutations at L212 and L213. Loss of a positively charged residue at position M231 (Arg → Leu), 15-20 Å away from the Q_B binding site (Class 3, Figure 1), is also capable of restoring photocompetence. The absence of M231Arg changes the charge balance in this region of the RC by freeing H125Glu and H232Glu for participation in other ionic interactions. Whether the effect of the absence of M231Arg in the revertant is direct or indirect, it influences the charge balance to restore as a whole the RC function of driving protons to Q_B in the absence of L212Glu and L213Asp (D. K. Hanson, L. Baciou, D. M. Tiede, S. L. Nance, M. Schiffer, and P. Sebban, unpublished observations).

The suppressor mutations which restore photocompetence should also affect the pH optima for these reactions. We expect that these "pseudo wild-type" suppressor strains will be instrumental in demonstrating the existence of alternative proton transfer pathways through the RC to the Q_B anion(s); the sequence changes found so far seem to indicate that protons travel to Q_B via a "directed diffusion" pathway rather than by a specific pathway involving a network of hydrogen bonds.

Double Mutant M246Ala-M247Ala → Glu-Asp

In continuing to probe the contribution of the protein side chains to the observed differences in the chemical properties of the two quinones, we symmetrized the quinone binding sites in the opposite way by constructing a double mutant which replaced the alanines in the Q_A binding pocket with the acidic groups found at the equivalent sites in the Q_B binding pocket (M246Ala-M247Ala → Glu-Asp). Thus, in this strain, the acidic groups are present in both the Q_A and Q_B binding sites. In addition, we reversed the symmetry relationships at the quinone sites by constructing the quadruple mutant L212Ala-L213Ala-M246Glu-M247Asp. Both of these mutants are PS^-. EPR spectra display significant amounts of both the P^+ and triplet states of the bacteriochlorophyll dimer in unreduced chromatophores isolated from both mutants (not shown).

Isolation and genetic characterization of photocompetent revertants. We selected photocompetent derivatives of the M246Ala-M247Ala double mutant strain. Genetic characterization determined that four of the six independent revertants isolated carried chromosomal suppressor mutations; the other two strains carried plasmid-borne mutations. In these two, sequence analysis determined that single base changes at the M246 codon restored the PS^+ phenotype. In one strain, M246Glu reverted to Ala, restoring the wild-type codon. In the other, M246Glu changed to Gly. Even though residue M247 is still an aspartic acid, each of these strains displays a wild-type phenotype and EPR spectra resemble that of the wild-type.

Summary and Discussion. The mutations made at M246-M247 change a small hydrophobic residue into a fairly bulky acidic residue. These substitutions alter the charge distribution at the Q_A site, and computer graphics modeling of the substitutions suggests that a glutamic acid residue at M246 could sterically hinder binding of UQ_{10} (Figure 3). The observation that the reversion events replace Glu with either Gly or Ala may suggest that a small, as well as nonpolar, group is required for function at this site. Although the Q_B site has acidic residues, there are no acidic residues in the Q_A site (see Table 1). However, several acidic residues are located in the second layer of amino acids surrounding those that actually form the Q_A binding pocket. None of the reversion events replaces the Asp at M247; note that in *C. aurantiacus*, M247 is Gln [13]. Asp at M247 must alter the pKs of protonatable residues nearby, among which are L6Glu and M244Glu.

The mechanism(s) by which chromosomal mutations compensate for the M246Glu-M247Asp mutations is more difficult to imagine. To relieve the steric interference, rearrangements of large segments would have to occur. The point mutation in the H chain that could cause large rearrangements is hard to predict. The suppression might occur in a non-RC gene, possibly resulting in the insertion of an unsubstituted benzoquinone into the double mutant Q_A site.

AMINO ACIDS THAT AFFECT THE PRIMARY REACTIONS

Although the RC structure shows considerable twofold symmetry in the arrangements of both the cofactors and the peptide chains, electron transfer occurs only through the pathway that is primarily associated with the L chain. Several amino acid side chains have the potential to contribute to the primary electron transfer reactions. A particularly striking symmetry-breaking pair is L181Phe-M208Tyr [23]. Each of these residues is positioned in proximity to the macrocycles of the dimeric and monomeric bacteriochlorophylls and the bacteriopheophytin molecule associated with its particular side of the RC.

We have constructed a series of mutants at L181 and M208, and find that they are involved in determining the rate of the initial electron transfer reaction, and also influence the redox potential of the bacteriochlorophyll dimer (Fleming et al., this volume). Amino acid substitutions at these sites did not change the direction of electron transfer. Mutations at both L181 and M208 can also affect the spectral properties of nearby cofactors. The shifts can be seen in room temperature RC spectra, and are more dramatic in RC spectra recorded at 4°K. In the following, we will discuss only the two mutants that cause the largest changes in the absorption spectra.

To study the effect of charged residues at the L181 position, Lys and Glu were substituted for the hydrophobic aromatic residue, Phe. To our surprise, both mutant RCs are functional, and the rate of the electron transfer is similar to that of the wild-type RC. The substitution of Lys for Phe at L181 splits the bacteriochlorophyll Q_x band into three

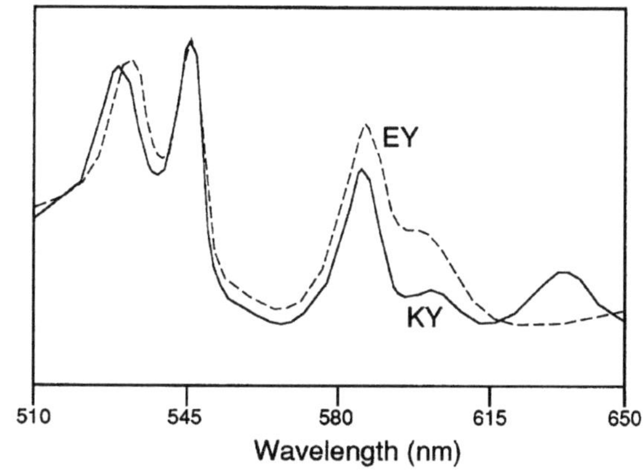

Figure 4. Absorption spectra of reaction centers measured at 4°K. In these RCs, Phe at L181 was substituted by Lys (KY) and by Glu (EY). The Q_x absorption band of the bacteriochlorophylls is split into two peaks in the EY and three in the KY mutants.

peaks (Figure 4). Modeling studies suggest that L181Lys can coordinate to the Mg of the M-side bacteriochlorophyll monomer, resulting in a hexacoordinated Mg. It is very unlikely that either Glu or Lys is ionized in the very hydrophobic environment in the interior of the protein. The Q_x absorption band contains absorption from the bacteriochlorophylls of the special pair and the monomeric bacteriochlorophylls. It is a single broad band in the wild-type and is split into two bands in the L181Glu mutant. The lower energy band at 601 nm in the L181Lys splits further into two bands (601 and 633 nm) with a frequency shift of $-833\ cm^{-1}$. This value is similar to the shift of $-813\ cm^{-1}$ that was observed for changing the penta coordinated monomeric bacteriochlorophyll with nitrogen ligands to a hexa coordinated bacteriochlorophyll [24].

When M208Tyr is replaced by His residue, red shift of the lower energy Q_x and Q_y bacteriopheophytin bands is observed (Figure 5); the shift of the Q_y band is very striking. Computer graphics modeling shows that by rotating the side chain of the His residue, it can be placed so that it forms a hydrogen bond (<3.0 Å) with the ring I acetyl carbonyl group of the L-side bacteriopheophytin molecule. No change in the acetyl's orientation is required. These observations therefore confirm that the lower energy bacteriopheophytin bands are that of the L-side bacteriopheophytin. In wild-type RCs, the red shift of the photoactive pheophytin in the Q_x region of the spectrum has been shown to be due to the formation of a hydrogen bond between L104Glu and the keto oxygen of ring V [25]. Our observation that hydrogen bond causes a red shift in the absorption spectra is in qualitative agreement in both direction and relative magnitudes of the shift with calculations [26]. Shifts in the Q_x and Q_y bands of $-51\ cm^{-1}$ and $-107\ cm^{-1}$, respectively, were predicted for water hydrogen bonded to the acetyl carbonyl. Experimentally, the Q_x band shifts $-101\ cm^{-1}$ from 545 to 548 nm; the Q_y band shifts $-224\ cm^{-1}$ from 755 to 768 nm.

The fact that the absorption changes can be rationalized based on the crystal structure of the *R. sphaeroides* RC suggests that the structure of the *R. capsulatus* RC is similar to it and that the mutants at positions L181 and M208 do not grossly affect the three-dimensional structure of the complex.

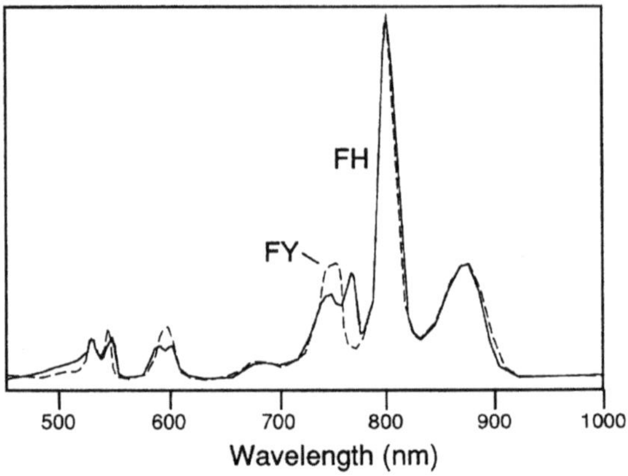

Figure 5. Absorption spectra of wild-type (FY) and M208His mutant (FH) reaction centers measured at 4°K. It shows the red shift of the Q_y absorption band of the L-side bacteriochlorophyll in the mutant M208His.

Acknowledgments. We thank D. C. Youvan for the gift of plasmids pU29 and pU2922, and deletion strain U43, G. Small for help with the low termperature absorption spectra, and J. Fajer for valuable discussions on the spectra. Supported by the U. S. Department of Energy, [1]Office of Health and Environmental Research and the [3]Office of Basic Energy Sciences, under Contract No. W-31-109-ENG-38, and the [2]National Science Foundation. M. S. and C.-H. C. are also supported by Public Health Service Grant GM36598.

REFERENCES

1. C.-H. Chang, O. El-Kabbani, D.M. Tiede, J.R. Norris, and M. Schiffer, *Biochemistry* 30:5352-5360 (1991).
2. J. Deisenhofer and H. Michel, *Science* 245:1463-1473 (1989).
3. G. Feher, J.P. Allen, M.Y. Okamura, and D.C. Rees, *Nature* 339:111-116 (1989).
4. J. Breton, *ISI Atlas of Science: BioChemistry* 1:323-328 (1988).
5. A.R. Crofts and C.A. Wraight, *Biochim. Biophys. Acta* 726:149-185 (1983).
6. P.H. McPherson, M.Y. Okamura, and G. Feher, *Biochim. Biophys. Acta* 1016:289-292 (1990).
7. H. Michel, K.A. Weyer, H. Gruenberg, I. Dunger, D. Oesterhelt, and F. Lottspeich, *EMBO J.* 5:1149-1158 (1986).
8. J.C. Williams, L.A. Steiner, G. Feher, and M.I. Simon, *Proc. Natl. Acad. Sci. USA* 81:7303-7307 (1984).
9. J.C. Williams, L.A. Steiner, R.C. Ogden, M.I. Simon, and G. Feher, *Proc. Natl. Acad. Sci. USA* 80:6505-6509 (1983).
10. D.C. Youvan, E.J. Bylina, M. Alberti, H. Begusch, and J.E. Hearst, *Cell* 37:949-957 (1984).
11. G. Belanger, J. Berard, P. Corriveau, and G. Gingras, *J. Biol. Chem.* 263:7632-7638 (1988).
12. Y.A. Ovchinnikov, N.G. Abdulaev, A.S. Zolotarev, B.E. Shmuckler, A.A. Zargarov, M.A. Kutuzov, I.N. Telezhinskaya, and N.B. Levina, *FEBS Lett.* 231:237-242 (1988a).

13. Y.A. Ovchinnikov, N.G. Abdulaev, B.E. Shmuckler, A.A. Zargarov, M.A. Kutuzov, I.N. Telezhinskaya, N.B. Levina, and A.S. Zolotarev, *FEBS Lett.* 232:364-368 (1988b).
14. D.C. Youvan, S. Ismail, and E.J. Bylina, *Gene* 38:19-30 (1985).
15. E.J. Bylina, S. Ismail, and D.C. Youvan, *Plasmid* 16:175-181 (1986).
16. E.J. Bylina, R.V.M. Jovine, and D.C. Youvan, *Bio/Technology* 7:69-74 (1989).
17. C.-K. Chan, L.X.-Q. Chen, T.J. DiMagno, D.K. Hanson, S.L. Nance, M. Schiffer, J.R. Norris, and G.R. Fleming, *Chem. Phys. Lett.* 176:366 (1991).
18. D.K. Hanson, S.L. Nance, and M. Schiffer, *Photosyn. Res.* 32:147 (1992).
19. A. Vermeglio and R.K. Clayton, *Biochim. et Biophys. Acta* 461:159 (1977).
20. M.L. Paddock, S.H. Rongey, G. Feher, and M.Y. Okamura, *Proc. Natl. Acad. Sci. USA* 86:6602-6606 (1989).
21. E. Takahashi and C.A. Wraight, *Biochemistry* 31:855-866 (1992).
22. E. Takahashi and C.A. Wraight, *Biochim. Biophys. Acta* 1020:107-111 (1990).
23. D.M. Tiede, D.E. Budil, J. Tang, O. El-Kabbani, J.R. Norris, C.-H. Chang, and M. Schiffer, *in*: "The Photosynthetic Bacterial Reaction Center," J. Breton and A. Vermeglio, eds, Plenum, New York, pp. 13-20 (1988).
24. P. M. Callahan and T. M. Cotton, *J. Am. Chem. Soc.* 109:7001-7007 (1987).
25. E.J. Bylina, C. Kirmaier, L. McDowell, D. Holten, and D.C. Youvan, *Nature* 336:182 (1988).
26. L. K. Hanson, M. A. Thompson, and J. Fajer, *in*: "Progress in Photosynthesis Research," J. Biggins, ed., Martinus Nijhoff Publishers, Dordrecht, Netherlands, pp. 1.3.311-1.3.314 (1987).

PROTON TRANSFER PATHWAYS IN THE REACTION CENTER OF *RHODOBACTER SPHAEROIDES*: A COMPUTATIONAL STUDY

P. Beroza, D. R. Fredkin, M. Y. Okamura, and G. Feher

Department of Physics, 0319
University of California San Diego
La Jolla, CA 92093-0319, USA

1. INTRODUCTION

We address the problem of determining proton transfer pathways from solution to the secondary quinone (Q_B) in the photosynthetic reaction center (RC) of *Rb. sphaeroides*. Based on the crystal structures for *Rb. sphaeroides*[1,2] and *Rps. viridis*,[3] the Q_B binding site is buried in the protein, out of contact with the aqueous solution. Thus, protons have to pass through the protein to reach the Q_B site. The mechanism of this transfer is important in understanding how reaction centers mediate the conversion of light energy into chemical energy.

The mechanism of proton transport in membrane proteins has been discussed extensively.[4–6] The general view is that protons move along a chain of proton donor and acceptor groups. These groups could be either side-chains of protonatable amino acids or bound water molecules.

In the reaction center, transfer of two protons from the cytoplasm to Q_B is coupled to the two-electron reduction of the quinone, resulting in dihydroquinone (see fig. 1). The

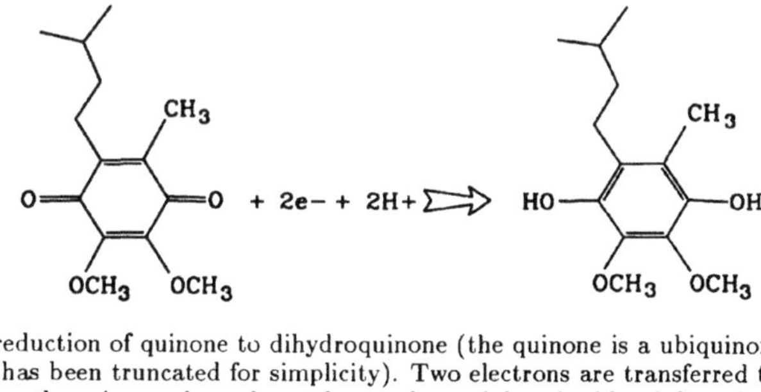

Figure 1. The reduction of quinone to dihydroquinone (the quinone is a ubiquinone-10; the isoprenoid chain has been truncated for simplicity). Two electrons are transferred to the secondary quinone (Q_B) from the primary donor located near the periplasmic side of the RC. After two protons are transferred from the cytoplasmic side, the dihydroquinone leaves the RC.

The Photosynthetic Bacterial Reaction Center II
Edited by J. Breton and A. Verméglio, Plenum Press, New York, 1992

two carbonyl oxygens of the quinone, which are hydrogen bonded to amino acid residues in the RC, are the ultimate proton binding sites. The proximal oxygen, defined as the one closer to the non-heme iron, is hydrogen bonded to His L190 (which is coordinated to the iron). The distal oxygen is hydrogen bonded to Ser L223 (see fig. 2B). After the transfer of two protons, the dihydroquinone leaves the RC and enters the membrane, where it is oxidized by the cytochrome bc_1 complex, which releases protons on the periplasmic side of the membrane. The reoxidized quinone binds to the RC and a new photochemical cycle is initiated. This cycle is part of a light-driven proton pump which generates the proton gradient that drives ATP synthesis.

The pathways of proton transfer in RCs from *Rb. sphaeroides* have been previously investigated experimentally by site-directed mutagenesis of protonatable residues near the Q_B binding site. The amino acid residues Glu L212, Asp L213, and Ser L223 have been shown to play important roles in proton transport to Q_B (for review, see ref. 7). Mutation of Asp L213 and Ser L223 blocked both the transfer of the second electron and the transfer of protons. Mutation of Glu L212 blocked only proton transfer. These results have been interpreted by the following model:[8] a) The first proton is transferred to the distal oxygen via a pathway involving Asp L213 and Ser L223. The binding of the first proton is required for the transfer of the second electron. b) The second proton is transferred to the proximal oxygen via a pathway involving Glu L212 and possibly Asp L213. The binding of the second proton occurs after the transfer of the second electron.

In this work we took a computational approach to investigate the problem of proton transfer pathways. We examined the X-ray structure of the RC from *Rb. sphaeroides* as determined by Allen et al.[1,2] * and searched for proton transfer pathways from external solvent to the internally bound quinone.

Water molecules may play an important role in proton transfer. Although water molecules were not resolved in the X-ray structure of *Rb. sphaeroides*, detailed examination revealed voids in the protein structure that could accommodate water molecules. † Water molecules were found in homologous internal cavities in the X-ray structure of *Rps. viridis*.[3] Following the method of Rashin et al.,[11] we predicted the positions of water molecules in the internal cavities in the RC. These water molecules were included in the RC structure to assess their potential role in proton transfer.

We present the results of our search for proton pathways with and without considering internally bound water molecules in the RC. We then discuss the general features of the pathways and compare our results with the experimental findings obtained from mutated RCs.

2. METHODS

2.1 Definition of a Proton Transfer Pathway

A pathway is a chain of proton donor and acceptor groups. These groups are either protonatable atoms from amino acid sidechains or water molecules whose positions were inferred from space-filling and hydrogen bonding considerations (see below). The groups considered as possible proton binding sites are listed in table 1; they include

*Comparison between the X-ray structure used in this work[1,2] and that reported by Chang, et al.[9] revealed significant discrepancies in the region of Q_B. Reconciliation of these discrepancies awaits more highly refined X-ray structures.

†Protein structures were examined visually by computer graphics using an Evans and Sutherland PS300 with the FRODO program[10] and an IBM/RS6000 model 540 with *Insight II*, a product from Biosym Technologies of San Diego, CA.

Table 1. Components of Proton Transfer Pathways

Group	Atom(s) †
Glutamic acid	Carboxyl oxygens
Aspartic acid	Carboxyl oxygens
Arginine	Terminal nitrogens
Lysine	Terminal nitrogen
Histidine	Imidazole nitrogens
Tyrosine	Hydroxyl oxygen
Cysteine	Sulfur
Serine	Hydroxyl oxygen
Threonine	Hydroxyl oxygen
Water	Oxygen

† Atoms allowed to receive and donate protons in computed proton pathways through the RC. Atoms on the same residue (i.e., the carboxyl oxygens in Glu or Asp, the terminal nitrogens in Arg, and the imidazole nitrogens in His) were assumed to be able to transfer a proton to each other.

acidic, basic, and hydroxyl residues and water. In defining a chain it was assumed that protons can readily exchange between protonatable atoms on the same residue (eg. in Glu or Asp, a proton can move between the two oxygen atoms of the carboxyl group).

Proton transfer between individual sites in a pathway may occur by either classical barrier crossing or quantum mechanical tunneling. In either case transfer will be likely only when the distance between proton donor and acceptor atoms is close to that of a hydrogen bond (~3 Å).[12,13] In identifying proton transfer pathways using the static X-ray structure of the protein, atomic motion is not taken into account. Thermal fluctuations and electrostatic forces that result from electron transfer to Q_B may bring donor and acceptor atoms to within hydrogen-bonding distance even though the distance between them determined by X-ray diffraction is too large for proton transfer.

In dynamical simulations of electron transfer in the RC of *Rps. viridis*, the RMS fluctuation of atomic positions was 0.52 Å.[14] Consequently, in our enumeration of pathways based on the static crystal structure, we expect a cutoff distance for proton transfer of ~4 Å. For larger distances between proton donor and acceptor groups, proton transfer would be likely only if significant motion (> 2 Å) of the groups occurs. Such motion is possible on the time scale of proton transfer (ms) which is much longer than the time scale of the simulations (ps).

Using a computer program, we enumerated all pathways from the quinone carbonyl oxygens to the external solvent. Two groups were considered to be connected with respect to proton transfer if the distance between them was less than a specified cutoff distance. By varying the cutoff distance, we obtained a different number of pathways for the transfer of protons from aqueous solvent to the quinone. Beginning at a quinone oxygen, connectivity was established to any neighboring proton donor/acceptor groups. The pathways were then extended from these groups to their neighbors, and so on.

When water molecules were included in the structure, the number of pathways from external solvent to the quinone carbonyl oxygens became very large ($> 10^4$). To limit the number, pathways with lengths exceeding ten groups were discarded. This removed from consideration long chains of water molecules joining internal cavities in the RC (see below).

A group was considered in contact with the external solvent when it had a protonatable atom that bordered the external solvent-accessible surface. Our convention is to call the solvent exposed residue the start of the proton transfer pathway and the carbonyl oxygen on the quinone the end of the pathway.

2.2 Solvent-Accessible Surfaces

We computed solvent-accessible surfaces following the procedure of Richards[15] (i.e., the surfaces are defined by the center of a spherical probe as it rolls along the van der Waals surface of the protein). In general, a protein will have multiple surfaces: one exterior surface and several interior surfaces.

In our algorithm, a surface is made up of a set of points that lie on a continuous surface (we used a density of 5 points per $Å^2$). Points that were close to each other were considered to lie on the same surface (for a detailed description of the distance criteria, see ref. 11). Thus, each surface was represented by a subset of the total solvent-accessible points. The largest subset of points was considered the external surface; a visual check was made to assure that this was the true external surface. Residues with a protonatable atom in contact with this surface were considered "external" and, therefore, capable of exchanging protons directly with the solvent. Atoms bordering internal surfaces were not allowed to exchange protons directly with the solvent.

2.3 Prediction of Bound Water Molecules

We applied the technique of Rashin et al.[11] to predict the positions of internally bound water molecules. We scanned the interior solvent-accessible surfaces for points that could form hydrogen bonds to protein atoms, using hydrogen bond length and bond angle criteria. For every point that could form simultaneous hydrogen bonds with three protein atoms we predicted a water molecule. The requirement of three hydrogen bonds is somewhat arbitrary and is intended to match the hydrogen bonds lost by a water molecule when it is removed from solution. We repeated the procedure to allow water molecules to form hydrogen bonds with previously predicted water molecules. Further details of the method appear in reference 11.

The method was tested on a set of eleven proteins that had water molecules resolved in their structures as determined by X-ray diffraction (X-ray coordinates were obtained from the Brookhaven Protein Data Bank[16]). We predicted water molecules using the protein coordinates without the bound waters and compared the predictions with the X-ray coordinates of the internal waters (see Table 2). The results were similar to

Table 2. Comparison of Predicted Water Molecules with X-ray Data

Protein §	Internal Waters ⋆ in X-ray Data	Predicted Waters † (total/X-ray)	RMS Deviation ‡ (Å)
Test set	69	81/63	0.61
1PRC	95	144/73	1.28
1PRC*	57	84/41	1.15
1RCR	—	67/—	—

§ Test Set The 11 protein structures used to test the method (see ref. 11). The numbers reported are the totals for all 11 proteins.

1PRC *Rps. viridis* reaction center

1PRC* *Rps. viridis* reaction center without the cytochrome subunit, shown for comparison with the *Rb. sphaeroides* RC, which has no bound cytochrome subunit.

1RCR *Rb. sphaeroides* reaction center

⋆ No water molecules are resolvable in the crystal structure of *Rb. sphaeroides*.

† The total number of predicted water molecules and the number of those that correspond to water molecules resolved in the X-ray structure.

‡ RMS deviation between the X-ray coordinates of the oxygen atoms of water molecules and the coordinates of the corresponding predictions.

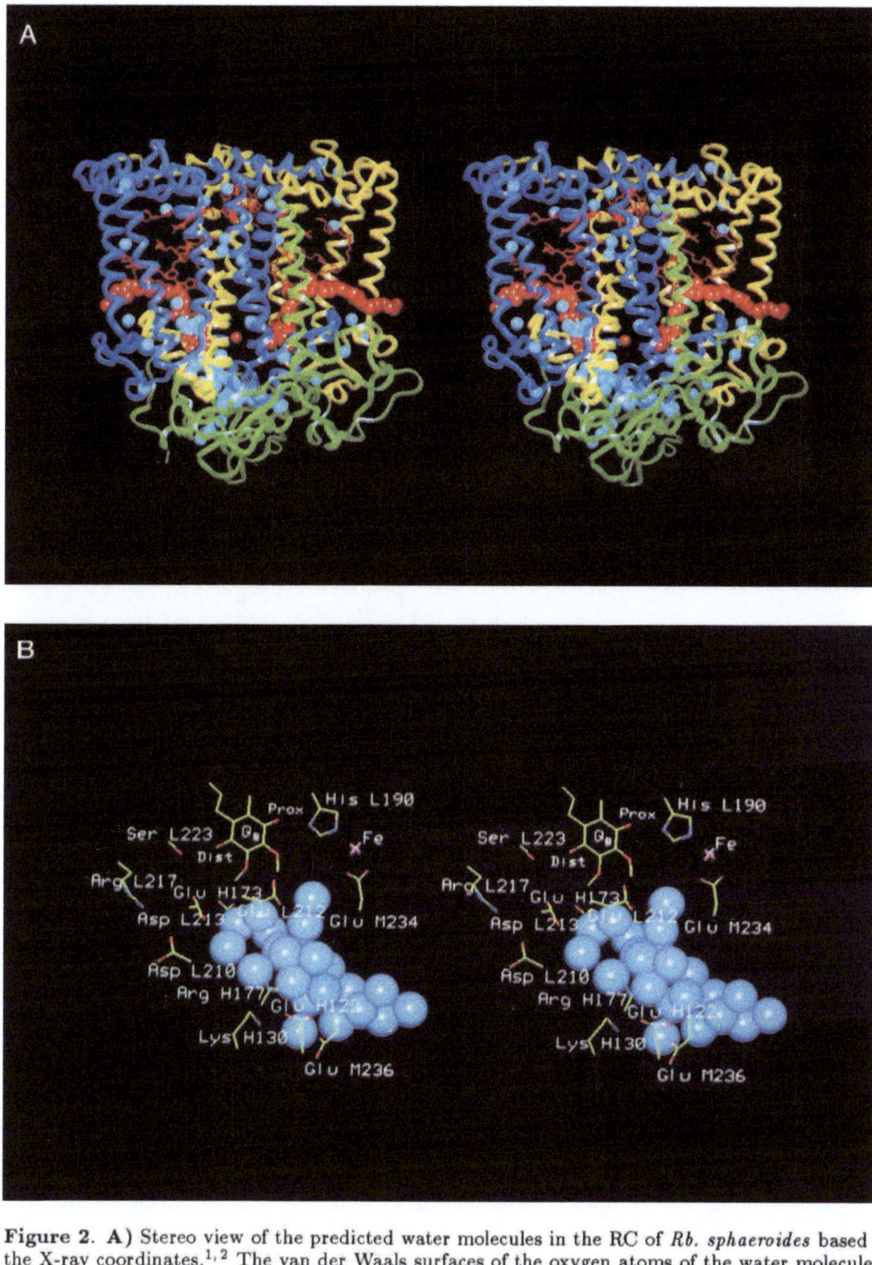

Figure 2. **A)** Stereo view of the predicted water molecules in the RC of ***Rb. sphaeroides*** based on the X-ray coordinates.[1,2] The van der Waals surfaces of the oxygen atoms of the water molecules (light blue), the iron, and the quinones are shown. Ribbons represent the secondary structure of the three protein subunits (L:yellow; M:blue; H:green). Cofactors are in red. Many water molecules were predicted to be near the Q_B site (on the left in each picture), and there are chains of water molecules that link the region directly below Q_B (the methoxy pocket) with solvated cavities much closer to the cytoplasm. **B)** An expanded stereo view of the Q_B binding site with nearby predicted water molecules and protonatable residues (Thr L226 is obscured by water molecules). The methoxy pocket (foreground) contains water molecules that are closest to Q_B. These predicted water molecules shorten the cutoff distances required to establish a proton transfer pathway from the external solvent to Q_B. Another cavity (background) contains water molecules that can form a chain from the methoxy pocket to the external solvent.

those of Rashin et al. About 90% of the internal water molecules in the crystal structures were predicted correctly, and about 20% of the predictions did not correspond to water molecules in the X-ray structure. Most of the discrepancy occurred in the the large proteins of the test set; in these, more water molecules were predicted than were observed in the X-ray structure. This trend continued with the RC from *Rps. viridis*, which is larger than any of the proteins in the test set. Twice as many water molecules were predicted than were resolved in the X-ray structure. However, it is not clear whether this over-prediction resulted from a failure of the predictive model or from loosely bound waters in the protein that did not show up in the X-ray structure.

3. RESULTS

3.1 Water Molecules in the Reaction Center of *Rb. Sphaeroides*

Using the algorithm described above, we predicted 67 internal water molecules in the reaction center of *Rb. sphaeroides* (see fig. 2A). Many bound water molecules were predicted to lie at the interface between the H-subunit and the membrane bound complex of the L and M subunits. This interface in the *Rps. viridis* structure has water molecules that were resolved by X-ray diffraction.

A hydrogen-bonded network of water molecules was predicted to fill a large cavity on the cytoplasmic side of Q_B. This cavity, which we have named the "methoxy pocket" because of its proximity to the methoxy groups in Q_B, contains water molecules that are closest to the Q_B carbonyl oxygens (fig. 2B). Although the methoxy pocket is deeply buried in the RC, it is near another solvated cavity that is closer to the cytoplasm. The closest distance between the water molecules in the two pockets is 3.9 Å. Both cavities are at the interface between the H-subunit and the L and M subunits.

In addition to the solvated internal cavities, there are also regions of external solvent-accessible surface that penetrate deep into the protein. It is likely that these invaginations contain ordered water molecules, but these molecules were not predicted by our algorithm because we searched only interior surfaces. Bound water molecules in this external region, which are not considered explicitly in this paper, are likely to play a role in proton transfer. In particular, there are clefts leading to Arg L217, Asp L210, and Glu H173. Although these residues appear to be deeply buried in the RC, they were considered external residues in our treatment.

3.2 Proton Transfer Pathways Without Water Molecules

The shortest pathways for proton transfer without water molecules are shown in figure 3. All pathways begin at residues that are below the cytoplasmic boundary of the membrane (i.e., since the iron is approximately at the level of the membrane surface, all residues with negative z values in the coordinate system defined in figures 3 and 4 are below the membrane).

The shortest pathways from external solvent to either carbonyl oxygen on Q_B begin at Arg L217, Asp L210, and Glu H173, which are accessible to external solvent by virtue of the narrow clefts discussed above. The proximal oxygen is more deeply buried in the protein than the distal oxygen. Therefore, it is not surprising that the pathways to the distal oxygen are shorter than those to the proximal oxygen.

Another feature of the computed pathways is the importance of Asp L213 for pathways leading to the distal oxygen. Even with cutoff distances as large as 5.2 Å all proton pathways to the distal oxygen include Asp L213. This result is independent of the presence of water molecules (see below).

The pathways originating farthest from Q_B (which converge with those shown in figure 3 at double circles) must bypass the methoxy pocket. To complete a pathway from one side of the pocket to the other, without the aid of water molecules, connections must be made around the outside of the pocket, and many of these connections require large cutoffs (> 5 Å). Because the methoxy pocket lies in the way of all pathways to the proximal oxygen, the cutoff distances for the pathways to the proximal oxygen are larger than those for the pathways to the distal oxygen.

3.3 Proton Transfer Pathways With Water Molecules

The shortest proton transfer pathways with predicted water molecules included in the structure are shown in figure 4. No water molecules were predicted to be within ~ 6 Å of either carbonyl oxygen of Q_B. As a result, the last proton transfers in the pathways shown in figures 3 and 4 are identical.

The predicted water molecules fill the methoxy pocket with a network of hydrogen-bonded water molecules, which provide many parallel pathways through the pocket

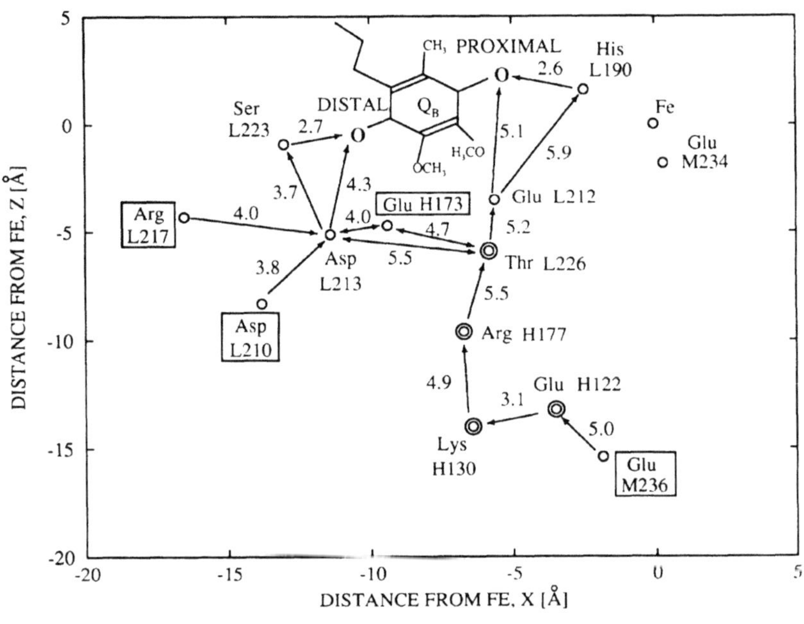

Figure 3. The shortest pathways from the external solvent to the Q_B carbonyl oxygens without water molecules. Numbers next to arrows indicate distances in Å. The origin of the coordinate system is at the position of the iron. The z axis is normal to the surface of the membrane, and the x axis is chosen so that the quinones and iron lie approximately in the xz-plane. Protonatable residues are shown as circles; those in contact with the external solvent are boxed. The coordinates of protonatable groups were projected onto the xz-plane (the coordinates of protonatable groups with more than one atom, Glu, Asp, His, and Arg, were taken as the average of all protonatable atoms in the group). Double-headed arrows indicate transfers that can participate in pathways to either Q_B carbonyl oxygen. Longer pathways from the external solvent (not shown) join these pathways at the residues with two circles. For example, the pathway shown from Arg H177 to the external solvent is the shortest of many. Short cutoff distances for transfer established pathways from the solvent to the distal oxygen, but longer cutoff distances were necessary to form a pathway from the solvent to the proximal oxygen.

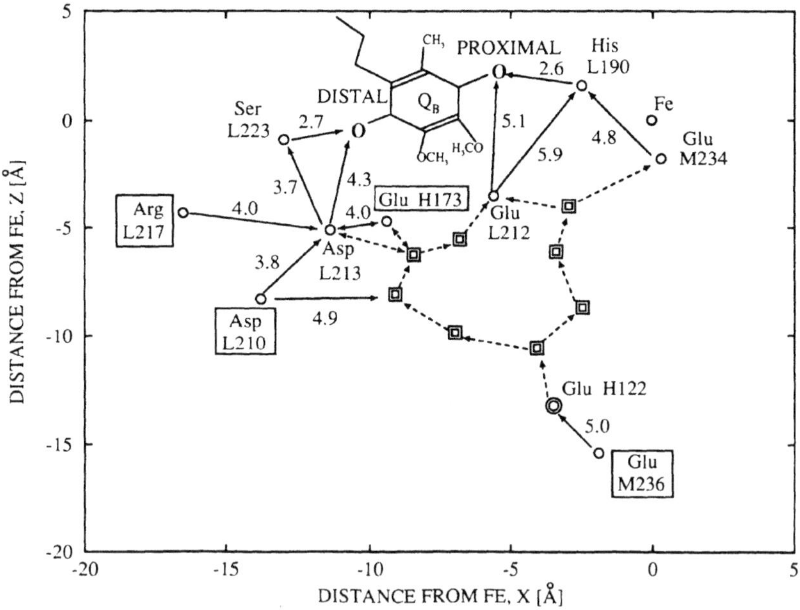

Figure 4. The shortest pathways from the external solvent to the Q_B carbonyl oxygens including predicted water molecules. Coordinate system and symbols as in figure 3. Predicted water molecules are shown as squares. Longer pathways from external solvent (not shown) converge with those shown at double squares and double circles. Dotted lines represent transfers that involve water molecules, all having distances under 4.0 Å. The shortest pathways from the solvent to the *distal oxygen* are not affected by considering water molecules. The pathways from the solvent to Glu L212 and His H190 are affected by the presence of water molecules, which reduce the cutoff distance from 5.2 Å to < 4 Å. Transfer distances near the proximal oxygen are still considerably longer than hydrogen bond length. This suggests that a structural change may be necessary for proton transfer to the *proximal oxygen*.

(only the shortest are shown in the figure). These water molecules significantly reduce the cutoff distances for proton transfer by allowing transfer through, rather than around, the pocket. Although there are more links necessary in some of the pathways across the pocket, the distances for individual proton transfers are significantly reduced.

For cutoff distances ≥ 3.9 Å, pathways were formed from the methoxy pocket to another solvated cavity that is closer to the cytoplasm. One water molecule in this cavity is 3.3 Å from the external surface. Thus, it is possible that proton transfer can occur from external solvent to the internal water molecules that are nearest to Q_B (those in the methoxy pocket) via a chain consisting entirely of water molecules.

4. DISCUSSION

Through a combination of computational work presented here and experimental work done on site-specifically mutated RCs, we identify two general properties of proton transfer in the RC: the *convergence of many pathways* from the external solvent on the cytoplasmic side of the RC, and the *role of internal water molecules* in these pathways. In addition, we compare the results of the *enumeration of pathways* to the individual Q_B oxygens with the results of experimental work done on mutated RCs.

4.1 Convergence of Pathways

There are a large number of pathways from the external solvent that converge on a few key residues near Q_B. The residues at the points of convergence of the pathways play a particularly important role in proton transfer. Consequently, site-directed mutagenesis of these residues to non-protonatable amino acids is expected to have a dramatic effect on the measured proton transfer rates. In contrast, site-directed mutation of residues that are farther removed from Q_B should have a relatively small effect, because other parallel proton pathways can compensate for the blockage introduced by the mutation.

The above argument is borne out by experiment. Of the four closest protonatable residues (His L190, Glu L212, Asp L213, and Ser L223), all but His L190 are essential for proton transfer.[8,17–21] Mutagenesis studies of the more distant residues show them to be less important in proton transfer. The mutation of two such residues, Arg L217 and Asp L210, to non-protonatable residues does not significantly affect proton transfer.[22] In the computed pathways, alternative pathways do not require these residues. In particular, if protons are transferred across the methoxy pocket by water molecules, many parallel pathways to the external solvent become available. These parallel pathways not only enhance proton transfer to the quinone, they also protect the RC from random mutations that might block a single pathway.

As a consequence of the large number of parallel pathways, it is difficult to identify effective candidates for site-directed mutagenesis. Aside from the residues near Q_B, which have already been mutated, none of the farther residues seems likely to block proton transfer to Q_B. After considering the energetics of proton transfer, the number of pathways may be reduced. In that case, mutations farther from Q_B may be more important than they appear based solely on pathway enumeration. It may also be possible to determine the effect of electrostatic changes introduced by mutation of charged residues, which may block proton transfer along a nearby pathway even though the mutated residue is not a member of the pathway.

4.2 Role of Internal Water Molecules

The presence of water molecules near the Q_B site suggests that water plays an important role in proton transfer. The positions of these waters are not established with certainty by the computational analysis. However, a strong indication of their existence is the presence in the X-ray structure of significant voids near Q_B that are bordered by polar groups in the protein. Because the distances between protonatable amino acid groups are, in some cases, significantly larger than the distances for proton transfer, internal water molecules are likely to play an important role as proton donors and acceptors.

Proton transfer along a hydrogen-bonded chain of water molecules has been explained, in analogy to proton transfer in ice, by the Grotthus mechanism.[4] In this mechanism, a concerted transfer of protons between adjacent members of the chain reduces the activation energy for transporting a charge in a low-dielectric medium. In the RC, this mechanism may occur for proton transfer across internal cavities as well as through the aqueous clefts leading from external solvent to the protein interior. Proton transfer in bacteriorhodopsin is inhibited by reducing the protein's water content;[23] this suggests that the importance of bound water molecules might be a general feature of proton transfer in membrane-bound proteins.

A hydrogen-bonded network of water molecules provides a natural framework for branched proton transfer pathways, thereby increasing the number of pathways from solvent to Q_B. Water is ideal for forming branches in pathways because of its ability to form up to four hydrogen bonds with other groups and because of its small size which enables it to change orientations easily.

4.3 Proton Transfer to the Distal Oxygen

Proton transfer pathways from the external solvent to the distal oxygen on Q_B were easy to identify. Paths were found that have cutoff distances as low as 4 Å, which, given the uncertainty in the coordinates and the presence of thermal motion of the protein, is probably small enough for proton transfer.

The shortest pathways (from Asp L210, Arg L217, and Glu H173) require no internally bound water molecules to transfer a proton from the external solution to the distal oxygen. The longer paths require water molecules to reduce cutoff distances needed to transfer a proton across the methoxy pocket. These longer pathways are probably less important in the wild type RCs but may offer alternative routes for proton transfer in mutant RCs in which the shorter pathways have been blocked.

For cutoff distances up to 5.2 Å, all pathways enumerated by our technique pass through Asp L213. Thus, Asp L213 appears to play a crucial role in the transfer of a proton to the distal oxygen. This is in agreement with experimental findings that showed that mutation of Asp L213 to its non-protonatable analog, Asn, blocks transfer of the first proton to Q_B.[17,18]

The central position of Asp L213 in proton transport in RCs from *Rb. sphaeroides* contrasts with its role in other photosynthetic bacteria. RCs from *Rps. viridis*,[24] *Rsp. rubrum*,[25] and *Cf. aurantiacus*[26,27] have Asn at the L213 position. These RCs have an Asp at the homologous position to Asn M44 in *Rb. sphaeroides*. When in the Asp L213 → Asn mutants in *Rb. sphaeroides*, Asn M44 was replaced by Asp, proton transfer was restored.[28] Thus, Asp M44 compensates for the mutated Asp L213. The results from pathway enumeration suggest that a new pathway is introduced by the mutation of Asn M44 to Asp, but the effect could be electrostatic– the result of restoring the negative charge that was lost in the mutation of Asp L213. RCs from *Rb. sphaeroides* and *Rb. capsulatis* in which Asp L213 was mutated to a non-protonatable residue had photosynthetic activity restored by mutations that appear to restore the net charge in the region (e.g., Asn M44 → Asp, Gly L225 → Asp, Arg M231 → Leu, or Arg M233 → Cys).[28,29] Combination of electrostatic energy calculations and pathway enumeration may help distinguish between electrostatic effects and the ability to donate (accept) protons.

Ser L223 is in the path with the shortest cutoff distance; this suggests its involvement in proton transfer to Q_B. Ser L223 is close to Asp L213 from which it presumably receives a proton (possible mechanisms for this transfer are discussed in ref. 7). Transfer by means of this pathway is supported by results from site-directed mutagenesis. Mutation of Ser L223 to Ala blocked electron and proton transfer, while mutation to the functionally similar threonine showed essentially no effect.[19]

An alternative pathway would involve the direct transfer of a proton from Asp L213 to the distal oxygen on the quinone. In the X-ray structure, one of the carboxyl oxygens on Asp L213 is 4.3 Å from the distal oxygen, which is not large enough to rule out direct transfer. Takahashi and Wraight have argued in favor of this mechanism,[20] ascribing results from mutagenesis of the serine to a conformational change.

4.4 Proton Transfer to the Proximal Oxygen

Proton transfer pathways to the proximal oxygen are more difficult to identify than those to the distal oxygen. The closest proton donor to the proximal oxygen, His L190 (a ligand to the non-heme iron), is within hydrogen-bonding distance. The shortest pathway from this residue to the external solvent involves Glu M234, another ligand to the iron. Whether this pathway is energetically feasible is unclear. The NH proton on the imidazole ring of the neutral His L190 may be considerably more acidic than it is in solution because of the strong electrostatic interaction with the positively charged iron, but protonation of Glu M234 seems unlikely for the same reason. Moreover, mutation of His L190 to Gln has little effect on the rates of electron and proton transfer.[21]

Glu L212 is an alternative proton donor either to His L190 (5.9 Å) or directly to the carbonyl oxygen of Q_B (5.1 Å). Although these distances are large, mutation of Glu L212 to Gln blocks proton transfer,[8] indicating its importance in proton transfer. Continuation of the proton pathways from Glu L212 to external solvent is influenced by the methoxy pocket. Bound water molecules in this pocket significantly shorten the cutoff distances needed to transfer a proton from the external solvent to Glu L212. Thus, water molecules play an important role in establishing proton transfer pathways to the proximal oxygen.

Asp L213, which is important for proton transfer to the distal oxygen, may also be important for proton transfer to the proximal oxygen. Experimental measurements of the rate of electron transfer to Q_B suggest that mutation of Asp L213 to Asn inhibits proton transfer to Glu L212.[30] Although our results identify Asp L213 in some of the shorter pathways, there are several other competing pathways from external solvent to Glu L212, which may circumvent Asp L213.

Despite the importance of water molecules discussed above, they do little to lower the large cutoff distances required for the *last transfers* in the pathways (to His L190 or the proximal oxygen). The closest an internal solvent-accessible surface comes to the proximal oxygen is ~6 Å, and the closest predicted water is 6.8 Å away. In *Rps. viridis*, the closest bound water molecule in a homologous position is 6.9 Å away from the corresponding carbonyl oxygen on the quinone. It appears, therefore, that proton transfer from Glu L212 to Q_B is not mediated by a bound water molecule.

The large distance between Glu L212 and the proximal oxygen suggests the possibility that amino acid residues, or the doubly reduced quinone, or both may move before the second proton is transferred to Q_B. This movement could bring either a protonatable residue or a bound water molecule closer to the proximal oxygen, thereby allowing proton transfer. The distance between Glu L212 and the proximal oxygen on the quinone in *Rps. viridis* is also large (5.8 Å), making it unlikely that the large distance is the result of an error in the X-ray structure.

5. SUMMARY

We have enumerated proton transfer pathways from the external solution to the two carbonyl oxygens on the quinones. The pathways appear to form a network in which many pathways converge on a few key residues in the vicinity of the quinone. For the protonation of the distal oxygen, Asp L213 appears crucial, although there is experimental evidence that RCs without an aspartic acid at this position can compensate for its absence by a mutation at another position.[28,29] For the protonation of the proximal oxygen, Glu L212 plays an important role, as shown by both the present computational study of pathways and by experimental work on mutated RCs.[8,18]

We have predicted a number of internally bound water molecules in the region of the Q_B binding site. These bound waters may be important for proton transfer, especially to the proximal oxygen, which is close to a large void believed to contain several water molecules (the methoxy pocket). Water molecules may also help funnel protons from different pathways to Q_B.

The distinction between internal surfaces and the external surface is somewhat artificial. The external surface has narrow clefts that penetrate to interior residues in the RC. It is likely that bound water molecules fill these clefts, as they fill the internal cavities. We are currently developing methods to predict the positions of these external water molecules. In addition, we are working on ways to remove the limitation on the maximum length for a pathway.

The relative contribution of each pathway to the final proton transfer rate depends not only on the geometry of the groups involved, but on the energetics of protonation at each donor/acceptor group along the pathway. Because protonatable amino acids are often charged, mutagenesis of these residues may block a pathway indirectly by changing the electrostatic energies along the pathway. The calculation of these energies is complicated by many strongly interacting titrating residues in the region, a problem that we have recently addressed.[31] However, calculation of energies of protonation of individual titrating sites is not sufficient to determine the energetics of proton transfer. The free energy change associated with proton transfer between a donor and acceptor site depends not only upon the energy required to protonate each site, but also on the interaction energy between them. In addition, there are activation energies required to transfer a proton between sites. We are currently investigating the energetics of proton transfer.

ACKNOWLEDGEMENTS

We thank J.P. Allen, D. Bashford, P. McPherson, M. Paddock, C. Perrin, A. Rashin, and D. Rees for helpful discussions. This work was supported by grants from the National Institutes for Health (NIH GM41637 and GM13191), the National Science Foundation (NSF DMB89-15631), and NIH Training Grant GM08326 (P.B.).

REFERENCES

1. J.P. Allen, G. Feher, T.O. Yeates, H. Komiya, and D.C. Rees, Structure of the reaction center from *Rhodobacter sphaeroides* R-26: the cofactors, *Proc. Natl. Acad. Sci. USA* 84:5730 (1987).
2. J.P. Allen, G. Feher, T.O. Yeates, H. Komiya, and D.C. Rees, Structure of the reaction center from *Rhodobacter sphaeroides* R-26: the protein subunits, *Proc. Natl. Acad. Sci. USA* 84:6162 (1987).
3. J. Deisenhofer, O. Epp, K. Miki, R. Huber, and H. Michel, Structure of the protein subunits in the photosynthetic reaction centre of *Rhodopseudomonas viridis* at 3 Å resolution, *Nature* 318:618 (1985)
4. J.F. Nagle and S. Tristram-Nagle, Hydrogen bonded chain mechanisms for proton conduction and proton pumping, *J. Membr. Biol.* 74:1 (1983).
5. Z. Schulten and K. Schulten, Proton conduction through proteins: an overview of theoretical principles and applications, *Methods Enzymol.* 127:419 (1986).
6. A. Warshel, Correlation between the structure and efficiency of light-induced proton pumps, *Methods Enzymol.* 127:578 (1986).
7. M.Y. Okamura and G. Feher, Proton transfer in reaction centers from photosynthetic bacteria, *Ann. Rev. of Biochem.* 61:861 (1992)
8. M.L. Paddock, S.H. Rongey, G. Feher, and M.Y. Okamura, Pathway of proton transfer in bacterial reaction centers: replacement of glutamic acid 212 in the L subunit by glutamine inhibits quinone (secondary acceptor) turnover, *Proc. Natl. Acad. Sci. USA* 86: 6602 (1989).
9. C.H. Chang, O. El-Kabbani, D. Tiede, J. Norris, and M. Schiffer, Structure of the membrane-bound protein photosynthetic reaction center from *Rhodobacter sphaeroides*, *Biochemistry* 30:5352 (1991).
10. T.A. Jones, Interactive computer graphics: FRODO, *Methods Enzymol.* 115:157 (1985).
11. A.A. Rashin, M. Iofin, and B. Honig, Internal cavities and buried waters in globular proteins, *Biochemistry* 25:3619 (1986).
12. J.B. Goodenough, Proton movements in inorganic materials, *Methods Enzymol.* 127:263 (1986).
13. S. Scheiner, Theoretical calculation of energetics of proton translocation through membranes, *Methods Enzymol.* 127:456 (1986).
14. H. Treutlein, K. Schulten, A.T. Brünger, M. Karplus, J. Deisenhofer, and H. Michel, Chromophore-protein interactions and the function of the photosynthetic reaction

center: a molecular dynamics study, *Proc. Natl. Acad. Sci. USA* 89:75 (1992).
15. F.M. Richards, Areas, volumes, packing, and protein structure, *Ann. Rev. Biophys. Bioeng.* 6:151 (1977).
16. F.C. Bernstein, T.F. Koetzle, G.J.B. Williams, E.F. Meyer Jr., M.D. Brice, J.R. Rodgers, O. Kennard, T. Shimanouchi, and M. Tasumi, The protein data bank: a computer-based archival file for macromolecular structure, *J. Mol. Biol.* 112:535 (1977).
17. M.L. Paddock, G. Feher, and M.Y. Okamura, pH dependence of charge recombination in RCs from *Rb. sphaeroides* in which Glu-L212 is replaced with Asp, *Biophys. J.* 57:569a (1990).
18. E. Takahashi and C.A. Wraight, Proton and electron transfer in the acceptor quinone complex of *Rhodobacter sphaeroides* reaction centers: characterization of site-directed mutants of the two ionizable residues, Glu^{L212} and Asp^{L213}, in the Q_B binding site, *Biochemistry* 31:855 (1992).
19. M.L. Paddock, P.H. McPherson, G. Feher, and M.Y. Okamura, Pathway of proton transfer in bacterial reaction centers: replacement of serine-L223 by alanine inhibits electron and proton transfers associated with reduction of quinone to dihydroquinone, *Proc. Natl. Acad. Sci. USA* 87:6803 (1990).
20. E. Takahashi and C.A. Wraight, A crucial role for Asp^{L213} in the proton transfer pathway to the secondary quinone of reaction centers from *Rhodobacter sphaeroides*, *Biochim. Biophys. Acta.* 1020:107 (1990).
21. J.C. Williams, private communication.
22. M.L. Paddock, S.H. Rongey, P.H. McPherson, A. Juth, G. Feher, and M.Y. Okamura, manuscript in preparation.
23. Y. Cao, G. Váró, M. Chang, B. Ni, R. Needleman, and J.K. Lanyi, Water is required for proton transfer from aspartate-96 to the bacteriorhodopsin Shiff base, *Biochemistry* 30:10972 (1991).
24. H. Michel, K.A. Weyer, H. Gruenberg, I. Dunger, D. Oesterhelt, and F. Lottspeich, The 'light' and 'medium' subunits of the photosynthetic reaction centre from *Rhodopseudomonas viridis*: isolation of the genes, nucleotide and amino acid sequence, *EMBO J.* 5:1149 (1986).
25. G. Bélanger, J. Bérard, P. Corriveau, and G. Gingras, The structural genes coding for the L and M subunits of *Rhodospirillum rubrum* photoreaction center, *J. Biol. Chem.* 236:7632 (1988).
26. Y.A. Ovchinnikov, N.G. Abdulaev, A.S. Zolotarev, B.E. Shmukler, A.A. Zargarov, M.A. Kutuzov, I.N. Telezhinshaya, and N.B. Levina, Photosynthetic reaction centre of *Chloroflexus aurantiacus*: I. primary structure of L-subunit *FEBS Lett.* 231:237 (1988).
27. Y.A. Ovchinnikov, N.G. Abdulaev, A.S. Zolotarev, B.E. Shmukler, A.A. Zargarov, M.A. Kutuzov, I.N. Telezhinshaya, and N.B. Levina, Photosynthetic reaction centre of *Chloroflexus aurantiacus*: primary structure of M-subunit, *FEBS Lett.* 232:364 (1988).
28. S.H. Rongey, M.L. Paddock, G. Feher, and M.Y. Okamura, manuscript in preparation.
29. M. Schiffer, C.K. Chan, C.H. Chang, T.J. DiMagno, G.R. Fleming, S.L. Nance, J.R. Norris, S.W. Snyder, M.C Thurnauer, D. Tiede, and D.K. Hanson, these proceedings.
30. P.H. McPherson, S.H. Rongey, M.L. Paddock, G. Feher, and M.Y. Okamura, The rate of electron transfer $Q_A^-Q_B \rightarrow Q_AQ_B^-$ in RCs from *Rb. Sphaeroides* in which Asp-L213 is replaced with Asn, *Biophys. J.* 59:142a (1991).
31. P. Beroza, D.R. Fredkin, M.Y. Okamura, and G. Feher, Protonation of interacting residues in a protein by a Monte Carlo method: application to lysozyme and the photosynthetic reaction center of *Rhodobacter sphaeroides*, *Proc. Natl. Acad. Sci. USA* 88:5804 (1991).

ELECTROSTATIC INTERACTIONS AND FLASH-INDUCED PROTON UPTAKE IN REACTION CENTERS FROM *RB. SPHAEROIDES*

Vladimir P. Shinkarev, Eiji Takahashi and Colin A. Wraight

Division of Biophysics and Department of Plant Biology
University of Illinois, Urbana, IL 61801 (U.S.A.)

INTRODUCTION

X-ray structural analysis of crystalline RCs from *Rhodopseudomonas viridis* [1,2] and *Rhodobacter sphaeroides* [3-5] has localized the cofactors and indicated the involvement of individual amino acid residues in cofactor binding. However the transient functional roles are still largely unclear. Following RC excitation by light, an electron is passed from the primary donor, a bacteriochlorophyll dimer (P), to the primary quinone (Q_A) and then to the secondary quinone (Q_B). In the absence of an electron donor to P^+ only one-electron activity is possible and the separated charges can recombine by two routes [6]:

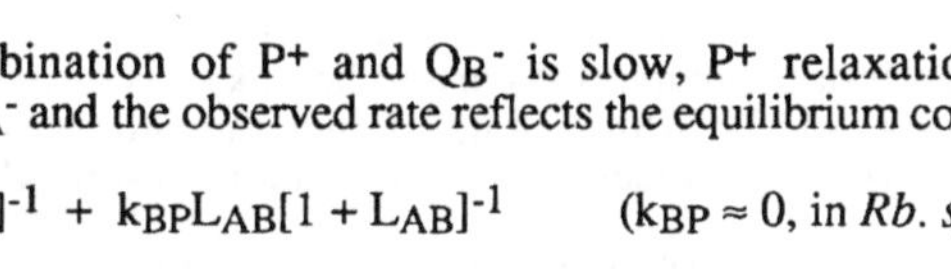

Scheme 1

If the direct recombination of P^+ and Q_B^- is slow, P^+ relaxation occurs largely by repopulation of $P^+Q_A^-$ and the observed rate reflects the equilibrium constant, L_{AB}:

$$k_{Q_B} = k_{AP}[1 + L_{AB}]^{-1} + k_{BP}L_{AB}[1 + L_{AB}]^{-1} \qquad (k_{BP} \approx 0, \text{ in } Rb.\ sphaeroides) \qquad (1)$$

Formation of the semiquinone states is accompanied by proton uptake due to protonation of amino acid residues influenced by the anionic charge [6-9], and the equilibrium and kinetics of the electron transfer from Q_A to Q_B are pH-dependent [10]. Several ionizable amino acid residues located in or near the Q_B site may be involved in this process [5]. Detailed kinetic analyses of *Rb. sphaeroides* mutants L212EQ and L213DN in which Glu^{L212} and Asp^{L213} were altered to Gln and Asn, respectively, have shown drastic changes of all characteristics of electron transfer and proton uptake [11-13]. It was suggested that the pH dependence of L_{AB} at alkaline pH is determined by protonation of Glu^{L212} [11] while the dependence at acidic pH is determined by protonation of Asp^{L213} or other amino acid residues [13,14].

The Photosynthetic Bacterial Reaction Center II
Edited by J. Breton and A. Verméglio, Plenum Press, New York, 1992

The details of the X-ray structure of the RC allow substantial modelling of the acceptor quinone functions, albeit limited by the static nature of the structural data, and some efforts have been put into calculation of the pK properties and protonation states of the ionizable residues of the protein [15,16]. The results are not inconsistent with experiment but consideration of all pairwise interactions, as required in these calculations, renders them largely inaccessible to any intuitive considerations. Here, we explore the use of a simple model to describe the behavior of the acceptor quinones in terms of pairwise electrostatic interactions of a small number of components, amenable to some intuitive conceptualization.

DEPENDENCE OF THE ONE-ELECTRON EQUILIBRIUM ON pH

To understand the nature of the pH dependence of L_{AB} on pH, we consider a simple phenomenological model in which we suppose only electrostatic interactions between reactants and in which only two amino acid residues, close to Q_B, are involved in the proton uptake by RCs after the first flash. The two amino acids, designated here as D (i.e., Asp) and E (i.e., Glu), generate four different protonation states: DHEH, D^-EH, DH E^- and D^-E^-. Electrostatic interactions between amino acid residues and quinone acceptors give rise to different pKs and rate constants for the various ionization states of D and E and the charge states of the quinones. Instead of one state of the quinone acceptors, for example Q_AQ_B, it is necessary to consider four: Q_AQ_B(DHEH), Q_AQ_B(D^-EH), Q_AQ_B(DHE^-) and Q_AQ_B(D^-E^-). One-electron transfer between the quinones involves 12 possible states, with 10 independent equilibrium constants and pKs [17]:

Scheme 2

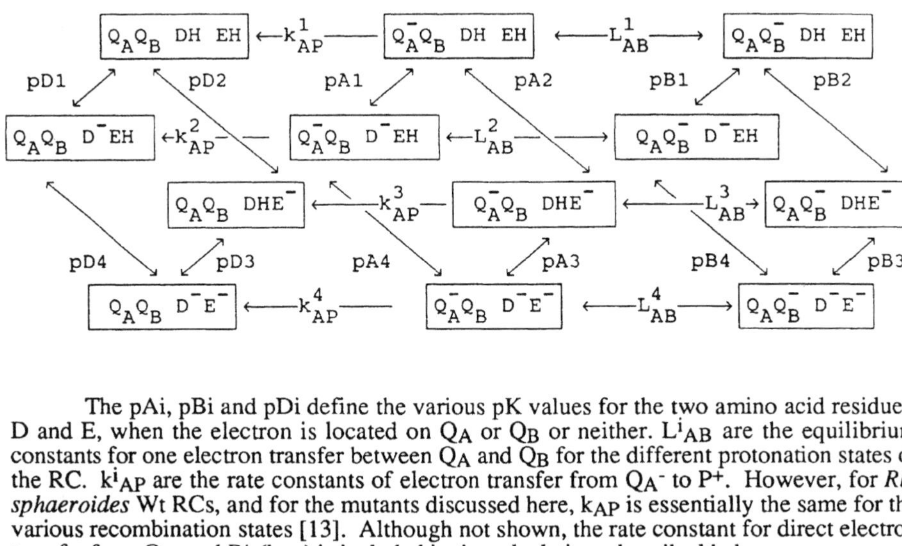

The pAi, pBi and pDi define the various pK values for the two amino acid residues, D and E, when the electron is located on Q_A or Q_B or neither. L^i_{AB} are the equilibrium constants for one electron transfer between Q_A and Q_B for the different protonation states of the RC. k^i_{AP} are the rate constants of electron transfer from Q_A^- to P^+. However, for *Rb. sphaeroides* Wt RCs, and for the mutants discussed here, k_{AP} is essentially the same for the various recombination states [13]. Although not shown, the rate constant for direct electron transfer from Q_B^- and P^+ (k_{BP}) is included in the calculations described below.

We suppose that the main effect of amino acid substitution on the first flash-induced electron transfer is due to altered electrostatic interactions. The electric potential difference ($\Delta\psi$, in mV) can be calculated from Coulomb's law: $\Delta\psi = 14387/(r\,\varepsilon)$, where r is the distance in Å between amino acid residues or between quinone and residue, and ε is the effective dielectric constant. For example, for r=10 Å and ε=10, the electric potential will be $\approx$ 144 mV. Differences in pK values for similar RC transitions in Scheme 2, arising from electrostatic interactions, can be calculated from the equation (T=295K): $\Delta pK = \Delta\psi F/2.3RT = \Delta\psi/60 \approx 240/(r\,\varepsilon)$.

The distances between Q_A, Q_B and amino acid residues D and E are, in Ångstroms:

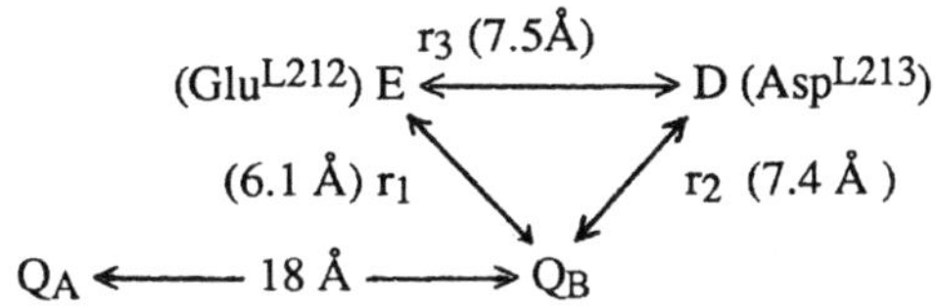

Scheme 3

(The distances are the average of the 4 pairs of carboxyl or carbonyl oxygen separations between amino acid residues and quinones). These distances were incorporated into the equations, with various assumed values of ε, to estimate the expected shifts in pK induced by electrostatic interactions.

The pH dependence of the P^+ dark relaxation is calculated using Eqn. 1, with a macroscopic equilibrium constant $L_{AB}{}^{app}$ that depends on pH, as indicated by the definition:

$L_{AB}{}^{app} = \Sigma_{H^+}[Q_AQ_B^-]/\Sigma_{H^+}[Q_A^-Q_B]$ $\qquad$ Σ_{H^+} indicates summation over all changing protonation states of the RC.

In Wt RCs, the acid and alkaline regions of pH dependence of P^+ dark relaxation are far enough apart that the ionization behavior of D and E can be treated independently. Thus, at alkaline pH the observed light-induced proton binding can be considered as arising from the interaction between a single ionizable group (EH/E^-) and Q_B^-, under conditions when D is fully deprotonated, and the scheme above can be simplified accordingly.

When the measured proton uptake reflects a change in the protonation state of a single group, $L_{AB}{}^{app}$ can be presented as:

$$L_{AB}{}^{app} = \{[Q_AQ_B^-(D^-E^-)] + [Q_AQ_B^-(D^-EH)]\}/\{[Q_A^-Q_B(D^-E^-)] + [Q_A^-Q_B(D^-EH)]\}$$

$$= L^2{}_{AB}(1+10^{pH-pB4})/(1+10^{pH-pA4}) = L^4{}_{AB}(1+10^{pB4-pH})/(1+10^{pA4-pH}) \qquad (2)$$

where $L^2{}_{AB}$, $L^4{}_{AB}$ are the microscopic equilibrium constants of electron transfer between Q_A and Q_B, and pA4 and pB4 the protonation equilibria, as defined in Scheme 2.

For low salt concentrations, in LDAO, RCs of *Rb. sphaeroides* are characterized by the following values: $pB4 \approx 11.2$, $pA4 \approx 9.8$, $L^4{}_{AB} \approx 0.5$, $L^2{}_{AB} \approx 14$ [10]. The shift in pK of 1.4 pH units for Glu^{L212} yields a change in electrical potential at Glu^{L212} of ≈ 85 mV. For the distance between Q_B and E (≈ 6 Å), this gives an effective dielectric constant, ε of ≈ 30.

At acid pH the observed proton binding by RCs can be considered as arising from the interaction between a single ionizable group (DH/D^-) and Q_B^-, under conditions when E is fully protonated. The analysis is similar to that for the alkaline region. In the low pH domain, RCs of *Rb. sphaeroides* are characterized by [13]: $pB1 \approx 5$-5.5, $pA1 \approx 4$, $L^2{}_{AB} \approx 14$ and $L^1{}_{AB} \approx 150$-450 for low salt concentrations. The shift in pK of ≈ 1 pH unit for Asp^{L213} (from ≈ 4-4.5 with the electron on Q_A to 5-5.5 with the electron on Q_B) gives a change in the electrical potential at Asp^{L213} of ≈ 60 mV. If we take the distance between Q_B and D as 4.5-10 Å (the Asp^{L213} is much closer to O5 of Q_B of the quinone ring than to O2, introducing large differences in distance depending on electron location in the quinone ring) then the effective dielectric constant, ε, is ≈ 20-50. Ionization of Asp^{L213} can also increase the pK of the Glu^{L212} [13]. Using $\varepsilon \approx 30$ predicts a shift of ≈ 1.1 pH units ($r \approx 7$ Å).

Fig.1A shows the pH-dependence of the time of P^+ dark relaxation in Wt RCs of *Rb. sphaeroides* [10,17]. With our simple model, the best fit of a theoretical curve to the

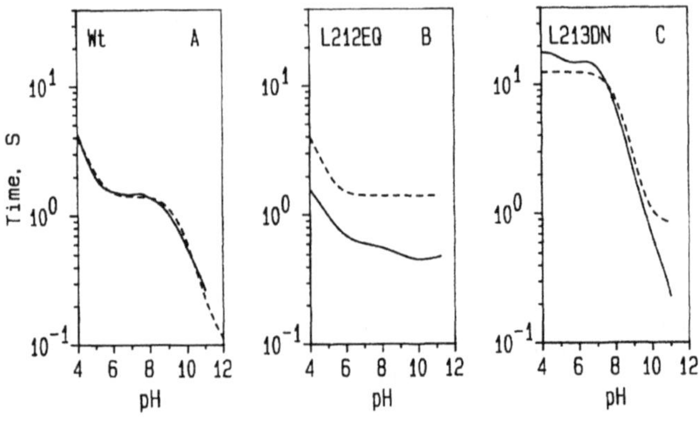

Figure 1. pH dependence of the $P^+Q_B^-$ charge recombination time in wildtype and mutant RCs of *Rb. sphaeroides*. Experimental data (solid lines) were taken from ref. 13, and theoretical dependences were calculated from Scheme 2, modified for the mutants, with ε = 25, k_{BP} = 0.03 s^{-1}, k_{AP} = 10 s^{-1}, $L^1{}_{AB}$ = 215, pD1 = 3.5 (for Wt and L212EQ), pD2 = 7.8 (for Wt and L213DN). All other parameters in the scheme were calculated from the electrostatic interactions between components.

experimental data is observed for a relatively high dielectric constant, $\varepsilon \approx$ 20-30, in agreement with the estimates above. With this value, the calculated pH dependence of the time of P^+ dark relaxation exhibits the observed acceleration with increasing pH from 4 to 6 and from 9 to 12, and pH independence between pH 6 and pH 9.

The qualitative validity of this model can be tested against the behavior of mutant RCs with the ionizable residues, Glu^{L212} and Asp^{L213}, substituted by neutral Gln and Asn.

L212EQ mutant RCs. L212EQ mutant RCs have the amino acid E changed to the neutral amide, Glu^{L212} --> Gln [11,13]. The flash-induced transitions of these mutant RCs are obtained from Scheme 2 by deleting the protonation state changes of E.

The kinetics of P^+ dark relaxation in mutant L212EQ RCs are characterized by the absence of acceleration of the dark relaxation seen in Wt RCs at alkaline conditions [11,13]. This, also, is reasonably well represented in Fig.1B by the theoretical curve, calculated for Scheme 2, using the same parameters as for the Wt at low pH, for a single ionizable group with pB1$\approx$ 4.8, $L^1{}_{AB} \approx$ 215 and $L^2{}_{AB} \approx$ 11. However, the reduction of P^+ in L212EQ mutant RCs is $\approx$ 2 times faster than in Wt RCs, indicating a small global change in the equilibrium constant L_{AB} for electron transfer. This has been speculated to reflect the difference in permanent dipole of glutamine compared to glutamic acid [13], although some effect from lack of saturation of quinone binding cannot be ruled out.

The general coincidence between the predictions of the simple electrostatic model and experimental results for Wt and L212EQ RCs supports the idea that the behavior of Wt RCs at alkaline conditions is mainly determined by electrostatic interactions between Q_B^- and negatively charged Glu^{L212}, which has an unusually high pK value [11,13].

L213DN mutant RCs. The expected one-electron equilibria in mutant RCs in which amino acid D is changed to a neutral species, Asp^{L213} --> Asn (mutant L213DN [12]), are obtained by deleting the protonation state changes of D in Scheme 2.

Fig. 1C shows the experimental data and theoretical curve for L213DN mutant RCs. There is qualitative correspondence between them and both show a characteristic slowdown of the kinetics of P^+ relaxation at lower pH. The model also predicts a shift in the pK of Glu^{L212} to more acidic values (pA4, pB4 in Wt RCs and pA2 and pB2 in L213DN RCs).

L212EQ/L213DN double mutant RCs. The time of $P^+Q_AQ_B^-$ charge recombination for L212EQ/213DN double mutant RCs is pH independent from pH 5 to 10 [13], consistent with the absence of amino acid residues around Q_B capable of protonation after electron transfer to Q_B. However, if the observed rate is dominated by k_{BP} in this mutant, then the kinetics are also not expected to be strongly modulated by the free energy of the state $P^+Q_B^-$.

pH DEPENDENCE OF THE RATE OF ELECTRON TRANSFER

The general Scheme 2 allows estimation of the pH dependence of the rate of electron transfer between Q_A and Q_B in Wt and mutant RCs. If the protolytic reactions are faster than the transfer of the electron then the rate of electron transfer should be described by a single exponential term with an apparent rate constant that is a weighted sum of the rate constants for all relevant protonated states of the RC with oxidized Q_B and semiquinone Q_A^- [17]:

$$k_{AB}^{app} = k^1_{AB}(Q_A^-Q_B\,EHDH) + k^2_{AB}(Q_A^-Q_B\,EHD^-) + k^3_{AB}(Q_A^-Q_B\,E^-DH) + k^4_{AB}(Q_A^-Q_B\,E^-D^-) \qquad (3)$$

This approximation is probably valid at neutral and low pH but it should be applied to alkaline conditions with caution. At high pH, the rate of proton uptake may be expected to become quite slow and for residues with high pK values the rate of deprotonation will also be slow. Taking the diffusion-limited rate constant of protonation to be $\approx 10^{11}$ M^{-1} s^{-1}, $k_{deprot} \approx 10^{(11\text{-}pK)}$ s^{-1} [18,19].

The pK of Asp^{L213} is < 5 (when Q_B is oxidized) and the estimated time for Asp^{L213} deprotonation is about 10 µs. So, the approximation of fast equilibrium can be applied to the protolytic reactions of Asp^{L213} at any pH. However, the pK of Glu^{L212} is ≈ 9.5 when Q_B is oxidized, so the time of deprotonation for Glu^{L212} could be as long as 10 ms. This is not faster than the measured rate of electron transfer at high pH, and the electron transfer between Q_A and Q_B will be non-exponential with terms determined by the dark distribution of RC states with protonated and deprotonated Glu^{L212} [17]. In fact, however, the net charge of the RC, which becomes more negative with increasing pH, will maintain a surface pH that is substantially lower than the bulk phase value. This may be taken into account by an effective rate constant of protonation greater than 10^{12} M^{-1} s^{-1} [20], or by correcting for the surface pH. Furthermore, both protonation and deprotonation are probably mediated by buffering species in solution which, at alkaline pH, are usually present at much higher concentrations than aqueous protons. H_2O itself can be an effective proton donor [19]. These circumstances could conspire to make proton delivery and exchange, per se, not rate limiting. This is not in contradiction with the observation of pH dependent kinetics of electron transfer which can arise from pH-dependent equilibrium effects [13,20]. Formally, this is similar to the descriptions of the P^+ relaxation in terms of the rapid one-electron equilibrium between Q_A and Q_B, binding equilibrium at the Q_B site, etc (Eqn. 1).

We will assume that the protonation equilibria occur faster than the electron transfer between Q_A and Q_B. The relevant equilibrium may, in fact, be one of internal (intraprotein) proton transfer. As a very simple approach to this problem, we consider the effect of an ionizable residue to arise from a smoothly varying charge as the pH is raised. The expected effect of a local charge on the rate of electron transfer between Q_A and Q_B can be qualitatively understood assuming that the free-energy of the electron transfer is linearly perturbed by the electric potential due to nearby charges:

$$\ln k_{AB} = \ln k^0_{AB} - 0.5\,\Delta\psi F/RT \qquad (4)$$

$$\ln k_{BA} = \ln k^0_{BA} + 0.5\,\Delta\psi F/RT$$

The factor 0.5 relates to identically shaped reactant and product potential energy curves. Thus, through perturbation of $\Delta\psi$, ionization of amino acids D and E will slow down the forward electron transfer and accelerate the reverse electron transfer. Because of the large distance between Q_A and Q_B (≈ 18 Å), we neglect the electrostatic interactions between D and E and Q_A, but some effect can be expected.

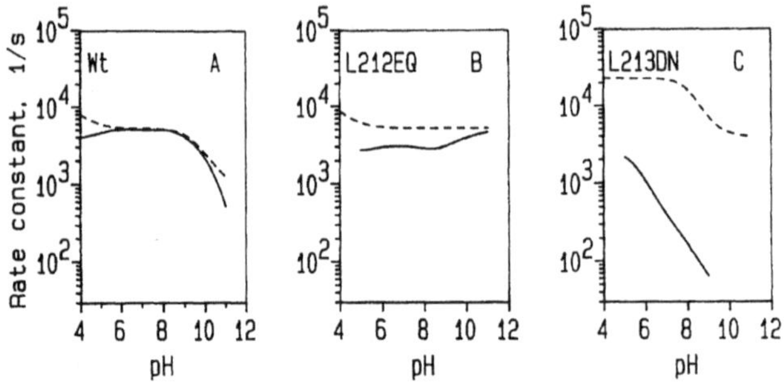

Figure 2. pH dependence of the rate constant (k_{AB}) of electron transfer between Q_A and Q_B in wildtype and mutant RCs of *Rb. sphaeroides*. Experimental data (solid lines) were taken from ref. 13, and theoretical dependences (dashed lines) were calculated from Scheme 2, modified for the mutants, with $\varepsilon = 25$, $k^1_{AB} = 2.3 \cdot 10^4\ s^{-1}$, $L^1_{AB} = 215$, pD1 = 3.5 (for Wt and L212EQ), pD2 = 7.8 (for Wt and L213DN). All other parameters in the scheme were calculated from the electrostatic interactions between components.

Wt RCs. Fig. 2A shows the pH dependence of the rate constant (k_{AB}^{app}) for electron transfer between Q_A and Q_B (solid line) in Wt RCs of *Rb. sphaeroides* [13]. k_{AB}^{app} depends slightly on pH in the range 5-8 and decreases more definitely at pH > 9. Theoretical curves were calculated at different values of effective dielectric constant, ε. For small dielectric constant, the theoretical curves predict a substantial increase in the rate constant at low pH which is not supported by the experimental data. A reasonable fit could be obtained using a relatively large effective dielectric constant, ε=20-30 (Fig.2, dashed line). The absence of any acceleration in the measured electron transfer rate at low pH may also reflect the asymmetric position of Asp^{L213} with respect to Q_B, giving rise to a substantially larger distance between it and the carbonyl O2 atom. Thus, the relevant distance between Asp^{L213} and Q_B may be different for the equilibrium and rate calculations of Figs 1 and 2. Yet another possibility is that electron transfer at low pH is rate limited by an entirely different process, such as a structural change.

L212EQ mutant RCs. Comparison of the theoretical and experimental dependences shows qualitative coincidence between them (Fig. 2B). The experimental data and the model both indicate that altering Glu^{L212} to Gln leads to disappearance of the pH-dependence of the rate constant of electron transfer between Q_A and Q_B, under alkaline conditions.

L213DN mutant RCs. Comparison of the theoretical and experimental pH dependencies of rate constant k_{AB} in L213DN RCs shows major discrepancies between them: (i) Alteration of Asp^{L213} to Asn leads to pH dependence of k_{AB} in all regions studied, from pH 5 to > 9, while calculations predict pH dependence only at pH > 8. (ii) The model predicts that protonation of Glu^{L212} at low pH leads to the disappearance of electrostatic retardation and therefore to an increase in the rate constant of electron transfer in mutant RCs in comparison with Wt RCs. However, the experimental data [12,13] do not support this prediction (Fig. 2C). The model calculations of the electron transfer rate and the experimental values differ by more than two orders of magnitude at high pH.

A possible explanation of the differences between L212EQ and L213DN mutants lies in the fact that Glu^{L212} is no nearer than 5 Å to any charged amino acid residues. On the other hand Asp^{L213} is surrounded by three ionizable amino acid residues (Asp^{L210}, Glu^{H173} and Arg^{L217}), all located closer than ≈ 4 Å and forming a cluster of strongly interacting charged amino acid residues. The interactions among them make it impossible to consider Asp^{L213} independently from the other amino acid residues of the cluster. Thus, the parameters obtained from fitting Wt data may not be applicable to the L213DN mutant, as we have assumed. The behavior of such a cluster may account for the broad profile of proton uptake, measured at pH <7 in Wt RCs [7,8], as well as the high value for the effective dielectric constant 20-50 suggested above.

Some inaccuracy is expected to arise from the use of point charges in the proposed model. The distances between the carboxylic oxygens of Glu^{L212} and the two carbonyl oxygens, O2 and O5, of Q are approximately the same, giving rise to only small differences in electrostatic energy depending on the location of the electron on O2 or O5. However, Asp^{L213} is much closer to O5 of the quinone ring than to O2, introducing large difference in the values of the electrostatic energy depending on electron location in the quinone ring. It is evident that our simple model does not consider these alternative distributions.

A more interesting source of the strong disagreement between the observed and calculated rate constants of electron transfer between Q_A and Q_B, is the possible absence of proton equilibration on the time scale of electron transfer between the quinones. This is consistent with a key position for Asp^{L213} in providing the path for all proton delivery to the Q_B binding domain. On the other hand, equilibration could easily occur on the time scale of electron transfer from Q_B^- to P^+, giving the qualitative coincidence between observed and calculated pH dependencies of the electron transfer equilibrium constant.

PROTON UPTAKE BY REACTION CENTERS

Following re-reduction of P^+ by a secondary electron donor, a second turnover of the RC results in the formation of quinol in the Q_B site. The quinol then exchanges for quinone in the pool:

$$Q_A Q_B \xrightarrow{h\nu_1} Q_A^- Q_B \xleftrightarrow{nH^+} Q_A Q_B^-(H^+)_n$$

$$Q_A Q_B^-(H^+)_n \xrightarrow{h\nu_2} Q_A^- Q_B^-(H^+)_n \xleftrightarrow{(2-n)H^+} Q_A Q_BH_2 \xleftrightarrow{Q \quad QH_2} Q_A Q_B$$

Scheme 4

After the first flash, reduction of Q_A and Q_B to the semiquinone anion state induces pK changes in several ionizable amino acids in the vicinity, resulting in a non-stoichiometric proton binding. After the second flash, further proton uptake satisfies the stoichiometry of quinol formation at the Q_B site. Because of the very high pKs of ubiquinol (> 12), two protons are expected to be taken up per QH_2 produced and released.

Proton uptake after the first flash reflects the protein's attempts to solvate, or charge compensate, the anionic semiquinones of Q_A and Q_B, and could involve both nearby and quite distant residues. The stoichiometry of proton uptake on the first flash is pH dependent, with a distinct peak at pH 9.5-10, and a less well defined peak at about pH 5.5-6 [8]. Site-directed mutation studies have strongly implicated Glu^{L212} in the proton binding behavior at high pH, and Asp^{L213} has been less conclusively implicated in the low pH range [11-13]. The X-ray data indicate a few other candidates around the Q_B head group binding site (Glu^{H173}, His^{L190}, Ser^{L223}), but with many more at greater distance [5].

We have investigated the relative contribution of distant residues through the salt dependence of the proton uptake in various pH regions. The salt dependence is strong near neutral pH, but is absent at alkaline pH and much less marked at acid pH. We suggest that this reflects the involvement of amino acid residues of varying exposure or accessibility to screening by counter ions. The observed dependence may explain some of the discrepancies in earlier measurements of the stoichiometry of proton uptake after the first flash [6-8].

pH dependence of proton uptake by RCs

Flash-induced proton uptake by RCs of *Rb. sphaeroides*, strain Ga, in the presence of ferrocene, was measured at various pH, at a salt concentration of 50 mM KCl (Fig. 3A). The proton uptake induced by the first flash is less than that induced by the second at all pH less than 9.5, and this pattern of odd versus even flashes is repeated in a longer series of flashes. The pH profile of proton uptake on the first flash is characterized by two main maxima - at pH ≈ 5 and ≈ 10, and a smaller additional maximum at pH ≈ 6.5 (Fig. 3B, and ref. 8). At the two main peaks approximately 0.7 and 1 proton per P^+ is taken up,

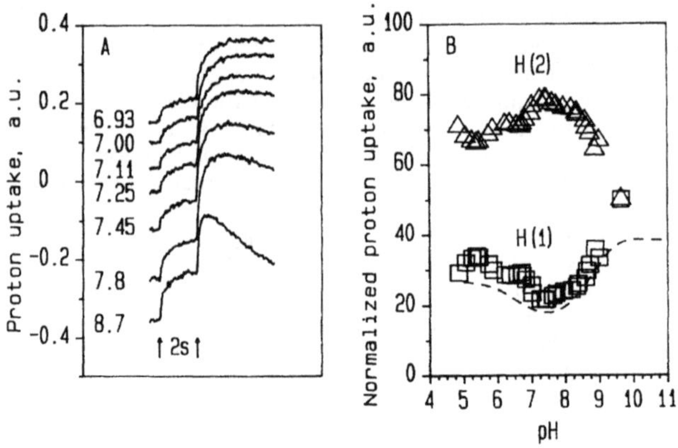

Figure 3. Flash-induced proton uptake by wild type RCs of *Rb. sphaeroides* (Ga). **A**. pH traces, measured with a glass electrode; 50 mM KCl, 100 μM ferrocene, 20 μM Q-10; flash period 2 s. **B**. pH dependence of the proton uptake stoichiometry on the first and second flash. The amplitudes were normalized by setting the sum equal to 100 (equivalent to 2 H^+). Conditions as for A. (----------) Data from McPherson et al. [8].

respectively. Between pH 6.5 and 8.5 the first flash induces uptake of 0.4-0.5 H^+ per P^+. Thus, the ratio of amplitudes for proton uptake on the first and second flash is substantially less than 1 at pH < 9.

The pH dependence of the first flash proton uptake (Fig. 3B) coincides well with the data of McPherson et al. [8], also obtained with a glass electrode for *Rb. sphaeroides* R-26 RCs (see dashed line), but differs from the data of Maróti and Wraight [7], obtained using pH-indicators and by a conductivity method. The reasons for this discrepancy are unclear, but the binary oscillations are very sensitive to the experimental conditions, including quinone, exogenous donor, intensity of the flash, dark adaptation of the sample, and, as shown here, salt concentration.

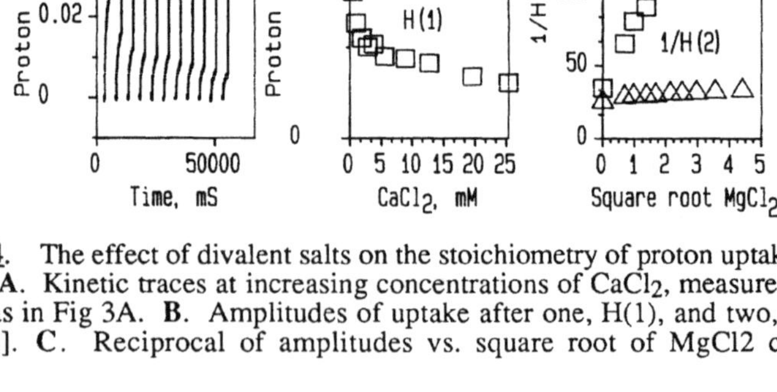

Figure 4. The effect of divalent salts on the stoichiometry of proton uptake by Wt RCs, at pH 7.5. **A**. Kinetic traces at increasing concentrations of $CaCl_2$, measured with a glass electrode, as in Fig 3A. **B**. Amplitudes of uptake after one, H(1), and two, H(2), flashes vs. [$CaCl_2$]. **C**. Reciprocal of amplitudes vs. square root of MgCl2 concentration.

After a second flash, approximately 1.2-1.6 protons are taken up, from pH 5 to 9. The amplitude declined steadily at pH above 8, and essentially equal stoichiometries, i.e., 1 H^+ per P^+, were taken up on each flash, at pH 9.5. At even higher pH, the net H^+ taken up after two flashes fell below 2, due to the unfavorable one-electron equilibrium, $Q_A^-Q_B$ <--> $Q_A Q_B^-$, blocking turnover of a fraction of the RCs on the second flash. Ubiquinone (Q-10) titration of the second flash proton binding gave a half saturating quinone concentration of about 4 μM. This is comparable to other titrations of Q_B activity under these conditions [21].

Effect of salts on proton uptake

In 1 M KCl, the maxima of proton uptake on the first flash occur at pH 6 and pH 9.5, where approximately 0.4 and 1 proton, respectively, are taken up per RC (not shown). The main difference compared to 50 mM KCl was a marked decrease in first flash proton uptake at pH ≈ 5 and pH 7.5, indicating a strongly differential effect of salt at different pH. This was further explored through salt titrations of proton uptake at selected pH values.

Neutral conditions. At pH 7.5, increasing concentrations of various salts lead to substantial changes in the flash-induced proton uptake pattern by RCs (Fig. 4A). The major effect is a decrease in the relative amplitude of the first flash-induced proton uptake as the salt concentration is increased (Fig. 4B). Initially the decrease is precipitous and is not fully compensated by an increase in proton uptake on the second flash; thus, the net uptake after two flashes is greater than $2H^+$. In very low salt, it appears that more than 1 H^+ may be taken up on the first flash, and preliminary measurements suggest that this may correspond to a slower phase (0.1-1 s) observed under these conditions. Some of this uptake must be re-released, also quite slowly, after the second flash. The first flash proton uptake depends inversely on the square root of the salt concentration, with salts of divalent cations, $CaCl_2$ and $MgCl_2$, effective at concentrations 6-7 times less than monovalents (Fig.4C). A convenient measure, which provides an internal calibration, is the ratio of second to first flash proton uptake - $H^+(2)/H^+(1)$. This was 1-1.5 at low salt concentrations and approximately 7-8 in 25 mM $CaCl_2$ or $MgCl_2$, or 1 M KCl. The linear dependence on $[c]^{-1/2}$ implies that the first flash proton uptake could be essentially eliminated in very high salt.

Alkaline conditions. Under alkaline conditions (pH 9), added salts had essentially no effect on the on the flash-induced proton uptake, even at high concentrations, e.g., $CaCl_2$ up to 25 mM (Fig. 5). The absence of salt effect on the ratio $H^+(2)/H^+(1)$ indicates that this proton uptake by RCs is due mainly to pK shifts of an amino acid residue(s) located so close to Q_B, that it cannot be screened by salts.

Acid conditions. The effect of salts on the relative amplitude of the first flash-induced proton uptake at acid pH was intermediate between neutral and alkaline conditions, and could not be eliminated (Fig. 5). At pH 5.6, the value of the ratio $H^+(2)/H^+(1)$ did not exceed 2.5, even at high $CaCl_2$ concentration. This indicates that it is attributable to both easily screened, and probably distant, residues, and to less accessible residues, presumably close to Q_B.

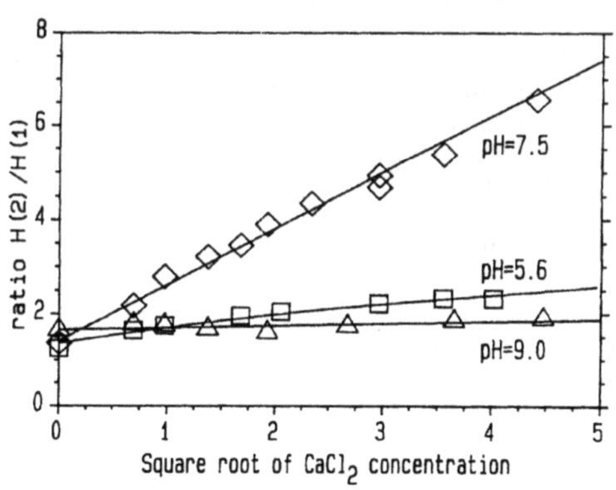

Figure 5. The effect of divalent salts on the stoichiometry of proton uptake by Wt RCs, at various pH. The ratio of proton uptake on the second and first flash, $H(2)/H(1)$, is plotted vs. the square root of $[CaCl_2]$. Measurements as in Fig. 3A, at pH 5.6, 7.5 and 9.0.

Ionic screening and mobile ion effects

The addition of salts can affect proton uptake by RCs through several related but distinct phenomena: a) Screening of the surface charges by mobile ions will cause changes in the surface pH. b) Changes in the surface pH will effect changes in the degree of dissociation of ionizable groups, thus altering the surface charge density. c) Screening of surface groups will also diminish their responsivity to the anionic semiquinone charge, i.e., will decrease the pK shifts responsible for the proton uptake.

The effect of KCl and others salts on this proton uptake in the presence of donor can be explained partially by the screening of flash-induced pK shifts of various amino acid residues. It can be qualitatively understood from the salt dependence of the Debye length, λ_D, which is a measure of the screening effect in liquids (for simplicity we consider here only the case of a symmetrical z-z valent salt):

$$\lambda_D = a/(z\sqrt{c}) \quad (5)$$

λ_D is expressed in Ångstroms, and the concentration of salt, c, in molarity (M). The coefficient of proportionality, $a = \sqrt{e^2/(\varepsilon kT)}$, is ≈ 3 in aqueous solution ($\varepsilon \approx 80$). For z=1 (KCl), the Debye length is ≈ 30 Å for 10 mM, and ≈ 3 Å for 1 M KCl. Increasing the salt concentration will progressively diminish the shift in pK of amino acid residues located further from Q_B. In other words, increasing the salt concentration leads to a shift from protons to cations as the main contributors to the compensation of the light induced extra charge on Q_B^-. Thus, increasing salt concentrations can be used to select the amount of amino acid residues involved in proton uptake. At high salt concentration the first flash induces proton uptake only by amino acid residues located near Q_B. The dependence of the amplitude of the first flash-induced proton uptake on salt concentration may be used for clarifying questions about the nature and number of amino acid residues closely involved in the proton uptake at a given pH.

The strong salt dependence of the relative amplitude of the first flash-induced proton uptake at neutral pH indicates that at low salt concentrations the compensation of the extra charge of the semiquinone is provided by the protonation-deprotonation reactions of amino acid residues that are readily screened by ions. Thus, at high salt concentration the charge compensation is provided largely by mobile cations.

The lack of effect of salts on proton uptake at alkaline pH indicates that the amino acid residue(s) involved are essentially inaccessible to mobile ions. Glu^{L212} is strongly implicated in this role [11,13], and the lack of a salt effect suggests this residue to be highly inaccessible. This is consistent with its high pK and with the X-ray structural data, which shows Glu^{L212} to be deeply sequestered.

At acid pH, the incomplete effect of high salt in suppressing proton uptake indicates a complex origin of the Q_B^- charge compensation. The involvement of many amino acid groups in the proton uptake at low pH implies that the pH-dependence of the equilibrium constant L_{AB} at low pH is determined by many protonatable amino acid residues. Site-directed mutagenesis studies have shown both Asp^{L213} [12,13] and Asp^{L210} [14] to contribute to the pH dependence of L_{AB}, and the X-ray stucture reveals these to be part of a cluster of strongly interacting residues.

Some salts also affect the second flash-induced proton uptake but this effect is smaller than that for the first flash. This may be explained by a scattering effect of RC aggregates generated at high concentration of the divalent salts and decreasing of actinic light. The slightly smaller amplitude of the proton uptake after the second flash observed at high salt concentration may also be due to a shift of the equilibrium $Q_A^-Q_B^- + 2H^+ \longleftrightarrow Q_AQ_BH_2$ in response to the decreasing surface proton concentration [17].

Penetration of the reaction center protein by salts

The pronounced effect of salt on the first flash-induced proton uptake suggests penetration of the protein by salts. Net penetration may not need to be very fast, although the formal description of the salt effect in terms of mobile counter ions does imply rapid microscopic redistribution. Experiments with azide and other compounds [22] indicate that penetration to the closest regions of the Q_B pocket occurs in the millisecond time range. There are many indications of the accessibility of interior regions of the RC protein to salts and water molecules, most notably the presence of water molecules and some ions inside proteins, determined by X-ray analysis of crystals. Both water molecules and ions are observed in the vicinity of the Q_B pocket of the *Rps. viridis* RC [2]. The resolution of the *Rb. sphaeroides* RC X-ray structure does not allow direct visualization of such entities, but the available space is similar to that of *Rps. viridis.*

Penetration of the polar portions of the RC protein by salts and their screening effect encourages the reinvestigation of all characteristics of the quinone acceptors as a function of salt concentration.

Ionic screening of charge interactions

The experimental pH-profile for proton uptake on the first flash, such as that presented in Fig.3B, can be approximated by the sum:

$$\Delta H^+ = \Sigma\,[\,1/(1+10^{pH_s-pK_i^d}) - 1/(1+10^{pH_s-pK_i^l})] \qquad (6)$$

where pK_i^d and pK_i^l are the pK values for i-th amino acid before the flash and after electron transfer to Q_B, respectively. For simplicity we do not consider the electrostatic interaction of different amino acid residues.

In Eqn. 6, pH_s is the surface pH, which is related to the bulk (volume) phase, pH_v, by the Boltzman equation:

$$pH_s = pH_v + \Delta\Psi_s/60 \qquad (7)$$

where $\Delta\Psi_s$ is the surface electric potential. The Gouy-Chapman equation relates $\Delta\Psi_s$ with the density of charges, σ, and allows the qualitative prediction of the dependence of the surface potential on salt (symmetrical z:z valent) concentration:

$$\sigma = (0.0073\sqrt{c})\,\sinh\,(z\,\Delta\Psi_s F/2RT) \approx 0.0073\sqrt{c}.z\Delta\Psi_s F/2RT \quad \text{(for small } \Delta\Psi_s)$$

or, $$\Delta\Psi_s = (2RT/zF)\,\ln\,[b + \sqrt{(1+b^2)}] \qquad (8)$$

where $\Delta\Psi_s$ is measured in mV, $b \approx 137\sigma/\sqrt{c}$, $RT/F \approx 25$ mV.

Using Eqns. 7 and 8, Eqn. 6 can be rearranged to:

$$\Delta H^+ = \Sigma\,[1/(1+10^{pH_v-\overline{pK}_i^d}) - 1/(1+10^{pH_v-\overline{pK}_i^l})] \qquad (9)$$

where: $\overline{pK}_i^d = (pK_i^d)_o + \Delta\Psi_s F/(2.3RT) = (pK_i^d)_o + (2/2.3)\,\ln[b+\sqrt{(1+b^2)}]$

and similarly for $\overline{pK}_i^l$. Eqn. 9 predicts that increasing the salt concentration will decrease $\Delta\Psi_s$ and shift the observed $\overline{pK}^d$ and $\overline{pK}^l$ values of protonatable amino acid residues to the acid direction (for pH_v above the isoelectric point).

For the i-th group of the protein, located at a distance r_i from Q_B, a shift of pK (ΔpK_i) due to the presence of the extra charge on the Q_B semiquinone can be approximated by the following formula:

$$\overline{pK}_i^l = \overline{pK}_i^d + \Delta\psi e/(2.3kT) = \overline{pK}_i^d + e^2\exp(-r_i/\lambda_D)/(2.3\;4\pi\varepsilon\varepsilon_o r_i kT) \qquad (10)$$

where $\Delta\psi$ is the electric potential difference at distance r_i from Q_B, e is the elementary charge, ε the dielectric constant and λ_D is the Debye length introduced above (Eqn. 5).

It is seen from Eqn. 10 that increasing the salt concentration leads to screening of the electrical potential of Q_B^- and a decrease in the magnitudes of the pK shifts ($\Delta pK_i = pK_i^l - pK_i^d$) for amino acid residues located around Q_B.

The contributions to the sum (Eqn. 9) can be approximately subdivided into two classes: (1) where the difference of pK_i^d and pK_i^l is large, i.e., the effective distance between Q_B and the protonatable amino acid residues is smaller than the Debye length (Σ_1), and (2) where the difference in pK_i is small, i.e., the distance between Q_B and the protonatable amino acid residues is larger than the Debye length (Σ_2). In the first case we can neglect the screening effect of salt, and in the second we can suppose that the shift of pK is small. Hence,

$$\Delta H^+ \approx \Sigma_1 + \Sigma_2 \approx \Sigma_1 + \sum_i \frac{\partial f}{\partial pK_i} \Delta pK_i \tag{11}$$

where

$$\frac{\partial f}{\partial pK_i} = \frac{10^{(pH_s - pK_i)} \cdot \ln 10}{(1 + 10^{(pH_s - pK_i)})^2} \approx \begin{cases} 10^{(pH_s - pK_i)} \cdot \ln 10, & \text{if } pH_s \ll pK_i \\ 10^{(pK_i - pH_s)} \cdot \ln 10, & \text{if } pK_i \ll pH_s \end{cases}$$

Taking into account Eqn. 10, the second sum can be rearranged:

$$\Sigma_2 = \sum_i \frac{\partial f}{\partial pK_i} \cdot \frac{e^2 \exp(-r_i/\lambda_D)}{4\pi\varepsilon\varepsilon_0 kT r_i \, 2.3} = \frac{e^2}{4\pi\varepsilon\varepsilon_0 kT \, 2.3} \cdot \sum_i \frac{\partial f}{\partial pK_i} \cdot \frac{\exp(-r_i/\lambda_D)}{r_i} \tag{12}$$

Eqn. 12 qualitatively explains the decrease in the first flash-induced protonation as a function of the square root of the salt concentration.

SUMMARY

1. A simple model is described for the pH dependence of the one-electron transfer between Q_A and Q_B, in RCs from *Rb. sphaeroides*, taking into account only the pairwise electrostatic interactions between Q_B^- and the two amino acid residues, Glu^{L212} and Asp^{L213}.

2. The electron transfer equilibrium between $Q_A^-Q_B$ and $Q_AQ_B^-$ is reasonably well described by the model, both for wild type RCs and for mutant RCs with either Glu^{L212} or Asp^{L213} altered to the corresponding amide - mutants L212EQ and L213DN.

3. The kinetics of the one-electron transfer between Q_A and Q_B are well described by the model, for Wt and L212EQ mutant RCs, but L213DN mutant RCs yield a very large discrepancy. An attractive explanation for this is that the absence of Asp^{L213} prevents attainment of protonation equilibrium on the timescale of the electron transfer kinetics, but does allow it on the much longer timescale of the electron transfer equilibrium measurements.

4. Proton uptake by RCs, after one and two flashes, was shown to have distinctive salt dependences it various pH. Increasing salt concentrations led to at least a 7-fold decrease in the relative amplitude of first flash-induced proton uptake by RCs at pH near neutral, but changed it much less at acid pH (< 6) and not at all at alkaline pH (≥ 9). Divalent cations were effective at much lower concentrations than monovalent cations.

5. The dependence of first flash proton uptake on salt concentration is explained by a salt screening effect of the Q_B semiquinone electric potential, leading to diminution of the pK shifts of amino acid residues located further from Q_B than the Debye screening length in the RC protein. Suppression of the contribution of amino acid residues located far from Q_B by the addition of high concentrations of salt reveals the contributions of those amino acid residues which are close to Q_B.

6. It is suggested that pH titration of proton uptake by RCs at high salt concentration is a convenient method for testing the role of amino acid residues in proton uptake in mutant reaction centers. The evidence to date supports the major involvement of a small number of residues in the H^+-binding stoichiometry of Q_B^- at alkaline (pH> 9.5) and acid (pH <6.5) and a more distributed response over many residues at neutral conditions (6.5 < pH < 8.5).

Acknowledgements

This work was supported by grants from the National Science Foundation (DMB 89-04991 and MCB 92-30660). V.P.S. gratefully acknowledges release time from the Biophysics Faculty, Moscow State University.

REFERENCES

1. Michel, H., Epp, O. and Deisenhofer, J. (1986) EMBO J. 5, 2445-2451.
2. Deisenhofer, J. and Michel, H. (1989) EMBO J., 8, 2149-2170.
3. Allen, J.P., Feher, G., Yeates, T., Komiya, H. and Rees, D.C. (1987) Proc. Natl. Acad. Sci. USA 84, 5730-5734.
4. Allen, J.P., Feher, G., Yeates, T., Komiya, H. and Rees, D.C. (1987) Proc. Natl. Acad. Sci. USA 84, 6162-6166.
5. Allen, J.P., Feher, G., Yeates, T., Komiya, H. and Rees, D.C. (1988) Proc. Natl. Acad. Sci. USA 85, 8487-8491.
6. Wraight, C.A. (1979) Biochim. Biophys. Acta 548, 309-327.
7. Maróti, P. and Wraight, C.A. (1988) Biochim. Biophys. Acta 934, 329-347.
8. McPherson, P.H., Okamura, M.Y. and Feher, G. (1988) Biochim. Biophys. Acta 934, 348-368.
9. Shinkarev, V.P., Verkhovsky, M.I. and Zakharova, N.I. (1989) Biochemistry (USSR) 54, 256-264.
10. Kleinfeld, D., Okamura, M.Y. and Feher, G. (1984) Biochim. Biophys. Acta 766, 126-140.
11. Paddock, M.L., Rongey, S.H., Feher, G. and Okamura, M.Y. (1989) Proc. Natl. Acad. Sci. USA, 86, 6602 - 6606.
12. Takahashi, E. and Wraight, C.A. (1990) Biochim. Biophys. Acta 1020, 107-111.
13. Takahashi, E. and Wraight, C.A. (1992) Biochemistry 31, 855-866.
14 Paddock, M.L., Juth, A., Feher, G. and Okamura, M.Y. (1992) Biophys. J. 61, 153a.
15. Beroza, P., Fredkin, D.R., Okamura, M.Y. and Feher, G. (1991) Proc. Natl. Acad. Sci. USA, 88, 5804-5808.
16. Gunner, M.R. (1992) This volume.
17. Shinkarev, V.P. and Wraight, C.A. (1993) in: "The Photosynthetic Reaction Center," J.R. Norris and J. Deisenhofer, eds, Academic Press. In press.
18. Bell, R.P. (1973) "The Proton in Chemistry," Cornell University Press, New York.
19. Gutman, M. and Nachliel, E. (1990) Biochim. Biophys. Acta 1015, 391-414.
20. Maróti, P. and Wraight, C.A. Biophys. J. Submitted.
21. McComb, J.C., Stein, R.R. and Wraight, C.A. (1990) Biochim. Biophys. Acta 1015, 156-171.
22. Takahashi, E. and Wraight, C.A. (1991) FEBS Lett. 283, 140-144.

INITIAL CHARACTERIZATION OF THE PROTON TRANSFER PATHWAY TO Q_B IN *RHODOPSEUDOMONAS VIRIDIS*: ELECTRON TRANSFER KINETICS IN HERBICIDE-RESISTANT MUTANTS

W. Leibl[1], I. Sinning[2], G. Ewald[2], H. Michel[2], and J. Breton[1]

[1]Section de Bioénergétique, DBCM, CE Saclay, 91191 Gif-sur-Yvette France

[2]Max-Planck-Institut für Biophysik, H.-Hoffmann Str. 7 6000 Frankfurt 71, Germany

INTRODUCTION

For the study of electron and proton transfer processes in proteins, photosynthetic reaction centers (RC) are of considerable interest as they provide a system that allows to trigger the reaction sequences in a natural way by light. Especially promising objects are the RC of purple bacteria where for two species the three dimensional structure of the RC is known in great detail. The use of single-site mutations adds a further powerful tool to shine light on the role of single amino acids in photosynthetic electron and proton transport.

The acceptor side of the reaction center of purple photosynthetic bacteria consists of a quinone-iron complex. The secondary quinone acceptor (Q_B) accepts in successive flashes two electrons from the first quinone acceptor (Q_A) and two protons from the external medium. The protein surrounding plays an important role in providing pathways for protons from the cytoplasmic phase to Q_B. For *Rps. viridis* as well as for *Rb. sphaeroides* the structure of the RC has been determined by X-ray crystallography (Deisenhofer et al., 1984; Allen et al., 1986; Chang et al., 1986). The availability of site-directed mutants in *Rb. sphaeroides* led to a detailed picture of the role of single amino acid residues in electron and proton transfer (Okamura and Feher, 1992) . Beside Ser L223 and Glu L212 which seem to be involved in the transfer of the first and second proton to Q_B respectively, Asp L213 was identified to be essential for the transfer of both protons (Paddock et al., 1989; Paddock et al., 1990; Takahashi and Wraight, 1990; Takahashi and Wraight, 1992). The equivalent residue in *Rps. viridis* is asparagine, thus the proton pathway could be quite different. Although a comparison of the pathways in the two otherwise very similar

The Photosynthetic Bacterial Reaction Center II
Edited by J. Breton and A. Verméglio, Plenum Press, New York, 1992

systems is highly desirable, there is in the case of *Rps. viridis* no information available about the involvement of individual residues in proton transfer.

In *Rps. viridis* no site-directed mutants are available. However, a number of herbicide-resistant mutants have been characterized (Sinning et al., 1989; Ewald et al., 1990). Herbicides of the triazine type compete with Q_B for the binding pocket and their binding is controlled by nearby residues which also are potential candidates for proton donors to reduced Q_B. Recently, we have characterized the kinetics and pH-dependence of the electron transfer in the quinone acceptor complex in *Rps. viridis* wild type (Wt). To get further insight into the mechanisms of quinone reduction we studied these reactions in herbicide-resistant mutants. In this contribution we present a first glance on these experimental results.

MATERIALS AND METHODS

Cultures of cells of Wt and herbicide-resistant mutants were grown as described (Ewald et al., 1990) and stored at −30 °C until use. For measurements the cells were thawed and treated essentially as described earlier (Leibl and Breton, 1991). A short incubation step in 50 mM buffer containing 2 μg/mL gramicidin, 200 μM TMPD or DAD and 3 mM ferricyanide was followed by two washing steps with the appropriate buffer to adjust the pH and to remove ferricyanide. TMPD or DAD were added as redox mediators in a sufficient concentration to reoxidize Q_B^- within a few seconds.

We used a photoelectric technique to detect the fraction of Q_A^- at a variable delay time after a saturating flash. In Fig. 1 an example of the basic photovoltage kinetics is given. The two traces are representative for the two extreme cases of a) RC with Q_A oxidized, and b) RC with Q_A reduced (either chemically or by a preflash).

From Fig. 1 it can be seen that the time-resolved photoelectric signal induced by a picosecond flash is quite different if Q_A is either oxidized or reduced. The difference is due

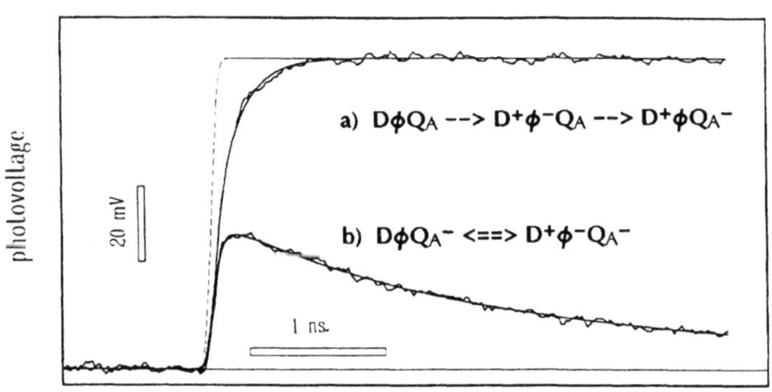

Figure 1. Kinetics of the photovoltage in whole cells of *Rps. viridis* (Wt) evoked by a picosecond flash (30 ps, 530 nm) in the presence of 10 mM *o*-phenanthroline, pH 9. (a) without preflash; (b) with a saturating preflash given 20 μs before the picosecond flash. An instrumental capacitative decay was deconvolved. Smooth lines are best fits according to the reaction scheme stated in the figure (D, primary donor; ϕ, intermediary acceptor). The dashed line shows the photovoltage kinetics from *Halobacterium halobium*, demonstrating the time resolution of the apparatus.

to the replacement of the positive phase due to charge stabilization on Q_A (200 ps) by a negative phase (2 ns) due to a back–reaction of the primary radical pair $D^+\phi^-$ when Q_A is reduced. Therefore the fraction of RC with reduced Q_A can be extracted from photovoltage kinetics. In a double flash experiment the first saturating flash creates the state $D^+Q_A^-$ in all RC. D^+ is rereduced in the sub–μs time range. The second (picosecond) flash given at a variable delay time (μs – s) tests the reoxidation of Q_A^-. By this double flash method the reoxidation kinetics of Q_A^- can be determined, which under normal conditions is due to electron transfer to Q_B.

The photovoltage kinetics were detected using the light–gradient method and a micro–coaxial measuring cell (Trissl et al., 1987). The excitation energy of the picosecond flash was ca. 250 μJ/cm^2 and 5 to 10 traces were averaged. Analysis of the kinetics was done by iterative convolution of a linear combination of the charge separation kinetics in the case of oxidized and reduced Q_A with the response function of the apparatus. The latter was independently determined by the ultrafast signal induced in oriented purple membranes from *Halobacterium halobium*.

RESULTS

We studied the kinetics of electron transfer in the quinone acceptor complex in Wt (Leibl and Breton, 1991) and in the following three herbicide–resistant mutants of *Rps. viridis*:

T1	:	Arg L217 –> His and Ser L223 –> Ala	(Sinning et al., 1989)
MAV5	:	Arg L217 –> His and Val L220 –> Leu	(Ewald et al., 1990)
MAV2	:	Glu L212 –> Lys	(Ewald et al., 1990).

The mutant T1 was selected by its resistance to terbutryn whereas the mutants MAV2 and MAV5 are resistant to atrazine. We measured the reoxidation kinetics of Q_A^- after one, two and three preflashes. In all cases at pH 7 a binary oscillation of these kinetics as a function of the number of preflashes was observed. This is taken as proof that the Q_A^- reoxidation kinetics can be assigned to electron transfer to Q_B and Q_B^- respectively, displaying the function of Q_B as a two–electron gate.

The values of the exponential rate constants of the first and second electron transfer to Q_B at pH 7 are listed in Table 1 for the mutants studied together with the values for Wt.

Table 1. Rate constants of the first (k_1) and second (k_2) electron transfer to Q_B at pH 7

	Wt	MAV5	T1	MAV2
k_1	$(18\ \mu s)^{-1}$	$(7\ \mu s)^{-1}$	$(3\ \mu s)^{-1}$	$(76\ ms)^{-1}$
k_2	$(65\ \mu s)^{-1}$	$(370\ \mu s)^{-1}$	$(40\ ms)^{-1}$	$(260\ \mu s)^{-1}$

In the two double mutants, **MAV5 and T1** the first electron transfer is accelerated and the second electron transfer is decelerated compared to Wt. The most striking effect observed in these mutants is the decrease of the rate of the second electron transfer by nearly three orders of magnitude in the mutant T1. In the mutant MAV5, which also shows the Arg L217 -> His mutation but has kept Ser L223, this rate is decreased only slightly by about a factor of 6. As in Wt the pH–dependence of the electron transfer is rather weak in the pH–range 7–9 and binary oscillations of the electron transfer kinetics are easily observed. At high pH, however, these oscillations are blocked in the mutant T1. Details of the characteristics of electron transfer at high pH in these two mutants will be treated in another publication.

In the mutant **MAV2** (Glu L212 -> Lys) at pH 7, the rate of the first electron transfer to Q_B is decreased by more than three orders of magnitude compared to Wt. Surprisingly, the kinetics of the second electron transfer are about 100–times faster than those of the first electron transfer, and not much slower than in the Wt. As in the other mutants the presence of binary oscillations in the electron transfer kinetics as a function of the number of preflashes clearly shows that Q_B functions as the electron acceptor and that the exchange Q_BH_2/Q_B is established within the time between the preflashes (200 ms).

There is no significant effect of pH on the first and second electron transfer in the pH range 7–10 in the mutant MAV2. However, at pH>10 a dramatic change in the kinetics is observed compared to lower pH. After the first flash a large fraction of Q_A^- is reoxidized very rapidly with a kinetics comparable to the one in Wt. After two flashes reoxidation is very slow and similar kinetics are observed after three flashes. Therefore at pH>10 a clear attribution of the slow kinetics to electron transfer to Q_B^- is not possible. Alternatively they are due to a charge recombination between Q_A^- and cyt+. The drastic effect induced by high pH is reversible.

DISCUSSION

All mutations studied are located in the Q_B binding region which overlaps with the binding region of herbicides. Ser L223 as well as Glu L212 are conserved in the RC of all purple bacteria. They have been shown to play in *Rb. sphaeroides* an essential role in the transfer to Q_B of the first and second proton, respectively (Paddock et al., 1990; Paddock et al., 1991).

In all mutants (except MAV2 and pH>10) as well as in Wt, the first electron transfer shows only a very weak pH–dependence. This is in line with the idea, that protonation events are not rate limiting for the stabilization of the first charge on Q_B. It is well established that the transfer of the second electron and of the first proton to Q_B are closely correlated. The second proton is not correlated to electron transfer but is essential for the replacement of the doubly reduced quinone by an oxidized one and thus for the observation of binary oscillations of the semiquinone state in a sequence of flashes. Therefore the loss of a residue which is essential in proton donation is expected either to strongly decrease the rate of the second electron transfer or to block the binary oscillations after the second flash.

The effect of Arg L217 –> His mutation

In the mutant MAV5 the second mutation (Val L220 -> Leu) can be assumed to be neutral with respect to electron and proton transfer. Therefore the differences with regard to Wt might largely be attributed to the replacement of Arg L217 by histidine. The small effects seen in this mutant suggest that this replacement is not introducing any major disturbances despite the presumably quite different pK's of both residues. This renders a crucial role of Arg L217 for proton supply to Q_B unlikely.

The effect of Ser L223 -> Ala mutation

In the mutant T1, a drastic decrease of the second electron transfer rate was found indicating that a block in the pathway of the first proton to Q_B has occurred. According to the results obtained in the mutant MAV5, this effect cannot be assigned to the mutation Arg L217 -> His, present in both RC. Therefore we attribute the inhibition of the second electron transfer in T1 to the loss of Ser L223 which is needed for an efficient donation of the first proton to Q_B. This attribution is in agreement with a very similar result obtained on a single Ser L223 -> Ala mutation in *Rb. sphaeroides*, where a 350-fold decrease of the rate of the second electron transfer has been measured (Paddock et al., 1990).

In Wt, Ser L223 was found to form a hydrogen bond to one of the carbonyl oxygens of the quinone both in *Rps. viridis* and in *Rb. sphaeroides*, making this residue a likely candidate for a direct proton donor. However, its high pK suggests that it can presumably only be deprotonated and reprotonated in a concerted manner (Paddock et al., 1990). In a recent model the (transiently) protonated hydroxyl group of Ser L223 was proposed as the actual proton donor (Okamura and Feher, 1992). It seems to us that a similar mechanism could be an attractive hypothesis in the case of *Rps. viridis*, regarding the fact, that the presence of Ser L223 leads to the formation of a cluster of three hydrogen bonds including Q_B, Ser L223, Asn L213, and Gly L225. In the mutant T1 only one of these hydrogen bonds (Q_B-Gly L225) is preserved (Sinning et al., 1990).

The effect of Glu L212 -> Lys mutation

This mutation leads to a decrease of the rate of the first electron transfer to Q_B by three orders of magnitude at pH 7. This is unexpected from an energetic point of view because the introduction of a positive charge should stabilize the state Q_B^-. Possible explanations for this effect could be a structural change and/or a decreased affinity for quinones on the Q_B site. A structural change could for example increase the distance between Q_A and Q_B and thereby strongly decrease the electron transfer rate. A decreased binding affinity could result from structural changes or be a pure electrostatic effect. In this case the electron transfer would be rate limited by quinone binding.

At present we are not able to distinguish between these or other possible explanations. In contrast to the mutants T1 and MAV5, the structure of the RC of MAV2 is not known. However, the conclusion that the positive charge on the lysine is responsible for the inhibition is supported by the observation that above pH 10 at least 60% of a fast rate of the first electron transfer, close to the one found in the Wt, is reconstituted. We suggest that the cause is a loss of the positive charge after deprotonation of Lys L212.

The very surprising result, that at pH<10 the second electron transfer is faster than the first one can easily be reconciled with either of the above interpretations. In the case of a structural change, the transfer of the first electron to Q_B might lead to a partial screening of the charge on the lysine and induce some relaxation of the protein into a more native conformation. In the case of a decreased affinity, it is well known that normally the semiquinone form of Q_B (formed after the first flash) is many orders of magnitude more strongly bound than the other species and for this reason the second electron transfer is not limited by the binding of the quinone.

Under high-pH conditions the binary oscillations seen at lower pH are lost. This could be related to a failure of supply of the second proton to Q_B indicating that an easily deprotonatable amino acid residue at position L212 is necessary for this reaction. Thus, in the Wt Glu L212 could be involved in the donation of the second proton as was proposed in *Rb. sphaeroides* (Paddock et al., 1989). This residue might be at least transiently protonated when Q_B is negatively charged delivering its proton to the state Q_BH^- which should have a very high pK. Another explanation that we cannot exclude presently would be an inhibition of the second electron transfer to Q_B above pH 10 in this mutant. The discrimination between these two possibilities and a definitive assignment of a proton donor role to Glu L212 awaits further experiments.

CONCLUSIONS

Photoelectric measurements were preformed to study the electron and proton transfer to Q_B in the RC of *Rps. viridis*. The advantage of this method is the possibility to use whole cells and thereby to avoid the isolation of RC, which could result in a modified or largely unoccupied Q_B-site.

The measurements show that Ser L223 most likely plays an important role in the transfer of the first proton, similar to the case of in *Rb. sphaeroides*. The interpretation is less clear in the case of Glu L212, which in our mutant unfortunately was replaced by another protonatable residue, although an important role in the transfer of the second proton to Q_B would be compatible with our data. It seems to be a reasonable hypothesis that these two highly conserved residues are designed as direct proton donors to Q_B in the RC of purple bacteria and that differences in the proton transfer pathway occur only on a later level.

REFERENCES

Allen, J.P., Feher, G., Yeates, T.O., Rees, D.C., Deisenhofer, J., Michel, H., and Huber, R. (1986) *Proc. Natl. Acad. Sci. USA* 83, 8589–8593.

Baciou, L., Sinning, I., and Sebban, P. (1991) *Biochemistry* 30, 9110–9116.

Chang, C.-H., Tiede, D., Tang, J., Smith, U., Norris, J., and Schiffer, M. (1986) *FEBS Lett.* 205, 82–86.

Deisenhofer, J., Epp, O., Miki, K., Huber, R., and Michel, H. (1984) *J. Mol.Biol.* 180, 385–398.

Ewald, G., Wiessner, C., and Michel, H. (1990) *Z. Naturforsch.* 45c, 459–462.

Leibl, W. and Breton, J. (1991) Biochemistry 30, 9634–9642.

Mathis, P., Sinning, I., and Michel, H. (1992) Biochim. Biophys. Acta 1098, 151–158.

Okamura, M.Y., and Feher, G. (1992) *Ann. Rev. Biochem.*, in press.

Paddock, M.L., McPherson, P.H., Feher, G., and Okamura, M.Y. (1990) *Proc. Natl. Acad. Sci. USA* 87, 6803–6807.

Paddock, M.L., Rongey, S.H., Feher, G., and Okamura, M.Y. (1989) *Proc. Natl. Acad. Sci. USA* 86, 6602–6606.

Sinning, I., Michel, H., Mathis, P., and Rutherford, A.W.R. (1989) *Biochemistry* 28, 5544–5553.

Sinning, I., Koepke, J., Schiller, B., and Michel, H. (1990) *Z. Naturforsch.* 45c, 455–458.

Takahashi, E., and Wraight, C.A. (1990) *Biochim. Biophys. Acta* 1020, 107–111.

Takahashi, E., and Wraight, C.A. (1992) *Biochemistry* 31, 855–866.

Trissl, H.-W., Leibl, W., Deprez, J., Dobek, A., and Breton, J. (1987) *Biochim. Biophys. Acta* 893, 320–332.

STUDY OF REACTION CENTERS FROM *RB. CAPSULATUS* MUTANTS MODIFIED IN THE Q_B BINDING SITE

Laura Baciou[1], Edward J. Bylina[2] and Pierre Sebban[1]

[1] UPR 407, Bat. 24, CNRS Gif/Yvette, 91198, FRANCE
[2] Pacific Biomedical Research Center, Honolulu, Hawaï, 96822 USA

INTRODUCTION

In photosynthetic reaction centers, herbicides block electron transfer between primary and secondary quinone acceptors, Q_A and Q_B, respectively, by competing with Q_B for its proteic binding site. The similitude between the reaction center from purple bacteria, which 3D structure is known at 3 Å resolution, and that of PSII of plants, still unknown, validates comparative studies concerning herbicides binding. Herbicide-resistant mutants from different bacterial species have been either designed by site directed mutagenesis or selected on their ability to grow in the presence of herbicides. This was done for Rb. capsulatus (Bylina and Youvan, 1987; Bylina et al., 1989), Rb.sphaeroides (Paddock et al., 1988) and Rps. viridis (Sinning and Michel, 1987). Among all these mutants, it has been pointed out that the replacement of Ile^{L229} in Rb. sphaeroides and Rb. capsulatus is of special interest. From the 3D structure of the reaction center from Rps. viridis with s-triazine herbicide terbutryn bound in the Q_B pocket, it appears that Ile^{L229} makes extensive contacts with that inhibitor and Q_B. Depending on the nature of the substitution in Rb. capsulatus, the mutations lead to various levels of photosynthetic growth and of terbutryn resistance compared to the wild type (Bylina and Youvan, 1987). Since assays have been done on whole cells it is likely that combination of different factors is at work in modulating the photosynthetic activities of the mutants. In order to explain the effects of the mutations, more precise studies are therefore required on isolated reaction centers. Two mutants of Rb. capsulatus are studied in this

The Photosynthetic Bacterial Reaction Center II
Edited by J. Breton and A. Verméglio, Plenum Press, New York, 1992

paper. The Ile^{L229} --> Ser mutant obtained by site directed mutagenesis and Thr^{L226} --> Ala, selected by resistance to atrazine. In Rps. viridis, Ala is naturally in the L226 position. It was then of interest to observe the consequences of that mutation on the binding of herbicides in Rb. capsulatus comparatively to Rps. viridis.

RESULTS

Q_B Binding

Endogenous UQ_{10} Titrations in the Wild Type and the Thr^{L226} --> Ala Mutant.

The Q_B pocket occupancy (% Act) is measured from the relative amplitude of the slow phase ($P^+Q_B^-$) of the charge recombination decay kinetics (data not shown, to be published). After purification, our reaction center preparations from the wild type, Thr^{L226} --> Ala and Ile^{L229} --> Ser respectively contain 30%, 30% and 10% of residual bound Q_B. In order to determine the dissociation constant (K_m) for the native UQ_{10}, we have titrated Q_B activity as a function of endogenous Q_B concentration by diluting the reaction center solution stock. This was done for the wild type and the Thr^{L226} --> Ala mutant, but not for Ile^{L229} --> Ser reaction centers in which the residual amount of Q_B was too low. When the total enzyme concentration ($[RC]_{total}$) is not negligible compared to the substrate ($[Q]_{total}$), % Act may be expressed as:

$$\% \text{ Act} = \frac{[RC]_{Q_B}}{[RC]_{total}} = \frac{\left(1 + r + \frac{K_m}{[RC]_{total}}\right) - \left(\left(1 + r + \frac{K_m}{[RC]_{total}}\right)^2 - 4r\right)^{1/2}}{2} \quad (1)$$

where r represents the amount of quinone per reaction center in the solution. The parameter r was biochemically determined by the method of Gast et al. (1985). It is about 1.5 Q_B per reaction center in the solution.

This analysis leads to K_m values of 0.85 ± 0.3 μM and 0.35 ± 0.15 μM for the wild type and the Thr^{L226} --> Ala mutant, respectively (Table 1). Thus, the affinity for the endogenous quinone is increased in that mutant compared to the wild type.

UQ_6 Titrations in the Wild Type, Thr^{L226} --> Ala and Ile^{L229} --> Ser mutants.

Assuming similar hypothesis as for the endogenous quinone, the UQ_6 titration curves of the Q_B activity in the wild type, the Thr^{L226} --> Ala and Ile^{L229} --> Ser mutants (data not shown) were fitted by equation 2.

$$\%Act = \frac{(K'_m + [RC]_{total} + [Q]_{total}) - \{(K'_m + [RC]_{total} + [Q]_{total})^2 - 4\,[RC]_{total} \cdot [Q]_{total}\}^{\frac{1}{2}}}{2\,[RC]_{total}} \quad (2)$$

where K'_m represents the apparent dissociation constant for UQ_6, and $[Q]_{total}$ includes the concentrations of added UQ_6 and residual UQ_{10}. The results are presented in Table 1. Confirming the above results obtained with the endogenous UQ_{10}, the binding affinity for UQ_6 is somewhat increased in the Thr^{L226} --> Ala mutant compared to wild type. That could result from steric reasons, the smaller side chain of Ala compared to Thr, allowing an easier binding of Q_B. In Ile^{L229} --> Ser, K'_m is decreased compared to the wild type. This may arise from the loss of van der Waals contacts between Q_B and Ser^{L229} compared to Ile at position L229.

Table 1. Dissociation constants (μM) for endogenous UQ_{10} and UQ_6 in the wild type Rb. capsulatus, the Thr^{L226} --> Ala and Ile^{L229} --> Ser mutants

	K_m endogenous UQ_{10}	K'_m UQ_6
Wild type	0.85 ± 0.3	3.7 ± 0.3
Thr^{L226} --> Ala	0.35 ± 0.15	0.88 ± 0.15
Ile^{L229} --> Ser		19.5 ± 0.5

K'_m values don't take into account the equilibrium constant between Q_A^- and Q_B^- states.

We may note that the ratio between the K'_m values of the wild type and the Thr^{L226} --> Ala mutant (3.7/0.88) for the UQ_6 titrations is consistent with that measured for the titrations with UQ_{10} endogenous quinone (0.85/0.35). In the wild type Rb. sphaeroides, Mc Comb et al. (1991) found dissociation constants of 1.7 μM and 4 μM for UQ_{10} and for UQ_6, respectively. Our results on Rb. capsulatus reaction centers are consistent with those findings.

Terbutryn and o-phenanthroline Binding Affinities

K_i's for Terbutryn

In order to obtain the dissociation constant values (K_i) for the herbicides terbutryn and o-phenanthroline, we have titrated the activity of reaction centers (% Act_i) as a function of the inhibitors concentration. These titration curves (data not shown, to be published) were fitted by equation (3) that accounts for a pure competitive process between Q_B and the inhibitor.

$$\% \; Act_i = \frac{[Q]}{[Q] + K_m^{'} + \frac{K_m^{'}}{K_i} \cdot [I]} \qquad (3)$$

where [Q] and [I] are the concentrations of free quinone and inhibitor, respectively. For the sake of simplicity we have identified these values to the total concentrations. The results are presented in Table 2.

The K_i values for terbutryn are essentially the same in the wild type and the two mutants. This result is logical for the Thr^{L226} --> Ala mutant considering that Thr^{L226} is not involved in the binding of that inhibitor. The growth resistance of Thr^{L226} --> Ala to terbutryn does not result from a decreased affinity of terbutryn, but rather from a larger quinone binding. The absence of effect of replacement of Ile^{L229} by Ser is surprising since Ile^{L229} makes extensive contacts with terbutryn. It seems therefore that these contacts are poorly affected when the side chain of Ile is replaced by the smaller side chain of Ser. Thus, the sensitivity of Ile^{L229} --> Ser towards terbutryn is likely to be due to a increase of the relative binding of terbutryn compared to quinone, and not to some steric hindrance of isoleucine compared to serine.

Table 2. Dissociation constants, K_i (μM) for terbutryn and o-phe in the wild type Rb. capsulatus, and the Thr^{L226} --> Ala and Ile^{L229} --> Ser mutants

	K_i terbutryn	K_i o-phe
Wild type	2.37 ± 0.3	146 ± 10
Thr^{L226} --> Ala	2.03 ± 0.3	6 ± 1.5
Ile^{L229} --> Ser	2 ± 0.2	242 ± 50

K_i's for o-phenanthroline

As for terbutryn, we have fitted the o-phenanthroline inhibition curves by equation 3. This leads to the values displayed in Table 2. The major observation is the greatly increased sensitivity of the Thr^{L226} --> Ala mutant compared to the wild type. The K_i value is decreased from 146 ± 10 μM in the wild type to 6 ± 1.5 μM in the Thr^{L226} --> Ala mutant.

In Rps. viridis, Ala is naturally in position L226. In that strain, the sensitivity to o-phenanthroline is much larger than that of Rb. capsulatus. This is readily apparent by comparing the I_{50}/Q_{50} ratio in the two strains. Q_{50} is the ubiquinone concentration for which half of the total activity is restored; at this quinone concentration, I_{50} represents the inhibitor concentration that inhibits half of initial activity. In Rps. viridis, I_{50}/Q_{50} ratios of 4 (Shopes and Wraight, 1987) and 6.5 (Sinning et al., 1989) were measured. These values are consistent with that of 8.5 obtained here in Thr^{L226} --> Ala. In contrast, it is about 125 in Rb. capsulatus wild type (this work) and about 240 in Rb. sphaeroides (Paddock et al., 1988). Thr L226 is conserved in all purple bacteria which are sequenced, except in Rps.

viridis. Taking into account the similar sensitivity to o-phenanthroline in wild type Rps. viridis and in the Thr^{L226} --> Ala mutant from Rb. capsulatus, it seems reasonable to propose that Ala L226 plays a role in the binding of that inhibitor or alternatively that the replacement of Thr^{L226} by Ala with a smaller side chain allows a much easier binding of o-phenanthroline, may be due to a local protein rearrangement.

pH Dependence of Q_B^- Stabilization

The effect of the mutations on the semiquinone stabilization can be observed on the rate of $P^+Q_B^-$ charge recombination (k_{BP}). The high sequence analogy of the quinone proteic pocket and the similar the nature of the primary and secondary quinones (UQ_{10}) in Rb. capsulatus and Rb. sphaeroides reaction centers lead to very close free energy gaps between the two states, $P^+Q_A^-$ and $P^+Q_B^-$, in both strains. It is thus very probable that charge recombination from $P^+Q_B^-$ occurs in Rb. capsulatus, as in Rb. sphaeroides (Kleinfeld et al., 1984), via $P^+Q_A^-$. Since $P^+Q_A^-$ and $P^+Q_B^-$ are in thermal equilibrium, k_{BP} is therefore sensitive to variations of the free energy level of $P^+Q_B^-$. It is thus of interest to measure the influence of pH on k_{BP} in Rb. capsulatus. The pH titrations of k_{BP} in the wild type and the Thr^{L226} --> Ala and Ile^{L229} --> Ser mutants are presented in Fig. 1.

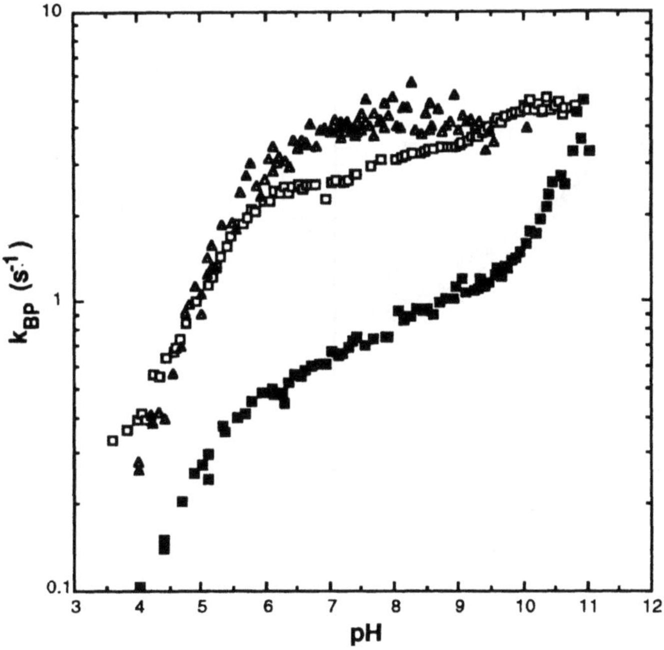

Figure 1. pH dependence of k_{BP}, the rate constant of $P^+Q_B^-$ charge recombination decays in the reaction centers from the wild type (■), Thr^{L226} --> Ala (□) and Ile^{L229} --> Ser mutants (▲).

The Wild Type

A smooth and regular increase of k_{BP} is observed in Rb. capsulatus from pH 7 ($k_{BP} \approx 0.65\ s^{-1}$) to pH 10 ($k_{BP} \approx 1.2\ s^{-1}$). Below pH 7, substantial decrease of k_{BP} is observed going from 0.65 s^{-1} at pH 7 to about 0.1 s^{-1} at pH 4. Above pH 10, k_{BP} accelerates from 1.5 s^{-1} at pH 10 to about 5 s^{-1} at pH 11. In Rb. sphaeroides, similar k_{BP} variations were observed at low and high pH and assigned to deprotonation of Asp^{L213} (Takahashi and Wraight, 1992) and Glu^{L212} (Paddock et al., 1989), respectively. In order to calculate the pH dependence of the $Q_A^- Q_B$ --> $Q_A Q_B^-$ equilibrium constant (K_2), we have also measured the pH dependence of the rate constant of $P^+Q_A^-$ charge recombination (k_{AP}) (data not shown). The equation derived by Wraight (1981): $K_2 + 1 = k_{AP}/k_{BP}$, leads to the pH titration curves for K_2 shown in Fig. 2. If only one protonation event occurs in Q_B vicinity in the pH range where the curves are analyzed, one can fit the K_2 curves as follows:

$$K_2\,(\mathrm{pH}) = K_2^{H^+}\,\frac{1 + 10^{\,(\mathrm{pH} - \mathrm{p}K_{Q_AQ_B^-})}}{1 + 10^{\,(\mathrm{pH} - \mathrm{p}K_{Q_A^-Q_B})}} \tag{4}$$

where $K_2^{H^+}$ represents the K_2 value at low pH, $pK_{Q_AQ_B^-}$ and $pK_{Q_A^-Q_B}$ the apparent pK's of the protonable group close to Q_B when the electron is on Q_B and on Q_A, respectively.

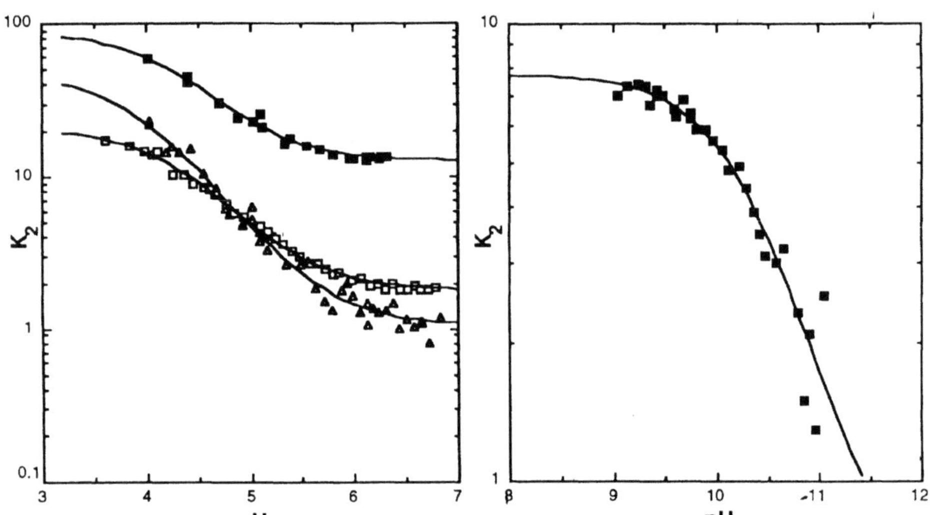

Figure 2. pH dependence of K_2, the $Q_A^-\,Q_B$ <--> $Q_A\,Q_B^-$ equilibrium constant, in the reaction centers from the wild type (■), the Thr^{L226} --> Ala mutant (□) and the Ile^{L229} --> Ser mutant (▲). The lines result from fitting the data with eq. 4. The derived pK's for the Q_B and Q_B^- states are shown in Table 3.

Fitting the K_2 curves of the left panel of Fig. 2 with eq. 4 for the wild type leads to $pK_{Q_A^- Q_B} = 4.2 \pm 0.2$ and $pK_{Q_A Q_B^-} = 5.1 \pm 0.1$ (Table 3).

These values are nearly the same as found for Rb. sphaeroides by Takahashi and Wraight: 4 and 5, respectively. That suggests similar distance and electrostatic environment between Asp^{L213} and Q_B in Rb. capsulatus and in Rb. sphaeroides. The right panel of Fig. 2 presents the pH dependence of K_2 for the wild type at high pH. Fitting this curve with eq. 4 leads to $pK_{Q_A^- Q_B}$ and $pK_{Q_A Q_B^-}$ values of 10.3 ± 0.1 and 11.6 ± 0.3, respectively (Table 3). These values are close to those attributed to deprotonation of Glu^{L212} by Mc Pherson et al. (1990).

Table 3. Protonation pK's related to the pH dependence of K_2

	Low pH		High pH	
	$pK_{Q_A^- Q_B}$	$pK_{Q_A Q_B^-}$	$pK_{Q_A^- Q_B}$	$pK_{Q_A Q_B^-}$
Wild type	4.2 ± 0.2	5.1 ± 0.1	10.3 ± 0.1	11.6 ± 0.3
Thr^{L226} --> Ala	4.3 ± 0.2	5.25 ± 0.1		
Ile^{L229} --> Ser	3.95 ± 0.2	5.5 ± 0.1		

Thr^{L226} --> Ala and Ile^{L229} --> Ser mutants

As displayed in Fig. 1 substantial destabilization of Q_B^- occurs in the mutants compared to the wild type. In the pH range 4-7, similar protonation events as in the wild type leads to decreases of k_{BP} values from 4.5 s^{-1} at pH 7 to 0.25 s^{-1} at pH 4 in Ile^{L229} --> Ser and from 2.5 s^{-1} at pH 7 to 0.33 s^{-1} at pH 3.7 in Thr^{L226} --> Ala. As shown in Table 3, the pK's derived from these pH titrations are close to those of the wild type suggesting that the mutations did not mainly affect that region of the Q_B pocket.

DISCUSSION

We have shown that in the wild type and the Thr^{L226} --> Ala mutant the dissociation constants for UQ_6 are 2 to 4 times higher than for endogenous UQ_{10}. The same result was found for Rb. sphaeroides reaction centers by Mc Comb et al (1990). This suggests that the parameters controling Q_B binding, are likely to be the same in Rb. sphaeroides and in Rb. capsulatus.

We found similar sensitivities to o-phenanthroline in Thr^{L226} --> Ala and in wild type Rps. viridis, where Ala is naturally in position L226. In wild type Rb. sphaeroides and Rb. capsulatus, where a Thr is in position 226, the sensitivity to that inhibitor is much lower. In Thr^{L226} --> Ala, the Thr -> Ala mutation mimics Rps. viridis Q_B pocket at the connection of the transmembrane helice E and the helice de either by opening a "high" affinity site or by favoring an easier binding of o-phenanthroline.

Substantial semiquinone destabilization is observed in mutants. As noted by Shopes and Wraight (1987), the midpoint redox potential (E_m) value for the couple Q_B/Q_B^- reflects the differential binding affinity for Q_B and Q_B^-. Our measure of an higher binding for Q_B in Thr^{L226} --> Ala is consistent with a lower E_m. This is not the case in Ile^{L229} --> Ser. We therefore have to postulate that in Ile^{L229} --> Ser, Q_B^- binding is more decreased compared to the wild type than it is for Q_B. That could be due to unfavorable electrostatic interaction between the negative charge present on the semiquinone and the partial negative charge present on the side chain of serine.

A significant difference is observed here between the patterns of the pH titration curves of k_{BP} in wild type Rb. capsulatus and Rb. sphaeroides, in the pH range 7-10. A marked increase of k_{BP} is displayed in the former strain but not in the latter. Tyr^{L215} is the only protonable amino acid present in the Q_B binding pocket of Rb. capsulatus which corresponds to a non-protonable residue in Rb. sphaeroides (Phe^{L215}). That could explain the different behavior of the two strains. Changing Tyr^{L215} to Phe in Rb. capsulatus reaction centers would greatly help to check that hypothesis.

To understand the differences of the spectroscopic properties from the two strains, it would be of interest to obtain crystals from Rb. capsulatus reaction centers and to resolve the 3D structure.

REFERENCES

Bylina, E. J., Jovine, R. V. M. and Youvan D. C. (1989)*Biotechnology 7, 69-74.*

Bylina, E. J. and Youvan D. C. (1987) Z. *Naturforsch.* 42c, 769-774.

Gast, P., Michalski, T. J. , Hunt, J. E. and Norris J. R. (1985) *FEBS Letters* 179, 2, 325-328.

Kleinfeld, D., Okamura, M. Y. and Feher G. (1984) *Biochim. Biophys. Acta* 766, 126-140.

Paddock, M. L., Rongey, S. H., Abresch, E. C., Feher, G. and Okamura M.Y. (1988) *Photosynt. Research* 17, 75-96.

Paddock, M. L., Rongey, S. H., Feher, G. and Okamura M. Y. (1989) *Proc. Natl. Acad. Sci.* 86, 6602-6606.

Shopes, R. J. and Wraight C. A. (1987) In the Proceedings of the VIIth International Congress on Photosynthesis. J. Biggins editor. Martinus Nijhoff/Holland. 397-400.

Sinning, I. and Michel H. (1987) Z. *Naturforsch.* 42c, 751-754.

Sinning, I., Michel, H., Mathis, P. and Rutherford B. (1989) *Biochemistry* 28, 5544-5553.

Takahashi, E. and Wraight C. A. (1990) *Biochim. Biophys. Acta* 1020, 107-111.

Takahashi, E. and Wraight C. A. (1992) *Biochemistry* 31, 855-866.

Wraight, C. A. (1981) *Israël Journal of Chemistry* 21, 348-354.

CALCULATIONS OF PROTON UPTAKE IN *RHODOBACTER SPHAEROIDES* REACTION CENTERS

M.R. Gunner[1] and Barry Honig[2]

[1]Department of Physics
City College of New York
New York, N.Y. 10031 U.S.A

[2]Department of Biochemistry and Molecular Biophysics
Columbia University
630 West 138th Street
New York, N.Y. 10032 U.S.A.

INTRODUCTION

In the first stages of photosynthesis carried out in proteins such as bacterial reaction centers (RCs) the energy of an absorbed photon is used to initiate a series of intra-protein electron transfers. These reactions have been well characterized (see reference 1 for a review). One important consequence of the resulting charge separation is proton uptake from the interior of the cell. The stoichiometry of proton uptake on formation of different redox states as a function of pH has been described in detail.[2-6] More recent studies have monitored the influence of specific amino acids on proton uptake stoichiometry and kinetics.[7-12] It is of interest to understand how the RCs three dimensional structure yields the observed coupling of electron and proton transfers.

The Photosynthetic Bacterial Reaction Center II
Edited by J. Breton and A. Verméglio, Plenum Press, New York, 1992

A plausible connection between proton motions and electron transfer reactions is simply the changed electrostatic field in different redox states of the protein. The calculation of electrostatic potentials throughout a protein has been attempted with increasing success in recent years (see references 13-15 for reviews). It has been shown to be possible to calculate electrochemical midpoints[16], and pK_a's.[17-19] The work presented here demonstrates that classical electrostatics can also provide insight into the coupling between proton and electron transfers.

METHODS

The heavy atom coordinates of the reaction centers from Rb. sphaeroides obtained by Allen et. al. were used in the calculations.[20-24] Comparisons were made to the Rb. sphaeroides structure of Chang et. al.[25,26] and to the Rps. viridis reaction center coordinates[27,28] (1PRC in the Brookhaven Data Bank). Protons were added with the program Discover and minimized with all heavy atoms fixed, assuming all Arg, Lys, Asp, and Glu are ionized and all other sites are neutral. No waters are localized in either Rb. sphaeroides structures.

The program DELPHI was used to solve the Poisson equation for the electrostatic potential in and around the RC as described in reference16. The electrostatic contributors to the free energy of a charge in the protein are separated into the reaction field energy and the interaction with other charges in the system.[29,30] The reaction field energy is the result of the polarization of electrons and dipoles in the surrounding media by a charge. This is treated in the Poisson equation by a dielectric constant. Water, with its dielectric constant of 80, is the most important contributor. However, there are also small scale rearrangements of the dipoles in the close packed protein, accounted for by an internal dielectric constant of 4.[16,31,32] The free energy of interactions between charges, modulated by the response of the media, is also provided by DELPHI.

A large number of simplifications were made in carrying out these calculations. The charge distribution on non-ionizable side chains and ionized Asp, Glu, Arg, and Lys were taken from the CHARMM charge set.[33] The change in charge when the acidic amino acids were neutralized was simply removal of -0.5 from each carboxylate oxygen. For Arg, 0.25 was removed from the 4 terminal protons while Lys had 0.33 removed from the three ε-protons. The reduced quinone was approximated by -0.5 on each carbonyl oxygen, while the oxidized primary donor had +0.5 on each Mg of the two bacteriochlorophylls of P. No membrane was added in the calculations reported here.

The Boltzmann distribution of different ionization states of the acidic and basic residues was calculated at different pHs in different redox states of the RC. The contribution of site i to the total free energy of a given ionization state at a particular pH is:

$$\Delta G(i) = \delta_i \left[\Delta G_{pol} + \Delta\Delta G_{rxn} + \frac{1}{2}\sum_{j \neq i}^{n} \delta_j \, \Delta G_{crg}(i,j) - \frac{c_i(pK_{soln} - pH)}{RT} \right] \quad (1)$$

ΔG_{pol} represents the free energy of interaction of i with charges in the protein that are not dependent on the pH such as the charge distribution on the backbone or on polar sidechains. $\Delta\Delta G_{rxn}$ is the difference in the reaction field energy for the residue in solution and in the protein. $\Delta G_{crg}(i,j)$ is the interaction energy between ionizable residues i and j when both are charged. pK_{soln} is the pK of this type of residue in solution. The delta function is 1 for an ionized residue and zero for a neutral group. The free energy of the system in a particular charge state is given by the sum of ΔG_i over all ionizable residues. The difficulty, as has been described previously,[19] is to obtain the fraction of each site that is ionized in a Boltzmann distribution given the extremely large number of possible states. A Monte Carlo sampling method was developed. A suggested by Beroza et. al[19] the algorithm incorporates internal proton transfer by allowing two sites that are strongly interacting to change state in a single Monte Carlo step 50% of the time. The single flips represent transfers to the external buffer.

RESULTS AND DISCUSSION

Transfer of an ionizable group from water into a low dielectric environment such as protein shifts the equilibrium between ionized and neutral form, favoring neutral species.[34] The loss of reaction field energy, often referred to as the desolvation penalty, for the ionizable amino acids in Rb. sphaeroides RCs is shown in figure 1a. Even without inclusion of a membrane, desolvation penalties range from 0 to 13 Kcal. In the absence of compensating interactions, this would shift ionizable residues pK_a's by as much as 10 pH units relative to those found in aqueous solution. Thus, the difference in reaction field energy in different positions in the protein makes an important contribution to the in situ ionization of acidic and basic sidechains. When the change in charge on ionization is well buried, as it is for liganded cofactors such as 6 coordinate hemes, the picture is quite different. It has been found that while placing a heme in a protein does make it harder to oxidize, there is little difference in desolvation penalty for hemes in different sites.[16]

One of the remarkable features of RCs is the cluster of acidic residues in the Q_B site. For example, Glu L 212, Glu M 234, and Asp L 213 form a pool of ionizable residues that in their charged state have a free energy of interaction with Q_B of greater than 5 Kcal. Asp L 210, Asp L 173, and Arg L 217 interact strongly (>5 Kcal) with at least one residue in the first group (see table 1). With 5 acids and only one base, ionization of these sites would appear to be unfavorable, since the cost of bringing negative charges close together is added to the cost of burying charged groups. However, it appears that ionization of the acids within this cluster is stabilized by the backbone. Although the ends of the transmembrane helices do contribute,

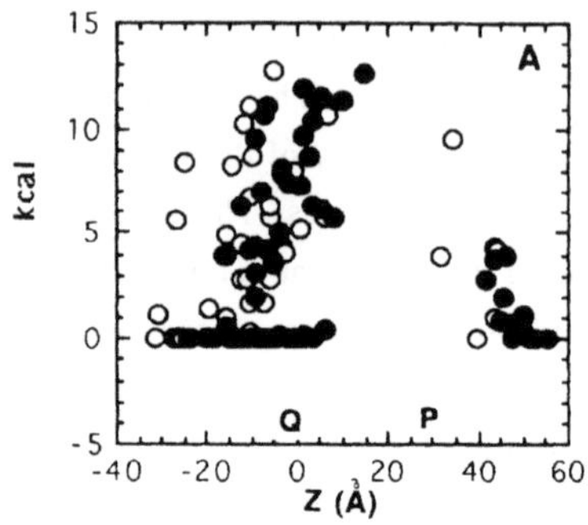

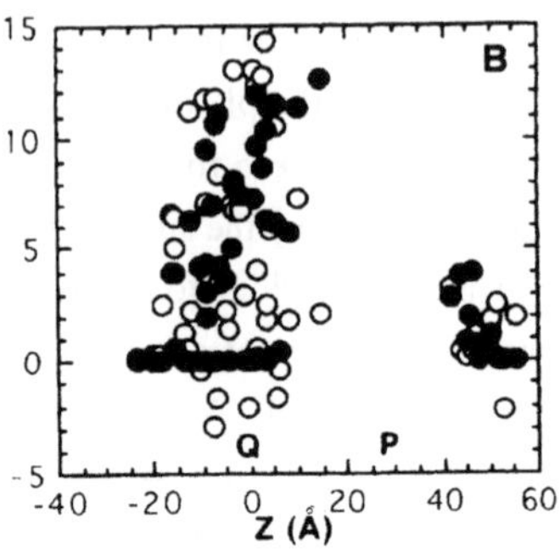

Figure 1. (A) Loss of reaction field energy for the acids (•) and bases (o) in RCs. (B) Comparison of the desolvation penalty of the acidic residues (•) and the stabilization of the charge on acidic residues by the partial charges on the backbone (presented as -ΔG for easier comparison) (o). Z is a line through a RC perpendicular to the membrane. The non-heme Fe is at Z=0. Approximate positions of the quinones and the magnesiums of the primary donor are shown.

Table 1. Interactions between acids and bases in the Q_B site (in Kcal). The residues in this table either have an free energy of interaction with Q_B,when both are charged, of greater than 5 Kcal or are strongly coupled to one of those residues

	Glu L 212	Glu M 234	Asp L 213	Asp L 210	Glu H 173	Arg L 217
Q_B	7.6	5.0	5.2	2.2	3.0	-2.8
L 212		6.2	5.2	2.6	3.0	-1.9
M 234			3.3	1.6	2.6	-1.2
L 213				8.5	5.2	-7.5
L 210					2.0	-5.7
H 173						-2.3

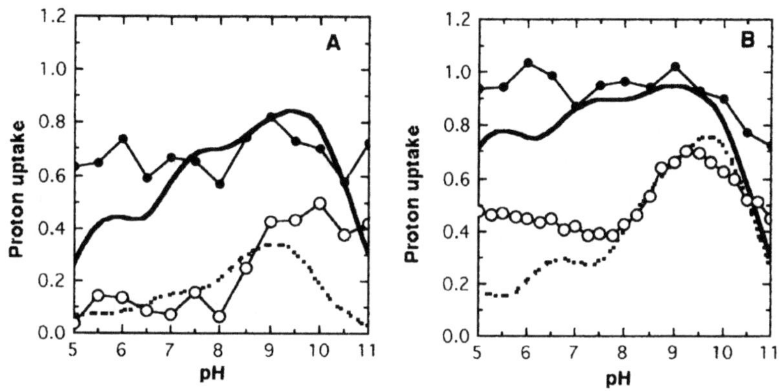

Figure 2. Comparison of the measured and calculated proton uptake stoichiometry as a function of pH. Lines without points are experimentally determined.[5] Points are the results of the Monte Carlo calculation. Formation of: (A) Q_A^- (Solid line, •), $P^+Q_A^-$ (dashed line, o) (B) Q_B^- (Solid line, •), $P^+Q_B^-$ (dashed line, o)

several loops in the cytoplasmic cap of the protein are even more important. Short runs of backbone at the turns of these loops are oriented to have their amide groups pointing inward towards the Q_B site. Each segment of the backbone adds to the field from its neighbor and the end result is a field that stabilizs acidic residues with an energy comparable to the desolvation penalty (see figure 1b). This same backbone arrangement destabilizes basic residues, balancing the favorable influence of the ionized acids.

It has long been of interest to understand the changes in ionization of specific sites that yield the observed proton uptake when RCs go from the ground state to different charge separated states.[2,5,6] This quantity was calculated from the difference in total charge on RCs, given a Boltzmann distribution of ionization states, in the ground state and the $P^+Q_A^-$, $P^+Q_B^-$, Q_A^-, or Q_B^- states. The results are shown in figure 2. The experimental and theoretical results agree remarkably well.

It is possible to see which residues contribute to the proton uptake shown in figure 2. While the identity of individual sites are expected to change as the calculations are refined, the correspondence between the data and calculations demonstrate that this picture provides at least a possible, if not a unique scenario. Figure 3a shows a consequence of the close coupling between acidic residues in the Q_B site. Asp L 210 and Asp L 213 perturb the pK_a of each other by 8.5 Kcals (6.2 pH units). Thus, over a wide pH range all RCs have one of these residues ionized, but almost no individual protein has both residues charged. In the calculation presented here, in the ground state at pH 8.0 41% of the RCs have L 210 ionized while 60% have L 213 ionized. As shown this balance is pH dependent, as the changing ionization of surrounding residues shifts the relative proton affinity of the two acids. $P^+Q_B^-$ formation is accompanied by a shift in average proton occupancy of ≈0.2 protons, with L 210 now 60% ionized while L 213 is 41% ionized. This will not be a direct response to the charge on Q_B^- which has a stronger interaction with L 213 (5.2 Kcal) than it does with L 210 (2.2 Kcal) (see table 1). Rather the ionization of Glu L 212 appears to be the cause. As shown in figure 3b, the calculation suggests Glu L 212 is ionized in the ground state. At low pH when Q_B is reduced L 212 picks up a proton, causing proton shifts in the surrounding residues, including those ascribed to L 210 and L 213 as well as final proton uptake from the buffer. The inability of L 212 to take up a proton at high pH correlates with the falloff in total proton uptake in the system (figure 2). The scenario presented here is neither unique or proved. Previous work suggested that L 212 is neutral in the ground state and that it is the loss of a proton in this species that leads to the falloff in proton uptake at high pH.[7] The picture presented here shows that a different, though equally important, role for this Glu is also consistent with the data for wild type RCs.

CONCLUSION

The work described here demonstrates how the protein structure can produce the observed coupling between electron transfer and proton uptake in RCs. The good match of theory and experiment for the pH dependence of proton uptake in different ionization states demonstrates this structure provides a very stable system. Thus, the occupancies of individual sites in a cluster of strongly interacting ionizable residues may be wrongly assigned in theory, or differently ionized in experiment, but the total charge on the system is little changed. The clusters of closely interacting residues therefore buffer the RC from the changes in redox state or pH. Thus, the relatively featureless dependence of the proton uptake with pH appears to be due to the buffering capacity of strongly coupled acids and bases. Ionization of the acids that provide this response is favored by loops in the backbone, a motif that orients runs of carbonyls away from the Q_B site.

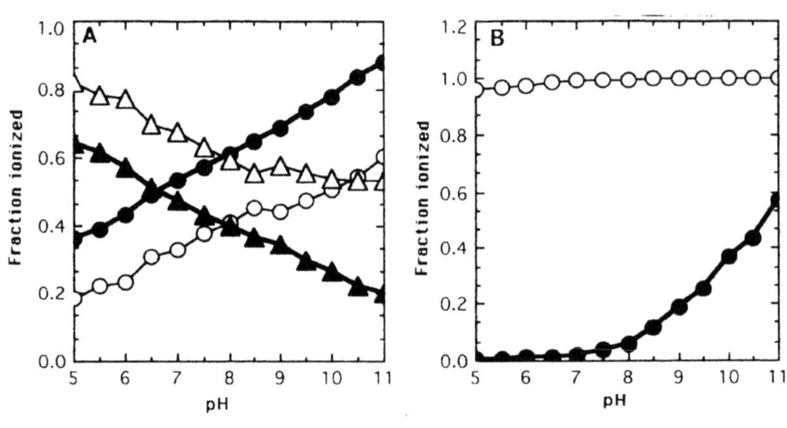

Figure 3. Fraction of Asp L 210 (A-o), Asp L 213 (A-Δ), and Glu L 212 (B-o) ionized at different pHs. The open points are the ground state, while solid points are for the state $P^+Q_B^-$. A and B have different vertical scales.

Observations of electron transfer kinetics[35] and the free energy of various redox states[36,37] have demonstrated that normal RC populations are heterogeneous,. The emerging view of RCs with many ionization states of similar energy provides one compelling reasons for heterogeneity of any population of protein.

ACKNOWLEDGEMENTS

We would like to thank Anthony Nicholls for continuing development of DELPHI, An-Suei Yang and Rosemary Samponga for helpful discussions regarding pK_a's and S. Sridharan for advice on Monte Carlo calculations. This work was supported in part by N.I.H. GM12897 (M.R.G) and N.S.F. DMB-03489.

REFERENCES

1. M.R. Gunner. *Current Topics in Bioenergetics*, 16:319 (1991).
2. C.A. Wraight. *Biochem. Biophys. Acta*, 548:309 (1979).
3. D. Kleinfeld, M.Y. Okamura and G. Feher. *Biochemistry*, 23:5780 (1984).
4. D. Kleinfeld, M.Y. Okamura and G. Feher. *Biochem. Biophys. Acta*, 809:291 (1985).
5. P. Maroti and C.A. Wraight. *Biochim. Biophys. Acta*, 934:329 (1988).
6. P.H. McPherson, M.Y. Okamura and G. Feher. *Biochim. Biophys. Acta*, 934:348 (1988).
7. M.L. Paddock, S.H. Rongey, G. Feher and M.Y. Okamura. *Proc. Natl. Acad. Sci. USA*, 86:6602 (1989).
8. I. Sinning, H. Michel, P. Mathis and A.W. Rutherford. *Biochem*, 28:5544 (1989).
9. M.L. Paddock, P.H. McPherson, G. Feher and M.Y. Okamura. *Proc. Natl. Acad. Sci.*, 87:6803 (1990).
10. L. Baciou, I. Sinning and P. Sebban. *Biochem.*, 30:9110 (1991).
11. E. Takahashi and C.A. Wraight. *Biochemistry*, 31:855 (1992).
12. E. Takahashi and C.A. Wraight. *Febs Lett*, 283:140 (1991).
13. A. Warshel and S.T. Russell. *Q. Rev. Biophys.*, 17:283 (1984).
14. K.A. Sharp and B. Honig. *Annu. Rev. Biophys. Biophys. Chem.*, 19:301 (1990).
15. M.R. Gunner and B. Honig. in *Cytochrome c Sourcebook* (eds. R.A. Scott & A.G. Mauk) in press (University Science Books, Mill Valley, 1992).
16. M.R. Gunner and B. Honig. *Proc. Natl. Acad. Sci. USA*, 88:9151 (1991).
17. D. Bashford and M. Karplus. *Biochemistry*, 29:10219 (1990).
18. D. Bashford and M. Karplus. *J. Phys. Chem.*, 95:9556 (1991).
19. P. Beroza, D.R. Fredkin, M.Y. Okamura and G. Feher. *Proc. Natl. Acad. Sci. USA*, 88:5804 (1991).

20. J.P. Allen, G. Feher, T.O. Yeates, H. Komiya and D.C. Rees. *Proc. Natl. Acad. Sci. USA*, 84:5730 (1987).
21. J.P. Allen, G. Feher, T.O. Yeates, H. Komiya and D.C. Rees. *Proc. Natl. Acad. Sci. USA*, 84:6162 (1987).
22. J.P. Allen, G. Feher, T.O. Yeates, H. Komiya and D.C. Rees. *Proc. Natl. Acad. Sci. USA*, 85:8487 (1988).
23. T.O. Yeates, H. Komiya, D.C. Rees, J.P. Allen and G. Feher. *Proc. Natl. Acad. Sci. USA*, 84:6438 (1987).
24. T.O. Yeates, H. Komiya, A. Chirino, D.C. Rees, J.P. Allen and G. Feher. *Proc. Natl. Acad. Sci. USA*, 85:7993 (1988).
25. C.H. Chang, D. Tiede, J. Tang, U. Smith, J. Norris and M. Schiffer. *FEBS Lett.*, 205:82 (1986).
26. O. El-Kabbani, C.-H. Chang, D. Tiede, J. Norris and M. Schiffer. *Biochemistry*, 30:5361 (1990).
27. J. Deisenhofer and H. Michel. *The EMBO Journal*, 8:2149 (1989).
28. J. Deisenhofer, O. Epp, R. Miki and H. Michel. *Nature*, 318:618 (1985).
29. M.K. Gilson and B. Honig. *Proteins*, 3:32 (1988).
30. M.K. Gilson and B. Honig. *Proteins*, 4:7 (1988).
31. M.K. Gilson and B. Honig. *Biopolymers*, 25:2097 (1986).
32. S. Harvey. *Proteins*, 5:78 (1989).
33. B.R. Brooks, R.E. Bruccoleri, B.D. Olafson, D.J. States, S. Swaminathan and M. Karplus. *J. Comput. Chem*, 4:187 (1983).
34. R.J. Kassner. *Proc. Natl. Acad. Sci. USA*, 69:2263 (1972).
35. C. Kirmaier and D. Holten. *Proc. Natl. Acad.USA*, 87:3552 (1990).
36. P. Sebban. *Biochim. Biophys. Acta*, 936:124 (1988).
37. P. Sebban and C.A. Wraight. *Biochim. Biophys. Acta*, 974:54 (1989).

CHLOROPHYLL TRIPLET STATES IN THE CP47-D1-D2-CYTOCHROME *b*-559 COMPLEX OF PHOTOSYSTEM II

Paul J.M. van Kan, Marloes L. Groot, Stefan L.S. Kwa, Jan P. Dekker and Rienk van Grondelle

Department of Physics and Astronomy, Free University
De Boelelaan 1081, 1081 HV Amsterdam, The Netherlands

ABSTRACT

Chlorophyll triplet states have been induced at cryogenic temperatures by illumination of isolated photosystem II CP47-D1-D2-Cytochrome *b*-559 (CP47-RC) and D1-D2-Cytochrome *b*-559 (RC) complexes with repetitive laser flashes. The triplet induced in RC and RC-CP47 complexes had a lifetime of 1.6 ms (at 77K) and was characterized by an absorbance difference spectrum peaking at 680 nm. In CP47-RC complexes, an antenna triplet on CP47 was found with a lifetime of 0.7 ms; the quantum yield of radical pair triplet formation was about 0.2-0.3 below 100K and decreased at higher temperatures. The temperature dependence was analyzed with an equilibrium model for the chlorophyll excited states and the radical pair; the free energy change (ΔG) for charge separation was found to be about -50 meV and independent of temperature.

INTRODUCTION

Considerable progress has been made with the isolation and purification of fully active, oxygen-evolving photosystem II complexes from higher plants and cyanobacteria. The most purified complexes contain the reaction center proteins D1 and D2, cytochrome *b*-559, the core antenna proteins CP47 and CP43, an extrinsic 33 kDa protein involved in the stabilization of oxygen-evolution and a number of small proteins with unknown functions.[1] These so-called "core"-preparations retain all primary and secondary electron transfer components, as well as the binding sites for the inorganic co-factors manganese, calcium and chloride. The minimum content of chlorophyll *a* was recently estimated to be about 30-35 per photoactive center.[2,3] About 6 of these chlorophylls are bound to the D1-D2-cytochrome *b*-559 complex, whereas about 13 chlorophylls may be bound to each of the core antenna proteins CP47 and CP43.[2,3]

The primary charge separation in such complexes involves the rapid oxidation of the primary electron donor P680, a chlorophyll *a* monomer or dimer, and the reduction of a

The Photosynthetic Bacterial Reaction Center II
Edited by J. Breton and A. Verméglio, Plenum Press, New York, 1992

pheophytin *a* molecule called I. The charge separation is stabilized by the transfer of the electron on the pheophytin *a* molecule to the secondary electron acceptor Q_A, a strongly bound plastoquinone A molecule. This reaction takes place with a lifetime of about 500 ps.[4] Further stabilization is achieved by reduction of $P680^+$ by a tyrosine residue of the protein backbone in 20-200 ns[5] and by a number of subsequent electron transfer reactions in the hundreds of microseconds and milliseconds time range involving a second plastoquinone molecule and a tetranuclear manganese cluster.[1]

The energy and primary electron transfer reactions in the photosystem II core complex have been described by the so-called exciton-radical pair equilibrium model.[6] This model supposes that the equilibration of excitation energy over all antenna pigments (including P680) is very fast and that the charge separation is trap-limited, i.e., that the formation of the primary radical pair is much slower than excitation equilibration. The model is attractive because it yields data on the thermodynamics of charge separation and its free energy change ΔG. For PS II core particles, Schatz et al.[4] have shown that the rate of formation of $P680^+ I^-$ takes about 100 ps and is fitted quite well by the exciton-radical pair equilibrium model. The extrapolated rate of the reaction $P680^* \ I \rightarrow P680^+ \ I^-$ was 370 ns^{-1}, corresponding to a ΔG of about -150 meV.

In photosystem II, the reaction center trap can be closed by reducing the plastoquinone acceptor Q_A in its singly reduced state. The amount of radical pair formation, however, is rather low and was estimated to be about 30%[7] or less[8,9] of that observed with oxidized Q_A. This low yield was explained by electrostatic interaction between the negative charge on Q_A and the charges of the radical pair, resulting in an increase of the free energy of the radical pair and a shift of the equilibrium between the radical pair and the excited antenna to the latter.[6] Van Mieghem et al.[8,9] provided experimental evidence for this suggestion by analyzing fluorescence decay kinetics in preparations in which Q_A was double reduced and protonated, due to which the electrostatic repulsion was strongly diminished. The results revealed a much larger amplitude of the about 100 ps phase attributed before to trapping,[4] suggested therefore a high efficiency of radical pair formation, and revealed a slow phase with a lifetime of about 7 ns. The latter phase was shown not be caused by unconnected chlorophyll and thought to be related to the relatively long-lived radical pair state.[9]

During the isolation of D1-D2-Cytochrome *b*-559 (RC)[10] and CP47-D1-D2-Cytochrome *b*-559 (CP47-RC)[11,12] complexes the plastoquinone Q_A is removed from its binding site. The primary reactants P680 and I, however, are present and the primary photochemistry proceeds efficiently in a manner that shows some analogies to the photochemistry in photosystem II core and membrane fractions with double reduced Q_A, because in both cases the negative charge of singly reduced Q_A is absent. In isolated RC complexes, the primary radical pair $P680^+ I^-$ is formed with high yield in about 20 ps or less[13-20] (depending on whether or not P680 is excited directly). Recombination of the primary radical pair occurs in 30-40 ns at room temperature,[21,22] is slower at very low temperatures[23] and might be multiphasic.[24,25,20] During the relatively long lifetime of the radical pair state, the coupling between the spins of the electrons on the $P680^+$ and I^- molecules gets lost, due to which after recombination of the charge separated state a spin polarized triplet of P680 is formed.[26] For isolated RC preparations, the yield was estimated to be about 23-30% at room temperature[22,27] and about 80% at 10K;[22] the lifetime was about 30 µs at 4°C and about a millisecond under anaerobic conditions.[27]

The exciton-radical pair equilibrium model predicts that the number of pigments coupled to P680 influences the ΔG values for the primary charge separation, trapping times and fluorescence lifetimes.[6] In addition, in preparations where the radical pair is transiently formed, the number of pigments coupled to P680 may have an effect on the triplet yield. In view of entropy effects, the ΔG decreases when the number of coupled antenna pigments increases.[6]

In this contribution, we report yields and lifetimes of flash-induced triplet states in

CP47-RC complexes as a function of temperature and excitation energy, and analyze the results in terms of the exciton-radical pair equilibrium model. The results provide new information on the thermodynamics of the system and suggest a temperature independent ΔG of charge separation of about -50 meV.

EXPERIMENTAL

CP47-D1-D2-Cytochrome *b*-559 (CP47-RC) complexes were isolated from spinach PS II membranes by incubation with the non-ionic detergent dodecyl-β,D-maltoside and separation by ion exchange chromatography.[11,12] Isolated D1-D2-Cytochrome *b*-559 (RC) complexes were prepared by a short incubation of CP47-RC complexes with Triton X-100, after which the samples were readjusted to dodecylmaltoside and purified by ion-exchange chromatography.[28,20] For low-temperature measurements, the samples were adjusted to BTT (20 mM BisTris, 20 mM NaCl, 10 mM $MgCl_2$, 1.5% taurine, pH 6.5) + 0.03% (w/v) dodecylmaltoside + 70% (w/v) glycerol and (in darkness) cooled to cryogenic temperatures in a nitrogen reservoir cryostat (Oxford) or a helium flow cryostat (Oxford). For the measurements an optical density at 675 nm (77K) of 0.6 (CP47-RC) or 0.3 (RC) cm^{-1} was used.

Steady-state absorption spectra were recorded with a Cary 219 spectrophotometer. Laser-flash induced absorption changes were recorded with equipment described before by van Mourik et al.[29] with some modifications. Excitation light was produced by a Quanta Ray PDL2 dye-laser pumped by a Quanta Ray DCR2A Nd-YAG laser. To obtain homogeneous illumination of the sample, a diffuser plate was placed in the beam in front of the cryostat entrance window. The excitation laser flashes (FWHM ~ 8 ns) were at 610 nm with 1 Hz frequency; for most of the experiments, the average of 32 shots was taken. Probe light was from a 150 W tungsten halogen lamp, filtered by a RG 610 filter. By using a shutter the sample was illuminated with the probe light during 10 ms before and after the laser flashes. A 1/5 m monochromator was used between the sample and the Si photodiode detector; the spectral bandwidth was 1.2 nm. The signals were amplified 100 times using an amplifier with a response time of 5 μs. Measured traces were digitized in a Gould 4500 transient recorder. It was checked that the amplitudes of the flash-induced absorbance changes did not decrease during the first 1000 illuminations (below 200K).

Absolute values for the triplet quantum yield were obtained by extrapolating the slope of the saturation curves to zero excitation intensity and converting the measured laser intensity to the number of incident photons per cm^2 using free chlorophyll *a* in Triton X-100 at 77K as a calibration standard. For the calculation the following values were assumed: triplet quantum yield $\Phi_{Chl} = 0.6$[30] and $\varepsilon_{Chl} = 100\ 000\ M^{-1}cm^{-1}$. The number of triplets per pigment were approximated by the ratio of the integrals of both the ΔA (at time zero) and A spectra from 648 to 690 nm for free chlorophyll and from 655 to 695 nm for the complexes.

RESULTS

CP47-RC complexes of photosystem II were isolated and purified in the presence of the non-ionic detergent dodecyl-β,D-maltoside. It was checked that the preparations were characterized by a red absorption maximum at 675 nm at room temperature.[31] Fig.1A shows the 77K absorbance spectrum of the CP47-RC complex. The peak at 542 nm and the relatively large amplitude of the contribution at 418 nm suggest the presence of RC-associated pheophytin *a*, while in the carotenoid region an RC-related contribution peaking at 491 nm is observed. The Chl Q_y absorption region shows considerable fine structure with peaks in the second derivative at 682, 677, 670 and 659.5 nm (Fig. 1B). This spectral fine structure is probably largely due to the chlorophylls bound to the CP47 protein, since in isolated CP47 a very similar fine structure was observed.[32]

The 77K and 4K triplet difference spectra of the CP47-RC and RC complexes are shown in Fig. 2. The spectra are characterized by a somewhat asymmetric bleaching at 680-681 nm. Apart for a contribution near 670 nm and a slight red-shift of the main bleaching in the CP47-RC complexes the spectra are very similar for both complexes. The 4K triplet difference spectrum of the RC preparation strongly resembles the spectrum reported earlier for long-time Triton X-100 exposed RC particles.[23] Both spectra reveal a positive absorbance change at 670-673 nm. Such positive features could not be detected with free

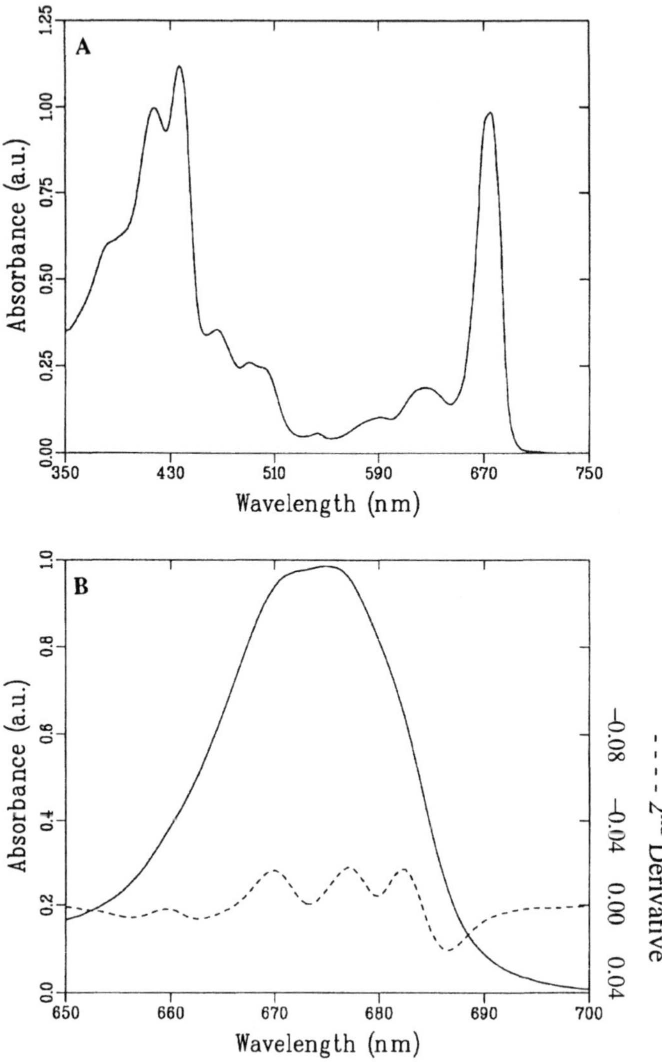

Figure 1. (A) 77K absorption spectrum of the CP47-RC complex, (B) solid line: enlargement of the Q_y-region, as in (A); dashed line: second derivative spectrum.

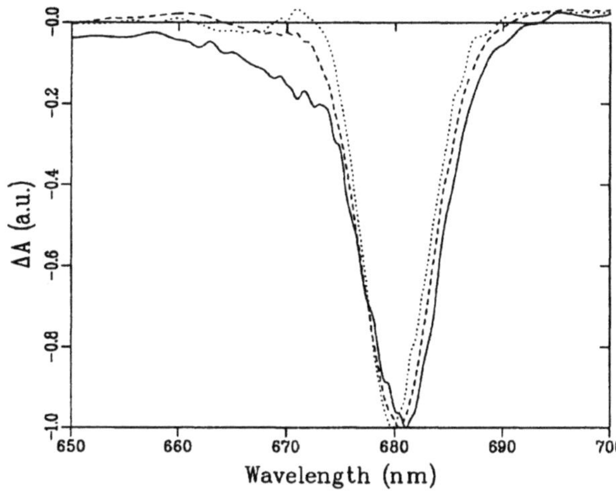

Figure 2. Laser-flash induced absorbance difference spectra of CP47-RC at 77K (full line), RC at 77K (dashed line) and RC at 4K (dotted line). The recordings are the differences of the average absorbances at t = 0-200 μs after the flashes and before the flashes.

chlorophyll in detergent[32] and suggest that triplet formation on one of the chlorophylls induces a shift of another chlorophyll to a somewhat shorter wavelength.The shoulder at 683 nm observed in Ref. 23 cannot be observed in Fig.2; the peak wavelengths of the bleachings (680 nm) and the wavelengths at the red flanks of the difference spectra at which

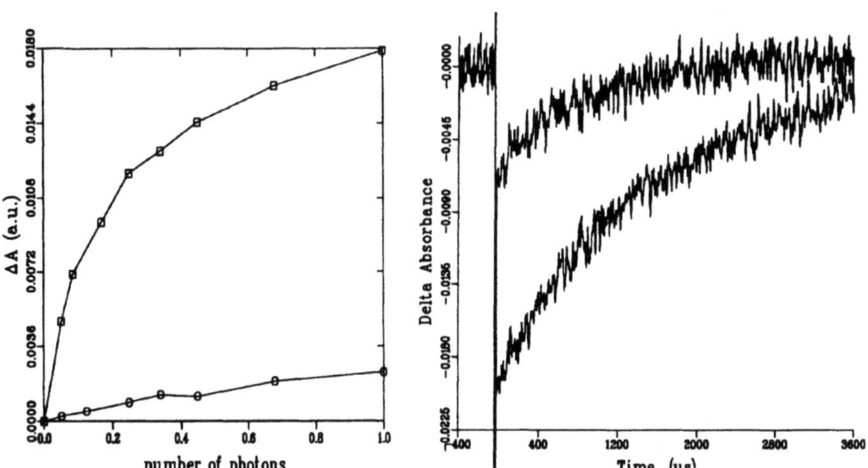

Figure 3. Left: laser-intensity dependence of the 77K absorbance changes at 680 nm (upper curve, squares) and 670 nm (lower curve, circles) in CP47-RC complexes (see Fig.2, full line). The number of 1.0 photons on the x-axis was calculated to correspond to 0.4 photons per chlorophyll. Right: laser flash induced absorbance changes at 77K in CP47-RC complexes at 680 nm (lower curve) and at 690 nm (upper curve). The trace at 690 nm was multiplied by five. The recordings are the average of 16 (680 nm) and 80 (690 nm) experiments.

50% of the bleachings occur (683.5 nm), however, are virtually identical. The lifetimes of the triplets in both complexes were calculated to be 1.6 ms at 77 K. This value is longer than the 0.7 ms lifetime calculated for the triplet at 683 nm in isolated CP47 complexes.[32] This value, however, is in the same order as found in solution at 4°C under anaerobic conditions,[27] suggesting that oxygen quenching does not occur in a glass.

The signal around 670 nm appeared to occur only at relatively high excitation energies. In Fig.3 (left), we compare the flash saturation curves measured at 680 nm and 670 nm. The results indicate that the triplet peaking at 680 nm saturates more easily than the triplet peaking at 670 nm and that both triplets are caused by different pigments.

The difference spectrum of the radical pair triplet appeared to be somewhat red-shifted in the CP47-RC complex. This could be due to a small redshift of P680 itself. It is, however, also possible that the red-shift is the result of formation to some extent of an antenna triplet on CP47. The absorbance difference spectrum of the latter triplet is characterized by a bleaching at 683 nm,[32] which is about 3 nm red-shifted compared to the reaction center triplet. Because the antenna triplet is characterized by a faster decay time than the reaction center triplet (0.7 ms[32] vs. 1.6 ms at 77K) we analyzed the decay time at the isosbestic point for the radical pair triplet formation (690 nm) in the CP47-RC complexes. The result is shown in Fig.3 (right) and reveals a clearly faster decay time at 690 nm than at 680 nm. This suggests that in the CP47-RC complex also a CP47 triplet is generated by flash illumination. The red-shift has been observed in all investigated batches of CP47-RC preparations. Therefore, it seems likely that the formation of a 683 nm CP47 antenna triplet in CP47-RC complexes is an intrinsic feature of the photochemistry in the CP47-RC complex of photosystem II.

If we assume that the rate of intersystem crossing from the radical pair to the triplet state of P680 is not dependent on temperature, measurement of the triplet (quantum) yield as a function of temperature gives information on the radical pair yield and on the thermodynamics of charge separation. We confirmed earlier results[22,23] that the triplet quantum yield is about 0.8 at 4-10 K for RC preparations (not shown). The results of experiments on the CP47-RC complexes are presented in Fig.4. The triplet yield appeared to be about 0.2-0.3 at low temperature and decreased above 120 K to less than 0.1 at 200 K.

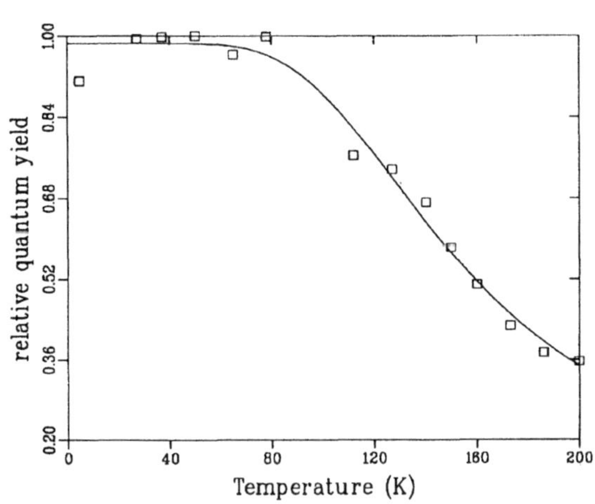

Figure 4. Squares: relative quantum yield of the radical pair triplet of CP47-RC complexes as a function of temperature. Line: Best fit of the relative quantum yield using the exciton-radical pair equilibrium model (Fig.5) and equations 1-3; the fit shown uses k_f = 20 ns^{-1}, k_{01} = 0.3 ns^{-1} and k_{02} = 0.01 ns^{-1} and yielded ΔG = - 48 meV

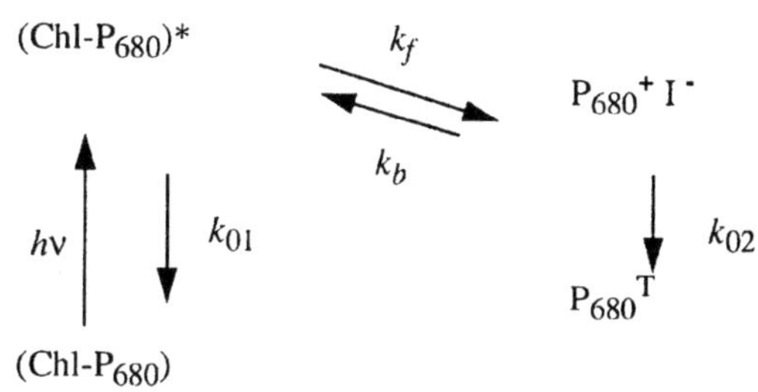

Figure 5. Schematic representation of the exciton-radical pair equilibrium model for energy transfer and primary charge separation in photosystem II

DISCUSSION

The triplet difference spectra measured in the CP47-RC complex are similar, but not identical, to the spectra obtained with isolated RC preparations. In these RC preparations the triplet signal is due to a spin polarized triplet state of the primary donor P680,[26,22] which presumably is formed from the radical pair state $P680^{+}I^{-}$. We conclude from our results that in the CP47-RC complex the overall signal is dominated by this radical pair triplet, but that to a smaller extent a CP47 antenna triplet is generated as well. The saturation curves in Fig.3 indicate that the radical pair triplet is initially produced at low excitation intensities; the signal around 670 nm has a lower quantum yield and probably represents formation of another antenna triplet state of perhaps some uncoupled chlorophyll.

The quantum yield of the radical pair triplet is also in the CP47-RC complexes clearly temperature dependent. We determined the yields for temperatures between 4 K and 200 K and modelled the yields by the exciton-radical pair equilibrium model of Schatz et al[6] (note that relative yields are more accurate, since all experiments have been performed in the same setup with identical probe pump geometry; for this reason, the shape of the temperature curve is more reliable than the absolute quantum yield values). The model is schematically reproduced in Fig.5.

The state (Chl-P680)* represents the equilibrium state of the excitations in the antenna formed on a timescale of < 20 ps; formation of the radical pair is trap-limited and takes up to 100 ps in core preparations.[4] P680 may be 'visited' many times by the excitation before charge separation takes place. The equilibrium between the excited and radical pair states can be described by the first-order rate constants k_f and k_b. This equilibrium decays in time; this effect is represented by the rate constants k_{01} and k_{02}. k_{01} is the sum of rate constants of fluorescence and internal conversion from the singlet excited state; k_{02} covers triplet formation and other decay processes of the radical pair $P680^{+}I^{-}$.

In the framework of this model (a standard two-compartment model), the quantum yield of the radical pair triplet formation Y can be described by:

$$\frac{Y}{Y_0} \approx \frac{k_f k_{02}}{k_f k_{02} + k_{01} k_{02} + k_{01} k_b} \qquad \text{if: } k_{01}\, k_{02} << k_f\, k_{02} \qquad (1)$$

in which Y_0 is the absolute yield at T = 0 K.

It is hardly feasible to vary all parameters in this expression, so we will discuss some estimations that may be used as fixed values. k_f determines the rate of charge separation.

The measured values for the decay of P* have been shown to increase by a factor of 2 upon cooling from room temperature to liquid helium temperatures.[13,14] We will in a first approximation neglect this linear dependence, because an activated back reaction k_b with an exponential temperature dependence will be introduced (see below, equation 3). In addition, for a perfect equilibrium model all pigments are coupled to P680. In this case, the forward rate constant k_f is approximated by:

$$k_f = \frac{k_{int}}{N} \qquad (2)$$

where N is the number of coupled Chl a. With $k_{int} \approx 350\ ns^{-1}$ (Refs. 6, 13-15, 18) and $N = 6$-8 this yields $k_f \approx 50\ ns^{-1}$ for the reaction center, which is well in accordance with recent results.[19] For CP47-RC (with $N \approx 20$) this would result in an estimated $k_f \approx 20\ ns^{-1}$ and for a core particle with $N \approx 40$ in a $k_f \approx 10\ ns^{-1}$, in agreement with the results in Ref. 4.

The value for the excited state lifetime of chlorophyll k_{01} is about $0.3\ ns^{-1}$ (see also Ref. 33). Estimation of k_{02} uses the notion that the lifetime of the radical pair in isolated RC preparations is 80-100 ns with a yield close to unity.[23] In a first approximation, we used this value ($k_{02} = 0.01\ ns^{-1}$) for the CP47-RC complexes; also this rate was approximated to be temperature independent.

With these estimations k_b is the only variable in expression (1). With:

$$k_b = k_f \cdot exp(\frac{\Delta G}{kT}) \qquad (3)$$

the free energy difference of the charge separation reaction ΔG is obtained. For $k_f = 20\ ns^{-1}$ (see above), a good fit to the data was obtained for ΔG = - (45-50) meV (Fig.4). Since we found ΔG to be independent of temperature, it is a reaction enthalpy and the entropic term in ΔG is negligible.

It is concluded that the simple exciton-radical pair equilibrium model fits the measured temperature dependence of relative triplet quantum yields quite well with reasonable parameters of k_f, k_{01} and k_{02}. However, is does not explain the decrease of the absolute triplet quantum yields in the CP47-RC complexes (20-30% vs. 65-80% for isolated RC complexes at low temperature). A possible explanation might be provided by the low-lying transitions at 683-690 nm in the CP47 antenna (see, e.g., Fig.1) that may form a 'sink' of excited states. An inital partition could take place between energy transfer to P680 and to these low-lying levels, which could result in reduced radical pair formation and, consequently, triplet quantum yield in the CP47-RC complexes.

According to our results, the temperature dependence of triplet yield can be fitted well by assuming a temperature independent free energy difference ΔG of charge separation. This result is in contrast with ΔG estimations from time-resolved fluorescence experiments on isolated RC complexes, in which an entropic ΔG linearly depending on temperature and a very small reaction enthalpy were found.[18,34] At present there is no good explanation for this paradoxal situation. It may be relevant to note here that a similar conflicting interpretation of fluorescence measurements and triplet yield data exists for the thermodynamics of the primary radical pair in purple bacteria.[35,36] Probably the models used are not sufficiently comprehensive to understand the detailed dynamics of singlet excited states, singlet and triplet states of the radical pair and the triplet state of the primary donor. We would not exclude the possibility of an equilibrium with entropic ΔG in the singlet state, an equilibrium with a constant enthalpy in the triplet state and a relatively slow intersystem crossing rate coupling these two equilibria. At the present stage of our experimental work on photosystem II complexes, the simple model we have used provides a satisfactory fit to the experimental results, and in addition provides a number of estimations (e.g., for k_f) that can be tested in future research.

ACKNOWLEDGMENTS

Thanks are due to Mrs. Ineke Vlaanderen for expert technical assistance with the isolation of the complexes. This research was supported in part by the Dutch Foundations for Chemical Research (SON) and for Biophysics (SvB), financed by the Netherlands Organization for Scientific Research (NWO). JPD was supported by a fellowship from the Royal Netherlands Academy of Arts and Sciences (KNAW).

REFERENCES

1. D.F. Ghanotakis and C.F. Yocum, Photosystem II and the oxygen-evolving complex, *Ann. Rev. Plant Physiol. Plant Mol. Biol.* 41:255 (1990)
2. P.J. van Leeuwen, M.C. Nieveen, E.J. van de Meent, J.P. Dekker and H.J. van Gorkom, Rapid and simple isolation of pure photosystem II core and reaction center complexes from spinach, *Photosynth. Res.* 28:149 (1991)
3. R. Barbato, H.L. Race, G. Friso and J. Barber, Chlorophyll levels in the pigment-binding proteins of photosystem II, *FEBS Lett.* 286:86 (1991)
4. G.H. Schatz, H. Brock and A.R. Holzwarth, Picosecond kinetics of fluorescence and absorbance changes in photosystem II particles excited at low photon density, *Proc. Natl. Acad. Sci. USA* 84:8414 (1987)
5. S. Gerken, K. Brettel, E. Schlodder and H.T. Witt, Optical characterization of the immediate electron donor to chlorophyll a_{II}^+ in O_2-evolving photosystem II complexes. Tyrosine as possible electron carrier between chlorophyll a_{II} and the water-oxidizing manganese complex, *FEBS Lett.* 237:69 (1988)
6. G.H. Schatz, H. Brock and A.R. Holzwarth, Kinetic and energetic model for the primary processes in photosystem II, *Biophys. J.* 54:397 (1988)
7. W. Leibl, J. Breton, J. Deprez and H.-W. Trissl, Photoelectric study on the kinetics of trapping and charge stabilization in oriented PS II membranes, *Photosynth. Res.* 22:257 (1989)
8. F.J.E. van Mieghem, W. Nitschke, P. Mathis and A.W. Rutherford, The influence of the quinone-iron electron acceptor complex on the reaction centre photochemistry of photosystem II, *Biochim. Biophys. Acta* 977:207 (1989)
9. F.J.E. van Mieghem, G.F.W. Searle, A.W. Rutherford and T.J. Schaafsma, The influence of the double reduction of Q_A on the fluorescence decay kinetics of photosystem II, *Biochim. Biophys. Acta* 1100:198 (1992)
10. O. Nanba and K. Satoh, Isolation of a photosystem II reaction center consisting of D-1 and D-2 polypeptides and cytochrome *b*-559, *Proc. Natl. Acad. Sci. USA* 84:109 (1987)
11. J.P. Dekker, N.R. Bowlby and C.F. Yocum CF, Chlorophyll and cytochrome *b*-559 content of the photochemical reaction center of photosystem II, *FEBS Lett.* 254:150 (1989)
12. J. Petersen, J.P. Dekker, N.R. Bowlby, D.F. Ghanotakis, C.F. Yocum and G.T. Babcock, EPR characterization of the CP47-D1-D2-cytochrome *b*-559 complex of photosystem II, *Biochemistry* 29:3226 (1990)
13. M.R. Wasielewski, D.G. Johnson, M. Seibert and Govindjee, Determination of the primary charge separation rate in isolated photosystem II reaction centers with 500-fs time resolution, *Proc. Natl. Acad. Sci. USA* 86:524 (1989)
14. M.R. Wasielewski, D.G. Johnson, Govindjee, C. Preston and M. Seibert, Determination of the primary charge separation rate in photosystem II reaction centers at 15K, *Photosynth. Res.* 22:89 (1989)
15. R. Jankowiak, D. Tang, G.J. Small and M. Seibert, Transient and persistent hole burning of the reaction center of photosystem II, *J. Phys. Chem.* 93:1649 (1989)
16. D. Tang, R. Jankowiak, M. Seibert, C.F. Yocum and G.J. Small, Excited-state structure and energy-transfer dynamics of two different preparations of the reaction center of photosystem II: a hole burning study, *J. Phys. Chem.* 94:6519 (1990)

17. D. Tang, R. Jankowiak, M. Seibert and G.J. Small, Effects of detergent on the excited state structure and relaxation dynamics of the photosystem II reaction center: a high resolution hole burning study, *Photosynth. Res.*27:19 (1991)
18. T.A. Roelofs, M. Gilbert, V.A. Shuvalov and A.R. Holzwarth, Picosecond fluorescence kinetics of the D1-D2-Cyt-b-559 photosystem II reaction center complex. Energy transfer and primary charge separation processes, *Biochim. Biophys. Acta* 1060:237 (1991)
19. J.R. Durrant, G. Hastings, Q. Hong, J. Barber, G. Porter and D.R. Klug, Determination of P680 singlet state lifetimes in photosystem two reaction centres, *Chem. Phys. Lett.* 188:54 (1992)
20. T.A. Roelofs, S.L.S. Kwa, R. van Grondelle, J.P. Dekker and A.R. Holzwarth, Primary processes and structure of the photosystem II reaction center: II. Low temperature picosecond fluorescence kinetics of a D1-D2-Cyt-b559 reaction center complex isolated by short Triton-exposure, *Biochim. Biophys. Acta*, submitted for publication
21. R.V. Danielius, K. Satoh, P.J.M. van Kan, J.J. Plijter, A.M. Nuijs and H.J. van Gorkom, The primary reaction of photosystem II in the D1-D2-cytochrome *b*-559 complex, *FEBS Lett.* 213:241 (1987)
22. Y. Takahashi, Ö. Hansson, P. Mathis and K. Satoh, Primary radical pair in the photosystem II reaction centre, *Biochim. Biophys. Acta* 893:49 (1987)
23. P.J.M. van Kan, S.C.M. Otte, F.A.M. Kleinherenbrink, M.C. Nieveen, T.J. Aartsma and H.J. van Gorkom, Time-resolved spectroscopy at 10K of the photosystem II reaction center; deconvolution of the red absorption band, *Biochim. Biophys. Acta* 1020:146 (1990)
24. Govindjee, M. van de Ven, C. Preston, M. Seibert and E. Gratton, Chlorophyll a fluorescence lifetime distributions in open and closed photosystem II reaction center preparations, *Biochim. Biophys. Acta* 1015:173 (1990)
25. P.J. Booth, B. Crystall, I. Ahmad, J. Barber, G. Porter and D.R. Klug, Observation of multiple radical pair states in photosystem 2 reaction centers, *Biochemistry* 30:7573 (1991)
26. M.Y. Okamura, K. Satoh, R.A. Isaacson and G. Feher, Evidence of the primary charge separation in the D_1D_2 complex of photosystem II from spinach: EPR of the triplet state, *in*: "Progress in Photosynthesis Research," J. Biggins, ed., Martinus Nijhoff Publishers, Dordrecht, The Netherlands, Vol. I, p. 379 (1987)
27. J.R. Durrant, L.B. Giorgi, J. Barber, D.R. Klug and G. Porter, Characterisation of triplet states in isolated photosystem II reaction centres: oxygen quenching as a mechanism for photodamage, *Biochim. Biophys. Acta* 1017:167 (1990)
28. S.L.S. Kwa, W.R. Newell, R. van Grondelle and J.P. Dekker, The reaction center of photosystem II studied with polarized fluorescence spectroscopy, *Biochim. Biophys. Acta* 1099:193 (1992)
29. F. van Mourik, C.J.R. van der Oord, K.J. Visscher, P.S. Parkes-Loach, P.A. Loach, R.W. Visschers and R. van Grondelle, Exciton interactions in the light-harvesting antenna of photosynthetic bacteria studied with triplet-singlet spectroscopy and singlet-triplet annihilation on the B820 subunit form of *Rhodospirillum rubrum*. *Biochim. Biophys. Acta* 1059:111 (1991)
30. A.J. Hoff, Triplets: phosphorescence and magnetic resonance, *in*: "Light Emission by Plants and Bacteria," Govindjee, J. Amesz and D.C. Fork, eds., Academic Press, New York, p. 225 (1986)
31. J.P. Dekker, S.D. Betts, C.F. Yocum and E.J. Boekema, Characterization by electron microscopy of isolated particles and two-dimensional crystals of the CP47-D1-D2-cytochrome *b*-559 complex of photosystem II, *Biochemistry* 29:3220 (1990)
32. P.J.M. van Kan, M.L. Groot, I.H.M. van Stokkum, S.L.S. Kwa, R. van Grondelle and J.P. Dekker, Chlorophyll triplet states in the CP47 core antenna protein of photosystem II, In Proc. IXth International Congress on Photosynthesis, Nagoya, Japan, in press
33. T.A. Roelofs and A.R. Holzwarth, In search of a putative long-lived relaxed radical pair state in closed photosystem II: kinetic modelling of picosecond fluorescence data, *Biophys. J.* 57:1143 (1990)
34. P.J. Booth, B. Crystall, L.B. Giorgi, J. Barber, D.R. Klug and G. Porter, Thermodynamic properties in D1/D2/cytochrome b-559 reaction centres investigated by time-resolved fluorescence spectroscopy, *Biochim. Biophys. Acta* 1016:141 (1990)
35. N.W.T. Woodbury and W.W. Parson, Nanosecond fluorescence from isolated photosynthetic reaction centers of *Rhodopseudomonas sphaeroides*, *Biochim. Biophys. Acta* 767:345 (1984)
36. A. Ogrodnik, M. Volk and M.E. Michel-Beyerle, On the energetics of the states $^1P^*$, $^3P^*$ and P^+H^- in reaction centers of *Rb. sphaeroides*, *in*: "The Photosynthetic Bacterial Reaction Center. Structure and Dynamics," J. Breton and A. Verméglio, eds., Plenum Press, New York, p. 177 (1988)

LIGHT REFLECTIONS II

G. Feher

University of California, San Diego
Department of Physics, 0319
La Jolla, California 92093-0319

I don't mind that you think slowly but I do mind that you are publishing faster than you think.

W. Pauli

At the last (1988) Cadarache meeting Jacques asked me to summarize the conference. Since we were all tired towards the end, he wanted me to do it with a light touch. I went overboard; it wasn't really a summary but a collection of anecdotes and flippant remarks. Jacques subsequently asked me to write it up, which I did, and it was included in the Proceedings under the title "Light Reflections". To my great surprise I received more favorable comments on that article than I did on any of the papers that I have published over the past 40 years. You can imagine the mixed feelings this elicited; did I perhaps pick the wrong profession? A few months ago Jacques asked me to give a "Light Reflections II" at this Cadarache meeting. I wrote him that I took a dim view of it, but would be willing to reconsider during the meeting. As the week of the meeting wore on and the number of lectures that were difficult to follow increased, I was leaning more and more towards giving one lecture that everybody could understand without too much effort. So yesterday I decided to give Light Reflections II and here I am.

First I would like to thank the organizers, Jacques and André, for inviting me to this beautiful place. Not only do I look back with pleasure on Cadarache I but I love to travel, in spite of the misgivings of my family who think that in view of my previous health problems I would be better off staying closer to home. Jacques himself seems to have been slightly worried. He very kindly provided me with a room on top of the castle with a beautiful view. But when he accompanied me to the room he was concerned. Will I be able to climb the stairs? Jacques, did it ever occur to you that you focused on the wrong problem? Perhaps it's my eyesight that has deteriorated so much that I can't enjoy the view. Where are you Jacques, I can't see you in the audience! The paternalistic attitude of well meaning people towards my state of health unfortunately often results in a not too diplomatic and rude reaction. I would like to take this opportunity to apologize to one of my colleagues who greeted me at the airport in Marseilles with what seemed to me a surprised 'Oh George you look so cool' to which I replied 'You don't look so hot either'. On another occasion, several years ago when somebody told me that I must see Dr. Zhivagho, my automatic reaction was 'Oh no I don't, I feel fine.' Let me assure you by paraphrasing George Burns: I look better, I feel better - and I lie better than ever before. Gee, what a weak reaction from you. When George Burns made his statement in Las Vegas he brought the house down. Perhaps it is telling me something besides the trivial that not all Georges are isomorphous. But enough of this morbid topic. Let me get back to travel and its enjoyments.

The Photosynthetic Bacterial Reaction Center II
Edited by J. Breton and A. Verméglio, Plenum Press, New York, 1992

Do you know what the most important scientific piece of equipment is? A Boeing 707. When one uses this piece of equipment one broadens one's horizon: one sees new places, new people, new cultures, and if one travels far enough, new civilizations. This brings to mind Gandhi's remark when he arrived in London after the 2nd World War and was asked what he thought about western civilization. 'I think it would be a good idea', he replied. A diametrically opposite attitude was expressed by Herman Göring of the 3rd Reich. 'When I hear the word "culture" I automatically reach for my revolver.' But let's not dwell on such topics; after all the title of the talk is "Light Reflections" and not "Gloomy Thoughts".

On the lighter side it is interesting to see how the English language is used and misused throughout the world, sometimes producing some funny results. Let me give you a few examples (modified from an Air France Bulletin of 12/1/89).

In a Copenhagen airline ticket office: *We take your bags and send them in all directions.*

At the reception desk in a Paris hotel: *Before proceeding to your room leave your values at the front desk.*

In a Japanese hotel: *You're invited to take advantage of the chambermaid.*

In a Moscow hotel room: *If this is your first visit to Russia, you're welcome to it.*

In a Moscow announcement: *There will be a Moscow exhibition of art by 10,000 Russian painters and sculptors; they were all executed over the past 5 years.*

Outside a Hong Kong tailor shop: *Customers may have a fit upstairs.*

In a Rome laundry: *Leave your clothes here and spend the afternoon having a good time.*

In a Norwegian cocktail lounge: *Ladies are requested not to have children in the bar.*

In an Acapulco hotel: *The manager has personally passed all the water served here.*

In an office of a Budapest doctor: *Specialist in women and other diseases.*

And to knock my own country of origin. In a Czechoslovak tourist agency: *Take one of our horsedriven city tours, we guarantee no miscarriages.*

Why am I dwelling so much on language? For the obvious reasons that it is our ultimate tool; we have to publish the results of our investigations. One scientist is as good as another until he/she has written a paper. And it is not the person with an idea, but the one who effectively communicates it in writing who gets the credit. And what difference in style and contents one finds!! There are erudite, articulate descriptions of nonsense, and there are descriptions of things that everybody knows in language that nobody understands.

And the syntax!! "Why did you bring these Proceedings to read from up for?" would not be an unusual sentence structure. And the ambiguities. For example: "The war will soon be all over" or the famous: "I received your reprint and shall waste no time reading it".

Unfortunately, when it comes to refereeing papers, strange criteria are often used. As F. Dyson put it: 'Most of the papers which are submitted to the Physical Review are rejected, not because it is impossible to understand them, but because it is possible. Those which are impossible to understand are usually published!!' Here is a ditty on refereeing, whose author's name I unfortunately have forgotten:

> There once was a fine referee
> who reviewed each paper with glee:
> 'What's new is not true
> what's true is not new
> unless it's been published by me'.

I guess the conclusion is that refereeing and peer review, like democracy and marriage, are poor institutions but the best that we have.

You all must have experienced at one time or another a sinking feeling when you received a manuscript from your students, postdocs or collaborators that you felt did not do justice to the work. I once had a student who wasn't too bad in the laboratory but he was close to being illiterate. I asked him whether he ever had a writing class. Yes, he had creating writing in high school, he said. Upon closer interrogation it turned out that the class spent most of the semester learning to write ransom notes. He must have been good at it because he was chosen valedictorian with the task of rewriting all the graffiti in the bathrooms. Then I had another student who was a real star in research. One of those rare fluctuations that make you happy to be at a university. But when it came to writing he was a total disaster. It prompted me to sit down and write a few pointers on how to write a manuscript:

SOME POINTS ON WRITING MANUSCRIPTS

1) *Good writing does not save bad science but good science can be wrecked by bad writing.*

2) *Never tell more than you know (or the reader wants to know).*

3) *Related to 2): It isn't what we don't know that hurts us but what we "know" and ain't so.*

4) *Make an outline before starting on the text.*

5) *Separate experimental findings from theory and speculations. (Reliable experimental facts remain valid forever, theories come and go).*

6) *On the science:*

   a) *Have clearly in mind what is new and what has been learned.*

   b) *If you use a theoretical model, state clearly the assumptions made.*

   c) *Think hard about the precision and accuracy of your results (if you can't make a reasonable estimate of your errors, you probably don't understand the procedure by which you arrived at the results).*

   d) *Weigh the strength of your conclusion by using appropriate words; in decreasing order of conviction: proves, shows, indicates, suggests and is consistent with.*

7) *If you encounter great difficulties in writing up a point - it is likely that you don't understand it well.*

8) *Try to be precise but keep in mind that the path between precision, i.e. rigor, and rigor mortis is a narrow one.*

9) *Avoid big words; they impress and intimidate only small minds.*

10) *Plain clarity is far better than ornate obscurity.*

11) *Don't use no double negatives.*

12) *Don't sloganize the english language.*

13) *Don't end a clear stream of DISCUSSION with a swampy delta.*

14) *Proofread carefully so you don't* leave *out words.*

15) *Be brief.*

LITERA SCRIPTA MANET

I have occasionally given these pointers to students and postdocs, but I am not convinced it has done much good. What a pity to see so much poor writing. Scientists often are the only people who have something to say and don't know how to say it.

A person who was a real expert of the English language was Noah Webster, of dictionary fame. He was a wordsmith who knew to use the right word for any occasion. The story goes that when Mrs. Webster came home and found Noah in bed with another woman she exclaimed: 'Oh Noah I am surprised'; 'No my dear, I am surprised, you're astonished' he replied.

In addition to language and style there are, of course, many other problems connected with publications, e.g. where, when, and how much to publish. When one publishes too late, one may get scooped, the work may be stale and obsolete. When one publishes too early the work may be incomplete or even wrong. On this, as on many other topics, Wolfgang Pauli made a succinct remark to one of his colleagues: 'I don't mind that you think slowly but I do mind that you are publishing faster than you think'. Another helpful admonition to people contemplating the submission of a paper is: Use your brain, it is the little things that count.

Concerning quantity of publications, an incident that occurred a few years ago comes to mind. I was on a committee for an award, and one of the candidates had a phenomenally long list of publications. All the committee members were terribly impressed. But how can anybody publish fifty meaningful papers a year? With difficulty, I convinced the committee to appoint one member who was an expert in the field to look the papers over and if possible to talk to the candidate himself. After a few months the expert reported back. He had told the candidate that the committee was worried about his prolific output. 'You must be publishing an article a week', he told him. 'Yes, on Tuesdays', the candidate replied.

I believe that, by and large, tenure and promotion committees make mistakes by weighing and counting publications rather than inquiring about the quality of the work. Let's take an example from history: the Ten Commandments that Moses brought down from Mount Sinai. The intellectuals of the time may well have been fretting that Moses came down after several days with only ten little sentences. Perhaps humble Moses stammered that he couldn't carry more than 2 stone tablets. As it turned out, the Ten Commandments became by far the most significant and often quoted publication per unit length.

But whether Moses could get tenure in a modern-day department is another question. Why, I believe not even God would make tenure. Can't you see the committee arguing that i) He didn't publish in a refereed journal, ii) nobody was able to repeat His experiments, and iii) what has He done recently? The last item is not only the nightmare of candidates for tenure but also of aging scientists. I suggest that we call it "the Prometheus syndrome". So Prometheus discovered fire but what has he done since? Hans (Deisenhofer) earlier this week showed a slide illustrating this point.

THE PROMETHEUS SYNDROME

'In reviewing your file, I noticed that you have not made any significant discoveries since you were awarded the Nobel Prize.'

This made me think of other great personalities in the past that were treated shabbily. Not only did they have tenure problems but in addition they were severely punished by the administration. Here is a partial list:

TENURE REVIEW		
	Tenure	*Action by Administration*
Adam & Eve	*denied*	*Cast out*
Pythagoras	*denied*	*Exile*
Socrates	*denied*	*Poisoning*
Jesus Christ	*denied*	*Crucifixion*
Alighieri Dante	*denied*	*Exile*
Giordano Bruno	*denied*	*Incineration*
Galileo Galilei	*denied*	*House arrest*
Karl Marx	*denied*	*Exile*

I hope that if any of you are up for tenure you will make it, but if for some reason you fail, you may take some solace of being in good company. And you may derive comfort in knowing that at least you won't be incinerated, crucified or exiled.

My time is up, and I want to conclude by thanking the organizers for having arranged such a successful conference. Some of us came confused on a specific topic, and after listening to the lectures and discussions are still confused on that topic, but let me assure you that our confusion is on a much higher level than when we came.

There are many topics that I did not have time to cover, e.g. grantsmanship, peer review, teaching, the relationships to students and postdocs, the use and misuse of computers, statistics modeling, etc. Enough material to last us through Cadarache VII. But by that time, as you have heard Werner (Mäntele) predict, the FTIR people will have taken over and they may not invite me to the conference. However, if you do, please reserve me a room - on the ground floor.

INDEX

MIX
Papier aus verantwortungsvollen Quellen
Paper from responsible sources
FSC® C105338

If you have any concerns about our products,
you can contact us on
ProductSafety@springernature.com

In case Publisher is established outside the EU,
the EU authorized representative is:
Springer Nature Customer Service Center GmbH
Europaplatz 3, 69115 Heidelberg, Germany

Printed by Libri Plureos GmbH
in Hamburg, Germany